HODDER

mathematics

GCSE IN A YEAR

SECOND EDITION

Series editor: **Roger Porkess**

Author team:

Catherine Berry
Pat Bryden
Diana Cowey
Dave Faulkner
Geoff Rigby
John Spencer
Julian Thomas
Christine Wood

MEI

Hodder & Stoughton

A MEMBER OF THE HODDER HEADLINE GROUP

Acknowledgements

The authors and publishers would like to thank the following companies, agencies and individuals who have given permission to reproduce copyright material. Every effort has been made to trace and acknowledge ownership of copyright. The publishers will be glad to make suitable arrangements with any copyright holder whom it has not been possible to contact.

Illustrations were drawn by Maggie Brand and Ann Kronheimer (with Maggie Mundy Illustrators Agency), Tom Cross, Bill Donohoe, Jeff Edwards, Phil Ford, Ian Foulis, Joe McEwan, Multiplex, Peters & Zabranski, Gary Rees (Linda Rogers Agency) and Tony Wilkins.

Photos supplied by Philip Gould/Corbis (page 86, left), David Turnley/Corbis (page 86, right), Dave Faulkner (page 191), Emma Lee/Life File (page 196), and Andrew Ward/Life File (page 197).

Page design and cover design by Lynda King.

Orders: please contact Bookpoint Ltd, 130 Milton Park, Abingdon, Oxon OX14 4SB.
Telephone: (44) 01235 827720, Fax: (44) 01235 400454. Lines are open from 9.00 – 6.00, Monday to Saturday, with a 24 hour message answering service. Email address: orders@bookpoint.co.uk

British Library Cataloguing in Publication Data

A catalogue record of this title is available from The British Library

ISBN 0 340 846909

First published 1999
Second edition 2002
Impression number 10 9 8 7 6 5 4 3 2 1

Year 2007 2006 2005 2004 2003 2002

Cover photo by John Ward.

Typeset by Pantek Arts Ltd, Maidstone, Kent, ME14 1NY.

Printed in Italy for Hodder & Stoughton Educational, a division of Hodder Headline Plc, 338 Euston Road, London NW1 3BH by Printer Trento.

Introduction

This textbook covers Intermediate Tier GCSE and is suitable for use with any specification. It also covers the mathematics requirements of Key Skills Application of Number at Level 3.

This is the second edition of the book. It has been adapted to take account of the new (2001) GCSE criteria. These first apply to one year courses for those starting in September 2002 and having final certification in 2003.

This book is designed for students following a one year course; these may be mature students preparing for higher education, evening class students or students retaking GCSE Mathematics aiming to achieve a grade C. It is assumed that the students are already able to add, subtract, multiply and divide whole numbers without the use of a calculator.

The book is divided into 33 chapters and arranged in four sections for easy reference: Number (Chapter 1–10); Algebra (Chapters 11–18); Shape, Space and Measures (Chapters 19–27); and Handling Data (Chapters 28–33).

Each chapter is divided into a number of double-page spreads, designed to be teaching units. The material to be taught is covered on the left-hand pages; the right-hand pages consist entirely of work for the student to do. Each chapter ends with a mixed exercise covering all of its content.

The left-hand pages have been designed to help teachers engage their students in whole class discussion. The symbol ? is used to indicate a Discussion Point; teachers should see it as an invitation.

On the right-hand pages are exercises of carefully graded questions and in some cases a practical activity or an investigation as well. The practical activities are suitable for both GCSE and Application of Number students; some can be used for portfolio tasks. These activities connect the mathematics classroom to the outside world and to other subjects.

This book provides plenty of opportunity to practise for both 'no calculator' and 'calculator' examination papers as the majority of questions are unmarked and can be done with or without a calculator. The no calculator icon indicates questions for which a calculator should definitely not be used. Where a question is marked with a calculator icon students are advised to use a calculator. Many questions have neither icon and these require a sensible judgement. Students should do as many of these as possible without a calculator in order to practise for the non-calculator GCSE questions. However, they also need to work through plenty of questions and using a calculator often allows them to work faster.

A final word to students: I hope this textbook will help you to understand mathematics, enjoy it and achieve success. Good luck!

Dave Faulkner 2002
GCSE in a Year editor

Information

How to use this book

This symbol next to a question means you need to use your calculator.

This symbol next to a question means you are not allowed to use your calculator.

This symbol means you will need to think carefully about a point and may want to discuss it.

Triangles

An **equilateral** triangle has 3 equal sides.

An **isosceles** triangle has 2 equal sides.

A **scalene** triangle has no equal sides.

A **right-angled** triangle has 1 right angle.

An **acuted-angled** triangle has 3 acute angles.

An **obtuse-angled** triangle has 1 obtuse angle.

$$\textbf{Area of a triangle} = \frac{1}{2} \times \textbf{base} \times \textbf{perpendicular height}$$

Quadrilaterals

square

rectangle

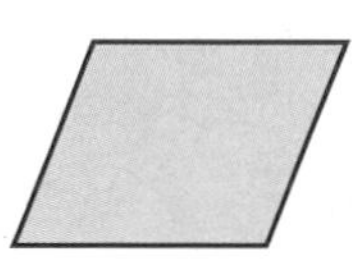

parallelogram

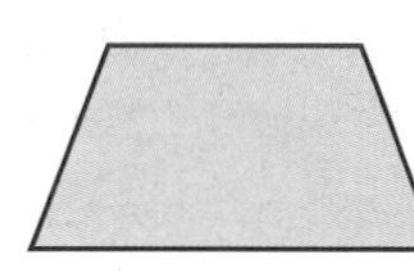

trapezium

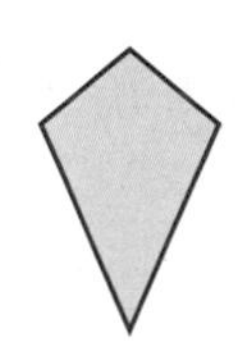

kite

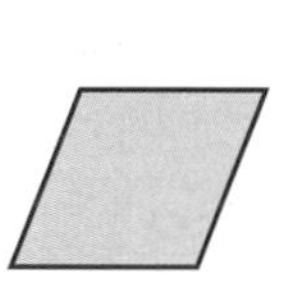

rhombus

$$\textbf{Area of a parallelogram} = \textbf{base} \times \textbf{vertical height}$$

vertical height

base

$$\textbf{Area of a trapezium} = \frac{1}{2}(a + b) \times h$$

a

h

b

Circles

Circumference of circle $= \pi \times$ diameter
$= 2 \times \pi \times$ radius

Area of circle $= \pi \times (\text{radius})^2$

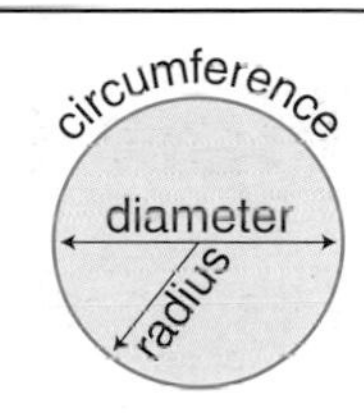

Solid figures

Volume of cuboid = length × width × height

Volume of prism = area of cross section × length

Volume of cylinder = πr^2 × length

Trigonometry

$\sin\theta = \dfrac{\text{opposite}}{\text{hypotenuse}}$

$\cos\theta = \dfrac{\text{adjacent}}{\text{hypotenuse}}$

$\tan\theta = \dfrac{\text{opposite}}{\text{adjacent}}$

Pythagoras' rule: $x^2 + y^2 = h^2$

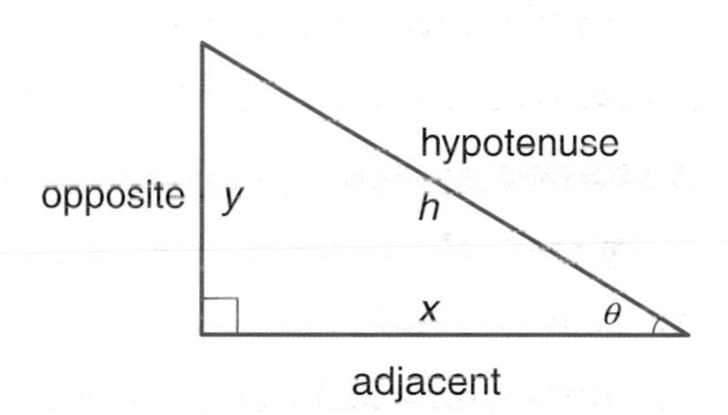

Units

Metric system

Length

k 1 kilometre $= 10^3$ metres = 1000 metres

h 1 hectometre $= 10^2$ metres = 100 metres

da 1 decametre $= 10^1$ metres = 10 metres

d 1 decimetre $= 10^{-1}$ metres $= \frac{1}{10}$ metre

c 1 centimetre $= 10^{-2}$ metres $= \frac{1}{100}$ metre : 100 centimetres = 1 metre

m 1 millimetre $= 10^{-3}$ metres $= \frac{1}{1000}$ metre : 1000 millimetres = 1 metre

The units for mass and capacity follow the same pattern. Notice that:
1 kilogram = 1000 grams
1 litre = 1000 millilitres

Notice also that: 1 tonne = 1000 kg

Imperial

12 inches = 1 foot
3 feet = 1 yard
1760 yards = 1 mile

16 ounces = 1 pound
14 pounds = 1 stone
8 stones = 1 hundredweight (cwt)
20 cwt = 1 ton

Contents

Number

Contents

Algebra

Contents

Shape, Space and Measures

Contents

Handling Data

One

Types of number

Squares and square roots

Anna designs these chocolate squares for a manufacturer.

$1 \times 1 = 1$

$2 \times 2 = 4$

$3 \times 3 = 9$

How many pieces are there in a 4×4 square?

The numbers 1, 4, 9, 16, … are called **squares** or **square numbers**.

A quick way of writing 4×4 is 4^2

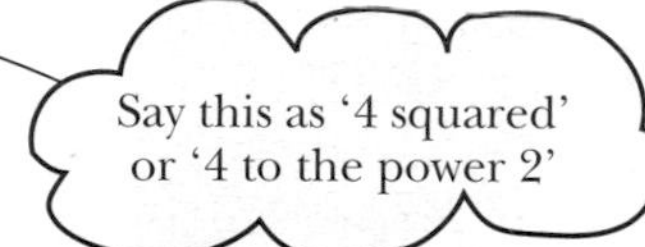

Anna's largest chocolate square has 25 pieces. You know that $5 \times 5 = 25$ so a square with 25 pieces has 5 pieces along each side.

We say that the **square root** of 25 is 5 and write $\sqrt{25} = 5$.

Cubes and cube roots

Anna designs these chocolate cubes for the Christmas market.

How many pieces are there in each cube?

The numbers 1, 8, 27, … are called **cubes** or **cube numbers**.

A quick way of writing $2 \times 2 \times 2$ is 2^3

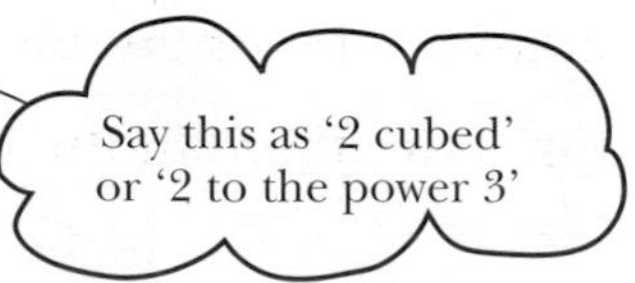

A chocolate cube with 8 pieces has 2 along each edge.

We say that the **cube root** of 8 is 2 and write $\sqrt[3]{8} = 2$.

1 Find out how to use the x^2 key on your calculator to work out

a) 7^2 b) 10^2 c) 15^2 d) 200^2

2 Find out how to use the $\sqrt{x}$ key on your calculator to work out

a) $\sqrt{36}$ b) $\sqrt{81}$ c) $\sqrt{144}$ d) $\sqrt{400}$

3 Find out how to use the x^3 key or x^y key on your calculator to work out

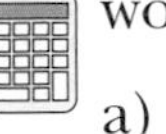

a) 4^3 b) 5^3 c) 10^3 d) 7^3

4 Find out how to use the $\sqrt[3]{x}$ or $x^{\frac{1}{y}}$ key on your calculator to work out

a) $\sqrt[3]{64}$ b) $\sqrt[3]{216}$ c) $\sqrt[3]{125}$ d) $\sqrt[3]{1000}$

5 Two square cakes are made in tins which are 30 cm by 30 cm.

The chocolate cake is cut into 5 cm by 5 cm pieces.

a) How many pieces is the chocolate cake divided into?

b) Now the lemon cake is cut into 3 cm by 3 cm pieces.

How many pieces does the lemon cake make?

6 Steven is designing a garden. He has 200 square slabs.
He uses all these slabs to make two identical square patios.

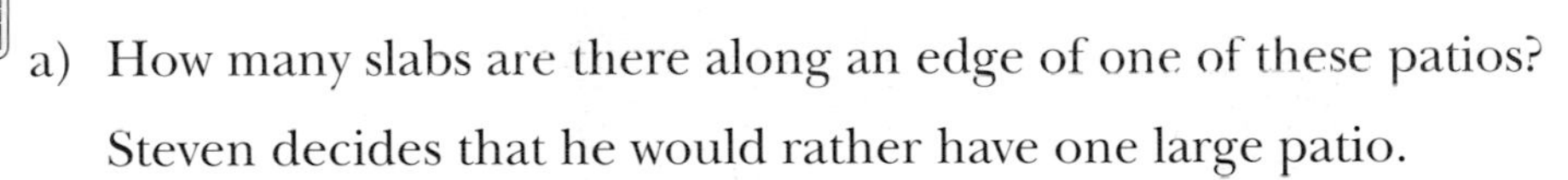

a) How many slabs are there along an edge of one of these patios?

Steven decides that he would rather have one large patio.

b) What size is the largest square patio that he can make?

c) How many slabs are left over?

7 Jody packs sugar cubes in boxes like this.

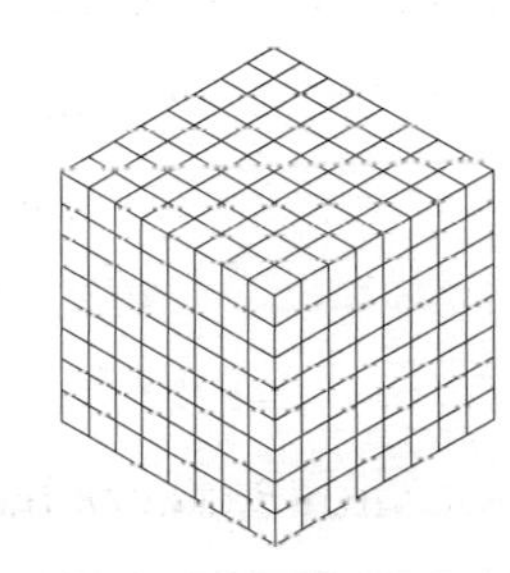

a) How many are there on the top layer?

b) How many are there in the box?

Multiples

Debbie works for a food and drink company. She decides how items are packaged together so that customers can buy in bulk.

Debbie decides that tins of tea should be packaged in pairs.

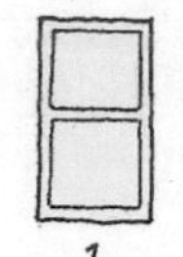

How many tins are there in 4 packages?

The numbers of tins, 2, 4, 6, 8, … are called the **multiples** of 2.

They are the answers to the 2-times table:

$1 \times 2 = 2$
$2 \times 2 = 4$
$3 \times 2 = 6$
and so on.

The multiples of 2 are the **even** numbers. The other numbers, 1, 3, 5, 7, …, are **odd**.

Factors

Debbie thinks that small bottles of Supa Juice will sell well in packages of 12.

She designs this package.
It has 3 rows of 4.

$3 \times 4 = 12$. We say that 3 and 4 are **factors** of 12.

Draw 2 other ways of arranging 12 bottles in a rectangular package.

The different arrangements tell you all the factors of 12.

You may have thought of 6 rows of 2. $6 \times 2 = 12$.

You may have thought of 1 row of 12. $1 \times 12 = 12$.

The factors of 12 are 1, 2, 3, 4, 6 and 12.

It is often useful to list the factors in order, like this.

Another way of saying this is that 12 is **divisible** by 1, 2, 3, 4, 6 and 12.

Primes

Next month there is going to be a special promotion on Supa Juice.

When you buy a pack of 12 you get one bottle free.

How can Debbie arrange 13 bottles in a rectangular package?

13 is an example of a **prime number**.

A prime number has just two different factors, 1 and itself.

The primes are 2, 3, 5, 7, 11, 13, …

Exercise

1 The first and last houses on each side of Trinity Road are numbered on this map.

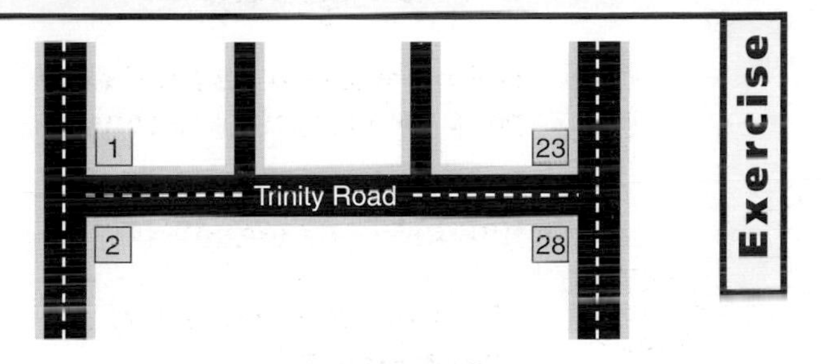

a) How many houses are there on the odd numbered side?

b) How many houses are there on the even numbered side?

2 Here is a box of chocolate eclairs.

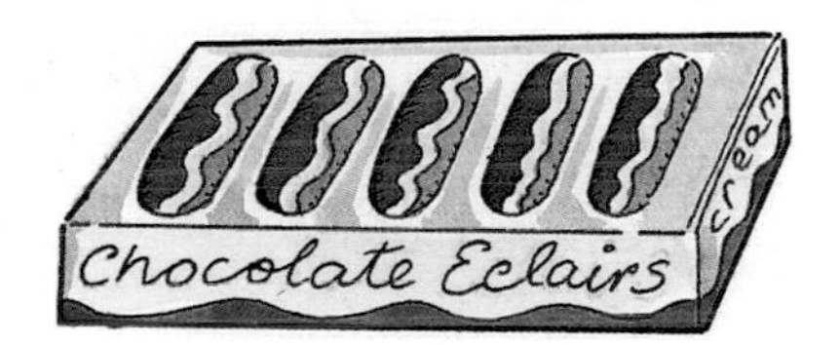

How many eclairs are there in

a) 2 boxes? b) 3 boxes? c) 4 boxes?

3 Look at this number pattern:
6, 12, 18, …, …, …, …, …, …, ….

a) Copy and complete the pattern.

b) Describe the numbers.

4 Lucy has 18 cork tiles.

She wants to use them to make a rectangular noticeboard in her room.

She sketches this arrangement.

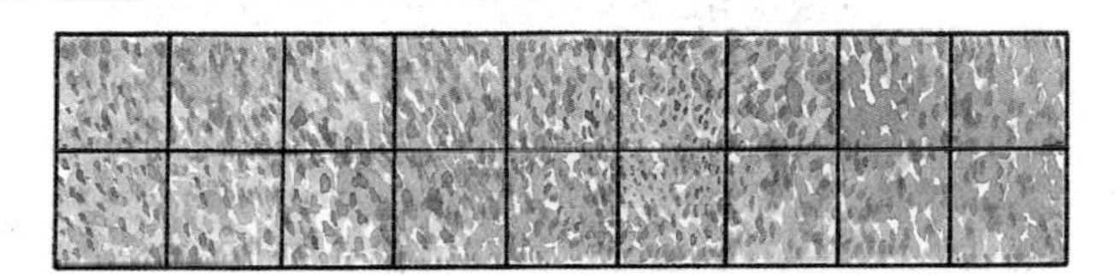

a) Draw all the other ways of making a rectangular noticeboard using exactly 18 tiles.

b) Now list all the factors of 18.

5 List all the factors of

a) 15 b) 8 c) 20 d) 7 e) 16 f) 35
g) 45 h) 36 i) 60 j) 72 k) 100 l) 144

6 List all the primes between 20 and 40.

Investigation

When two even numbers are added the answer is always even.

What happens when you add

a) two odd numbers?

b) an even and an odd number?

c) three odd numbers?

Work out some similar rules for multiplication.

Draw all the different ways of arranging 18 yoghurts in a rectangular pattern

a) using one layer

b) using more than one layer

Which of these arrangements do you think a customer would prefer?

Prime factorisation

Which of these are prime numbers?

19, 20, 43, 84

If a number is not itself prime then it can be written as the product of primes.

e.g. $20 = 2 \times 2 \times 5$ ← Each of these is a prime

and $84 = 2 \times 2 \times 3 \times 7$ ← Each of these is a prime

This is called **prime factorisation**.

Ian and Lin are both finding the prime factorisation of 20.

Ian writes this. Lin writes this.

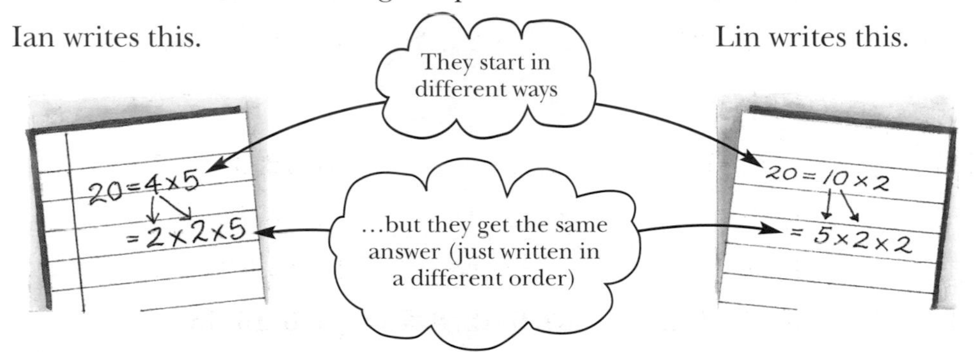

Remember that you must go on factorising until all the numbers are primes. Sometimes it may take several lines of working.

What is the prime factorisation of 360 *?*

Highest common factor (HCF)

You can find the HCF of 12 and 20 like this:

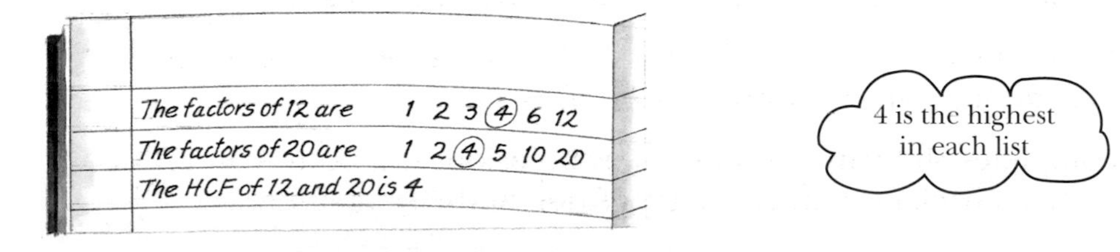

The HCF of 12 and 20 is 4.

Lowest common multiple (LCM)

You can find the LCM of 9 and 6 like this:

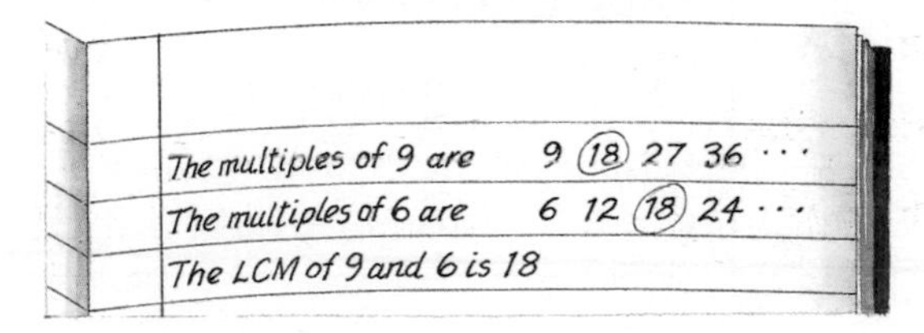

18 is the lowest in each list

The LCM of 9 and 6 is 18.

Exercise

1 Find the prime factorisation of each of these numbers.

a) 14 b) 15 c) 28 d) 36

e) 30 f) 27 g) 90 h) 126

i) 150 j) 210 k) 539 l) 1540

2 Find the HCF of each of these.

a) 6, 4 b) 6, 15 c) 18, 12 d) 12, 4

e) 10, 25 f) 3, 8 g) 18, 45 h) 14, 10

i) 21, 49 j) 22, 33 k) 63, 36 l) 56, 126

m) 12, 24, 54 n) 25, 35, 75 o) 56, 24, 32 p) 60, 80, 100

3 Find the LCM of each of these.

a) 10, 4 b) 5, 6 c) 4, 8 d) 12, 9

e) 6, 10 f) 3, 7 g) 27, 18 h) 16, 8

i) 14, 35 j) 8, 20 k) 20, 30 l) 45, 10

m) 2, 3, 4 n) 5, 15, 2 o) 9, 12, 8 p) 6, 10, 15

4 Look at these gears.

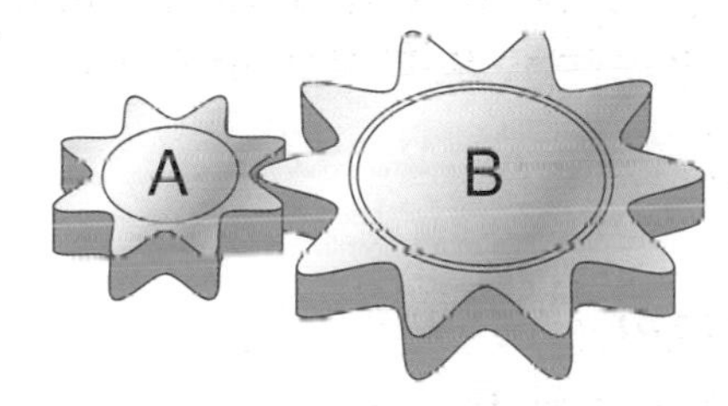

a) A completes 10 turns. How many turns does B complete?

b) What is the least number of turns that A can complete so that B also completes an exact number of turns?

c) B completes 30 turns. On how many occasions will both A and B have been back in their starting position at the same time?

5 Graham's three children, Angela, Bob and Cathy, visit him on Sunday afternoons. Angela visits every 2 weeks, Bob visits every 3 weeks and Cathy visits every 4 weeks. All three children visited Graham last Sunday.

How often do

a) Angela and Bob visit on the same day?

b) Angela and Cathy visit on the same day?

c) Bob and Cathy visit on the same day?

Finishing off

Now that you have finished this chapter you should be able to:

★ work out squares and find square roots

★ work out cubes and find cube roots

★ recognise odd and even numbers

★ work out multiples

★ find factors

★ recognise primes

★ work out prime factorisations

★ find the HCF of two or more numbers

★ find the LCM of two or more numbers.

Use the questions in the next exercise to check that you understand everything.

Mixed exercise

1 Work out

a) 9^2 b) $\sqrt{100}$ c) 6^3 d) $\sqrt[3]{27}$

e) $\sqrt{49}$ f) 12^3 g) 25^2 h) $\sqrt[3]{1000\,000}$

2 A chessboard is a large square divided into 64 smaller squares.

How many squares has it along each side?

3 How many small cubes are there in each of these large cubes?

a)

b)

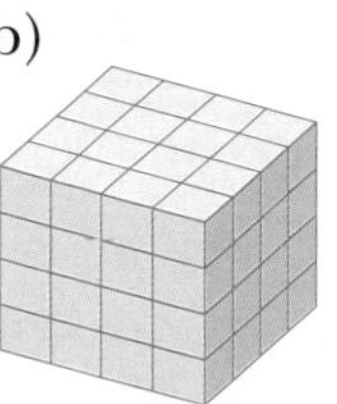

c)

4 A newspaper is made up of large sheets of paper. Each sheet is folded in half, to make 4 pages of newspaper.

a) How many pages are there in a newspaper made up of 8 sheets?

b) How many pages are there in a newspaper made up of 15 sheets?

c) Can a newspaper made up in this way have 42 pages?

5 a) Write down the first 10 multiples of 5.

b) Is 95 a multiple of 5?

c) Is 107 a multiple of 5?

d) Explain how you can tell whether or not a number is a multiple of 5.

Mixed exercise

6 Phil has 24 square slabs to make a rectangular patio.
He could arrange them like this:

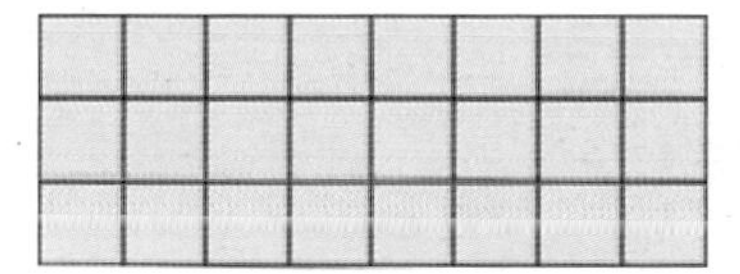

a) Draw all the other ways of arranging 24 slabs in a rectangle.

b) List all the factors of 24.

7 Work out which of these numbers are primes.

a) 17 b) 26 c) 31 d) 39 e) 73 f) 91

8 Look at this list of numbers.

3, 6, 9, 27, 29, 36

Write down the ones that are

a) factors of 27
b) multiples of 12
c) odd
d) squares
e) cubes
f) primes

9 Kelly goes to a disco every fourth Saturday, and ten-pin bowling every third Sunday. She does both during the weekend of 1 and 2 March.

a) How many weeks pass before the two outings again occur in the same weekend?

b) What will be the dates?

c) How many more times before the year end will Kelly have the two outings in the same weekend?

10 Work out the prime factorisation of

a) 18 b) 48 c) 100 d) 120

11 Write down the highest common factor of

a) 15, 20 b) 16, 36 c) 80, 30 d) 24, 36, 60

12 Write down the lowest common multiple of

a) 4, 6 b) 24, 8 c) 20, 50 d) 6, 10, 18

Investigation

a) Find the squares of 1, 2, 3, 4, ... up to 20.

Look at the last digit of the squares. e.g. for 13: $13 \times 13 = 169$, so the last digit of the square is 9.

b) Do you think that a square number could end with the digit 2?

c) Could 213 643 be a square number?

d) The last digit of a square number is 5.

What does this tell you about the original number?

e) Work out some rules about numbers and the last digit of their squares.

Two

Using numbers

Brackets

What is 1+ 2 × 3*?*

Did you add or multiply first?

John does it like this:

Lisa does it like this:

It seems that 1 + 2 × 3 can either be 9 or 7!

Using brackets avoids this confusion.

You always work out brackets first.

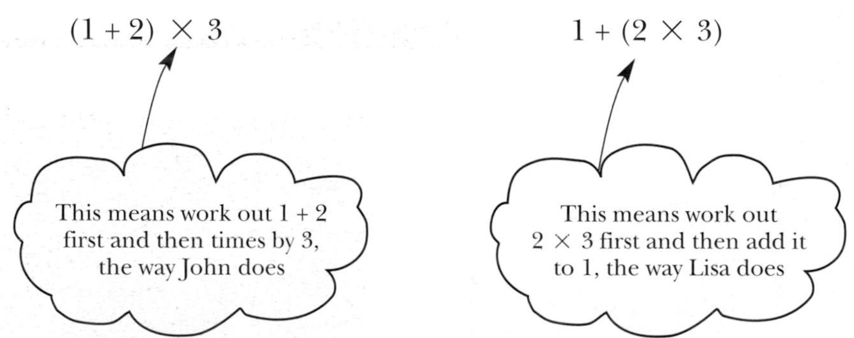

You should always work out operations in this order:

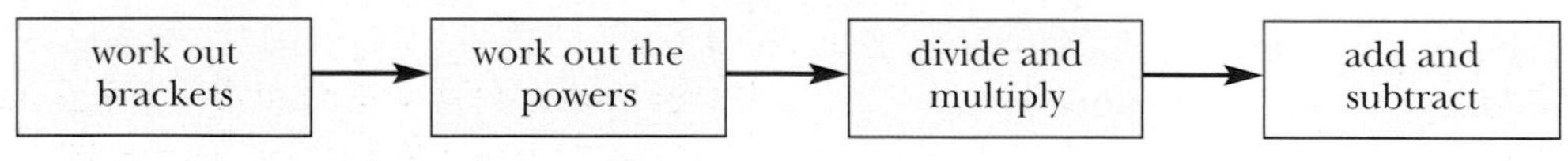

This is how a scientific calculator works out 1 + 2 × 3:

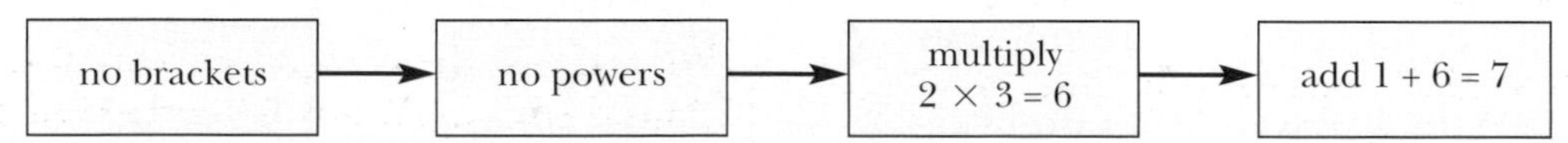

Using brackets, key in 1 + (2 × 3) = and check that you get 7.

Using brackets, key in (1 + 2) × 3 = and check that you get 9.

Exercise

1 Work out the value of

a) $4 + (5 \times 3)$ b) $(4 + 5) \times 3$ c) $(11 - 2) \times 4$

d) $11 - (2 \times 4)$ e) $2 \times (7 - 3)$ f) $(2 \times 7) - 3$

2 Work out the value of

a) $5 + 3 \times 4$ b) $8 \times 3 - 1$ c) $12 \div 2 + 4$ d) $16 - 2 \times 3$

3 Work out the value of

a) $(3 + 2) \times (4 - 1)$ b) $3 + (2 \times 4) - 1$

c) $(5 \times 2) + (3 \times 4)$ d) $(20 \div 4) - 2$

4 Complete the following by using +, –, × or ÷ in each box.

You can use brackets as well to show which operation is done first.

For example: 3 □ 4 □ 2 = 14.

One solution is $(3 + 4) \times 2 = 14$.

Another is $(3 \times 4) + 2 = 14$.

a) 4 □ 5 □ 3 = 6 b) 7 □ 5 □ 3 = 4 c) 2 □ 6 □ 3 = 15

d) 5 □ 2 □ 2 = 14 e) 3 □ 5 □ 1 = 12 f) 3 □ 2 □ 2 = 8

g) 1 □ 6 □ 2 = 4 h) 6 □ 6 □ 4 = 9 i) 8 □ 2 □ 3 = 7

j) 10 □ 8 □ 2 = 9

Investigation

Look at these calculations.

$4 + 4 + \frac{4}{4} = 9$ $\quad 4 - \frac{(4 + 4)}{4} = 2$

$\frac{4}{4} = 1$ $\quad \frac{4+4}{4} = \frac{8}{4} = 2$

$(4 \times 4) + (4 \times 4) = 32$

Each calculation uses exactly four 4s.

Using exactly four 4s invent 10 calculations of your own which give different answers.

Buy a meal for your family or a group of friends. Write down the total cost using the signs ×, +, and ().

Write down the cost per person, if it is shared equally. You will need to use the ÷ sign now as well.

Using the number line

Shola has £50 in her bank account. She has these bills to pay:

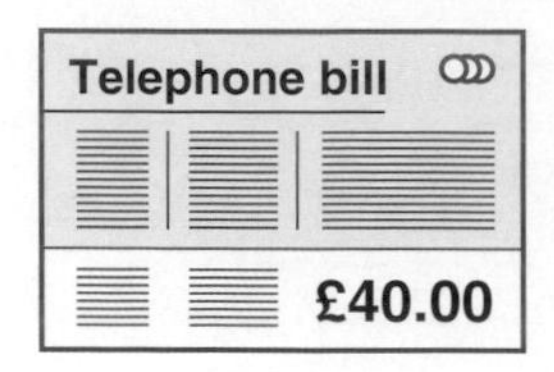

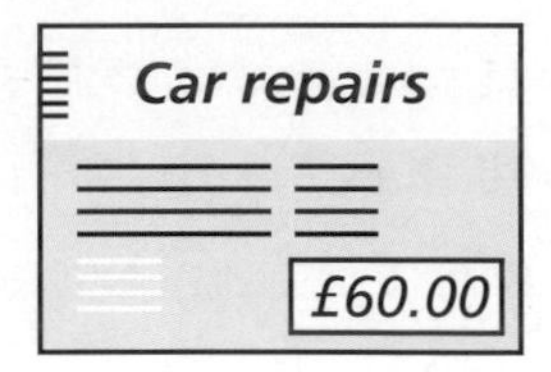

How much is left in her account if she pays the telephone bill?

How much is left if she pays the electricity bill?

How much is left if she pays the car repairs bill?

The bank might not pay a £60 cheque when Shola has only £50 left in her account. Alternatively they might pay the cheque and let Shola be overdrawn by £10. In this case, her balance is –£10.

You can see this on the **number line**.

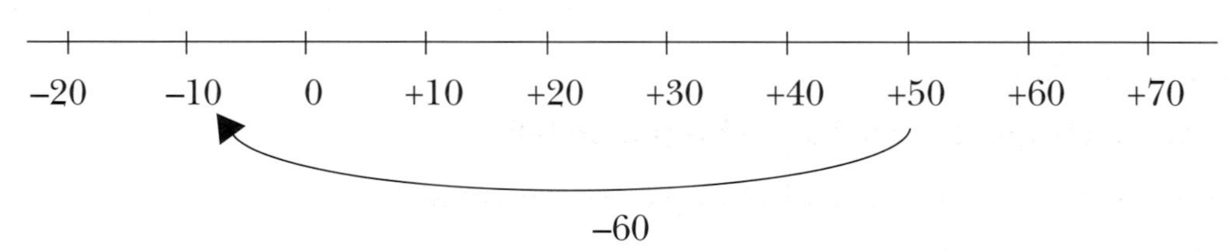

This number line is horizontal. It can also be vertical.

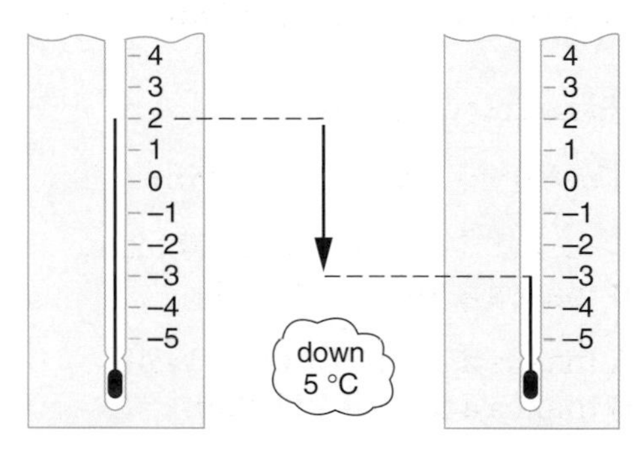

In winter, the temperature is often near zero.

The temperature one day is 2 °C. At nightfall it drops by 5 °C.

The new temperature is

$$2\text{ °C} - 5\text{ °C} = -3\text{ °C}$$

Which is the warmer 2 °C or –3 °C?

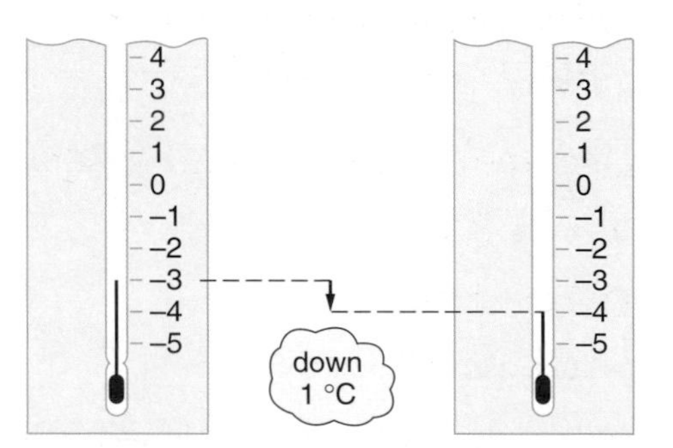

During the night, the temperature drops by another 1 °C.

The new temperature is

$$-3\text{ °C} - 1\text{ °C} = -4\text{ °C}$$

Which is the warmer –3 °C or –4 °C?

The next morning, the temperature rises by 7 °C. What is the new temperature?

1 Use a number line to work out

a) $5 - 1$ b) $0 - 4$ c) $1 - 7$ d) $-1 + 2$

e) $5 + 2$ f) $-5 - 2$ g) $8 - 5$ h) $-8 + 5$

i) $5 - 2$ j) $-5 + 2$ k) $3 + 4$ l) $-3 - 4$

2 Work out

a) $20 - 21$ b) $-1 + 101$ c) $-50 + 30$ d) $1000 - 2000$

3 Copy and complete this table, using the information below.

The temperature in London is 4 °C.

Manchester is 5 °C colder than London.

Leeds is 2 °C colder than Manchester.

Inverness is 3 °C colder than Leeds.

Accra is 36 °C hotter than Inverness.

City	Temperature (°C)
London	
Manchester	
Leeds	
Inverness	
Accra	

4 Oliver has £125 in his bank account.

He writes a cheque for £58 to pay his phone bill.

Then he writes a cheque for £104 to pay for car repairs.

Finally he pays in a cheque for £29.

What is the new balance of his account?

5 A and B are two shops.

Find the total profit or loss when

a) A makes a profit of £28 000 and B makes a loss of £15 000.

b) A makes a profit of £6000 and B makes a loss of £11 000.

c) A makes a loss of £2000 and B makes a loss of £9000.

6 Which of these temperatures is colder:

a) −2 °C or 3°C?

b) 0 °C or −1 °C?

c) −2 °C or −4 °C?

This diagram shows the leader board in a golf competition.

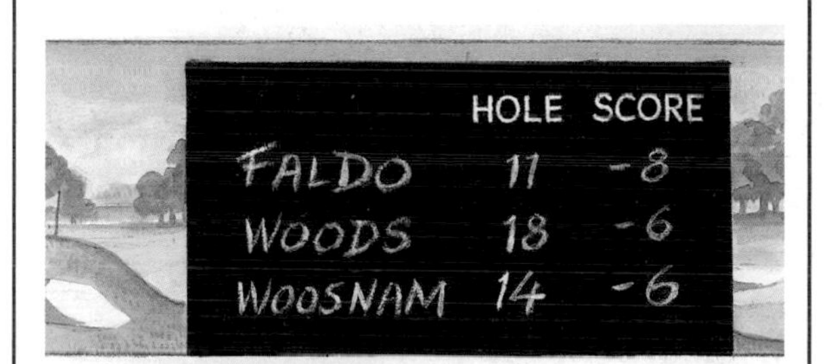

Find out and explain what the numbers in the score column mean.

Explain the meaning of words such as ‘Birdie’ and ‘Bogey’.

Using negative numbers

By how many degrees did the temperature rise?

Minneapolis Herald
Thursday 8th March
BIG THAW
From -8°C to +12°C in 12 hours

A change in temperature is given by:

new temperature – old temperature

In this case, the change (in degrees Celsius) is 12 – (–8).

But what is the answer to 12 – (–8)?

Look at the thermometer.

+12 Up 20 °C 0 −8 °C

You can see by counting that the change from –8 °C to +12 °C is +20 °C.

So 12 – (–8) = 20 = 12 + 8.

You can see that –(–8) = +8.

Where there are two signs before one number, the signs follow these rules.

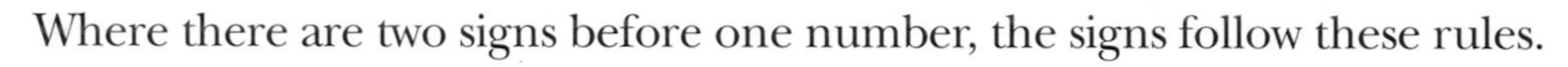

+ (+8) = +8   + (–8) = –8   – (+8) = –8   – (–8) = +8

What happens when you multiply a negative number by a positive one, for example (–3) × 2?

You can see this on the number line. Multiplying by 2 doubles the distance from zero.

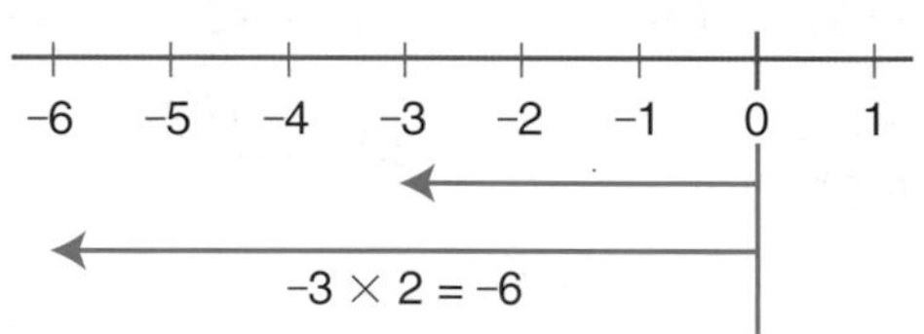

You can write this the other way round too, as 2 × (–3) = (–6).

In this, 2 means (+2), so you can see that

– × + → –

and

+ × – → –

What happens when two negative numbers are multiplied together?

When you multiply two negative numbers together the answer is positive.

– × – → +

This is the same for positive numbers.

+ × + → +

The rules for division follow the same pattern.

– ÷ + → –

+ ÷ – → –

and

– ÷ – → +

+ ÷ + → +

Exercise

1 Work out

a) $(+2) + (+5)$ b) $(+9) - (+4)$ c) $(+6) + (-2)$
d) $(+5) - (-3)$ e) $(-3) + (+7)$ f) $(-1) - (+2)$
g) $(+2) + (-7)$ h) $(-5) - (-1)$ i) $(-6) + (-2)$
j) $(-1) + (+5) - (+2)$ k) $-(+4) - (-4) + (+3)$ l) $(-3) + (+6) - (-5)$

2 Work out

a) $(+5) \times (+4)$ b) $(+5) \times (-4)$ c) $(-5) \times (+4)$
d) $(-5) \times (-4)$ e) $(+3) \times (-7)$ f) $(-4) \times (+8)$
g) $(-2) \times (-4)$ h) $(+8) \times (+7)$ i) $(+2) \times (-10) \times (-3)$
j) $(-3) \times (+5) \times (+2)$ k) $(-4) \times (-3) \times (-2)$ l) $(+5) \times (-5) \times (+5)$

3 Work out

a) $(+18) \div (+3)$ b) $(+18) \div (-3)$ c) $(-18) \div (+3)$
d) $(-18) \div (-3)$ e) $(-12) \div (+4)$ f) $(+35) \div (-7)$
g) $(+24) \div (+6)$ h) $(-42) \div (-6)$ i) $(+36) \div (-9)$
j) $\dfrac{(-16)}{(+2)}$ k) $\dfrac{(+48)}{(+8)}$ l) $\dfrac{(-55)}{(-5)}$

4 Work out

a) $\dfrac{(+9)-(+3)}{(+2)}$ b) $\dfrac{(-36)}{(-3)+(-9)}$ c) $\dfrac{(+2)-(-5)+(-1)}{(-3)}$

d) $\dfrac{(+5)\times(-8)}{(+10)}$ e) $\dfrac{(+56)}{(-2)\times(-7)}$ f) $\dfrac{(+9)\times(-20)}{(-5)\times(+6)}$

g) $\dfrac{(-4)\times(+3)\times(+6)}{(+8)}$ h) $\dfrac{(-54)}{(-2)\times(-3)\times(+3)}$ i) $\dfrac{(-12)\times(-2)\times(-8)}{(-3)\times(-4)}$

Investigation

What happens to the sign when you keep multiplying negative numbers?

Write down the value of $(-1)^2$, $(-1)^3$, $(-1)^4$, and so on.

What is the value of $(-1)^{213}$?

Find out how to do calculations like $(-5) \times (-4)$ on your calculator.

Write some instructions to enable a friend to do them.

Finishing off

Now that you have finished this chapter you should be able to:

- ★ work out operations in the correct order
- ★ multiply and divide positive and negative numbers
- ★ use a number line to add and subtract positive and negative numbers.

Use the questions in the next exercise to check that you understand everything.

Mixed exercise

1 Work out

a) $9 - 2 + 3$ b) $44 - 20 \div 4$ c) $11 \times 3 - 1$

d) $(14 - 10) \div (5 - 1)$ e) $[8 - (2 + 1)] \times 4$ f) $24 \div [(3 \times (5 - 3)]$

g) $(\sqrt{16})^2$ h) $\sqrt{(10^2)}$

2 Work out

a) $2 - 7$ b) $-20 + 24$ c) $-13 - 9$

d) $4 - 11 + 4$ e) $12 - 18 - 5$ f) $-10 + 45 - 35$

3 The temperature is –2 °C at midnight tonight. It is forecast to rise by 3 °C by midday and then fall 5 °C by midnight tomorrow.

a) What temperature is forecast for midday?

b) What temperature is forecast for midnight tomorrow?

4 Here is the control panel for the lift in a large store. For each of these trips say how many floors the lift goes up or down.

Example: 1 to 3 is up 2 floors.

a) 3 to 4 b) 4 to 3

c) 3 to 0 d) 0 to –2

e) –2 to 2 f) 2 to –1

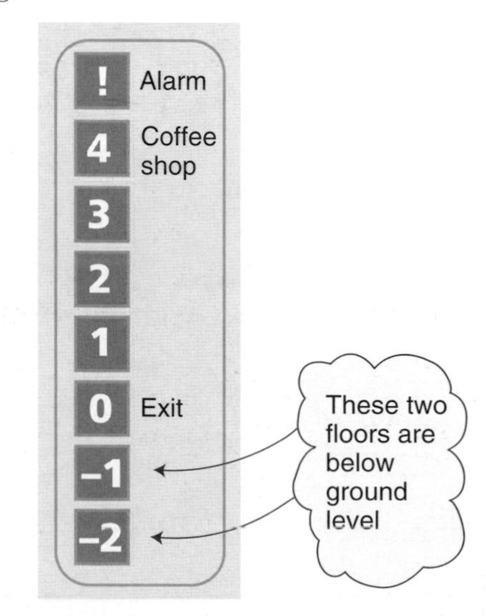

5 Victoria has £140 in her bank account on 8 June.

She keeps this record of her money:

Victoria's bank lets her be overdrawn. Work out her balance after each of the transactions.

6 Local time in Calais is 1 hour ahead of local time in Dover.

The ferry crossing from one to the other takes 90 minutes.

a) A ferry leaves Dover at 0945 local time.

What will be the time in Calais when it arrives?

b) A ferry leaves Calais at 1635 local time.

What will be the time in Dover when it arrives?

7 Work out

a) $(-5) + (+7)$ b) $(+4) + (-9)$ c) $(-6) - (-5)$

d) $(-9) \times (+5)$ e) $(-2) \times (-4)$ f) $(-3)^2$

g) $(+28) \div (-7)$ h) $(-54) \div (+6)$ i) $(-75) \div (-5)$

j) $\dfrac{(+3)-(+11)}{(+2)}$ k) $\dfrac{(-48)}{(+4)\times(-3)}$ l) $\dfrac{(+2)^2\times(-5)\times(-3)}{(+10)\times(+6)}$

Investigation

Make ten cards with numbers and signs on like this:

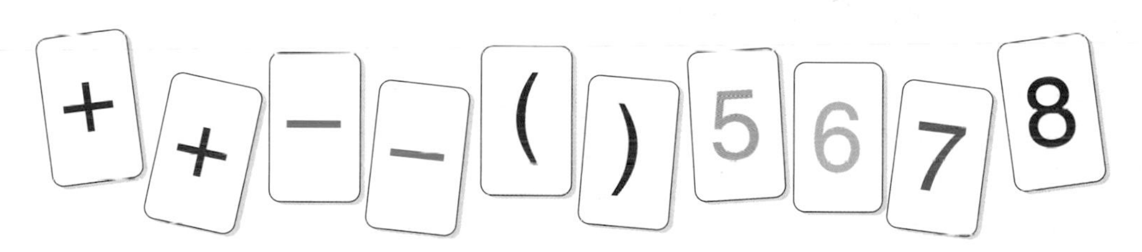

Using the numbers as single digits (i.e. not as 56, 75…), combine some or all of the cards to make different expressions.

What are the largest and smallest numbers you can make?

Write down as many combinations as you can which give the answers

a) 0 b) 14

Three

Fractions

Equivalent fractions

You can write 1 out of 4 as $\frac{1}{4}$. We call it one quarter. $\frac{1}{4}$ is a **fraction**.

The top, sometimes called the **numerator**, is 1.

The bottom, sometimes called the **denominator**, is 4.

Look at these diagrams:

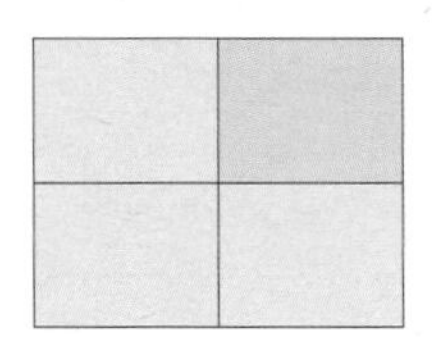

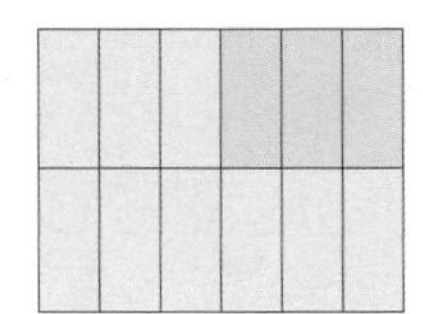

You can see that $\frac{1}{4}$, $\frac{2}{8}$ and $\frac{3}{12}$ all mean the same thing.

They are **equivalent fractions**. They can all be written as $\frac{1}{4}$.

$\frac{1}{4}$ is the **simplest form** because it has the smallest numbers.

To find an equivalent fraction, you multiply (or divide) the top and bottom by the same number

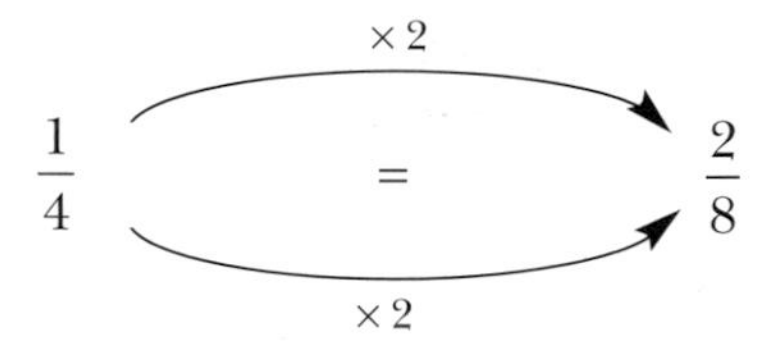

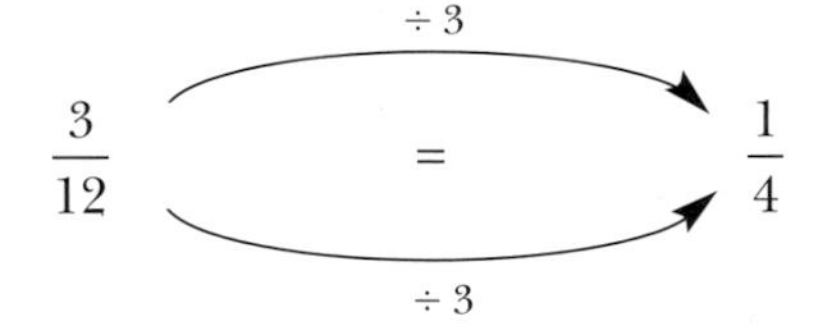

Adding and subtracting

Glyn serves at the delicatessen. There is a large pork pie on display. The first customer buys $\frac{1}{4}$ of the pie, and the second customer buys $\frac{3}{8}$ of it.

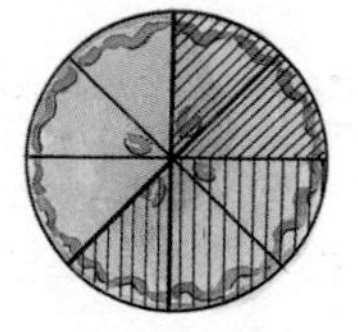

How much of the pie do they buy altogether?

You need to add 1 quarter to 3 eighths. But you cannot add quarters to eighths. You need to change the quarters into eighths first.

You have already seen that $\frac{2}{8}$ is equivalent to $\frac{1}{4}$. You can now write:

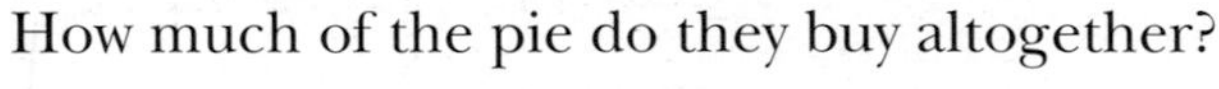

$$\frac{1}{4} + \frac{3}{8} = \frac{2}{8} + \frac{3}{8} = \frac{5}{8}$$

Exercise

1 This chart shows when the sports hall is booked (/).

Time	7	8	9	10	11	12	1	2	3	4	5	6	7	8	9	10	11
Sat	/		/	/	/	/		/	/		/	/	/	/	/	/	
Sun			/	/		/	/	/	/	/		/	/	/	/	/	

a) What fraction of the sessions on Saturday are booked?

b) What fraction of the sessions on Sunday are booked?

c) For what fraction of the whole weekend is the hall booked?

2 Find the missing number in each of these.

a) $\frac{1}{2} = \frac{?}{4}$ b) $\frac{1}{3} = \frac{?}{6}$ c) $\frac{6}{8} = \frac{3}{?}$ d) $\frac{10}{15} = \frac{2}{?}$

e) $\frac{1}{4} = \frac{?}{12}$ f) $\frac{6}{9} = \frac{2}{?}$ g) $\frac{3}{4} = \frac{?}{16}$ h) $\frac{6}{10} = \frac{?}{5}$

i) $\frac{30}{40} = \frac{3}{?}$ j) $\frac{7}{8} = \frac{?}{16}$ k) $\frac{1}{6} = \frac{?}{30}$ l) $\frac{7}{10} = \frac{21}{?}$

3 Write down three fractions equivalent to $\frac{1}{2}$.

4 Write these fractions in their simplest form.

a) $\frac{4}{8}$ b) $\frac{9}{12}$ c) $\frac{6}{16}$ d) $\frac{4}{12}$

e) $\frac{10}{15}$ f) $\frac{4}{20}$ g) $\frac{18}{24}$ h) $\frac{6}{30}$

i) $\frac{20}{80}$ j) $\frac{40}{56}$ k) $\frac{48}{72}$ l) $\frac{90}{225}$

5 Work out

a) $\frac{1}{2} + \frac{3}{8}$ b) $\frac{11}{16} - \frac{3}{16}$ c) $\frac{7}{8} - \frac{1}{4}$ d) $\frac{7}{16} + \frac{3}{8}$

e) $\frac{1}{3} + \frac{1}{6}$ f) $\frac{9}{10} - \frac{1}{5}$ g) $\frac{2}{3} + \frac{5}{6}$ h) $\frac{1}{2} + \frac{1}{4} + \frac{1}{8}$

6 These boards are made up of 2 parts.

The total thickness and the thickness of the top layer are shown (in inches). Find the thickness of the bottom layer.

a) $\frac{1}{8}$ $\frac{1}{2}$

b)

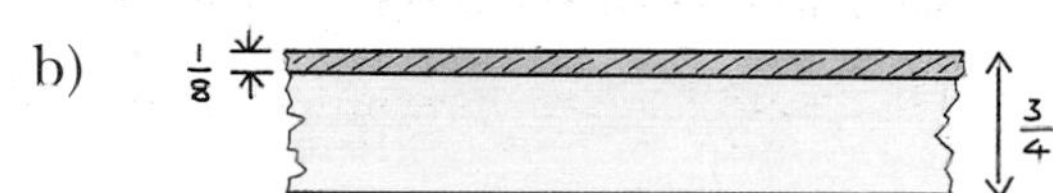

c) $\frac{5}{16}$ 1

d)

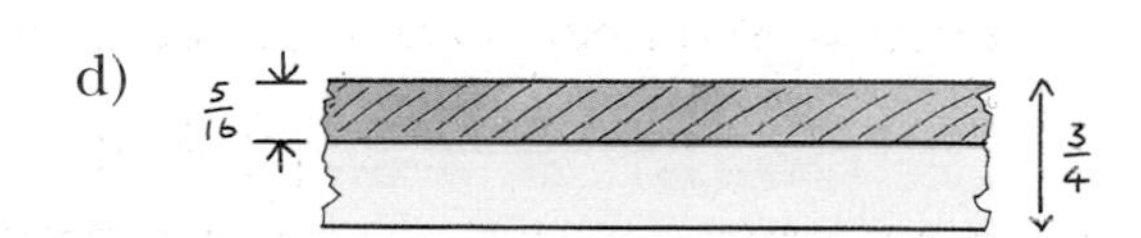

Improper fractions & mixed numbers

George and Marie are counsellors.

They have these afternoon appointments.

	1.00	1.30	2.00	2.30	3.00	3.30	4.00
GEORGE	Ron	Liz	Ian	Pat	Tim	Con	Bill
MARIE	Sal	Vic	Jim	Uma	Col		

How long is George booked for?

George has 7 appointments, each $\frac{1}{2}$ hour long.

7 ÷ 2 = 3 remainder 1

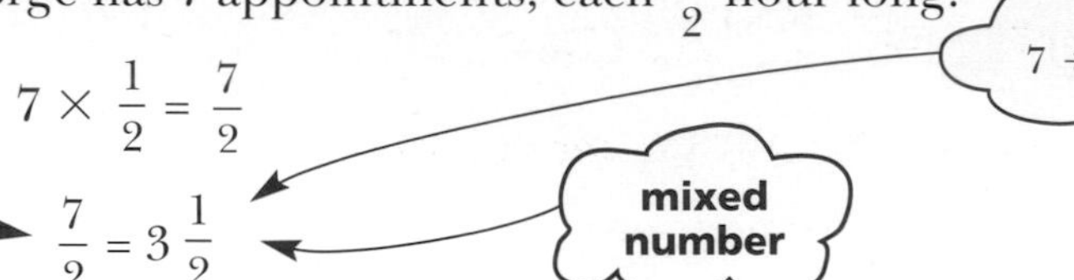

$7 \times \frac{1}{2} = \frac{7}{2}$

$\frac{7}{2} = 3\frac{1}{2}$

improper fraction

George is booked for $3\frac{1}{2}$ hours.

How do you change $3\frac{1}{2}$ back into $\frac{7}{2}$?

3 is 6 halves ($\frac{6}{2}$)

$3\frac{1}{2} = 3 + \frac{1}{2} = \frac{6}{2} + \frac{1}{2} = \frac{7}{2}$

Adding and subtracting

This diagram shows how far Ben, Chloe and Joel live from school (in miles).

Ben — Chloe — School — Joel

Ben to School: $2\frac{1}{4}$; Chloe to School: $1\frac{3}{4}$; School to Joel: $2\frac{1}{2}$

How many miles is Joel from Chloe? You need to work out $1\frac{3}{4} + 2\frac{1}{2}$

First add the whole numbers: $1 + 2 = 3$

Then add the fractions: $\frac{3}{4} + \frac{1}{2} = \frac{3}{4} + \frac{2}{4}$

You have to change the half into quarters before you can add them

$= \frac{5}{4} = 1\frac{1}{4}$

The whole numbers

So $1\frac{3}{4} + 2\frac{1}{2} = 3 + 1\frac{1}{4} = 4\frac{1}{4}$

The fractions

So Joel is $4\frac{1}{4}$ miles from Chloe.

How many miles is Chloe from Ben? You need to work out $2\frac{1}{4} - 1\frac{3}{4}$

$2\frac{1}{4} - 1\frac{3}{4} = \frac{9}{4} - \frac{7}{4} = \frac{2}{4} = \frac{1}{2}$

Change them both into improper fractions

Chloe is $\frac{1}{2}$ mile away from Ben.

Exercise

1 Change these improper fractions to mixed numbers.

a) $\frac{9}{2}$ b) $\frac{13}{8}$ c) $\frac{12}{5}$ d) $\frac{11}{3}$ e) $\frac{15}{4}$ f) $\frac{13}{6}$

2 Brian works at a health centre. He sees 9 people for half an hour each.

How many hours does it take him?

3 Tess records 13 programmes each lasting quarter of an hour.
How many hours does it take?

4 Change these mixed numbers into improper fractions.

a) $3\frac{1}{2}$ b) $4\frac{3}{8}$ c) $1\frac{7}{16}$ d) $2\frac{3}{4}$ e) $5\frac{1}{3}$ f) $3\frac{11}{16}$

5 A doctor has a $2\frac{1}{2}$ hour clinic. How many $\frac{1}{2}$ hour appointments can be fitted in?

6 Joanna has $2\frac{3}{4}$ hours left of a videotape.

How many $\frac{1}{4}$ hour programmes can she record?

7 This map shows the distances, in miles, between six stations.

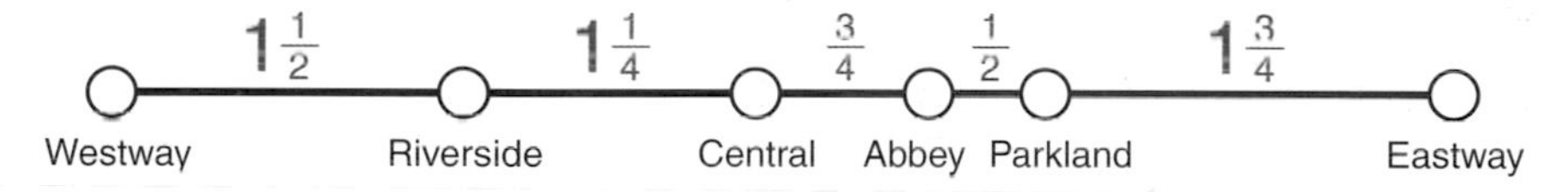

What is the distance from

a) Central to Parkland?
b) Westway to Central?
c) Riverside to Abbey?
d) Westway to Eastway?

8 Work out

a) $2\frac{7}{8} + \frac{5}{8}$ b) $1\frac{3}{4} + 3\frac{3}{8}$ c) $3\frac{7}{16} - 2\frac{1}{8}$ d) $5\frac{1}{4} - 3\frac{11}{16}$

e) $2 - \frac{13}{16}$ f) $4\frac{5}{8} + 1\frac{1}{2}$ g) $4\frac{3}{4} - 1\frac{7}{8}$ h) $1\frac{1}{2} + 2\frac{2}{3}$

i) $4\frac{7}{12} - 2\frac{1}{4}$ j) $4\frac{2}{3} + 2\frac{5}{6}$ k) $3\frac{7}{10} - 2\frac{1}{5}$ l) $5\frac{3}{5} + 2\frac{1}{4}$

Find out how to use a scientific calculator to add and subtract fractions. Then use it to check your answers to question 8.

Multiplying fractions

Dave, Becky and Ravi share a pizza.

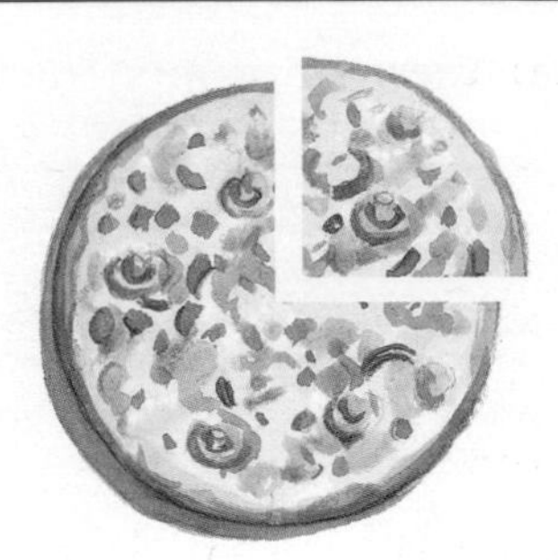

Dave has a quarter of it. How much is left?

Becky and Ravi share the other three quarters.

Each has half of it.

Each gets $\frac{1}{2}$ of $\frac{3}{4}$:

$$\frac{1}{2} \times \frac{3}{4} = \frac{3}{8}$$

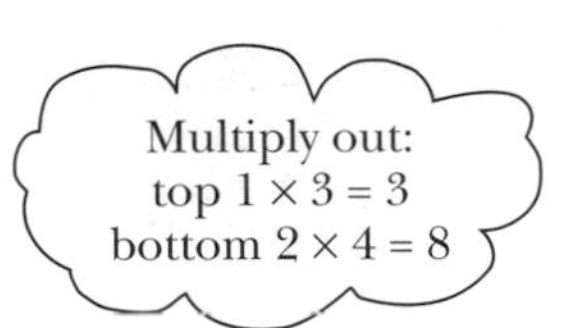

Dave only eats half of his piece. What is $\frac{1}{2} \times \frac{1}{4}$*?*

The pizza costs £8.00, but Ravi has a voucher that means he only pays for three quarters of the price.

Dave and Becky each work out $\frac{3}{4}$ of £8.

Dave writes $\frac{3}{4}$ of $8 = {}_1\frac{3}{\cancel{4}} \times \frac{\cancel{8}^2}{1} = \frac{6}{1} = 6$

Becky writes one quarter of 8 is $8 \div 4 = 2$
so three quarters of 8 is $2 \times 3 = 6$.

How much does Ravi pay for the pizza?

The next two examples show you how to multiply mixed numbers.

Change $4\frac{1}{2}$ into an improper fraction

$$\frac{2}{3} \times 4\frac{1}{2}$$

Cancel

$$= \frac{{}^1\cancel{2}}{{}_1\cancel{3}} \times \frac{\cancel{9}^3}{\cancel{2}_1}$$

Multiply out and change back into a mixed number

$$= \frac{3}{1} = 3$$

Change the mixed numbers into improper fractions

$$2\frac{1}{3} \times 3\frac{3}{4}$$

Cancel

$$= \frac{7}{{}_1\cancel{3}} \times \frac{\cancel{15}^5}{4}$$

Multiply out and change back to a mixed number

$$= \frac{35}{4}$$

$$= 8\frac{3}{4}$$

The **reciprocal** of 2 is $\frac{1}{2}$, the reciprocal of 3 is $\frac{1}{3}$, …

What is the reciprocal of 4?

What happens when you multiply a number by its reciprocal?

Exercise

1 Work out

a) $\frac{1}{2} \times \frac{1}{3}$ b) $\frac{1}{2} \times \frac{3}{8}$ c) $\frac{1}{4} \times \frac{3}{5}$ d) $\frac{3}{4} \times \frac{5}{6}$

e) $\frac{3}{8} \times \frac{2}{3}$ f) $\frac{6}{7} \times \frac{7}{10}$ g) $\frac{3}{8} \times \frac{5}{8}$ h) $\frac{3}{4} \times \frac{10}{1}$

2 Work out

a) $\frac{1}{2}$ of 7 b) $\frac{3}{4}$ of 6 c) $\frac{1}{3}$ of 8 d) $\frac{2}{5}$ of 4

e) $\frac{5}{8}$ of 20 f) $\frac{2}{3}$ of 14 g) $\frac{3}{8}$ of 10 h) $\frac{5}{6}$ of 9

3 Work out

a) $\frac{1}{2} \times 6\frac{1}{2}$ b) $\frac{3}{4} \times 4\frac{1}{2}$ c) $\frac{1}{3} \times 2\frac{5}{8}$ d) $2\frac{1}{2} \times \frac{7}{10}$

e) $2\frac{4}{5} \times \frac{5}{8}$ f) $\frac{3}{4} \times 5\frac{1}{3}$ g) $1\frac{1}{2} \times 2\frac{1}{2}$ h) $2\frac{1}{4} \times 3\frac{1}{2}$

i) $3\frac{2}{3} \times 1\frac{1}{2}$ j) $5\frac{1}{3} \times 3\frac{3}{4}$ k) $1\frac{3}{8} \times 3\frac{1}{2}$ l) $6\frac{2}{5} \times 1\frac{7}{8}$

4 Amanda lives $2\frac{3}{4}$ miles from work. She works 5 days a week.
How many miles does she cover, travelling to and from work, in a week?

5 Janine goes shopping for a pair of shoes.
How much does Janine save if she buys these shoes in the sale?

6 Jam is on special offer at the supermarket.
How much do 2 jars cost?

7 A magazine has 60 pages. Nine tenths of these pages have photos.

a) How many pages have photos?

b) There are 80 photos in the magazine.
Three quarters of them are in colour.
How many are in black and white?

8 Paula buys 60 lbs of boiled sweets.
She makes up twenty $\frac{1}{4}$ lb bags,
fifteen $\frac{1}{2}$ lb bags and fifteen $\frac{3}{4}$ lb bags.
How much has she left over?

9 a) Write down the reciprocal of
(i) 5 (ii) $\frac{1}{5}$. What is $5 \times \frac{1}{5}$?

b) Write down the reciprocal of 1.

c) Does 0 have a reciprocal?

Dividing fractions

Becky and Ravi share $\frac{3}{4}$ of a pizza.

$$\frac{1}{2} \text{ of } \frac{3}{4} = \frac{1}{2} \times \frac{3}{4} = \frac{3}{8}$$

They each have $\frac{3}{8}$ of the pizza.

Another way of working this out is to say $\frac{3}{4}$ of a pizza is divided between 2 people.

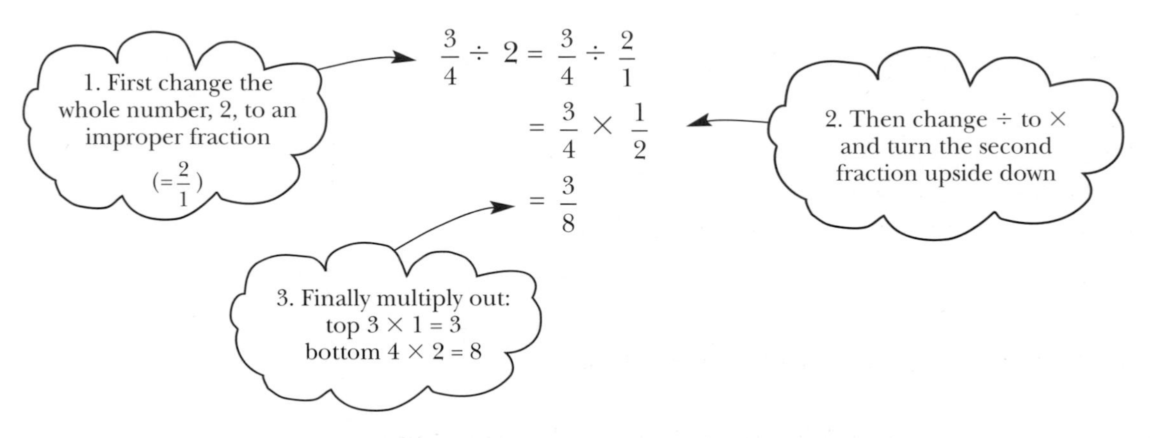

Dave, Becky and Ravi share $4\frac{1}{2}$ chocolate bars equally.

How much does each person get?

Ravi works it out like this:

Dave does it like this:

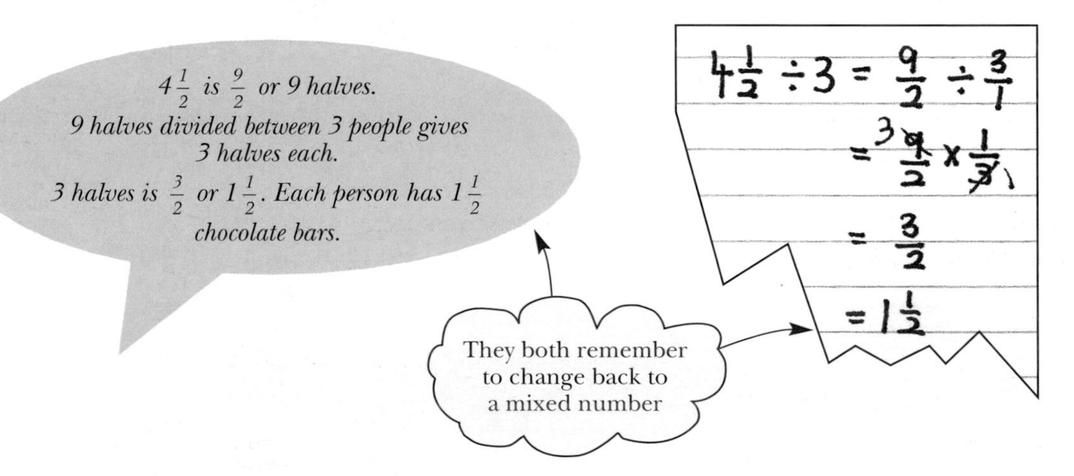

Here are two more examples of division.

$$5 \div \frac{3}{4} = \frac{5}{1} \div \frac{3}{4}$$
$$= \frac{5}{1} \times \frac{4}{3}$$
$$= \frac{20}{3} = 6\frac{2}{3}$$

$$4\frac{1}{2} \div 1\frac{1}{4} = \frac{9}{2} \div \frac{5}{4}$$
$$= \frac{9}{{}_1\not{2}} \times \frac{\not{4}^2}{5}$$
$$= \frac{18}{5} = 3\frac{3}{5}$$

Exercise

1 Work out

a) $5 \div 4$　b) $\frac{1}{5} \div 2$　c) $1\frac{3}{5} \div 4$　d) $\frac{1}{3} \div 3$

e) $2\frac{1}{2} \div 5$　f) $2\frac{1}{4} \div 3$　g) $\frac{5}{8} \div 2$　h) $1\frac{1}{2} \div 6$

2 Work out

a) $4 \div \frac{1}{3}$　b) $3 \div \frac{1}{2}$　c) $12 \div \frac{3}{4}$　d) $12 \div \frac{2}{5}$

e) $3\frac{1}{2} \div 4$　f) $2\frac{1}{4} \div 1\frac{1}{4}$　g) $3\frac{3}{4} \div \frac{3}{8}$　h) $2\frac{3}{16} \div 1\frac{1}{4}$

i) $2\frac{5}{8} \div 3\frac{1}{2}$　j) $8\frac{3}{4} \div 1\frac{1}{4}$　k) $6\frac{7}{8} \div 2\frac{3}{4}$　l) $12 \div 3\frac{1}{3}$

3 A grocer buys pieces of cheese weighing 5 kg.

a) How many $\frac{1}{2}$ kg pieces can he get from this?

b) How many $\frac{1}{4}$ kg pieces can he get from this?

4 A box is $12\frac{1}{2}$ inches long, 5 inches wide and $1\frac{1}{4}$ inches high.

Toy bricks are cubes with edges $1\frac{1}{4}$ inches long.

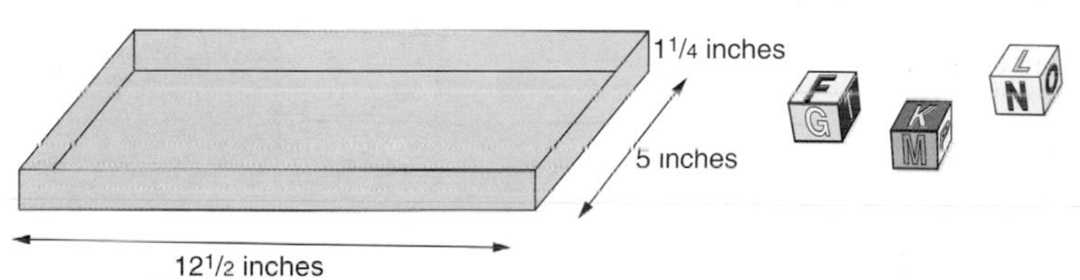

How many toy bricks can fit into the box?

5 Jermaine's car does 35 miles per gallon and he has 6 gallons of petrol in the tank.

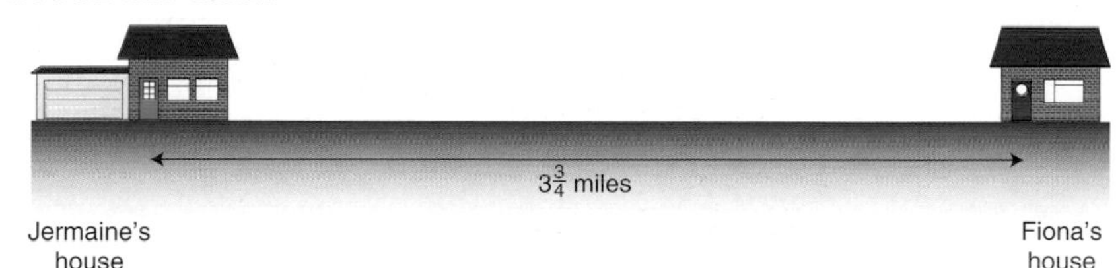

How many times can he go to Fiona's house and back?

6 A bookshelf is $29\frac{1}{4}$ inches long.

How many books can fit on the shelf if each book is

a) $\frac{3}{4}$ inch thick?　b) $1\frac{1}{8}$ inches thick?　c) $1\frac{5}{8}$ inches thick?

Finishing off

Now that you have finished this chapter you should be able to:

- ★ find equivalent fractions and simplest form
- ★ change improper fractions and mixed numbers
- ★ add, subtract, multiply and divide fractions
- ★ find a fraction of a quantity
- ★ write down a reciprocal.

Use the questions in the next exercise to check that you understand everything.

Mixed exercise

1 Look at this room booking chart for Kathryn's hotel.

Kathryn puts an x in the box for a booked room.

What fraction (in its simplest form) of rooms are booked on

a) Monday? b) Tuesday?

c) Wednesday? d) Thursday?

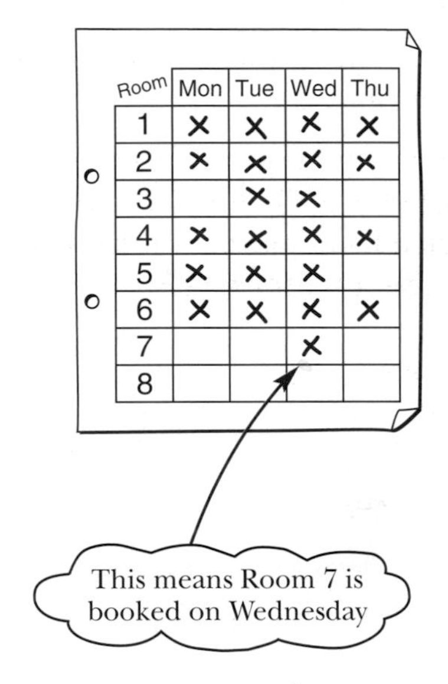

Room	Mon	Tue	Wed	Thu
1	×	×	×	×
2	×	×	×	×
3		×	×	
4	×	×	×	×
5	×	×	×	
6	×	×	×	×
7			×	
8				

2 Martyn, Chris and Rebecca set up a business.

Martyn gives one eighth of the money.

Chris gives three eighths.

Rebecca gives the rest.

What fraction of the money does Rebecca give?

3 Parvez is a chef. He allows 1 kg of rice for 8 people.

How many people can he serve with $2\frac{1}{4}$ kg?

4 Bottles of wine are packed in boxes of 6.

Diana has $3\frac{1}{2}$ boxes. How many bottles of wine does Diana have?

5 Lyn is orienteering. Here is her map.

She has reached Checkpoint 2.

a) What distance has she travelled?

b) How far has she still to go?

c) What distance will she have travelled when she reaches the finish?

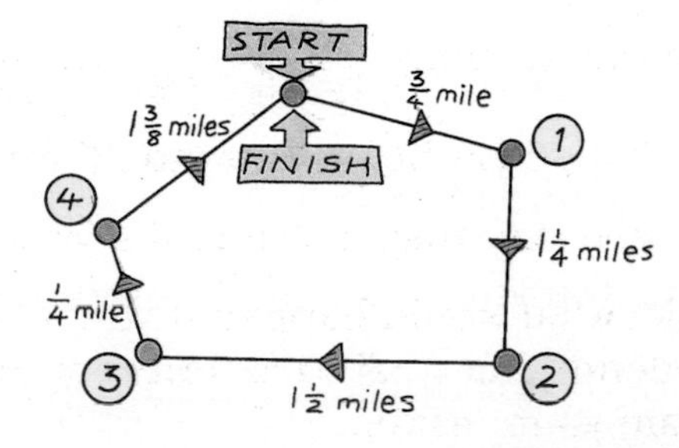

6 Arrange these numbers in order of size, smallest first.

a) $4 \quad \frac{11}{3} \quad \frac{17}{4} \quad 4\frac{3}{16}$ b) $\frac{11}{4} \quad 3 \quad 2\frac{13}{16} \quad \frac{23}{8}$

7 a) Write down the reciprocal of (i) 8 (ii) $\frac{1}{4}$ (iii) $1\frac{1}{2}$.

b) What is the result when you multiply a number by its reciprocal?

8 Hightown's annual rainfall last year was 96 cm.

a) One sixth ($\frac{1}{6}$) of the rain fell in January. How many cm is this?

b) One eighth ($\frac{1}{8}$) of the rain fell in April. How many cm is this?

9 Elizabeth buys this jacket.

a) How much does she save by buying it in the sale?

b) Another shop prices the same jacket at £65, and reduces the price by a quarter in the sale. Is this a better deal?

10 Work out

a) $2\frac{7}{8} + 1\frac{3}{4}$ b) $5\frac{1}{16} - 4\frac{1}{4}$ c) $5\frac{2}{3} - 1\frac{1}{6}$ d) $2\frac{4}{5} + 6\frac{7}{10}$

e) $\frac{1}{4} \times 2\frac{2}{3}$ f) $6\frac{3}{4} \times 1\frac{1}{2}$ g) $4\frac{5}{8} \times 1\frac{1}{4}$ h) $1\frac{1}{3} \times 3\frac{3}{4}$

i) $1\frac{1}{2} \div 2$ j) $2\frac{3}{4} \div \frac{1}{4}$ k) $1\frac{1}{8} \div 4\frac{1}{2}$ l) $6\frac{1}{4} \div 1\frac{2}{3}$

11 Work out

a) $5\frac{1}{2} - 3\frac{3}{4} - 1\frac{1}{8}$ b) $(\frac{2}{3} - \frac{1}{6}) \div 5$ c) $6 \div (\frac{1}{2} + \frac{1}{4})$

d) $2\frac{1}{2} \times 1\frac{3}{5} \times 1\frac{1}{4}$ e) $2\frac{1}{4} + 3\frac{2}{3} + 4\frac{1}{2}$ f) $(4\frac{1}{2} - 1\frac{1}{6}) \div \frac{3}{4}$

Investigation

Write the fractions $\frac{1}{12}, \frac{2}{12}, \frac{3}{12}, \ldots, \frac{11}{12}$, in their simplest form.

What happens when you try to do this for elevenths?

What about ninths, tenths and thirteenths?

Predict what would happen if you tried to simplify the set of fractions with denominator 29 (i.e. different numbers of twenty-ninths). Explain your answer.

Four

Decimals

Tenths and hundredths

This number line between 2 and 3 is split into **tenths**.

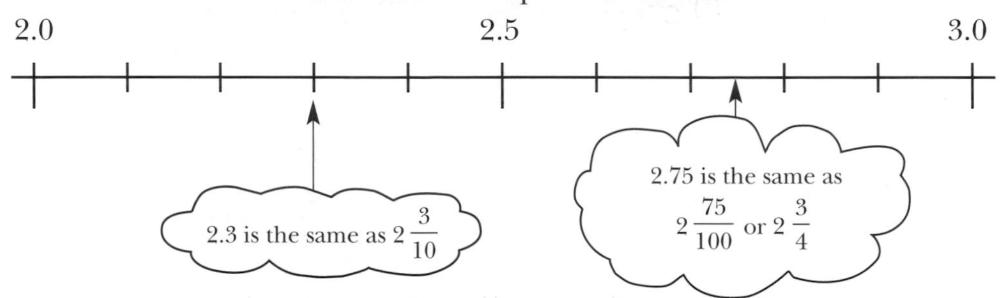

When a tenth is divided into ten parts each part is a **hundredth**.

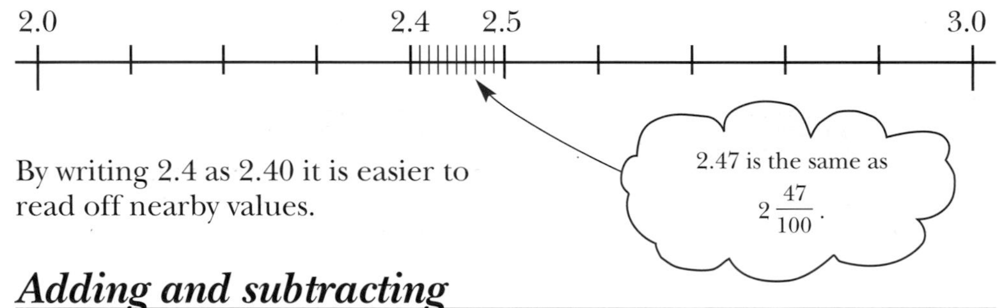

By writing 2.4 as 2.40 it is easier to read off nearby values.

Adding and subtracting

This is the design of Sophie's garden

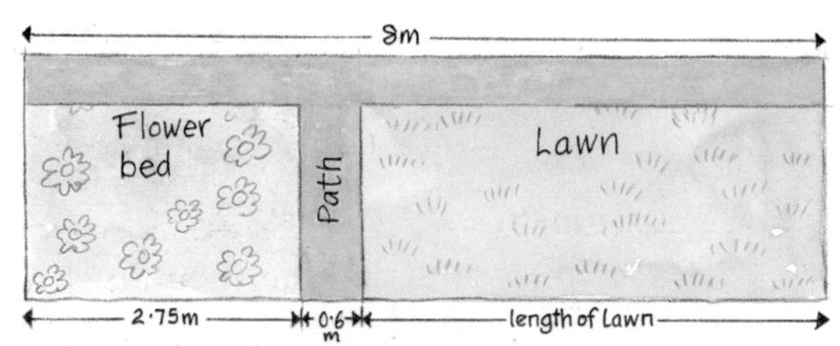

She wants to know how long her lawn is.

She works it out like this:

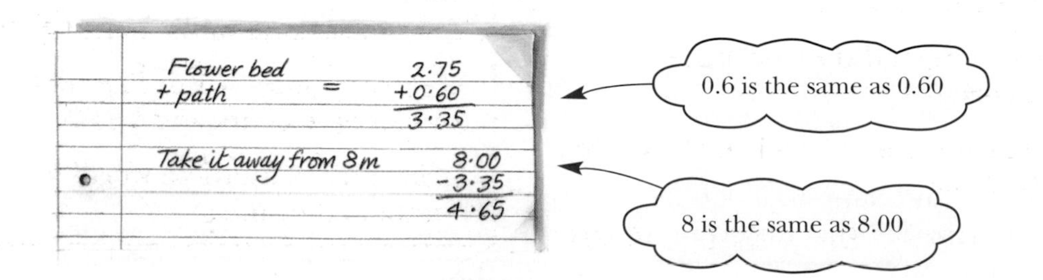

How long is the lawn?

Exercise

1 Write down a fraction equal to each of these decimals.

a) 0.3 b) 0.43 c) 9.2 d) 7.09

2 Write down a decimal equal to each of these fractions.

a) $\frac{7}{10}$ b) $2\frac{3}{10}$ c) $\frac{47}{100}$ d) $3\frac{7}{100}$

3 Ria takes the temperature of 3 patients. The readings are shown below (in degrees Celsius). Write down each reading.

a) 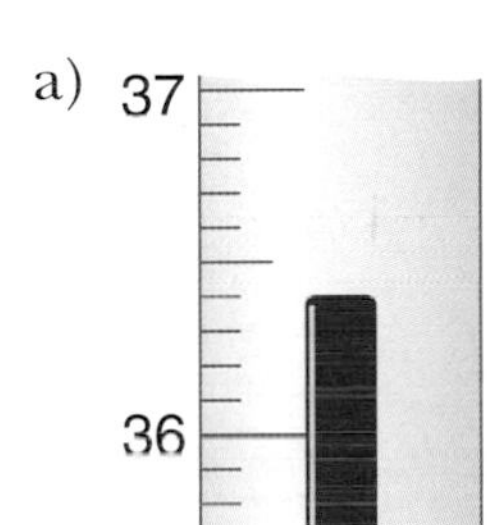

b) 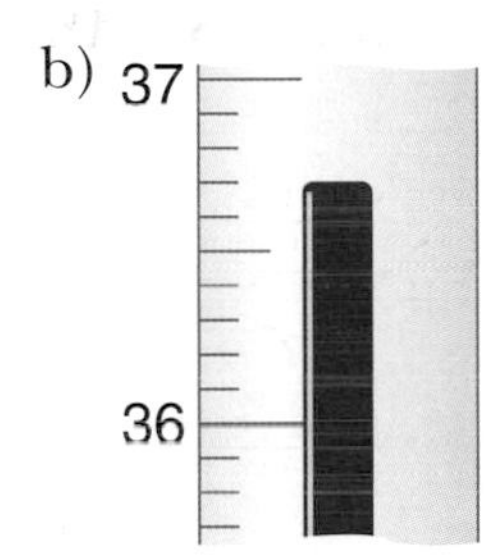

c) 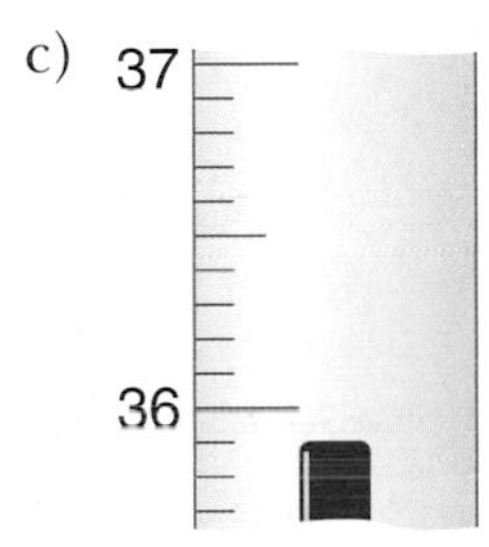

4 The diagrams show the height of a high jump bar (in metres). Write down the height of the top of each bar.

a) 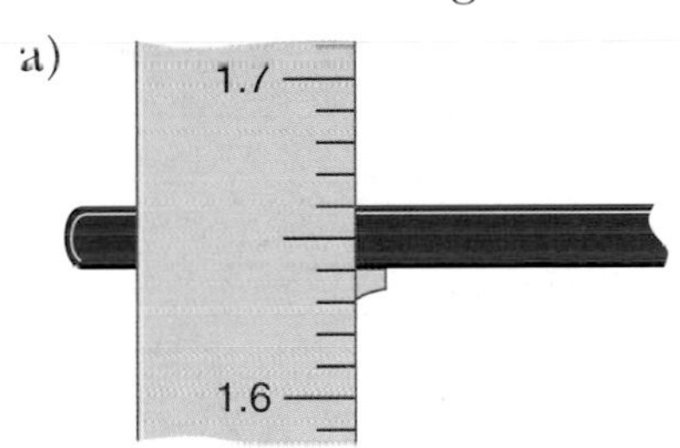

b) 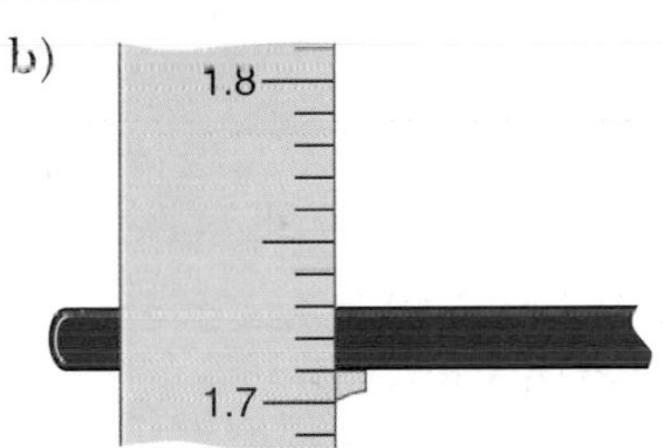

5 Work out

a) 4.6 + 9.3 b) 18.3 + 5.26 c) 4.008 + 1.9
d) 10.9 – 5.4 e) 12 – 7.9 f) 21.3 – 4.26

6 Gemma buys a birthday card for £1.80, a newspaper for £0.45 and a bag of crisps for £0.36. How much change does she get from £5?

7 The school discus record is 39.24 metres.

On his first throw Mick throws 38.16 metres.

a) How far short of the record is he?

b) With his second throw Mick throws 41.02 metres.

How much did he beat the record by?

8 Arrange these numbers in order of size, smallest first.

0.8 0.08 0.1 0.81 0.009

In metric units, 1 millimetre is $\frac{1}{1000}$ of a metre or 0.001 metres.
In the same way, 1 milligram is $\frac{1}{1000}$ of a gram or 0.001 grams.
You can say that milli- means 0.001.

There are other words like milli-, for example centi- and kilo-.

Make a list of all these words and what numbers they mean.

Multiplying and dividing

Multiplying or dividing by 10, 100, … can be carried out quickly as follows:

multiplying by 10 moves the decimal point one place to the right

multiplying by 100 moves the decimal point two places to the right

- $3.5 \times 10 = 35$ (3.5 becomes 35) • $7.4 \times 100 = 740$

dividing by 10 moves the decimal point one place to the left

dividing by 100 moves the decimal point two places to the left

- $29 \div 10 = 2.9$ (29 becomes 2.9) • $1.2 \div 100 = 0.012$

Remember: multiplying by 10, 100, … makes a number bigger
dividing by 10, 100, … makes a number smaller

Paul is leading a party of 12 people (including himself) on a hiking weekend. They stay one night at a hostel that charges £11.35 per person.

How much does it cost the group?

The cost is $12 \times £11.35$. Paul works it out like this:

He makes the 11.35 into a whole number by moving the decimal point 2 places to the right…

…then he multiplies the whole numbers

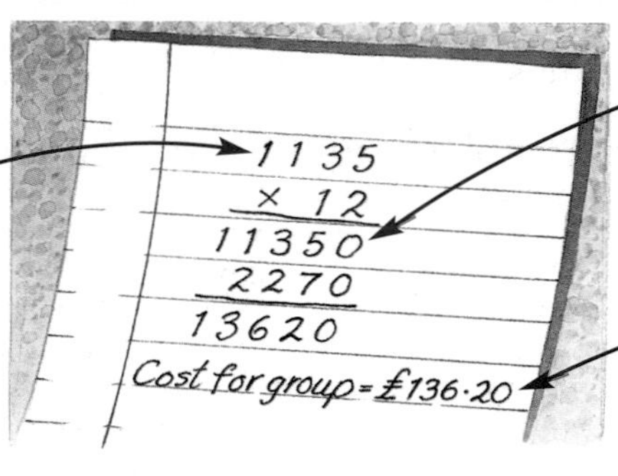

He moves the decimal point 2 places to the left to get his final answer

Why does Paul's method work?

Paul and Angela share the driving. They drive the minibus 186 miles during the weekend. It does 6.2 miles per litre of petrol.
How many litres of petrol does it use?

Paul works it out like this:

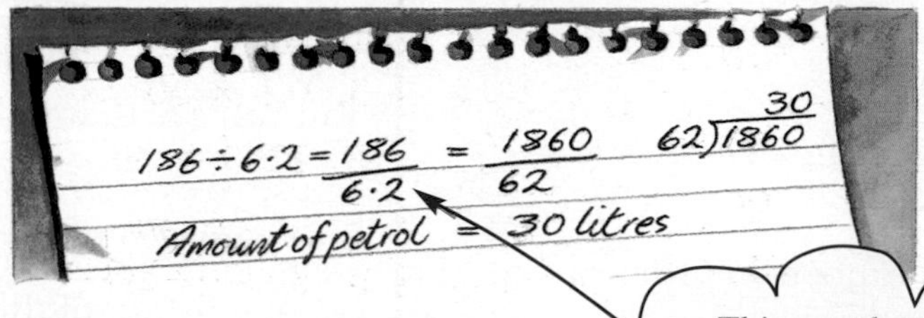

This must be a whole number before you divide. The top and bottom of the fraction have been multiplied by 10

The minibus uses 30 litres of petrol.

Exercise

1
a) 8.4×10 b) 3.9×100 c) $56 \div 10$ d) $7.3 \div 100$
e) 72×10 f) 17.5×100 g) $240 \div 10$ h) $3000 \div 100$
i) 0.6×1000 j) $0.5 \div 10$ k) 4.375×100 l) $82 \div 1000$

2
a) 3×1.45 b) 3.2×1.5 c) 0.6×5 d) 6×0.568
e) 1.2×3.5 f) 4.16×3 g) 0.2×5.2 h) 2.5^2

3
a) $6 \div 0.3$ b) $16.8 \div 1.2$ c) $15.4 \div 0.35$ d) $5 \div 0.8$
e) $4 \div 0.5$ f) $7.2 \div 2$ g) $5.6 \div 1.2$ h) $0.27 \div 0.9$

4 You can convert temperatures from Celsius to Fahrenheit by multiplying by 1.8 and adding 32. Convert each of these, giving your answer to the nearest degree.

Malta	27 °C
Cyprus	29 °C
Tunisia	36 °C

5 Hana's car travels 7.2 miles on one litre of petrol.

Work out, to the nearest litre, how much petrol she needs for a journey of

a) 115 miles b) 295 miles

c) Petrol costs 71.5p per litre. Work out the cost of petrol to the nearest 10p for each journey.

6 A nursing home has 20 residents each paying £400 a week.

a) What is the total income per week?

b) What is the total income per year?

7 Western Bank made a profit of eighty million pounds last year.

Approximately how much profit is made each week?
(Take 1 year = 50 weeks)

8 Work out the total cost of this stationery order.

Item	Unit cost (£)	Quantity	Cost (£)
Pens	0·16	300	
Pads of paper	0·80	50	
Files	1·20	200	
Envelopes	0·05	4000	
		Total	

Try out the following mental arithmetic test on your friends.

1. 30×10
2. $600 \div 20$
3. 0.1×50
4. 40×700
5. $1200 \div 30$
6. $\frac{1}{10} \times 900$
7. 250×40
8. 0.01×600
9. $8000 \div 20$
10. $10 \times 20 \times 50$

Now make up a test of your own to try out on your friends.

Fractions to decimals

You know that $\frac{1}{4} = 0.25$, $\frac{1}{2} = 0.5$ and $\frac{3}{4} = 0.75$.

What about $\frac{1}{8}$?

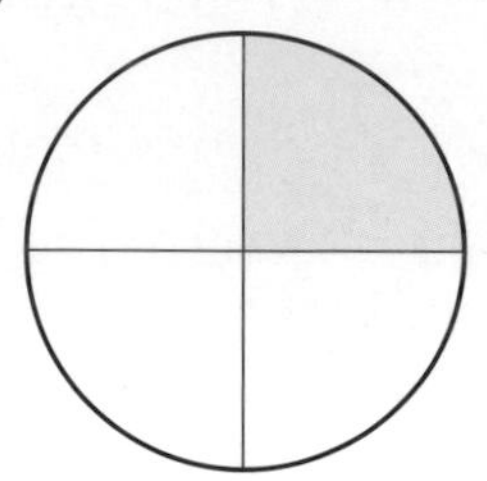

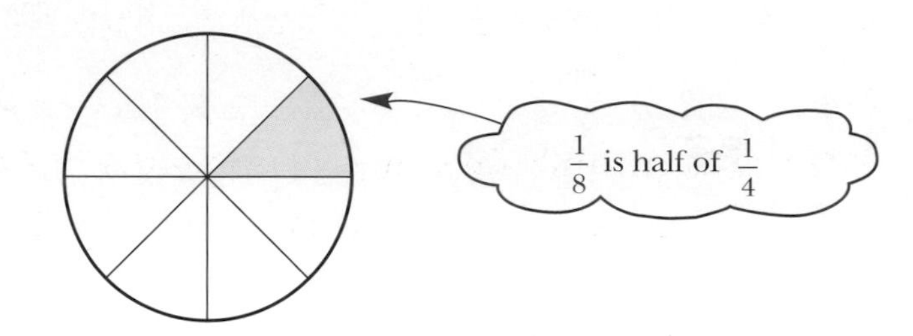

What decimal is half of 0.25?

$0.25 \div 2 = 0.125$

So $\frac{1}{8} = 0.125$

Also, $\frac{1}{8}$ is $1 \div 8 = 0.125$

$8\overline{)1.000}$ = 0.125

So again, $\frac{1}{8} = 0.125$.

We can do this to change any fraction into a decimal.

What is $\frac{1}{5}$ as a decimal?

Something interesting happens when we work out $\frac{1}{3}$ as a decimal.

$\frac{1}{3} = 1 \div 3 = 0.333\ldots$

the 3s go on forever!

$3\overline{)1.000\ldots}$ = 0.333...

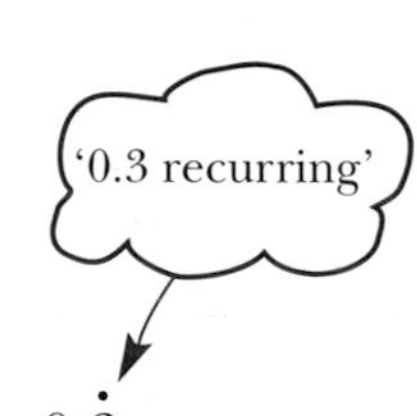

This is a **recurring decimal**.

Instead of writing out all the 3s (that would take forever!) we write $0.\dot{3}$.

The same thing happens for $\frac{1}{6}$.

$\frac{1}{6} = 1 \div 6 = 0.166\ldots$

$6\overline{)1.000\ldots}$ = 0.166...

So $\frac{1}{6} = 0.1\dot{6}$

'0.16 recurring'

Changing fractions to decimals lets you make comparisons.

Which is larger, $\frac{1}{6}$ or 0.17?

Exercise

1 Change the fraction to a decimal:

a) $\frac{3}{5}$ b) $\frac{3}{8}$ c) $\frac{9}{20}$ d) $\frac{5}{4}$

e) $\frac{8}{25}$ f) $\frac{1}{16}$ g) $\frac{7}{40}$ h) $\frac{61}{50}$

2 Change the fraction to a recurring decimal:

a) $\frac{4}{3}$ b) $\frac{1}{9}$ c) $\frac{5}{6}$ d) $\frac{5}{12}$

3 Arrange these numbers in order of size, smallest first.

0.66 0.07 $\frac{3}{5}$ 0.1 $\frac{2}{3}$

Investigation 1

What is $\frac{1}{11}$ as a decimal?

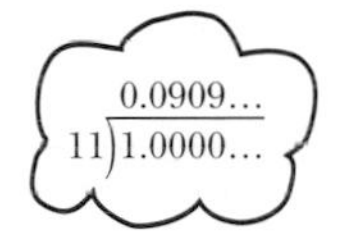

So $\frac{1}{11} = 0.\dot{0}\dot{9}$

a) Work out the decimal form of

(i) $\frac{2}{11}$ (ii) $\frac{6}{11}$

These results have been put into this table.

Fraction	Decimal
$\frac{1}{11}$	0.0909...
$\frac{2}{11}$	0.1818...
$\frac{3}{11}$	
$\frac{4}{11}$	
$\frac{5}{11}$	
$\frac{6}{11}$	0.5454...
$\frac{7}{11}$	
$\frac{8}{11}$	
$\frac{9}{11}$	
$\frac{10}{11}$	

b) What do you think the missing decimals are?

Investigation 2

What is $\frac{1}{7}$ as a decimal?

So $\frac{1}{7} = 0.\dot{1}4285\dot{7}$

a) Copy and complete this table.

Fraction	Decimal
$\frac{1}{7}$	$0.\dot{1}4285\dot{7}$
$\frac{2}{7}$	
$\frac{3}{7}$	
$\frac{4}{7}$	
$\frac{5}{7}$	
$\frac{6}{7}$	

b) Comment on your results.

Investigation 3

Some fractions such as $\frac{1}{8} = 0.125$ have a decimal form that terminates.

Others such as $\frac{1}{6} = 0.166\ldots$ have a decimal form that recurs.

Investigate which fractions have a decimal form that terminates and which recur.

Finishing off

Now that you have finished this chapter you should be able to:

- ★ change between decimal form and fractions
- ★ arrange numbers in size order
- ★ add and subtract decimals
- ★ multiply decimals (including multiplying by 10, 100, etc.)
- ★ divide decimals (including dividing by 10, 100, etc.)
- ★ work out squares and square roots of decimals.

Use the questions in the next exercise to check that you understand everything.

Mixed exercise

1 How tall are these people?

a)

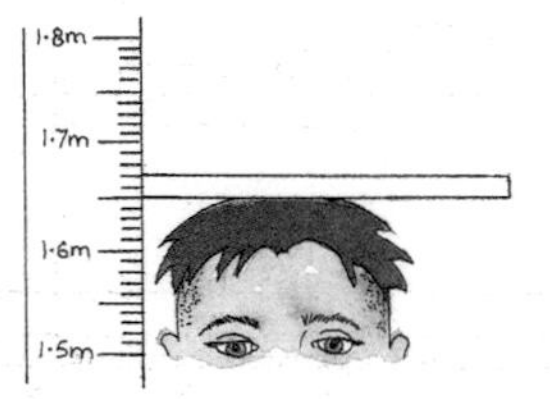

b)

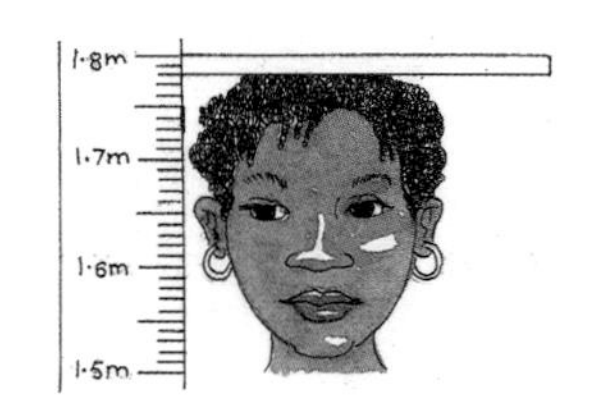

2 Change the decimal to a fraction:

a) 0.7 b) 5.4 c) 1.83 d) 6.371

3 Change the fraction to a decimal:

a) $2\frac{3}{10}$ b) $3\frac{17}{100}$ c) $6\frac{7}{100}$ d) $\frac{141}{1000}$

4 Arrange the following in order of size, smallest first.

$\frac{7}{3}$ 2.04 $2\frac{3}{10}$ 2 2.35

5 A pencil is 5 inches long. How many centimetres is it?
(Remember 1 inch = 2.54 cm.)

6 This chart shows Hana's temperature when she was in hospital.
From Monday to Tuesday, Hana's temperature falls from 38.5 to 37.9 °C.

This is a 0.6 °C fall.

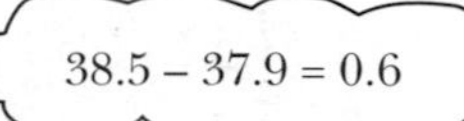

What rise or fall is there

a) from Tuesday to Wednesday?

b) from Wednesday to Thursday?

c) from Thursday to Friday?

Day	Temperature (°C)
Mon	38·5
Tue	37·9
Wed	37·3
Thu	36·9
Fri	37·0

7 Veena gets these quotes for her treasure hunt packs.

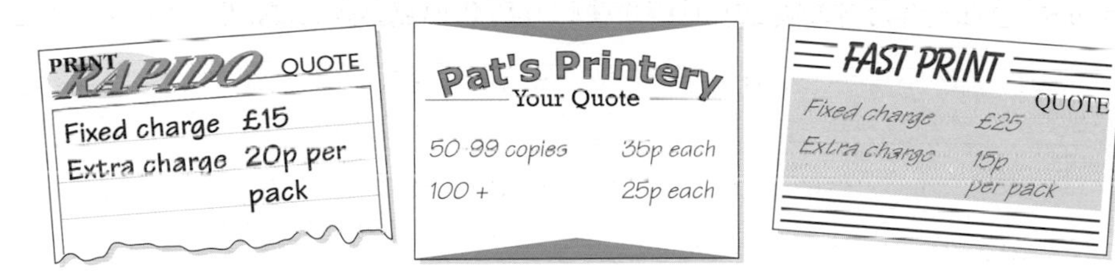

a) What is the lowest cost for 80 packs?

b) What is the lowest cost for 275 packs?

c) When does Print Rapido become cheaper than Pat's Printery?

8 A path is 9 metres long. How many paving slabs are needed to make the path if each slab is

a) 0.9 metres long? b) 0.6 metres long?

9 This is Shamir's bookcase. The wood is 1.6 cm thick.

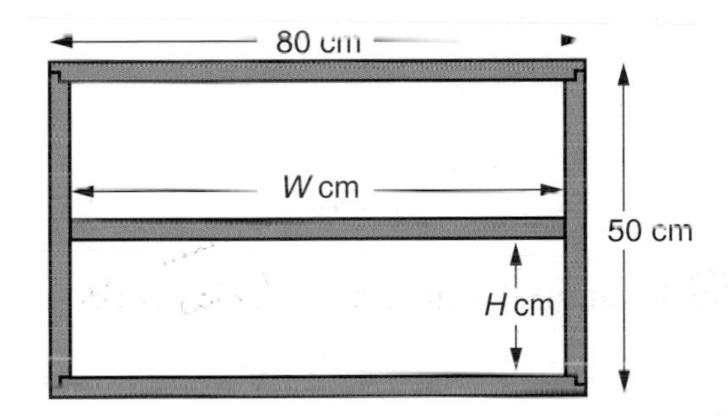

a) Work out the inside width, W cm, of the shelves.

b) The shelves are of equal height.

Work out the inside height, H cm, of each shelf.

10 50 000 people each pay £30 to attend a pop concert.

a) How much money is collected?

b) 4000 T-shirts are sold at the concert. They cost £52 000 altogether.

How much does one T-shirt cost?

11 Find the value of

a) 5.3^2 b) $\sqrt{8.41}$

c) 6.8^2 d) $\sqrt{15.21}$

12 Change these fractions into decimals.

a) $\frac{4}{5}$ b) $\frac{5}{8}$

c) $\frac{7}{20}$ d) $\frac{7}{6}$

e) $\frac{5}{16}$ f) $\frac{2}{7}$

Find examples of measuring instruments with scales in

a) whole b) 0.5

c) 0.2 d) 0.1 units.

Can you find examples of any other divisions?

What does each instrument measure?

Five

Percentages

25%, 50% and 75%

Student Survey findings

FACT **50% of Year 10 have part-time jobs**

The survey says that 50% of students have a part-time job.

That means 50 out of every 100.

You can show this in a 10×10 square like this:

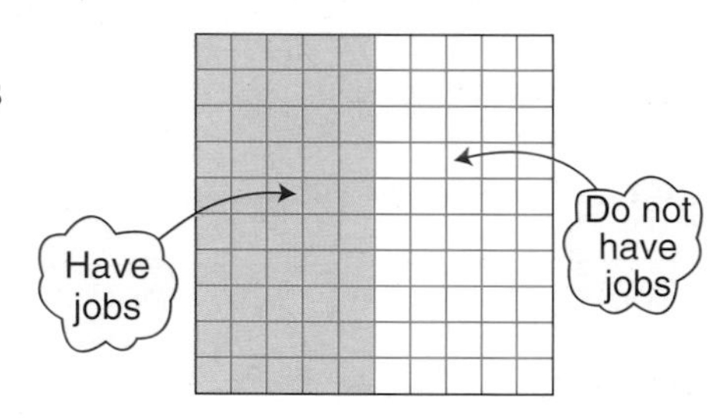

What fraction is shaded? Write the fraction in its simplest form.

Write this fraction as a decimal.

You can see that 50% is the same as $\frac{1}{2}$ or 0.5. So half of Year 10 students have part-time jobs.

FACT **25% of Year 10 cannot swim**

Again you can show this in a 10×10 square:

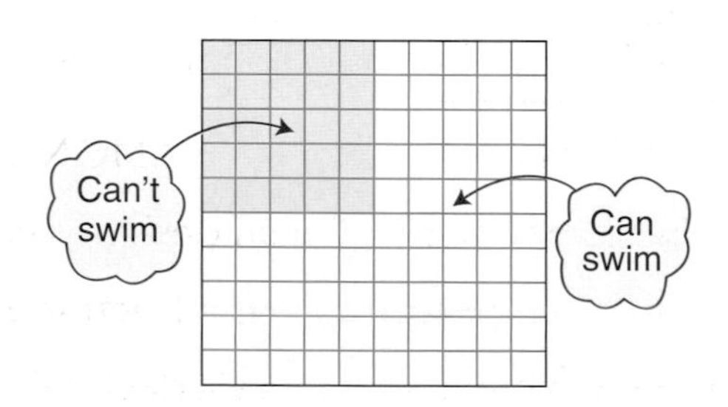

What fraction of the students can't swim? Write the fraction in its simplest form.

You can see that 25% is the same as $\frac{1}{4}$ or 0.25.

You can also see from the diagram that 75% of the students can swim.

What fraction of the students can swim?

75% is the same as $\frac{3}{4}$ or 0.75.

1 Look at this pie chart.

It shows the amount of sales a company makes in different parts of the world.

Other
UK
USA
Rest of Europe

a) Which of these is the correct answer?

(i) The amount of sales in the UK is

A less than 25% B 25% C more than 25%

(ii) The amount of sales in the Rest of Europe is

A less than 50% B 50% C more than 50%

(iii) The amount of sales in the USA is

A less than 25% B 25% C more than 25%

b) Estimate the percentage of sales in other countries.

2 Ben scores 11 marks out of 20 in a test. Is this

A less than 50% B 50% or C more than 50%?

3 Pat's food intake is 28% fat, 15% protein and the rest is carbohydrates.

What percentage is carbohydrates?

4 Each of these floor designs is made of 100 tiles.

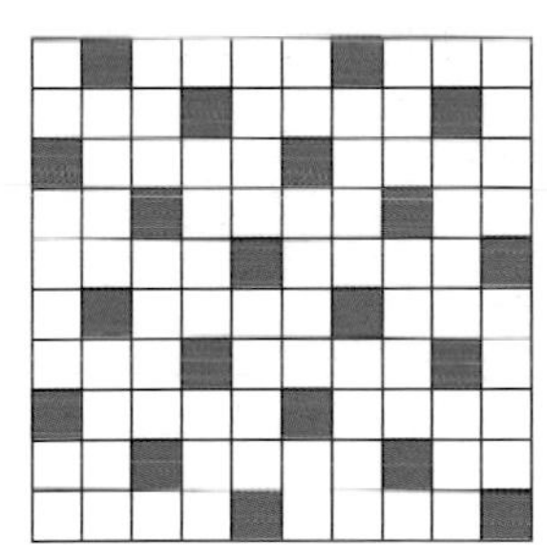

KNIGHT

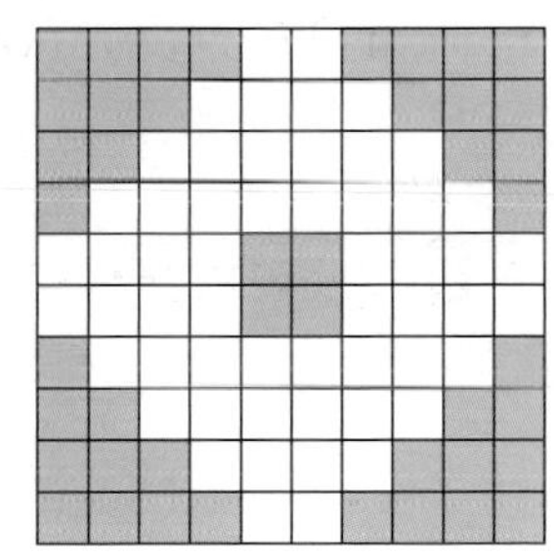

DIAMOND

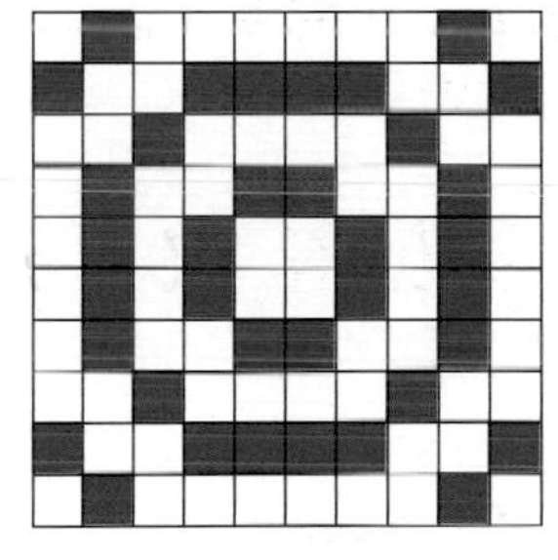

PATHWAYS

a) What percentage of each design is coloured?

b) Write each percentage as a fraction in its simplest form.

c) Write each percentage as a decimal.

Make two floor designs of your own where 20% of the tiles are white and the rest are black.

Look at real floors and tiles for ideas.

Percentage calculations

Stephen runs a fashion business.

Mark works for Stephen.

He earns £170 a week.

He gets a 3% pay rise.

Mark works out how much extra money he will get like this:

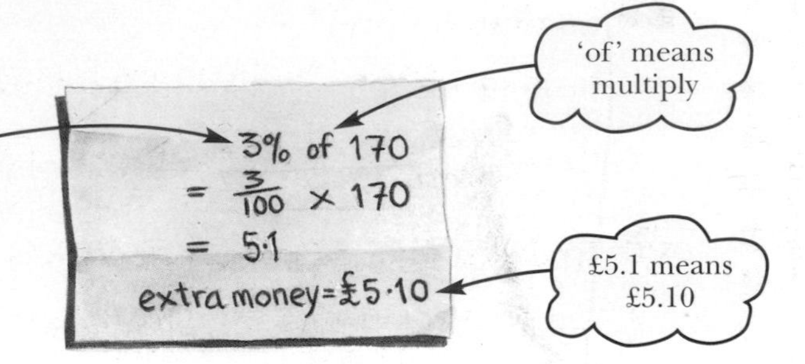

How much does Mark earn after the pay rise?

You can work it out as £170 + £5.10 = £175.10

100% + 3% = 103%

Alternatively you can find 103% directly.

$$\mathbf{103\% = \frac{103}{100} = 1.03 \text{ and } £170 \times 1.03 = £175.10}$$

Use whichever method you feel confident with.

Staff get 15% discount when they buy clothes from Stephen.

Mark buys clothes worth £125.60.

His discount is 15% of £125.60.

He works it out like this:

How much does he pay for the clothes?

You can work it out as £125.60 – £18.84 = £106.76

100% – 15% = 85%

Alternatively you can find 85% directly.

$$\mathbf{85\% = \frac{85}{100} = 0.85 \text{ and } £125.60 \times 0.85 = £106.76}$$

In the first year Stephen's business made a profit of £40 000.

Stephen expects his profit to rise by 7% each year.

His expected profit for the second year can be written £40 000 × 1.07.

What does £40 000 × $(1.07)^2$ *represent?*

What does £40 000 × $(1.07)^3$ *represent?*

Exercise

1 Work out

a) 80% of 300 b) 60% of 250 c) 50% of 631

d) 15% of 580 e) $12\frac{1}{2}$% of 200 f) 3.6% of 775.

2 Jordan is doing a survey by post.

He sends out 250 questionnaires and expects to get 30% back.

How many replies does he expect?

3 Emily sells her house for £75 000. The estate agent charges her 2% of the selling price.

a) How much does she pay the agent?

b) Emily later finds out that another agent is charging only 1.75%.

How much would she have saved?

4 Neil earns £8000 a year. He is given a pay rise of 3%.

a) How much does he earn after the rise?

b) The following year Neil gets a 4% pay rise.

How much does he earn after this rise?

5 This year 650 full-time and 5000 part-time students enrolled at Avonford College. A 6% rise in full-time enrolments and a 8% rise in part-time enrolments is planned for next year. How many students in total does the college expect next year?

6 A holiday costs £400. This price is increased by 10%, then the price is reduced by 10% for last-minute bookings.

How much does a last-minute booking cost?

7 Bob runs a travel company. He expects to increase prices by 5% next year.

A Mediterranean cruise costs £1500 this year.

a) How much will the cruise cost next year?

b) He expects to increase prices the following year by 4%.

How much will the cruise cost the following year?

c) How would the answer to b) be affected if the prices were increased by 4% next year and 5% the following year?

8 The population of an island is 100 000.

It is predicted that in each year its population will rise by 5% of the figure at the start of the year.

a) Work out the predicted population after 1 year.

b) Work out the predicted population after 2 years.

c) What does $100\ 000 \times (1.05)^{10}$ represent?

d) Explain whether or not the predicted population will have exceeded 200 000 after 20 years.

Further percentage problems

Josh works for a charity that wants to buy this computer. The price includes VAT at 17.5%.

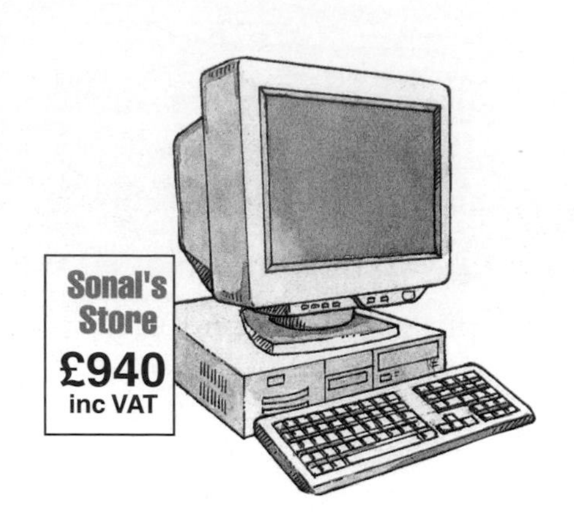

The charity does not pay VAT.

He works out how much the charity pays like this:

117.5% is 940

1% is $\frac{940}{117.5}$

100% is $\frac{940}{117.5} \times 100 = 800$

The charity pays £800 for the computer.

How much does the charity save by not paying VAT?

Josh goes to another store and buys this printer. There is a special promotion today.

He pays £170 for the printer.

The usual price is 100% and the discount is 15% so he pays 100% – 15% = 85%

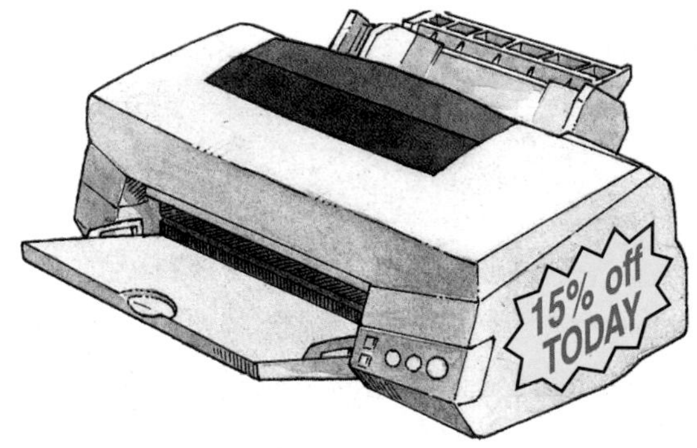

Josh works out the usual price like this:

85% is 170

1% is $\frac{170}{85}$

100% is $\frac{170}{85} \times 100 = 200$

The usual price of this printer is £200.

How much does Josh save by buying it on the special promotion?

Exercise

1 These prices are inclusive of VAT at 17.5%.

Work out the price exclusive of VAT.

a) 

b) £150

c) 

2 Jenna has just had a 4% pay rise. Her salary is now £13 000.

a) What was her salary before the rise?

b) Twelve months later she gets another 4% rise.

What is her salary after this rise?

3 

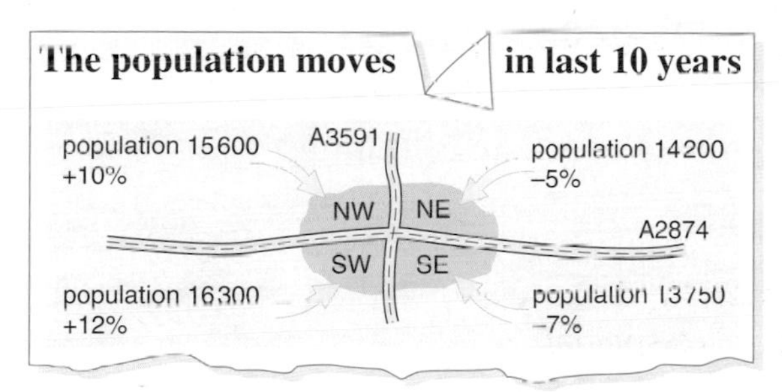

The newspaper gives the present population of each quarter of the town. The percentage change that has taken place in the last 10 years is also given.

a) What is the missing word in the headline? Choose North, South, East or West.

b) Work out the population of each quarter as it was 10 years ago.

c) Work out the percentage change in the number of people living North of the A2874.

4 Henry is a salesman. His sales are £350 000 this year.

This is 25% more than 3 years ago and 85% more than 7 years ago.

Work out his sales figures (to the nearest thousand pounds)

a) three years ago

b) seven years ago.

c) His target sales for next year is an 8% increase, with a further 8% increase the year after.

Work out his target sales figures for the next two years.

Fractions to percentages

Ali is doing a survey for a travel agent.

He asks 50 people

20 people say 'yes'.

What is $\frac{20}{50}$ as a percentage?

You need to change $\frac{20}{50}$ into an equivalent fraction with 100 on the bottom.

$$\frac{20}{50} \overset{\times 2}{=} \frac{40}{100} = 40\%$$

(numerator and denominator both ×2)

So $\frac{20}{50}$ is the same as 40% or 0.4.

Another way to obtain the percentage is to multiply the fraction by 100.

$$\frac{20}{50} \times 100 = 40$$

So $\frac{20}{50}$ is the same as 40%

What is $\frac{9}{20}$ as a percentage?

Work it out both ways to check you get the same answer.

Percentage change

Helen buys a new mountain bike and in the next month she cycles 600 miles on it. The previous month she did only 400 miles on her old one.

This is an increase of 200 miles.

You can write it as a percentage:

$$\textbf{Percentage increase} = \frac{\textbf{increase}}{\textbf{original}} \times \textbf{100}$$

In this case, percentage increase $= \frac{200}{400} \times 100$

$= 50$

Notice that a percentage increase (or decrease) is always based on the original value, not the new.

Exercise

1 Write down a percentage equal to each of these fractions.

a) $\frac{4}{5}$ b) $\frac{9}{25}$ c) $\frac{1}{8}$ d) $\frac{17}{40}$ e) $\frac{36}{120}$ f) $\frac{2}{3}$

2 Sam's survey of 80 people found that 38 read Daily News, 25 read News Today and the rest read neither paper. Nobody reads both.

What percentage of people read

a) Daily News? b) News Today? c) neither paper?

3 This bar chart shows the number of male and female employees at each of a firm's two sites.

a) How many employees does the firm have altogether?

b) What percentage of the employees are male?

c) What percentage work at Parkway?

4 Peter currently earns £16 000 a year. He is interested in these jobs.

What percentage increase in salary would Peter get as

a) a programmer?

b) a manager?

5 Nita and Mark run a business. They have drawn up this table of their profits last year.

Quarter	1	2	3	4
Profit (£ thousands)	32	44	50	34

a) What percentage of the total profit was made in the first quarter?

b) What percentage of the total profit was made in the second half of the year?

c) The expected profit for this year is £200 000. Work out the expected increase in profit as a percentage of last year's profit.

6 Jo compares her current sales figures with those of last year.

Month	May	June	July	August
Last year	£4000	£5000	£4800	£5600
This year	£4600	£4800	£5040	£5796

a) For each month, calculate the percentage increase or decrease in sales compared with last year.

b) Calculate the percentage increase in sales during the whole four month period.

Making comparisons

Ranjit does a survey of people's opinions of their local bus and train services. Here are his results.

Service	Number Satisfied	Number questioned
Bus	79	111
Train	37	59

Which service is satisfying more of its customers?

Ranjit works out what proportion of the customers he asked are satisfied with each service.

Bus: $\frac{79}{111}$ Train: $\frac{37}{59}$

These figures are still not easy to compare, so he writes them as percentages.

Bus: $\frac{79}{111} = 0.711\ldots = 71\%$ (to nearest 1%)

Train: $\frac{37}{59} = 0.627\ldots = 63\%$ (to nearest 1%)

The bus service satisfies more of its customers.

Sometimes the proportions you need to compare might be very close together. For example, which is larger, $\frac{95}{212}$ or $\frac{47}{105}$?

$\frac{95}{212} = 0.4481132\ldots$

$\frac{47}{105} = 0.447619\ldots$

You can see from the third decimal place that $\frac{95}{212}$ is slightly larger.

Writing these numbers as whole number percentages would not show the difference – they both round to 45%.

Which is larger, $\frac{1}{3}$ *or* 0.3?

Which is larger, $\frac{2}{11}$ *or* 0.182?

Exercise

1 Ella does a survey to find out people's opinions on two brands of personal stereo. These are her results.

Brand	Number questioned	Number reporting faults
A	129	13
B	186	21

Which brand is the more reliable?

2 This chart shows a manufacturer's daily output of bicycle frames from 3 production lines.

Line	Total output	Rejects
A	800	36
B	840	41
C	625	29

a) Work out the percentage of rejects for each production line.

b) Which production line is the most efficient?

c) Suggest a possible reason why the output of machine C was lower than A or B.

3 This table shows the numbers of votes cast in 4 parish council elections.

Parish	Votes cast	No. of voters
Waterbeach	105	267
Oakington	154	309
Witchford	187	393
Northwood	137	291

Which parish had

a) the highest percentage turnout?

b) the lowest percentage turnout?

c) a turnout of approximately 2 voters in 5?

4 Harriet manages her company's training centres. She draws up this table to compare the success of each centre.

Centre	No. of passes	No. of recruits
Northhill	35	44
Heartland	39	71
Southdown	41	52

a) Work out the percentage of recruits at each centre who passed.

b) How would you interpret these results?

Collect five newspaper articles containing expressions like '1 in 3' or '20%'.

a) Work out each fraction as a percentage.

b) Work out each percentage as a fraction in its simplest form.

c) Explain whether you would use '44%' or '11 in 25' in a heading.

Finishing off

Now that you have finished this chapter you should be able to:

★ find the fraction and decimal equivalents of simple percentages

★ work out percentage problems when you don't know the original amount

★ calculate the outcome of a percentage increase or decrease

★ change a fraction into a percentage

★ use percentages to calculate proportions.

Use the questions in the next exercise to check that you understand everything.

Mixed exercise

1 What percentage of the costs is

a) labour?

b) overheads?

c) materials?

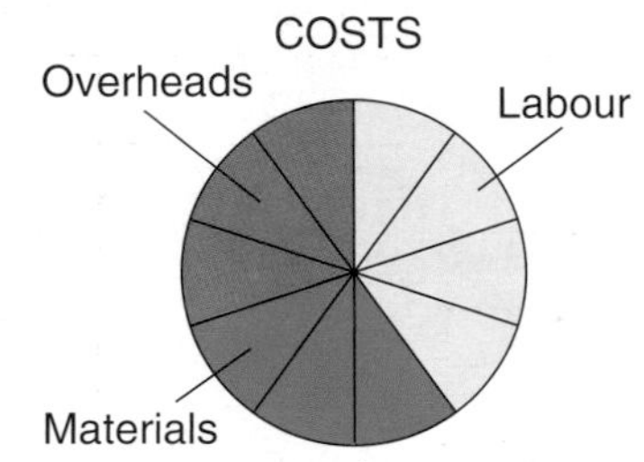

2 Anna does an experiment on the air someone breathes out.

Here are her results in litres.

a) What percentage is oxygen?

b) What percentage is nitrogen?

c) What percentage is carbon dioxide?

Total	Oxygen	Nitrogen	Carbon dioxide
200	32	160	8

3 Ben usually buys a 250 ml can of orange at lunchtime.

One day the can is larger and is marked '20% extra'.

How much orange does this larger can contain?

4 Keith sees the same rucksack on sale in 2 shops.

Which shop is cheaper?

5 Each day 4500 cars and 325 lorries cross this bridge.

a) How much is paid in tolls each day?

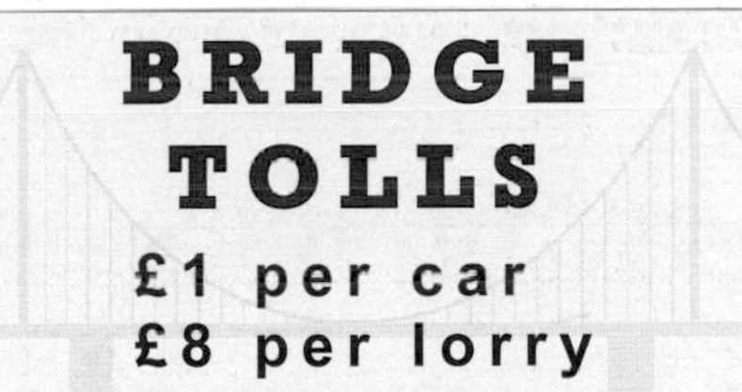

b) The car toll price is now increased by 30% and the lorry toll by 25%.

The number of cars crossing decreases by 5% and the number of lorries decreases by 8%.

How much is paid in tolls now?

c) Work out the percentage increase in tolls each day.

6 Quickcash Bank made a profit of £2.25 billion last year. This year it is expected that their profit will increase by 3.3%.

Work out the expected increase in profit, giving you answer to the nearest million pounds.

7 Ranjit is looking at how prices have changed in the last 3 years.

Item	Price now	% change over last 3 years
House	£75 200	up 14%
Car	£11 400	up 11%
Computer	£1199	down 30%
Calculator	£7.50	down 25%

Work out the prices 3 years ago giving your answers to 3 significant figures.

8 Tim's house is worth £90 000.

He expects the value to increase by 6% each year.

Find, to the nearest £100, the expected value of his house in

a) 1 year

b) 3 years

c) 12 years.

9 Isabel is testing children's mental arithmetic skills using a 20 question test.

a) The pass mark is 16. What percentage is this?

Isabel tests children in 3 schools and gets these results.

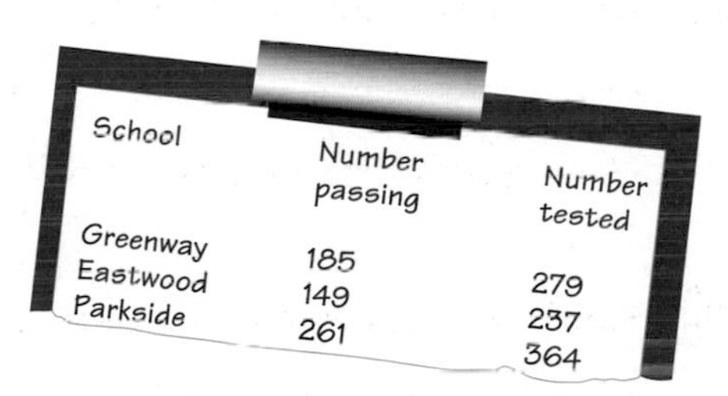

School	Number passing	Number tested
Greenway	185	279
Eastwood	149	237
Parkside	261	364

b) Which school has the highest pass rate?

c) Which school has the lowest pass rate?

10 Jo sells 80 tennis rackets in May. She expects sales to increase by 20% in June and then fall by a third in July. How many tennis rackets does she expect to sell in July?

Six

Units

Length, mass, capacity and time

What is the length of a) your arm? b) a football pitch?

In the **Imperial system** you can measure lengths in inches, feet, yards and miles.

12 inches = 1 foot
3 feet = 1 yard
1760 yards = 1 mile

In the **metric system** you can measure lengths in millimetres (mm), centimetres (cm), metres (m) or kilometres (km).

10 mm = 1 cm
100 cm = 1 m
1000 m = 1 km

Sandy's cat is 3 kilograms. In every day English this is called the cat's **weight**. In science the word **mass** is used instead.

What is the mass of a) a leaf? b) a car?

In the Imperial system you can measure mass in ounces (oz), pounds (lb), stones and tons.

16 oz = 1 lb
14 lb = 1 stone
160 stone = 1 ton

In the metric system you can measure mass in grams (g), kilograms(kg) and tonnes.

1000 g = 1 kg
1000 kg = 1 tonne

The petrol tank in Sam's car has a **capacity** of 50 litres.

What is the capacity of a) a teacup? b) a bucket?

In the Imperial system you can measure capacity in pints and gallons.

8 pints = 1 gallon

In the metric system you can measure capacity in millilitres (ml), centilitres (cl) and litres (l).

1000 ml = 1 l
100 cl = 1 l

How long does it take to a) travel to college? b) walk a mile?

Time is often measured in seconds (s), minutes and hours (h). Longer periods of time can be measured in days, weeks and years.

60 seconds = 1 minute
60 minutes = 1 hour
24 hours = 1 day

Timetables use the 24 hour clock so that, for example, 7.00 a.m. is 0700, 2.30 p.m. is 1430, 5.45 p.m. is 1745, and so on.

Exercise

1 Convert

a) 2 m into cm
b) 3 km into m
c) 40 mm into cm
d) 6 feet into inches
e) 1.3 m into mm
f) 18 yards into feet

2 Convert

a) 1.2 kg into g
b) 10 stones into pounds
c) 250 g into kg
d) half a pound into ounces

3 Convert

a) 1.5 litres into ml
b) 10 pints into gallons
c) 70 cl into litres
d) half a gallon into pints

4 Convert

a) 120 minutes into hours
b) 3 minutes into seconds
c) 90 seconds into minutes
d) 1 hour into seconds

5 a) How long does the News last?

b) How long does the Local news last?

c) What programme is on at quarter to seven?

d) How long does Tennis highlights last?

Evening TV

1800 The News
1830 Local news
1840 Tennis highlights
1915 Top of the Pops

6 Paul buys three 330 ml cans and Emily buys a litre bottle.

a) Who gets more and by how much?

b) Give two possible reasons why Paul bought the cans rather than a bottle.

7 Annabel needs 1 kg of margarine.

The shop has no 1 kg tubs but it does have 500 g tubs and 250 g tubs.

Give three ways in which Annabel can make up 1 kg of margarine.

8 Hugh's office has a shelf 1.2 m long.

Hugh has 16 box files.

Each box file is 7 cm thick.

Will they all fit on the shelf?

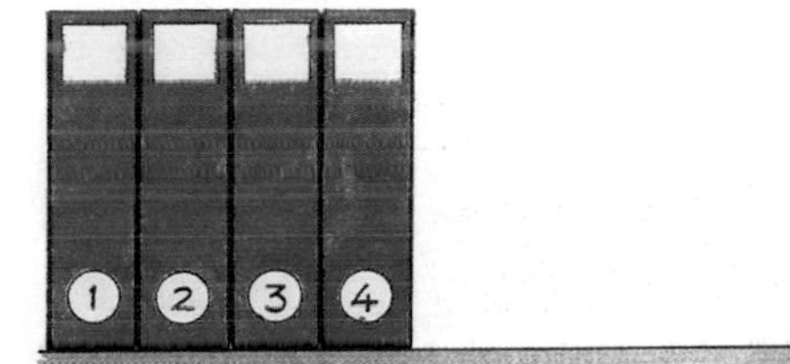

Measure at least 8 everyday objects to find things

a) about 1 foot long

b) about 1 metre long.

Now weigh at least 8 everyday objects to find things

c) which weigh about 1 kilogram

d) which weigh about 5 pounds.

Imperial and metric units

It is useful to be able to make quick conversions between different types of units. These conversions are worth remembering.

Length

1 inch is about 2.54 cm

39 inches is about 1 m

1 foot is about 30.5 cm

$\frac{5}{8}$ mile is about 1 km

Mass

1 ounce is about 28 grams

2.2 pounds is about 1 kilogram

Capacity

$1\frac{3}{4}$ pints is about 1 litre

1 gallon is about 4.5 litres

Calum's grandmother has given him this cake recipe.

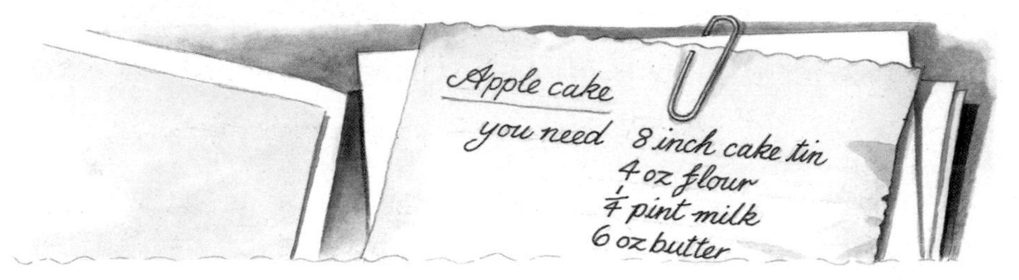

All Calum's cookery equipment is modern and uses metric units.

He works out the size of the tin in centimetres.

He works out how much flour he needs.

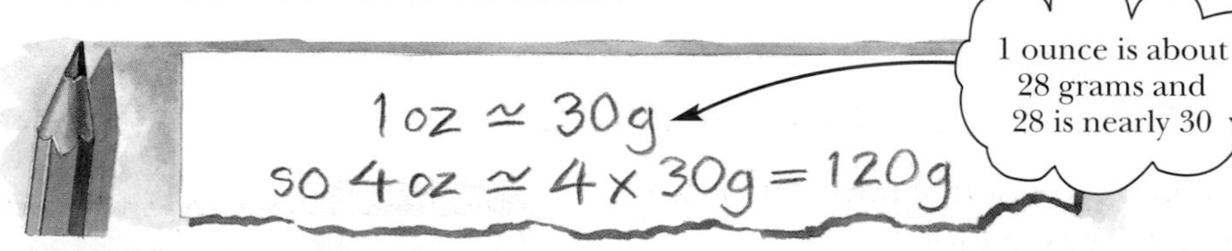

How much butter does he need?

Next Calum works out how much milk he needs.

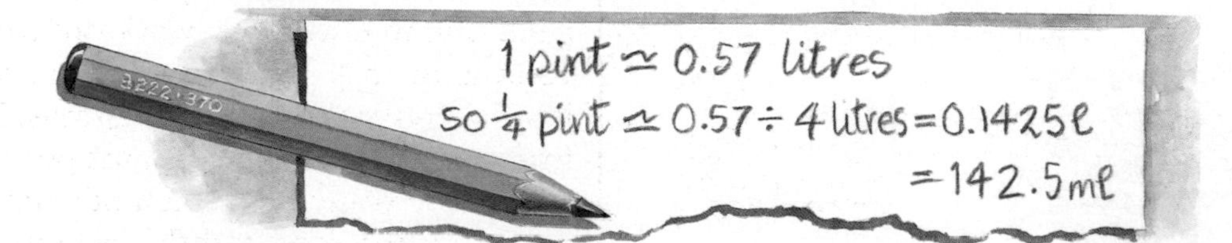

Many recipe books have the quantities listed in both Imperial and metric units. Usually the books advise you to use either Imperial or metric, not a mixture of both. Why?

Exercise

1 Using the rough conversions on the opposite page, convert

a) 60 gallons into litres b) 12 inches into centimetres
c) 80 km into miles d) 44 lb into kg
e) 30 litres into gallons f) 2 m into feet and inches
g) 150 g into ounces h) 6 mm into inches.

2 Ewan drives out of Aberdeen and sees this distance sign (in miles).

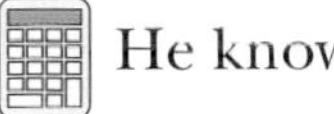

He knows that 5 miles is 8 km.

How far in kilometres is

a) Edinburgh? b) Glasgow?

Edinburgh	130
Glasgow	145

3 a) Gill weighs 9 stone 10 pounds. How many kilograms is this?

b) Jeff weighs 11 stone 6 pounds. How many kilograms is this?

4 Convert these quantities as shown.

a)

b)

c)
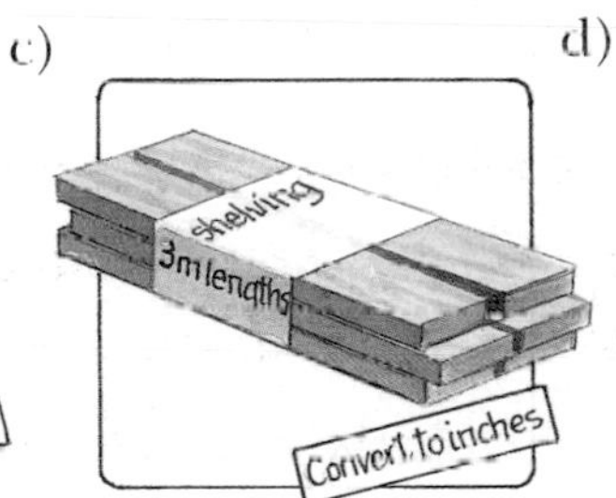

d)

5 These are some of the specifications for Neil's car.

Length	4.3 m	Petrol tank capacity	50 litres
Width	1.7 m	Fuel consumption	8 km per litre

a) Will Neil's car fit into a garage 18 feet long?

b) What is the width of Neil's car in feet and inches?

c) What is the petrol tank capacity to the nearest gallon?

d) What is the fuel consumption in miles per gallon?

6

a)

b)
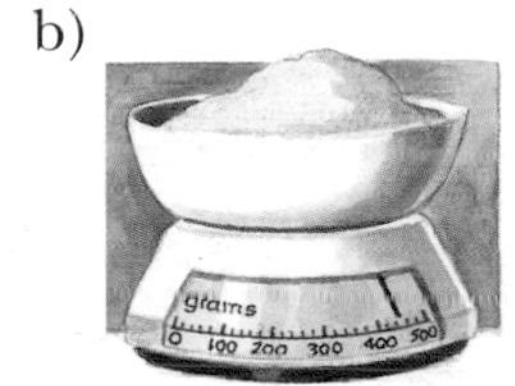

Write down each of these readings in grams, then convert it to pounds.

Some people say that certain cereal packets could be made smaller and so be less wasteful of the environment's resources.

Take 5 different cereal packets. Measure each one in cm and so work out the volume in cm^3. The weight of the cereal is marked on the packet in grams. Now work out the density in grams per cm^3 (weight ÷ volume). Which of your packets is the most tightly packed, and so the most environmentally friendly?

Finishing off

Now that you have finished this chapter you should be able to:

★ change units of length, mass and capacity within the Imperial system and within the metric system

★ change units of time

★ use the 24-hour clock

★ change between Imperial and metric units.

Use the questions in the next exercise to check that you understand everything.

Mixed exercise

1 Convert

a) 2.4 cm into mm

b) 750 g into kg

c) half a mile into yards

d) 13.6 km into m

e) 135 lb into stones and lb

f) 1.5 litre into cl.

2

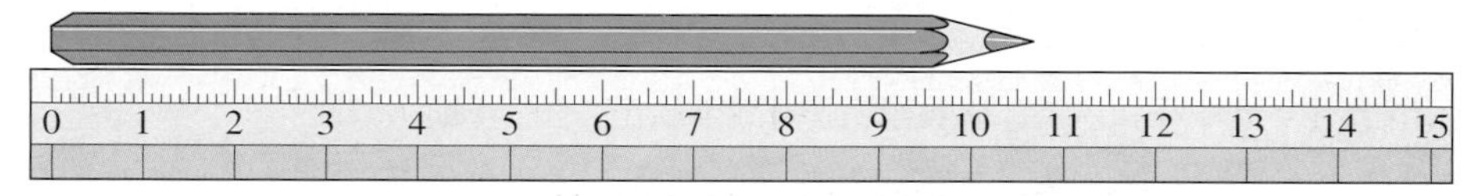

This is a 15 cm ruler. Write down the length of the pencil in

a) centimetres

b) millimetres.

3 Connor buys 5 pieces of curtain material each 120 cm long.

a) How many metres of curtain material does he buy?

b) A metre of material costs £4.50. How much does Connor pay?

c) The width of this material is $1\frac{1}{2}$ metres.

What is this in inches? (Give your answer to the nearest inch.)

4 A container weighs 400 kg.

a) How much do 5 containers weigh?

b) Write this amount in tonnes.

c) A truck weighs 3.5 tonnes. It is loaded with 5 containers. What is the total weight of the truck and its load?

d) The truck costs thirty thousand six hundred and nine pounds. Write this cost in figures.

5 Mandy's baby needs 200 ml of milk at each feed. How many feeds can Mandy get out of

a) a 1-litre carton of ready-to-drink baby milk?

b) a tin of baby milk powder that makes 5 litres?

6 Using the 24 hour clock, what time is 6.30 p.m.?

7 Convert

a) 6 inches into centimetres
b) 36 litres into gallons
c) 60 miles into kilometres
d) 55 kg into pounds
e) 4 ounces into grams
f) 2 pints into litres

8 Convert these Imperial measures into metric ones.

a)

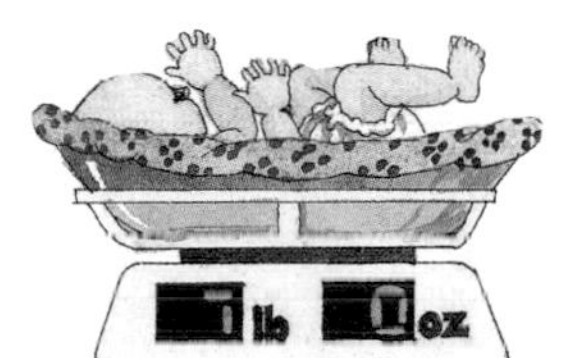

b)

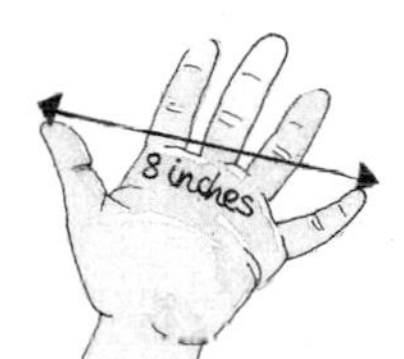

c)

9 A water tank has a capacity of 200 gallons. It is three quarters full.
How many litres does it contain?

10 a) What is 40 miles per hour in kilometres per hour?
b) What is 110 kilometres per hour in miles per hour?

11 In one hour Alice travels 90 km.

a) How many metres does she travel in one hour?
b) How many metres does she travel in one second?
c) What is 90 kilometres per hour in metres per second?

This chart from a road atlas shows the distances in miles between 6 cities.

a) You can see that the distance from Southampton to Newcastle is 323 miles, but you need to travel from Southampton to Newcastle passing through all the other cities on the way.

Cambridge is 61 miles from London

Birmingham					
101	Cambridge				
108	207	Cardiff			
120	61	155	London		
202	230	318	285	Newcastle	
129	132	141	80	323	Southampton

Look at a map and decide which orders are sensible.

Find the one with the shortest total distance.

b) Make your own distance chart for 6 towns or villages near you.

Seven

Ratio and proportion

Ratio

Leaf green paint is made by mixing yellow and blue.

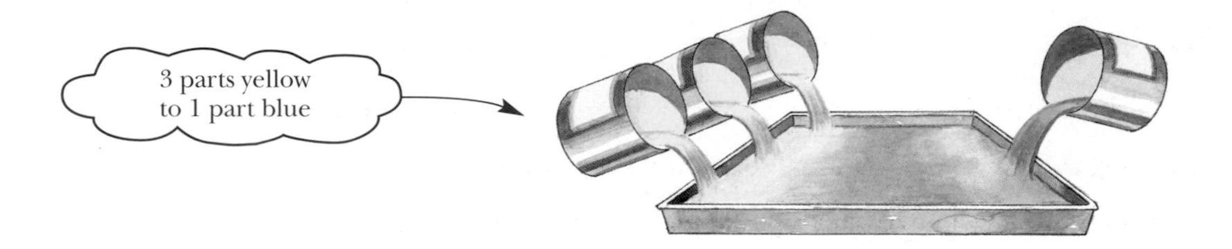

For example, 3 tins of yellow are mixed with one tin of blue.

The **ratio** 3 parts to 1 part is written 3:1. (It is read as 'three to one'.)

How many tins of each colour do you need for twice as much Leaf green?

You can multiply (or divide) both parts of a ratio by any number and get an **equivalent ratio**: 3:1 is the same as 6:2.

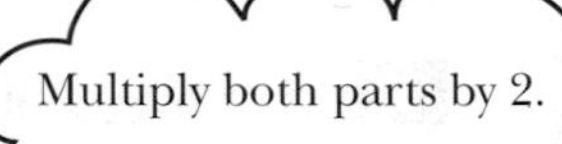

What other ratios are equivalent to 3:1 ?

Josh and Caitlin buy some Leaf green paint and share the cost 60:40.

You can simplify the ratio 60:40 as follows:
divide both sides by 10 to get 6:4
divide both sides by 2 to get 3:2

3:2 is in simplest form. It has the smallest whole numbers.

(You could also divide by 20 straight away to get 3:2.)

The cost of the paint is £40. Josh and Caitlin share this cost in the ratio 3:2.

How much does each pay?

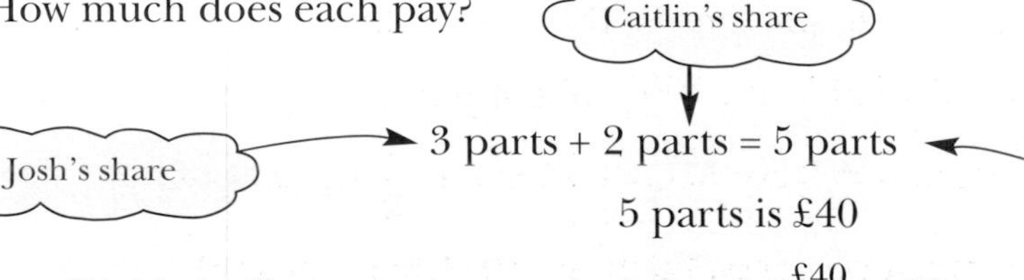

3 parts + 2 parts = 5 parts

Total

5 parts is £40

Divide by 5 1 part is $\frac{£40}{5} = £8$

Josh pays 3 parts which is $3 \times £8 = £24$

Caitlin pays 2 parts which is $2 \times £8 = £16$

Check: the £40 cost is shared and £24 + £16 = £40

What fraction of the cost does Josh pay?

What percentage of the cost does Caitlin pay?

Exercise

1 Write each ratio in its simplest form.

a) 8:4 b) 15:20 c) 60:40 d) 12:36

e) 14:35 f) 8:12 g) 50:20 h) 96:36

i) 6:8:10 j) 40:60:80 k) 6:12:15 l) 48:60:108

2 Write each ratio in a common unit and then put it in its simplest form.

a) 15 minutes:2 hours b) 30 mm:2 cm c) 750 g:2 kg

d) £5:125p e) 300 ml:1.5 l f) 1500 m:10 km

3 One unit of cleaning fluid is mixed with 25 units of water before use.

a) How much water is mixed with 4 units of cleaning fluid?

b) How much cleaning fluid is mixed with 150 units of water?

4 a) How much of this plant food would you mix with 3 litres of water?

b) What is the correct mixing ratio of plant food to water?

Write this ratio in its simplest form.

5 a) Share £600 in the ratio 3:2. b) Share £540 in the ratio 2:7.

c) Share £2250 in the ratio 2:3:4. d) Share £3200 in the ratio 4:5:1.

6 Rosie and Natasha share a flat.

a) The rent is £110 per week and they pay it in the ratio 2:3.

How much does each of them pay?

b) The phone bill is £91 and they pay it in the ratio 3:4.

How much does each of them pay?

7 Brass is made by mixing copper and zinc in the ratio 7:3.

a) How much brass can be made with 280 g of copper?

b) How many grams of each metal are needed to make $\frac{1}{2}$ kg of brass?

8 Charlotte and Thomas start a business. Each month Charlotte and Thomas share profits in the ratio 3:5.

a) In May the profit is £400. How much does each get?

b) In June Charlotte gets £240. Work out the total profit.

c) In July Charlotte gets £156 less than Thomas. How much does each receive?

Unitary method

Jade needs some mineral water.

Her local shop sells these two sizes.

Which do you think is the better buy?

Jade compares the prices by working out the price of 1 litre from each container.

The 2 litre container costs 60p:

$$\text{price of 1 litre} = \frac{60\text{p}}{2} = 30\text{p}$$

The 5 litre container costs 140p:

$$\text{price of 1 litre} = \frac{140\text{p}}{5} = 28\text{p}$$

The 5-litre container is the better buy: it has the lower price per litre.

This method of comparing prices, by finding the cost of one unit, can be extended to work out the expected cost of a different number of units.

Example

200 g of meat costs £1.44. How much does 340 g cost?

Solution

200 g costs £1.44

$$1 \text{ g costs } \frac{£1.44}{200}$$

$$340 \text{ g costs } \frac{£1.44}{200} \times 340 = £2.448$$

This has to be rounded to give a whole number of pence

340 g of meat costs £2.45 to the nearest penny

This method, of first finding what one unit costs, is called the **unitary method**.

You have already used it to solve percentage problems (see page 40).

Exercise

1 Work out which is the better/best buy.

a)

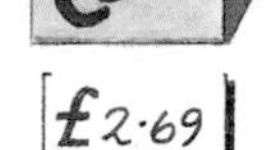

b)

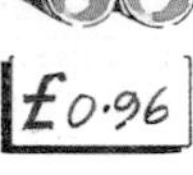

c)

2 The mass of 3 cm^3 of gold is 57.9 g. Work out the mass of

a) 5 cm^3 of gold b) 30 cm^3 of gold.

3 Verity, Misha and Kay are all paid at the same hourly rate.

Verity earns £168 for a 35-hour week.

a) Work out the hourly rate of pay.

b) Misha works a 38-hour week. How much does he earn?

c) Kay earns £201.60 a week. How many hours does she work?

4 Ian can drive 153 miles on 18 litres of petrol.

a) How far can he drive on 30 litres of petrol?

b) How many litres of petrol does he need to drive 119 miles?

c) He has 25 litres of petrol in the tank. Will this be enough to drive to see his sister, who lives 110 miles away, and back?

5 100 g of cereal contains 2.3 milligrams of vitamin B_6.

a) How much vitamin B_6 is there in a 30 g serving of cereal?

b) The amount of vitamin B_6 in a 30 g serving of cereal is 40% of Chloe's recommended daily allowance.

Work out Chloe's recommended daily allowance of vitamin B_6.

6 Luke makes 12 buns. He uses 300 g of flour, 120 g of sugar, and 120 g of butter.

a) How much of each ingredient would he use to make 4 buns?

b) How much flour would he need for 20 buns?

c) What is the greatest number of buns he can make with 100 g of butter?

7 Abigail takes four minutes and forty seconds to type a document containing 252 words.

What is her speed in words per minute?

> Go to a supermarket and find an item which you can buy in at least 3 sizes.
>
> Is it the case that the larger the quantity the lower the unit cost?

Changing money

How do you get foreign money?

Lisa is going on holiday to the USA. She wants to change £400 into dollars. She goes to this bureau de change.

Rates	£1=
Australia	2.65 dollars
Japan	170 yen
S. Africa	13.6 rand
USA	$1.58

Each pound is worth $1.58, so £400 is worth

$400 \times 1.58 = 632$

Lisa receives $632 for her £400.

How many dollars would Lisa get for a) £200*? b)* £750*?*

Note: The bureau de change will probably charge commission when changing Lisa's £400 into dollars. A typical rate is 2%, and 2% of £400 is £8. So Lisa would actually pay £408 for her $632.

Lisa arrives in the USA and buys a meal costing $22.

Each $1.58 is worth £1. Lisa works out the price in pounds by finding out how many times $1.58 goes into $22:

$22 \div 1.58 = 13.92$

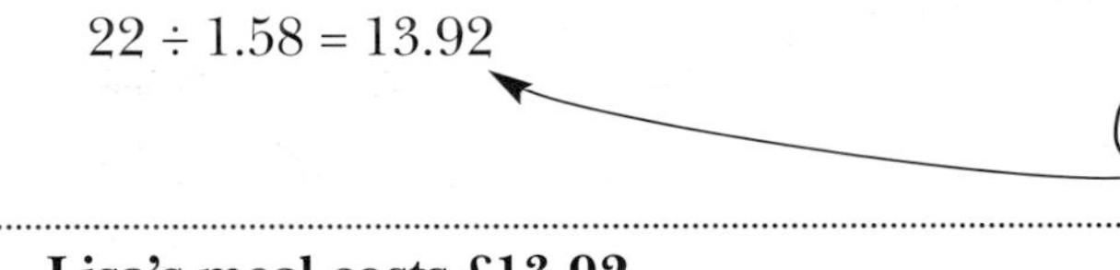

This is rounded to 2 decimal places so it is correct to the nearest penny

Lisa's meal costs £13.92.

Find the cost in pounds of a) a jacket costing $55, *b) a coach ticket costing* $96.

Lisa makes this conversion table to help her to convert dollars into pounds. She has worked out some of the conversions (to the nearest penny).

$1	£0.63	$15	
$2	£1.27	$20	
$3		$30	
$4		$40	
$5		$50	
$10	£6.33	$100	

Copy and complete this conversion table.

How can she use it to work out $90 *in pounds?*

Exercise

Use the exchange rates opposite for each of these questions.

1 Rea, Kate and Callum go to Japan.

a) How many yen does Rea get for £40?

b) How many yen does Callum get for £95?

c) How much does Kate pay for 18 500 yen? (Give your answer to the nearest pound.)

d) A trip to Tokyo costs 8000 yen. How much is this in pounds and pence?

e) A meal costs 2800 yen. How much is this in pounds and pence to the nearest 50 pence?

2 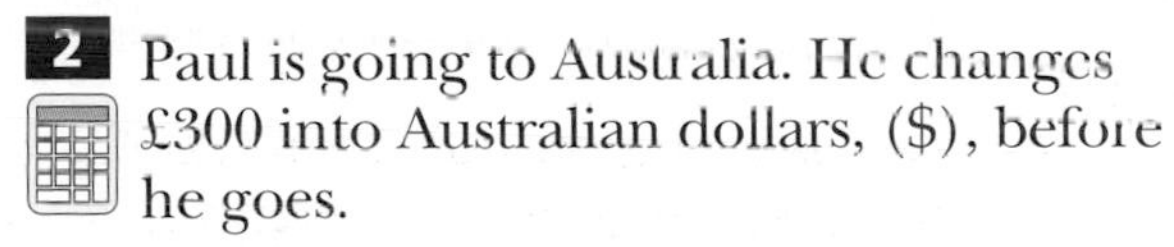

Paul is going to Australia. He changes £300 into Australian dollars, ($), before he goes.

a) How many dollars does he get?

b) He pays 2% commission. How much does he pay in total?

c) Paul buys these items whilst in Australia.

What is the cost of each item in pounds and pence?

d) At the end of his holiday Paul changes 130 dollars back into pounds. He pays no commission. How much does he get?

3 Lucy goes to South Africa. She makes this conversion table.

a) Copy and complete the table.

b) Explain how she would use it to find the cost in pounds and pence of an item costing (i) 70 rand (ii) 450 rand.

c) Work out the cost of these items in pounds and pence.

Rand	£	Rand	£
10	£0.74	150	
20		200	
50		500	
100	£7.35	1000	

Choose a country that you would like to visit.

Find out the name of its currency, and the current exchange rate against the pound.

Draw up a table to help you to convert prices quickly into pounds whilst you visit the country.

Distance, speed and time

What is the speed limit on motorways?

How fast does an aeroplane fly?

How many metres per second can you run?

Speed is usually measured in miles per hour (m.p.h.), kilometres per hour (km/h) or metres per second (m/s).

Tim drives at 100 km/h on the motorway.

Do you think that he drives at exactly 100 km/h *all the time?*

Tim drives 90 km on country roads in $1\frac{1}{2}$ hours.

What is his average speed in km/h?

$$\textbf{average speed} = \frac{\textbf{distance covered}}{\textbf{time taken}}$$

$$= \frac{90}{1\frac{1}{2}} = 60$$

Tim's average speed is 60 km/h.

Kanwal does the same journey in $1\frac{1}{4}$ *hours.*

What is her average speed?

Tim's average speed on a motorway is 100 km/h.

How long does a motorway journey of 225 km take?

$$\textbf{time taken} = \frac{\textbf{distance covered}}{\textbf{average speed}}$$

$$\frac{225}{100} = 2.25 \text{ or } 2\frac{1}{4}$$

Be careful here!
.25 hours is not 25 minutes.
To change 0.25 hours into minutes you multiply by 60
0.25 × 60 = 15 so
0.25 hours = 15 minutes

The time taken is $2\frac{1}{4}$ hours (or 2 hours 15 minutes).

How long does it take Tim to travel 60 km *on the motorway?*

How far does Tim travel in $1\frac{3}{4}$ hours?

$$\textbf{distance covered} = \textbf{average speed} \times \textbf{time taken}$$

$$= 100 \times 1\frac{3}{4} = 175$$

Tim travels 175 km in $1\frac{3}{4}$ hours.

How far does Tim travel in 45 *minutes?*

Exercise

1 What distance is covered by

a) Linton cycling at 50 km/h for 2 hours?

b) Jovanka driving at 70 km/h for $1\frac{1}{2}$ hours?

c) Philip flying at 880 km/h for $2\frac{1}{4}$ hours?

d) Liz running at 15 km/h for 1 hour 20 minutes?

2 Hana has four meetings today.
This is her schedule.

Work out the average speed

a) between London and Milton Keynes (84 km apart)

b) between Milton Keynes and Leicester (81 km apart)

c) between Leicester and Sheffield (105 km apart).

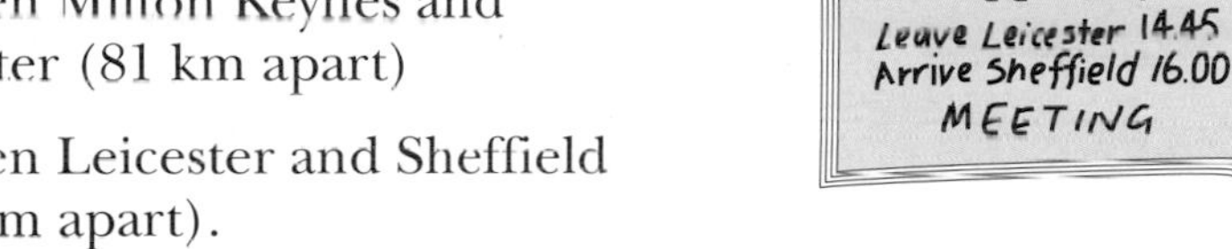

3 Work out the time it takes to

a) cycle 75 km at 30 km/h

b) fly 2250 km at 600 km/h

c) drive 100 km at 60 km/h

d) run a marathon (42.2 km) at 20 km/h.

4 Darren from Liverpool and Vicky from Hull drive to Manchester to meet for lunch.

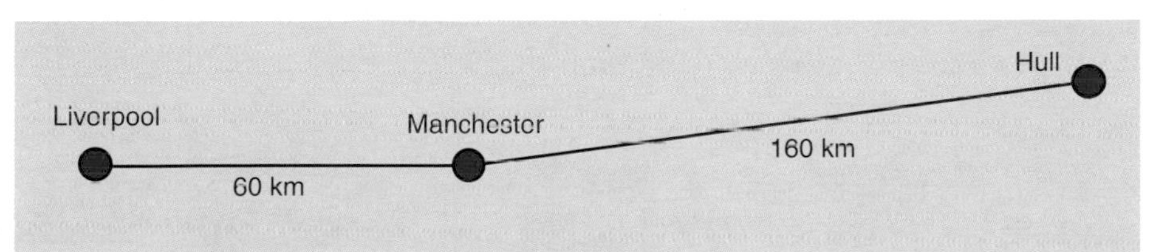

a) Darren leaves home at 1140 and his average speed is 60 km/h.
What time does Darren arrive in Manchester?

b) Vicky leaves at 1045 and expects to take 2 hours.
What will her average speed be?

c) Vicky's journey takes 30 minutes longer than planned.
What is her average speed?

d) How long did Darren have to wait on his own?

Finishing off

Now that you have finished this chapter you should be able to:

★ solve problems using ratio
★ use the unitary method
★ change money
★ solve problems with distance, speed and time.

Use the questions in the next exercise to check that you understand everything.

Mixed exercise

1 Write each ratio in its simplest form.

a) 25 : 40 b) 56 : 35 c) 16 : 32 : 8
d) 450 g : 2 kg e) 5 cm : 1 km f) 1.5 litres : 150 ml

2 £20 000 is shared in the following way. Andrew gets one quarter of it. Harriet gets one fifth of the remainder. Justin and Mel share the rest in the ratio 3 : 2.

a) How much does each person get?

b) What is the ratio of Andrew's share to Harriet's share in its simplest form?

c) What percentage of the £20 000 does Justin get?

d) What fraction of the money does Mel receive?
Write your answer in its simplest form.

3 Kathryn wants some hot chocolate. Which of these is the best value?

4 George makes concrete by mixing sand, gravel and cement in the ratio 2 : 4 : 1.

a) How much of each does George need to make 10.5 m^3 of concrete?

b) He has only 2.5 m^3 of sand. What is the greatest amount of concrete that he can make?

5 Sue and Bob are going on holiday to the USA. The exchange rate is £1 = $1.40.

a) How many dollars will Sue get for £110?

b) Sue pays $30 for a coach ticket to New York. How much is this in pounds and pence?

c) Bob buys $200 for a ticket and pays 2% commission charge. Work out the cost in pounds and pence.

6 8.4 g of iron combines with 3.6 g of oxygen to form 12 g of iron oxide.

a) How much oxygen combines with 14 g of iron?

b) How much iron combines with 13.2 g of oxygen?

c) How much of each element is required to produce 50 g of iron oxide?

7 a) Lucy cycles for 45 minutes at an average speed of 30 km/h. How far does she travel?

b) Anant is an airline pilot. He has 900 km to cover in $1\frac{1}{4}$ hours.

What average speed does he need to maintain?

c) Tracey has 50 km to drive. She thinks her average speed will be 65 km/h. How long, to the nearest 5 minutes, will it take her?

8 Asif plots the distance his coach has travelled against time. This journey is broken into 3 stages by 2 breaks. This is his graph.

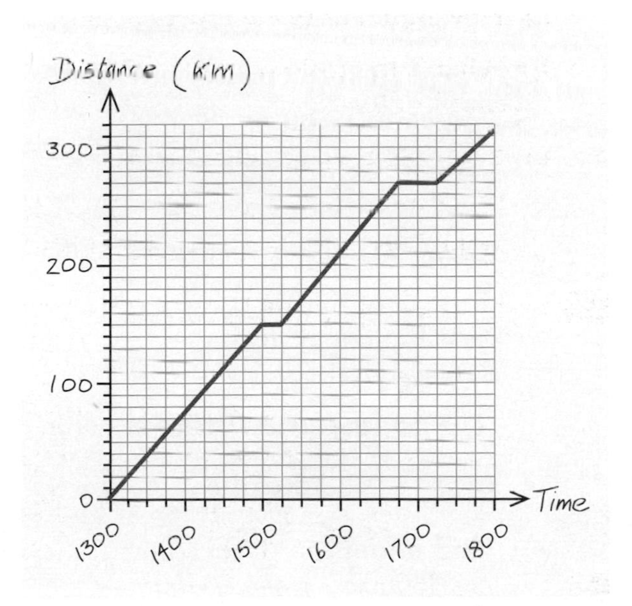

a) What is the average speed for the first stage of the journey?

b) How long was the first break?

c) What is the average speed for the second stage of the journey?

The total distance covered is 315 km.

d) What is the total time taken including breaks?

e) What is the average speed for the whole journey?

Eight

Money

Simple and compound interest

Katie has saved £3000 and wants to invest it for two years in Avonford Building Society. The rate of interest is 5% p.a. (per annum or each year).

Katie works out her yearly interest like this:

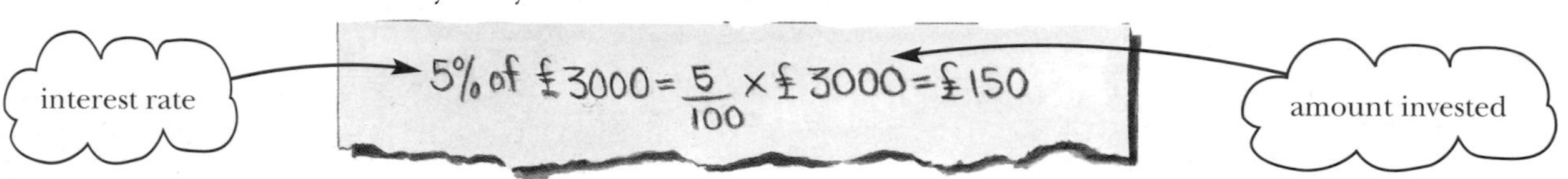

The £150 interest is sent to Katie at the end of the first year.

The £3000 in the account will earn another £150 interest during the second year.

The **simple interest** Katie earns over 2 years will be

$2 \times £150 = £300$

You can use this formula for calculating the simple interest, I:

$$I = \frac{P \times R \times T}{100} = \frac{PRT}{100}$$

where P is the money invested (sometimes called the principal),

R is the rate in % p.a.,

T is the time in years.

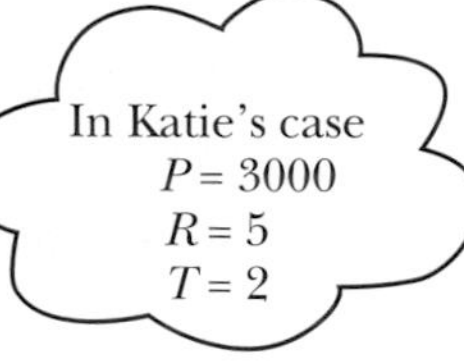

Check that this formula gives £300 *for Katie's simple interest over* 2 *years.*

Instead of having the interest sent to her at the end of the first year Katie may decide to have it added to her account to earn extra interest in the second year.

The **compound interest** Katie will earn is worked out as follows:

Year 1: As before, the interest is 5% of £3000 = $\frac{5}{100} \times £3000 = £150$

The amount in the account at the end of Year 1 is
£3000 + £150 = £3150.

Year 2: The interest is 5% of £3150 = $\frac{5}{100} \times £3150 = £157.50$.

The amount in the account at the end of Year 2 is
£3150 + £157.50 = £3307.50

The amount of compound interest that Katie gets is £307.50.

You can work this out as either £3307.50 – £3000 = £307.50 or £150 + £157.50 = £307.50.

Why does $£3000 \times (1.05)^2$ *give the amount in the account at the end of Year* 2?

Exercise

1 Calculate the simple interest on each of these investments.

a) £600 at 5% p.a. for 1 year;

b) £3000 at 5.5% p.a. for 2 years;

c) £5000 at $6\frac{1}{4}$% p.a. for 3 years;

d) £850 at 4% p.a. for 6 months.

2 Calculate the compound interest on each of these investments.

a) £500 at 4% p.a. for 2 years;

b) £2000 at 6% p.a. for 3 years;

c) £1000 at 5.5% p.a. for 4 years;

d) £15 000 at $6\frac{1}{4}$% p.a. for 3 years.

3 The members of the Lewis family want to maximise their investment income. They have collected these leaflets.

Rosie has £800 to invest and Scott has £1250.

a) Say where each person should invest, and what simple interest he or she will receive over 2 years.

b) How much interest would Scott lose by investing in the Southern Building Society instead?

c) Mr Lewis has £7000 to invest. He decides to have the interest added to his account each year to earn extra interest in the following year. How much is in his account after 4 years?

4 Mohan and Alka see this advert.

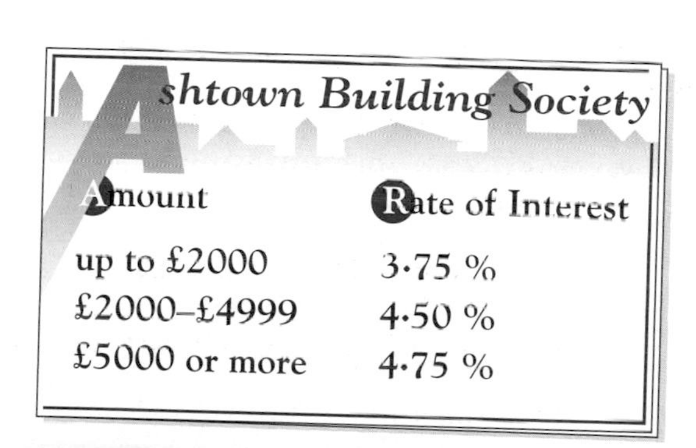

Ashtown Building Society

Amount	Rate of Interest
up to £2000	3·75 %
£2000–£4999	4·50 %
£5000 or more	4·75 %

How much simple interest will be earned by

a) Mohan investing £1500 for 3 years?

b) Alka investing £750 for 3 years?

Mohan and Alka decide to invest all their money in a joint account.

They share the interest in proportion to the money they invested.

c) What rate of interest will they earn now?

d) How much extra interest does each get over the 3-year period?

Go to 4 building societies or banks and find out the rates of interest they offer on 'instant access' accounts.

What is the highest rate you can get on
a) £200? b) £1500? c) £6000?

Income and tax

Emma has a holiday job at her local store earning £4.60 an hour for a 37-hour week. She works 3 hours' overtime which is paid at 'time and a half'. She works out her pay like this:

time and a half is 1.5 times

37 × £4·60 = £170·20 (basic pay)
3 × 1·5 × £4·60 = £20·70 (overtime pay)
£190·90 (total pay)

Emma leaves school and gets a job in fashion. Her salary is £8000 plus 20% commission. She sells £15 000 of goods in her first year. She works out her income like this:

£8000 (basic salary)
20% of 15000 = $\frac{20}{100}$ × 15000 = 3000 → £3000 (commission)
£11000 (total salary)

What are the advantages and disadvantages of being on commission?

Tax

Emma's **gross salary** is £11 000. She pays tax on part of her income.

This part is her **tax free allowance**, the part of her income on which she doesn't pay tax.

In Emma's case it is £4000

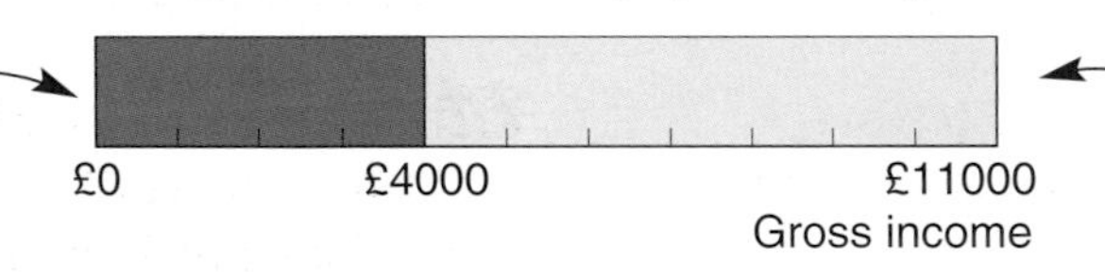

This £7000 is called her **taxable income**. Emma pays tax on it at a rate of 20p in the £

Emma works out her tax like this:

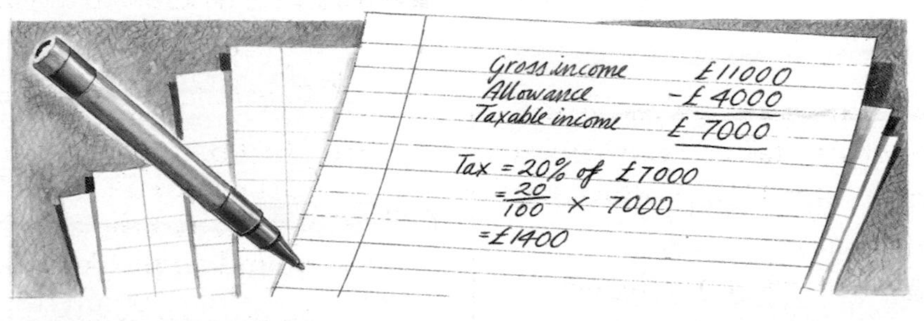

Emma's **net income** is the amount that she actually receives.

It is worked out by subtracting tax (and any other deductions) from gross income.

What is Emma's net income?

Exercise

1 Work out how much each of these people earns per week.

a) Josie is paid £5.00 an hour and works 37 hours a week.

b) Warren works Monday morning, Tuesday afternoon, Thursday morning and all day Friday. He is paid £32 for half a day.

c) Marie earns £5.80 an hour for a 36-hour week. She does 4 hours' overtime at 'time and a half'.

d) Richie earns £6.50 an hour for a 38-hour week. He does 3 hours' overtime at 'double time'.

2 These are Phil's sales figures for July to September.

Month	July	August	September
Sales (£)	2000	2240	1880

Phil's basic salary is £1100 a month and he gets 15% commission on the value of his sales.

a) Calculate his total pay for each month.

b) His pay for October is £1454. Calculate the value of his sales.

3 Sharon earns £12 500 a year. Her tax free allowance is £3900.

a) Calculate her taxable income.

b) She pays tax at 25p in the pound.

Calculate the amount of tax that she pays.

c) Calculate her net income.

4 Hayley's gross salary is £16 350. She has a tax free allowance of £4120. She pays tax at 25p in the pound.

a) Calculate her net salary.

b) Hayley receives a 3% pay rise. Her allowance remains the same.

Calculate the increase in her net salary.

In questions 5 and 6 tax is paid at 20p in the pound on the first £5000 of taxable income, and at 24p in the pound on the remainder.

5 Jo's gross salary is £17 800. Her tax free allowance is £4300. Work out her net salary.

6 Kelly earns £26 400 a year. Her allowance is £4160.

a) Calculate her net income.

b) The tax rates are changed to 19p in the pound on the first £6000 of taxable income and 25p in the pound on the remainder.

How much better or worse off is Kelly under the new rates?

Find out the present tax allowance for a single person and the rates at which tax is paid.

Use these figures to work out the net salary of

a) Samit who earns £13 500 a year,

b) Malcolm who earns £65 400 a year.

Bills

Emily moved into a flat on 21 July. Three months later she receives this electricity bill.

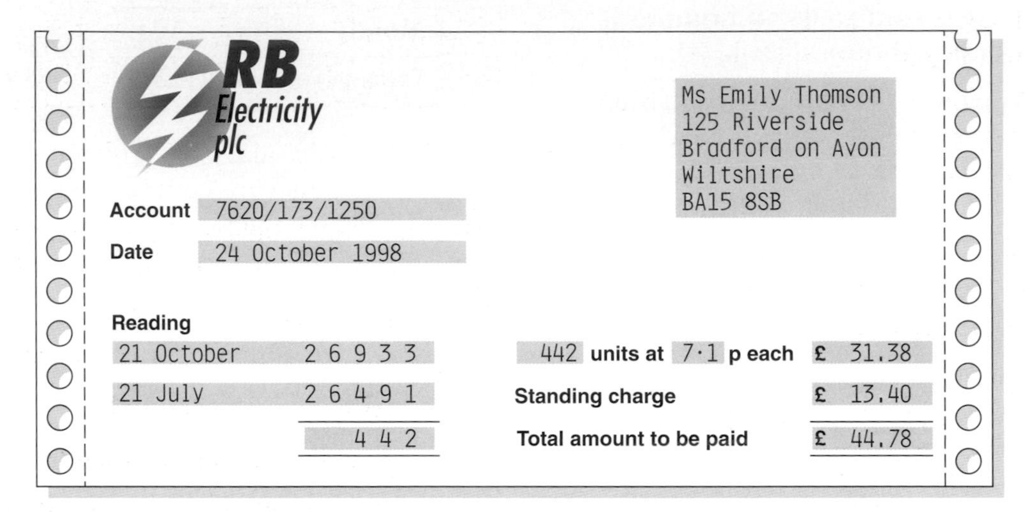

RB Electricity plc

Ms Emily Thomson
125 Riverside
Bradford on Avon
Wiltshire
BA15 8SB

Account 7620/173/1250

Date 24 October 1998

Reading	
21 October	2 6 9 3 3
21 July	2 6 4 9 1
	4 4 2

442 units at 7·1 p each	£ 31.38
Standing charge	£ 13.40
Total amount to be paid	£ 44.78

How is this worked out?

The number of units Emily has used is the difference between the two readings.

Emily has used 442 units. The cost of units is 7.1 pence each.

You work out the total cost of units like this:

$$442 \times 7.1\text{p} = 3138.2\text{p}$$

This is 3138p (or £31.38) to the nearest penny

The standing charge is the same every quarter: it is £13.40 however much electricity Emily uses. The standing charge is added to the cost of the units to work out the total amount to be paid.

Do you think that Emily will use the same amount of electricity in the next three months? Explain your answer.

Emily insures the contents of her flat with Wright Insurers. Emily lists her contents and a value of £6500 is agreed.

The insurance costs £4.50 for each £1000 of contents value and lasts for a year.

The amount she pays is £4.50 × 6.5 = £29.25

The contents value is six and a half (6.5) thousand

Why does Emily take out insurance?

Exercise

1 Here are two of Meryl's gas meter readings.

a) Each unit costs 13.2p. How much will the units cost in total?

b) Meryl pays a standing charge of £14.80 a quarter. Work out her total bill.

29th October

29th January

2 On 5 November the reading on Kerry's electricity meter is 6295.

On 5 February it is 6959.

a) Each unit costs 7.6p. How much will these units cost in total.

b) Kerry pays a quarterly standing charge of £12.50. Work out her total bill.

c) For the next bill, due in May, Kerry has set aside £50.50. How many units does she expect to use?

3 On 2 April Steve's water meter reading is 15726. On 2 October it is 16404.

Each unit costs 13.8p. Steve pays a half yearly standing charge of £12.40.

a) Work out his total bill.

b) A new tariff is introduced.

The standing change is reduced by £2.80 but the cost of a unit is increased by 0.1p.

Would you expect Steve to be better off under the new tariff?

4 Joe is an insurance broker. The table below shows the cost, per person, of his holiday insurance. It depends on the destination and the number of nights away.

	U.K.	Europe	USA	Far East
Up to 8 days	8	13	18	22
9–15 days	10	16	22	28
16–30 days	12	19	26	34
31–91 days	16	26	36	44

Write down what Joe charges each of these people for holiday insurance.

a) Natalie going to the USA for 17 nights.

b) Oliver going to the Far East for 4 weeks.

c) Jack taking his three children on a holiday in the U.K. for a week.

d) Mandy taking her son on a European tour for 10 days.

Find six appliances that use electricity.

How much does it cost to use each one for an hour?

Hidden extras

Buy now – pay later

Poppy wants to buy this television but she does not have £475.

She works out how much it will cost to buy it over 12 months.

How much more is this than the cash price?

Caution: Poppy must be very careful about doing a deal like this. If she fails to make the monthly payments, she may lose both the television and the money she has already paid.

Value added tax (VAT)

VAT is a tax which you pay when you buy either goods or services.

What is the present rate of VAT?

Steve sees the same fax machine in two different stores.

At Bill's Bargains the VAT to be added is 17.5% of £210.

$$17.5\% \text{ of } 210 = \frac{17.5}{100} \times 210 = 36.75$$

The VAT is £36.75.

The full price is £210 + £36.75 = £246.75.

Alternatively you can work this out as £210 × 1.175 = £246.75

Which store is offering the better deal?

Write down two reasons why Steve might still decide to buy the fax machine from Ceri's Corner.

The price including VAT (117.5%) *is* £246.75. *By using the unitary method, you should check that the price ex VAT* (100%) *works out as* £210.

Exercise

1 Work out the cost of each item spread over 36 months and the extra amount paid.

a) b)

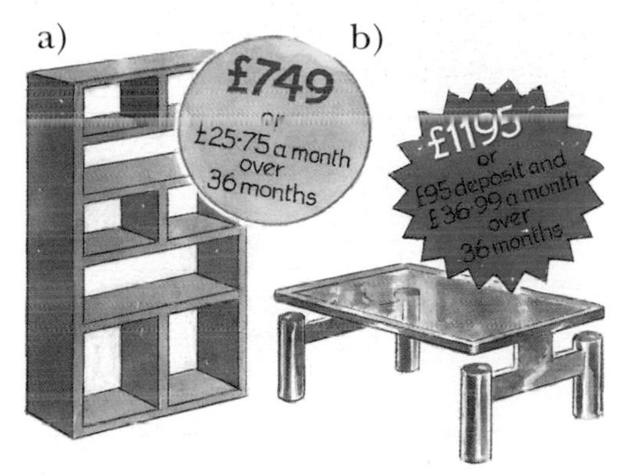

2 This table gives the repayments for each £100 of a loan.

Time in months	12	18	24	36
Repayment per month	£11.00	£8.10	£6.60	£5.00

How much is repaid in total on a loan of

a) £600 over 24 months?

b) £800 over 36 months?

c) £250 over 12 months?

d) £375 over 18 months?

3 An office chair costs £130 ex VAT and an office desk costs £370 ex VAT.

a) Work out the price of each item including VAT at 17.5%.

b) The addition of VAT at 17.5% increases the price by about one sixth. Use this approximation to make a rough check of your answers to part a).

4

Store A's prices exclude VAT at 17.5%.
Store B's prices include VAT.

Which store offers the better deal and by how much for

a) the filing cabinet? b) the computer?

5 Lauren works for a charity which can reclaim VAT. She buys these items:

1 printer cartridge @ £16.75.

2 rolls of fax paper @ £3.19 per roll.

4 reams of paper @ £5.29 per ream.

These prices include VAT at 17.5%.
How much can Lauren reclaim?

Use a spreadsheet to make a table with these headings.

Item	Price (ex VAT)	VAT	Price (inc VAT)

Find out the price (exclusive of VAT) of 6 items used in an office (e.g. stationery, computer equipment) and add on VAT at the current rate to get the price (inclusive of VAT).

Profit and loss

Andy owns this sports clothing shop.

He buys the golf sweaters at £50 each and the ski-suits at £200 each.

Which item is more profitable for him?

At first glance the ski-suits look more profitable: Andy makes £20 profit on each suit, and only £10 on each sweater. But businesses are interested in profit as a percentage of cost:

$$\% \text{ profit} = \frac{\text{profit}}{\text{cost}} \times 100$$

$$\text{So for the sweaters, } \% \text{ profit} = \frac{10}{50} \times 100 = 20$$

profit on one sweater (10); *cost of one sweater* (50)

$$\text{For the ski-suits, } \% \text{ profit} = \frac{20}{200} \times 100 = 10$$

Why do businesses need to be concerned about percentage profit?

Andy does not always make a profit. He sells the last sweater for £48 in a sale. It cost him £50 so he makes a loss of £2.

Work out this loss as a percentage of the cost price.

Andy buys track suits for £40 and prices them so as to make a 15% profit.

He works out the selling price like this:

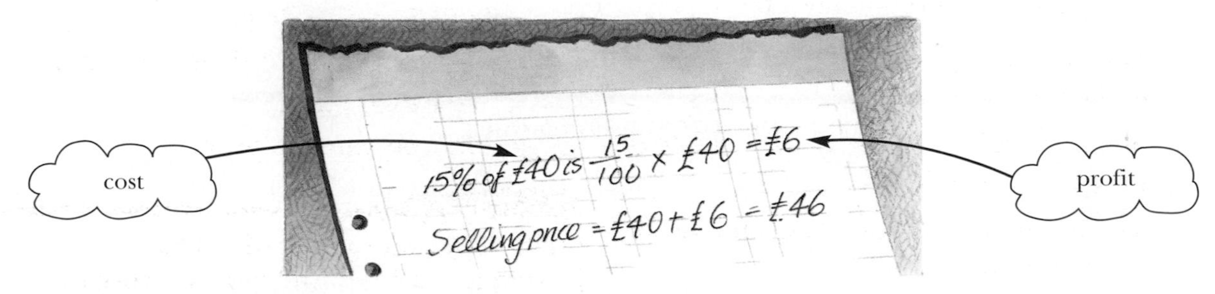

Note that the percentage profit is a percentage of the *cost price.*

The selling price (115 %) is £46. By using the unitary method, you should check that the cost price (100 %) works out as £40.

Exercise

1 Calculate the percentage profit or loss on

a) a table bought for £250 and sold for £380;

b) a coat bought for £80 and sold for £110;

c) a brief-case bought for £25 and sold for £17.50;

d) a tent bought for £35 and sold for £29.95.

2 Work out the selling price of

a) a clock bought for £60 and sold at a 20% profit;

b) a lamp bought for £62 and sold at a 40% profit;

c) a roof-rack bought for £80 and sold at a 15% loss;

d) a CD bought for £15 and sold at a 5% loss.

3 Carl buys 20 items for £600. He sells them at £39 each. Calculate

a) his total profit on the deal;

b) his percentage profit on the deal.

4 Sam buys jackets for £50 each and prices them at £70. He buys trousers for £20 each and prices them at £30. In a sale prices are reduced by 20%.

a) How much profit does Sam make on each item when he sells them in the sale?

b) Work out each profit as a percentage of the cost price.

c) Sam sells coats at £94.50 each and makes a profit of 40%. How much does he pay for a coat?

5 Esme agrees to landscape a garden for £1000. She pays 4 other people to do the job for her.

She pays Mark £160 to lay a path.

She pays George £330 to install a pond and a fountain.

She pays Della and Kirsty £90 each to plant shrubs and bulbs.

a) How much money does Esme make on the deal?

b) Express your answer to a) as a percentage of Esme's outlay.

6 Chloe sells hats. She buys 12 of them for £7.50 each. She prices them so as to make a 40% profit and sells 10 of them at this price. The remaining 2 are sold off at £5 each. Calculate

a) her total receipts from the sales

b) her total profit

c) her percentage profit.

Find the share prices of 5 major companies 6 months ago and work out the percentage profit/loss a shareholder would have made by buying then and selling today.

Finishing off

Now that you have finished this chapter you should be able to:

- ★ work out simple and compound interest
- ★ work out wages, commission and taxation
- ★ work out household bills
- ★ work out the costs of 'buy now – pay later' deals
- ★ work out VAT
- ★ calculate profit or loss.

Use the questions in the next exercise to check that you understand everything.

Mixed exercise

1 Judy invests £4000 in a building society for 2 years at 5.5% p.a.

a) How much simple interest does she earn?

b) How much extra interest would she receive by investing in a building society offering her 5.75% p.a.?

2 Mary invests £900 for 3 years at 6% p.a. The interest at the end of each year is added to the account to increase the principal.

How much is in the account at the end of 3 years?

3 Nina and Salil get these jobs.

a) How much does Nina earn for a 37-hour week?

b) How much does Salil earn for a 41-hour week?

c) How many hours does Nina work in her second week to earn £266?

4 Colin is paid a basic salary of £9600, plus 20% commission on the value of his sales. His first year's sales are £60 000.

a) Work out his total salary.

b) How much better off would he be with a basic salary of £12 750 and 15% commission?

5 Sara's gross income is £14 000 a year. She has a tax-free allowance of £4400. She pays tax on the rest at a rate of 20p in the pound.

a) Calculate Sara's net income.

b) How much extra tax will she pay per year if her salary goes up by £900?

6 Harry sells house insurance in Avonford. Harry explains to clients that house insurance depends on many factors but as a rough guide it works out at £3.50 for each £1000 of house value.

Work out an estimate of the house insurance for a house valued at

a) £60 000 b) £125 000

7 Tammy's gas meter reading was 4539 on 19 September and 4701 on 19 December. Units cost 48.2p and she pays a standing charge of £13.70 a quarter.

Work out her bill for this quarter.

8 Sal buys this settee over 30 months.

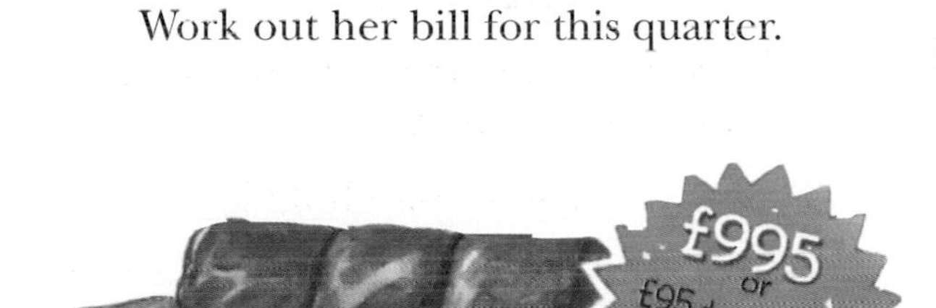

Work out

a) how much Sal pays;

b) how much above the cash price she pays.

9 Namita buys a CD player for £314.90. This includes VAT at 17.5%.

a) How much of the cost is VAT?

b) VAT at 17.5% accounts for roughly one seventh of the total cost. Use this approximation to make a rough check on your answer to a).

c) The VAT is reduced to 15%. Work out the new price inclusive of VAT.

10

a) Emma pays £20 for a bowl. Work out the percentage profit.

b) Emma makes a 40% profit on a vase. Work out how much it cost her.

c) Emma buys 30 bowls and 25 vases and sells them all.

Work out her total profit on the deal.

11 Terry manages a clothing store. He buys shirts at £8 each and trousers at £12 a pair.

a) He prices a shirt at £11.20. Work out his percentage profit on each one.

b) He makes a 50% profit on the trousers. Work out his selling price.

c) Terry buys 50 shirts and 40 pairs of trousers.

Calculate the total cost.

d) He sells 44 shirts and 35 pairs of trousers at the prices in a) and b). He sells the remainder at '25% off' in the sale.

Calculate Terry's total receipts.

e) What is Terry's percentage profit on the whole deal?

Nine

Indices and standard form

Powers

You have already met squares and cubes on page 2.

For example: $5 \times 5 = 5^2$

the value of 5^2 is 25

and $4 \times 4 \times 4 = 4^3$

the value of 4^3 is 64

2 is the **power** (or **index**). We say '5 to the power 2' or '5 squared'

We say '4 to the power 3' or '4 cubed'

When 64 is written as 4^3, then 4^3 is called **index form**.

Index form extends beyond squares and cubes in the natural way so

$$6 \times 6 \times 6 \times 6 = 6^4$$

Find out how to use the x^y key on your calculator.

Use it to check that $6^4 = 1296$.

You have already seen the meaning of positive powers.

What is the meaning of a zero or a negative power?

Look at this table.

$16 = 2 \times 2 \times 2 \times 2$	2^4
$8 = 2 \times 2 \times 2$	2^3
$4 = 2 \times 2$	2^2
$2 = 2$	2^1
$1 = 1$	2^0
$\frac{1}{2} = \frac{1}{2}$	2^{-1}
$\frac{1}{4} = \frac{1}{2 \times 2} = \frac{1}{2^2}$	2^{-2}
$\frac{1}{8} = \frac{1}{2 \times 2 \times 2} = \frac{1}{2^3}$	2^{-3}

(Each row is obtained from the one above by $\div 2$ on both sides.)

You can see the meaning of 2 to the power zero and to a negative power.
Notice that

$$2^0 = 1 \qquad \text{and} \qquad 2^{-3} = \frac{1}{2^3}$$

You can write the rules as general laws.

$$\mathbf{a^0 = 1} \qquad \mathbf{a^{-n} = \frac{1}{a^n}}$$

Exercise

1 Work out the value of

a) 6^2 b) 5^3 c) 3^4 d) 2^6 e) 4^4 f) 3^5

g) 7^3 h) 10^4 i) 11^2 j) 5^1 k) 2^9 l) 1.5^3

2 Write each value as a fraction. For example $2^{-2} = \frac{1}{4}$.

a) 4^{-2} b) 10^{-3} c) 5^{-2} d) 8^{-1} e) 3^{-3} f) 6^{-2}

g) 10^{-2} h) 3^{-4} i) 4^{-1} j) 2^{-4} k) 6^{-3} l) 10^{-4}

3 Work out the value of

a) 7^2 b) 3^{-2} c) 10^{-1} d) 4^3 e) 9^{-2} f) 6^0

g) 2^5 h) 5^{-3} i) 10^6 j) 9^1 k) 5^{-1} l) 7^0

m) 6^3 n) 8^{-2} o) 4^1 p) 5^4

4 Write down the following numbers as powers of ten.

a) one million b) one hundredth

c) one hundred d) one thousandth

e) one tenth f) one thousand

Investigation

Copy and complete this table.

$2^1 = 2$	$4^1 =$	$6^1 =$	$8^1 = 8$
$2^2 =$	$4^2 =$	$6^2 = 36$	$8^2 =$
$2^3 =$	$4^3 = 64$	$6^3 =$	$8^3 =$
$2^4 = 16$	$4^4 =$	$6^4 =$	$8^4 = 4096$
$2^5 =$	$4^5 =$	$6^5 = 7776$	$8^5 =$

Look at the pattern of last digits of the numbers in each column.

(i) What do you think the last digit is for each number to the power 6?

(ii) Write down what you think the last digit is in each of these numbers.

a) 4^9 b) 6^8 c) 2^{11} d) 8^7 e) 2^9 f) 8^{10}

(iii) Work out rules to predict the last digit of even numbers given in index form. (4^9, 6^8, … are examples of index form.)

(iv) Carry out a similar investigation with the odd numbers 1, 3, 5, 7 and 9 and work out rules which predict the last digit of odd numbers given in index form.

Using index form

When you multiply two numbers given in index form you add the powers.

$$2^4 \times 2^3 = 2^{4+3} = 2^7$$

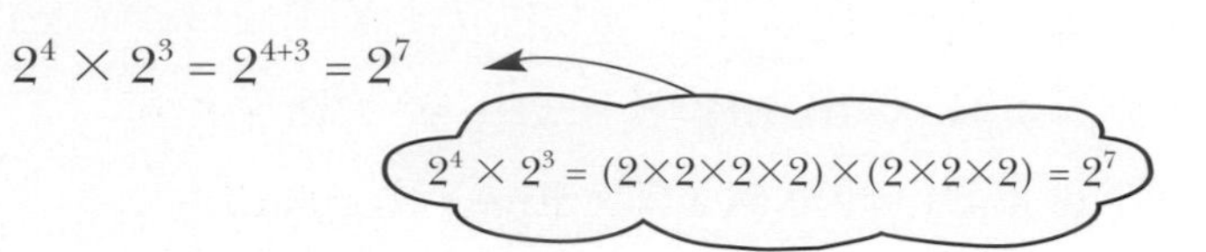

What is $2^2 \times 2^3 \times 2^4$*?*

When you divide one number by another you subtract the powers.

$$2^5 \div 2^3 = 2^{5-3} = 2^2$$

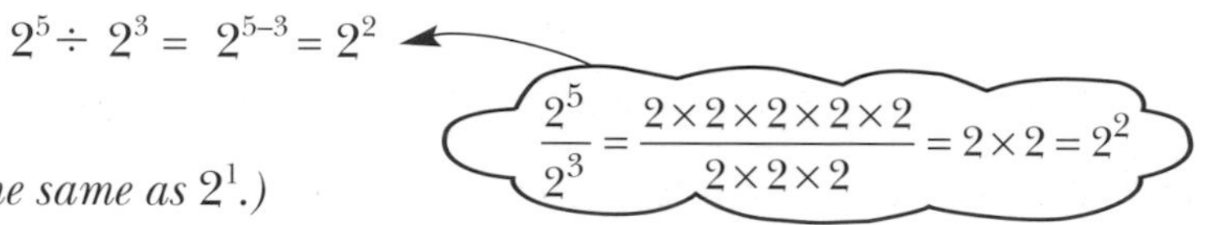

What is $2^4 \div 2$*? (Hint: 2 is the same as* 2^1*.)*

What is $(2^3)^4$ *in index form?*

You can write these rules as general laws.

$$\mathbf{a^m \times a^n = a^{m+n}} \qquad \mathbf{a^m \div a^n = a^{m-n}} \qquad \mathbf{(a^m)^n = a^{m\times n}}$$

Calculators and standard form

Without using a calculator work out 5 000 000 × 80 000.

(Hint: 80 000 = 8 × 10 × 10 × 10 × 10.)

Now work out 5 000 000 × 80 000 *on your calculator.*

Without a calculator 5 000 000 × 80 000 gives 400 000 000 000 but your calculator probably displays something like one of these

4 11

Your calculator uses this to mean 4×10^{11} It does not mean 4^{11}.

Your calculator can only show a limited number of digits (probably 8 or 10).

It does not have enough space to display 400 000 000 000.

The results must be equal so

$$400\ 000\ 000\ 000 = 4 \times 10^{11}$$

4×10^{11} is an example of **standard form**.

The leading number, 4, is between 1 and 10.

Standard form is a tidy way of writing large (or small) numbers.

Exercise

1 Work these out giving your answer in index form.

For example $4^3 \times 4^2 = 4^5$

a) $5^2 \times 5^4$ b) $2^6 \times 2^3 \times 2^2$ c) 6×6^3 d) $10^6 \div 10^3$

e) $2^8 \div 2$ f) $(10^2)^3$ g) $(3^4)^2$ h) $4^5 \times 4^{-3}$

i) $10^3 \div 10^{-1}$ j) $(5^{-1})^2$ k) $3^4 \times 3^{-2} \times 3$ l) $2^{-1} \div 2^{-2}$

m) $\dfrac{4^9}{4^3 \times 4^4}$ n) $\dfrac{3^4 \times 3^2}{3^8}$ o) $\dfrac{(2^3)^2 \times 2}{2^7}$ p) $\dfrac{10^4 \times 10^6 \times 10}{10^5 \times 10^3}$

Question 2 prepares you for the work on standard form that follows.

2 Without using a calculator work out

a) 52×10 b) $83 \div 10$ c) $6.9 \div 10$ d) 23×100

e) $47 \div 100$ f) 6.4×1000 g) 0.7×10 h) $528 \div 100$

i) 93.45×100 j) 0.8×10 k) $9.12 \div 100$ l) 573×1000

3 Work out the value of the numbers displayed here.

a) 

b) 5.72 13

4 Sara, Majid and Tom are each working out $20\ 000\ 001 \times 40\ 000\ 003$.

Sara uses her calculator and writes down 8×10^{14}.

Majid uses his calculator and writes down $8\ 000\ 001 \times 10^{14}$.

Tom forgot his calculator so he works it out by long multiplication.

Tom claims that both Sara and Majid are wrong.

Is Tom's claim correct? Explain your answer.

Investigation

(i) Use your calculator to work out

a) 7×10 b) $7 \times 10 \times 10$ c) $7 \times 10 \times 10 \times 10$

... and so on until you get an answer in standard form.

How many digits can your calculator display?

(ii) Use your calculator to work out

a) $7 \div 10$ b) $7 \div 10 \div 10$ c) $7 \div 10 \div 10 \div 10$

... until the form of the answer changes.

For which calculation does the form of your answer change?

What does the display show and what do you think this display means?

Standard form

Large numbers

The speed of light is 300 000 000 m/s.

Large numbers like this are rather untidy.

300 000 000 is the same as

$3 \times 100\,000\,000$

or $3 \times 10 \times 10 \times 10 \times 10 \times 10 \times 10 \times 10 \times 10$

This can be written as 3×10^8.

This is an example of **standard form**. The leading number, 3, is between 1 and 10.

The speed of sound in standard form is 3.3×10^2 m/s.

What is 3.3×10^2 in decimal form?

$$3.3 \times 10^2 = 3.3 \times 10 \times 10 = 330$$

Small numbers

The wavelength of yellow light is 0.000 000 6 metres.

Small numbers like this are also untidy.

$$0.000\,000\,6 = \frac{6}{10\,000\,000} \text{ or } \frac{6}{10 \times 10 \times 10 \times 10 \times 10 \times 10 \times 10}$$

This can be written as $\frac{6}{10^7}$ or 6×10^{-7}.

This is in standard form.

The wavelength of mercury green light is 5.4×10^{-7} m.

$$5.4 \times 10^{-7} = \frac{5.4}{10^7} = \frac{5.4}{10 \times 10 \times 10 \times 10 \times 10 \times 10 \times 10} = 0.000\,000\,54$$

The wavelength of mercury green light is 0.000 000 54 m.

This number, 0.000 000 54, can also be written as 54×10^{-8}.

This is not in standard form because the leading number, 54, is not between 1 and 10. You can convert it like this

$$54 \times 10^{-8} = 5.4 \times 10 \times 10^{-8} \quad \text{(because } 54 = 5.4 \times 10\text{)}$$

$$= 5.4 \times 10^{-7} \quad \text{(because } 10 \times 10^{-8} = 10^{-7}\text{)}$$

Exercise

1 These numbers are in standard form.

Write them out in full.

a) 6×10^2 b) 3×10^4 c) 7×10^{-3} d) 4×10^{-5}

e) 4.5×10^6 f) 5.4×10^{-3} g) 9.4×10^3 h) 8.75×10^{-4}

i) 1.6×10^{-2} j) 2.75×10^6 k) 8.3×10^{-5} l) 1.05×10^4

m) 7.3×10^3 n) 8×10^{-9} o) 4×10^{-1} p) 8.25×10^{10}

2 Write these numbers in standard form.

a) 4000 b) 800 000 c) 0.003 d) 0.0009

e) 26 000 f) 0.025 g) 7 500 000 h) 0.000 037

i) 810 j) 0.005 43 k) 0.93 l) 64 000

m) 0.016 n) 147 000 000 o) 0.507 p) 9040

3 Alison is using this spreadsheet for her science assignment.

	D	E	F
1	637 983	5E–03	4.38E+09
2	7.25E–04	9.42E+08	694
3	4.6E+12	83 926	7.5E–06

Large and small numbers are displayed using an E.

So in D2, 7.25E–04 means 7.25×10^{-4}.

Work out the value of the entry in

a) E1 b) D3 c) E2 d) F3

4 This table shows the approximate diameter of each planet.

Planet	**Diameter (m)**
Mercury	4.88×10^6
Venus	1.21×10^7
Earth	1.28×10^7
Mars	6.79×10^6
Jupiter	1.44×10^8
Saturn	1.21×10^8
Uranus	5.08×10^7
Neptune	4.95×10^7
Pluto	2.30×10^6

Arrange them in order of size.

The population of Sweden is about 8 800 000 or 8.8×10^6.

Find out the approximate population of nine other countries.

Present your results in a table, showing the populations in both standard form and as ordinary numbers.

Calculations using standard form

This diagram shows the distances of the planets from the Sun.

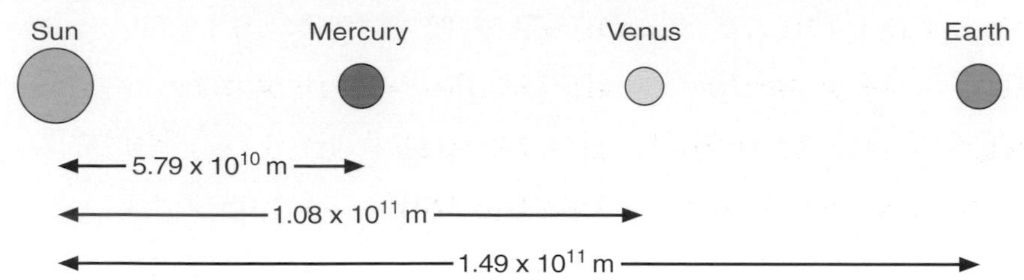

How far is Earth from Mercury when they are at their closest?

It is $(1.49 \times 10^{11} - 5.79 \times 10^{10})$ metres.

You work it out on your calculator like this.

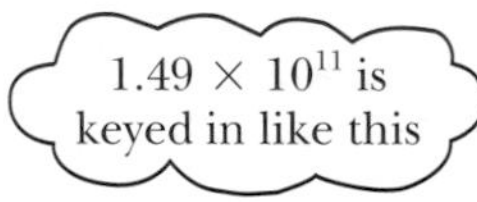

5.79 × 10^{10} is keyed in like this

Earth is 9.11×10^{10} m from Mercury when they are at their closest.

The volume of the Earth is 1.08×10^{21} m^3.

Each cubic metre has, on average a mass of 5.52×10^3 kg.

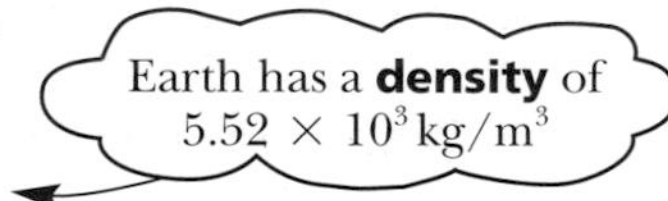

What is the mass of the Earth?

$$\textbf{mass} = \textbf{volume} \times \textbf{density}$$

$$= (1.08 \times 10^{21}) \times (5.52 \times 10^3)$$

Work this out on your calculator. Check you get 5.96×10^{24}.

The mass of the Earth is 5.96×10^{24} kg.

5.96×10^{24} *is in standard form. What is its value?*

The Earth is about 1.5×10^{11} metres from the Sun.

Light travels at a speed of 3.0×10^8 metres per second.

How long does it take light to travel from the Sun to the Earth?

Kylie works it out like this:

$$\text{Time in seconds} = \frac{\text{distance}}{\text{speed}} = \frac{1.5 \times 10^{11}}{3 \times 10^8} = \frac{1.5 \times 10^3}{3} = 500$$

Check this calculation on your calculator.

What is 500 *seconds in minutes and seconds?*

1 Work out these calculations, and give your answers in standard form.

a) $(4 \times 10^3) + (8 \times 10^4)$ b) $(7 \times 10^6) - (2 \times 10^5)$

c) $(3 \times 10^5) \times (5 \times 10^6)$ d) $(8 \times 10^{11}) \div (5 \times 10^4)$

e) $(2.5 \times 10^{-4}) \times (6 \times 10^{12})$ f) $(7.4 \times 10^{-2}) - (3.8 \times 10^{-3})$

g) $(4.5 \times 10^{-7}) \times (3.2 \times 10^{-11})$ h) $(4 \times 10^7)^2$

2 Complete this table by multiplying volume by density to get mass.

Planet	Volume (m^3)	Density (kg/m^3)	Mass (kg)
Mercury	1.45×10^{19}	5.42×10^3	
Venus	9.29×10^{20}	5.25×10^3	
Mars	5.21×10^{19}	3.94×10^3	
Jupiter	1.56×10^{24}	1.31×10^3	

3 The kinetic energy, E, of an electron is $\frac{1}{2}mv^2$.

When $m = 9 \times 10^{-31}$ kg and $v = 2 \times 10^7$ metres per second, what is E?

4 This table shows the population and area of different continents.

Continent	Population	Area (km^2)
Europe (incl Russia)	4.95×10^8	4.94×10^6
Africa	7.43×10^8	3.03×10^7
Oceania	8.51×10^6	2.5×10^7

a) Population density = population ÷ area.

Which of these continents has the highest population density?

b) What is the population of Oceania to the nearest million?

c) What is the area of Africa to the nearest million km^2?

5 The moon is a sphere whose radius, r, is 1.7×10^6 metres. Its volume, V, is $\frac{4}{3}\pi r^3$.

a) What is the volume of the moon?

b) The mass of the moon is 7.4×10^{22} kg.

What is the density of the moon in kg/m^3?

Find out the radius of each planet in metres.
Set up a spreadsheet to work out the volume of each planet in m^3.

You may assume that each planet is a sphere and its volume, V, is $\frac{4}{3}\pi r^3$.

Finishing off

Now that you have finished this chapter you should be able to:

- ★ work out positive, zero and negative powers of numbers
- ★ work with numbers in index form
- ★ write numbers in standard form
- ★ work out the value of numbers in standard form including those given on a calculator display
- ★ do calculations with numbers in standard form.

Use the questions in the next exercise to check that you understand everything.

Mixed exercise

1 Work out

a) 8^2 b) 10^3 c) 3^0 d) 2^7

e) 4^{-2} f) 12^2 g) 5^{-3} h) 10^1

i) 7^{-1} j) 6^0 k) 2^{-5} l) 2^{10}

2 Work these out, giving your answer in index form.

For example $3^2 \times 3^5 = 3^7$

a) $2^5 \times 2^3$ b) $7^4 \div 7$ c) $(5^3)^2$ d) $4^2 \times 4^6$

e) $6^5 \div 6^2$ f) $(4^2)^2$ g) $3^5 \times 3 \times 3^{-2}$ h) $\sqrt{3^2}$

3 The numbers in this question are in standard form.
Write them out in full.

a) The radius of the Earth is $\mathbf{6.4 \times 10^6}$ metres.

b) A capillary tube has radius $\mathbf{2 \times 10^{-4}}$ metres.

c) A red blood cell has a diameter of $\mathbf{7.5 \times 10^{-6}}$ metres.

d) The density of mercury is $\mathbf{1.36 \times 10^4}$ kg/m^3.

4 Write these numbers in standard form.

a) A train has a mass of **200 000** kg.

b) The thickness of a piece of cardboard is **0.0015** metres.

c) Thorium-230 has a half life of **83 000** years.

d) The linear expansivity of aluminium is **0.000 026** per degree Kelvin.

5 Work out the value of these calculations, and give your answer in standard form.

a) $(5 \times 10^4) + (8 \times 10^5)$
b) $(3.1 \times 10^{-2}) - (7 \times 10^{-3})$
c) $(4 \times 10^8) \times (9.7 \times 10^{13})$
d) $(3.6 \times 10^{12}) \div (9 \times 10^4)$
e) $(3.2 \times 10^{14}) \times (7.5 \times 10^{-9})$
f) $(4.9 \times 10^{11}) \div (2.8 \times 10^{-5})$
g) $(4.5 \times 10^7)^2$
h) $\sqrt{1.6 \times 10^{13}}$

6 The diameter of the Earth is 1.28×10^7 metres.
Work out the distance around the equator in metres.

7 a) Calculate the number of seconds in a year.

b) Light travels at 3×10^8 metres per second.
How many metres does light travel in a year?

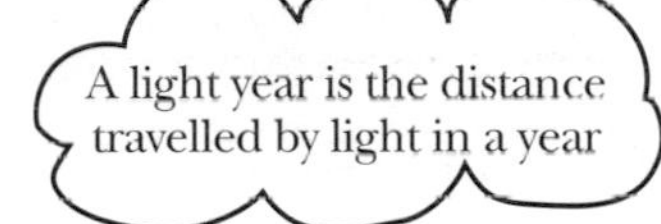

c) How many kilometres is this?

d) Sirius A is a star 4.2 light years away from Earth.
How many kilometres is this?

Investigation

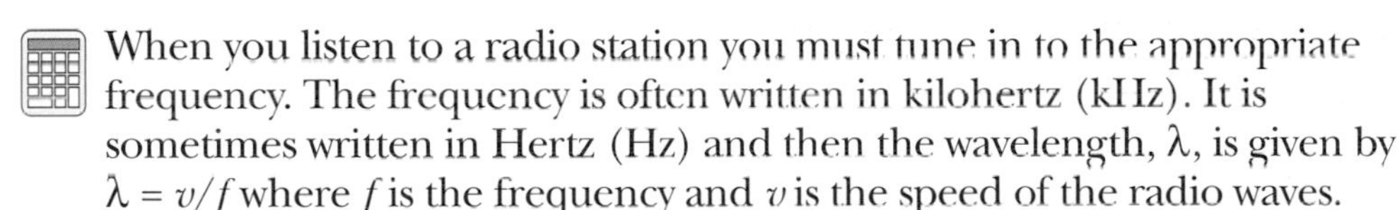

When you listen to a radio station you must tune in to the appropriate frequency. The frequency is often written in kilohertz (kHz). It is sometimes written in Hertz (Hz) and then the wavelength, λ, is given by $\lambda = v/f$ where f is the frequency and v is the speed of the radio waves.

Copy and complete this table. (Note: f is required in kHz in the table but must be in Hz to use in the formula, and $v = 3 \times 10^8$ ms^{-1}.)

Station	f (kHz)	f (Hz)	λ (m)
Radio 4	198		
Radio 4			417
Talk Radio	1089		
Radio 5 Live			433
Radio 5 Live	909		
Virgin			247
World Service	648		

Find out what these prefixes mean:

a) mega-
b) micro-
c) giga-
d) nano-

Try to find examples of when they are used in real life.

Ten

Approximations

Decimal places

Connor is doing a survey to find out how well known leading twentieth century figures are. He asks 35 people to identify these two.

(1)

(2)

Of the people he asks, 19 recognise (1) to be Mother Theresa.

Connor works this out as a percentage.

$$\frac{19}{35} \times 100 = 54 \cdot 2857\ldots$$

He has to decide how accurately to state this in his report.

How many decimal places, if any, do you think Connor should use?

Look at this number line.

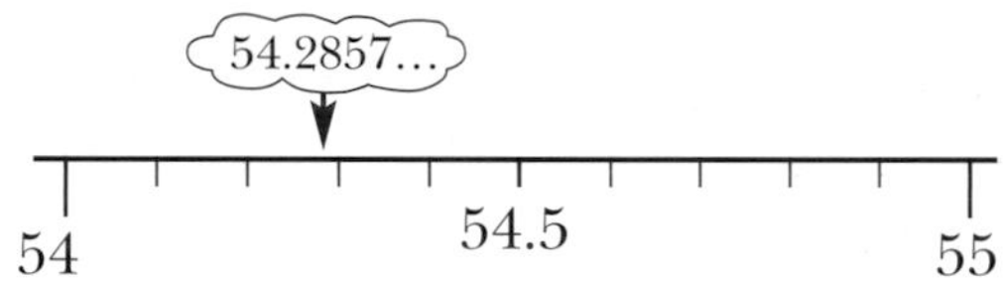

54.2857... is nearer to 54 than to 55. It is 54 correct to the nearest whole number.

Now look at this number line.

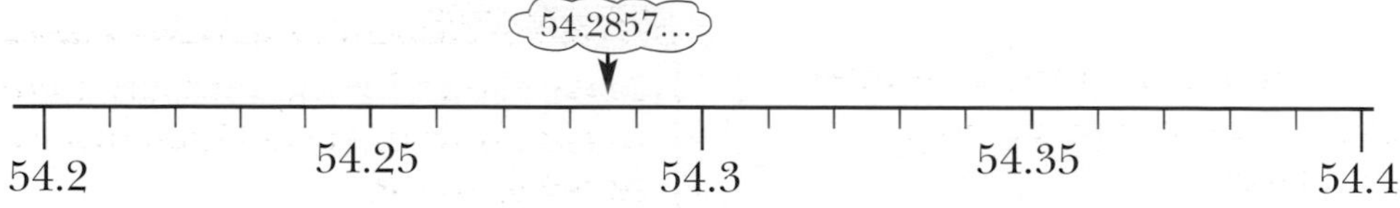

54.2857... is nearer to 54.3 than to 54.2. It is 54.3 correct to 1 decimal place.

Copy the last number line. Colour the range of numbers which would be written as 54.3 correct to 1 decimal place.

Out of 35 people, 15 recognise (2) to be Nelson Mandela.
What is this as a percentage

a) correct to the nearest whole number? *b) correct to 1 decimal place?*

Exercise

1 The value of π is 3.141592… Write this value correct to

a) 1 decimal place b) 2 decimal places c) 3 decimal places.

2 Use your calculator to find the square root of 20.

Write this value correct to

a) 1 decimal place b) 2 decimal places c) 3 decimal places.

3 How many decimal places do you give if your answer is correct to the nearest

a) tenth? b) hundredth?

4 For each of these, write down your estimate of the reading to the nearest hundredth, and then write the reading correct to the nearest tenth.

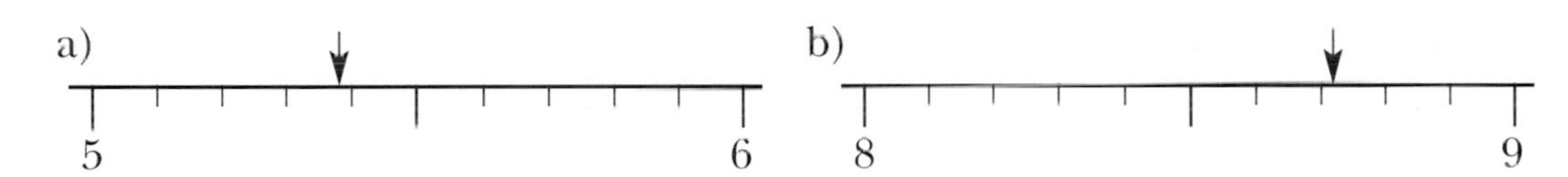

5 Write these fractions as decimals correct to 3 decimal places.

a) $\frac{1}{6}$ b) $\frac{1}{12}$ c) $\frac{2}{3}$ d) $\frac{3}{7}$

6 Calculate 12.6% of 43.8, giving your answer correct to 2 decimal places.

7 a) Measure the length and width of this rectangle in centimetres giving your answers correct to 1 decimal place.

b) Use these values to calculate the area of the rectangle giving your answer correct to 1 decimal place.

8 Work out the mean of these numbers giving your answer correct to 2 decimal places.

4, 4, 5, 7, 8, 12, 15

9 A circle has radius 7.3 m. Using $\pi = 3.14$ calculate, correct to 1 decimal place,

a) its circumference b) its area.

Make the following measurements and decide how many decimal places it is sensible to use.

a) The height of a friend in metres.

b) The length and width of a sheet of paper in centimetres.

c) The length and width of a car in metres.

Significant figures

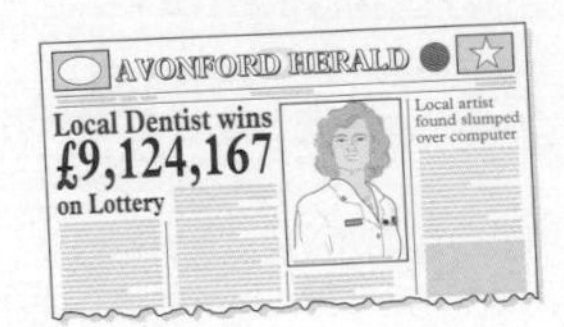

Look at these two newspaper headlines. Which do you think is better?

Sam has actually won £9 124 167, but the Avonford Star describes this as £9 million, which is £9 000 000. They decided their readers were not interested in the exact amount and so they rounded it to 1 **significant figure**.

£9 million	or £9 000 000	has 1 significant figure: it is the 9.
£9.1 million	or £9 100 000	has 2 significant figures: 9 and 1.
£9.12 million	or £9 120 000	has 3 significant figures: 9, 1 and 2.

Example

Write a) 294 217 to 3 significant figures (3 s.f.),

b) 0.004 297 to 2 significant figures.

Solution

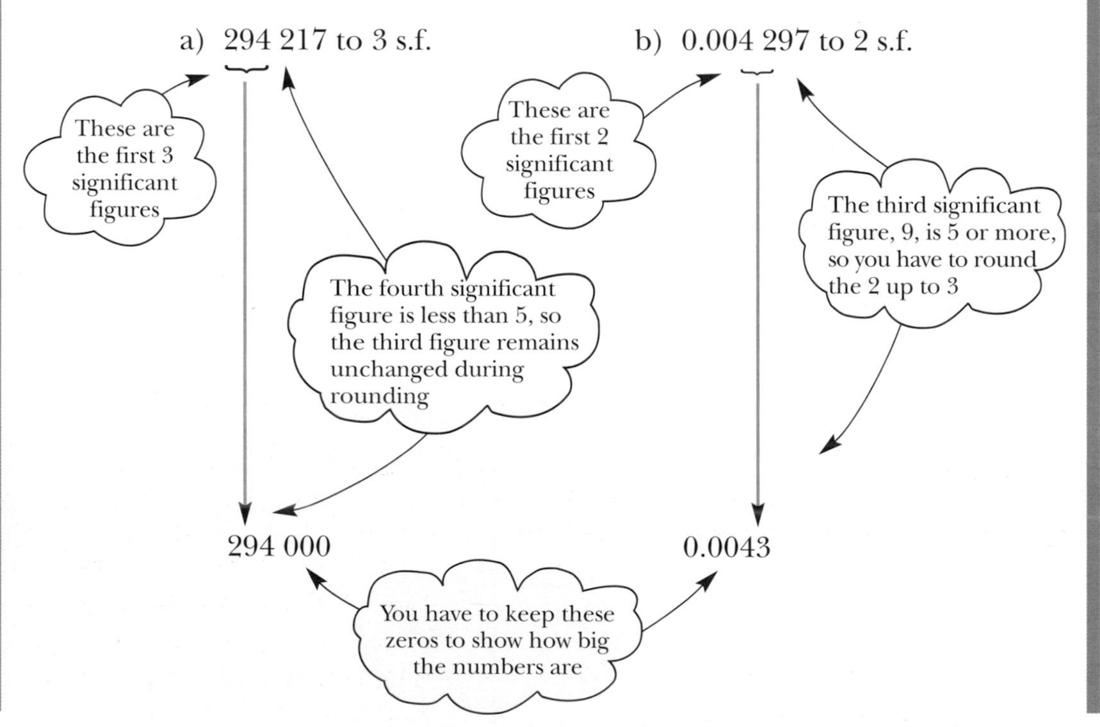

In the example above, 0.004 297 *is rounded to* 2 *significant figures.*

What do you get if you round it to 2 *decimal places?*

Which way of rounding do you think is better?

Write 0.004 297 *correct to* 3 *significant figures.*

1 Write each of these to the number of significant figures (s.f.) shown.

a) 77 328 to 2 s.f.

b) 42.195 to 3 s.f.

c) 5372 to 2 s.f.

d) 758 423 to 3 s.f.

e) 3780 to 1 s.f.

f) 61.977 to 3 s.f.

g) 6.7394 to 2 s.f.

h) 53 660 to 1 s.f.

i) 33.830 7 to 3 s.f.

j) 0.005 38 to 1 s.f.

2 For each of these, write down your estimate of the reading to 4 significant figures, then write the reading correct to 3 significant figures.

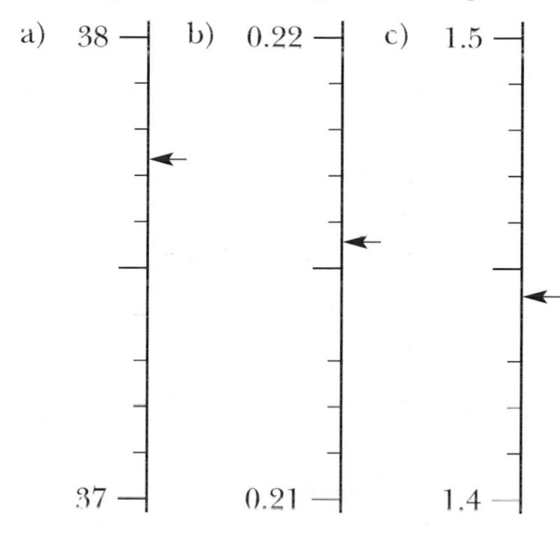

3 Repeat question 2 giving your answers correct to 2 significant figures.

4 Work out 15% of 175 000 correct to 2 significant figures.

5 A football pitch is 114 m long and 73 m wide. Work out, correct to 2 significant figures

a) the perimeter b) the area.

6 The profits of a business for the four quarters of a year are

£55 000 £79 000
£127 000 £98 000

a) Work out the average quarterly profit correct to 2 significant figures.

b) Work out the total profit for the year correct to 2 significant figures.

7 Tickets for a play are £4.25 each. This table shows the number of tickets sold for each of the three performances.

Day	Thursday	Friday	Saturday
Tickets sold	257	319	348
Income			

a) Copy and complete the table, writing each day's ticket income correct to 3 significant figures.

b) Write down the total income correct to 2 significant figures.

Rough checks

You should always do a rough check to make sure that the answer you get on your calculator is sensible.

A rough check does not enable you to pick up errors that only make a small difference to your answer but it does stop you getting it completely wrong.

Mr Harris wants to reseed his lawn.

He asks his four children to work out the area of the lawn.

19.8 m

10.3 m

They each try to work out 19.8 × 10.3 on their calculators. Here are their answers.

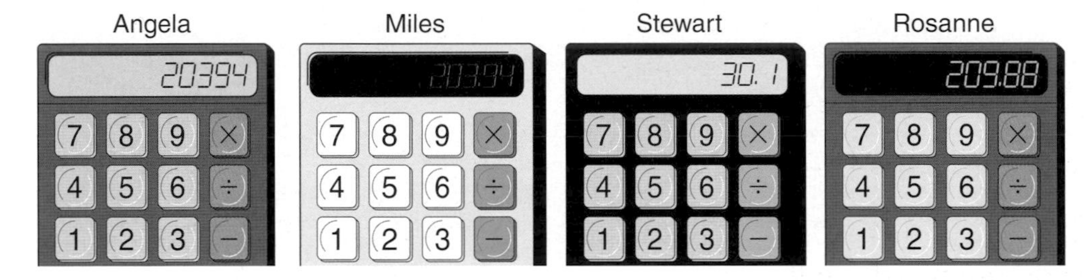

Which of these answers seem sensible?

One way to decide is to say 19.8 is about 20 and 10.3 is about 10, so the answer should be roughly $20 \times 10 = 200$.

Miles and Rosanne both have sensible answers. It is not easy to judge which of them is correct without doing an accurate calculation.

Do the calculation on your calculator and find out which is correct.

Elliott estimates the saving on these trainers.

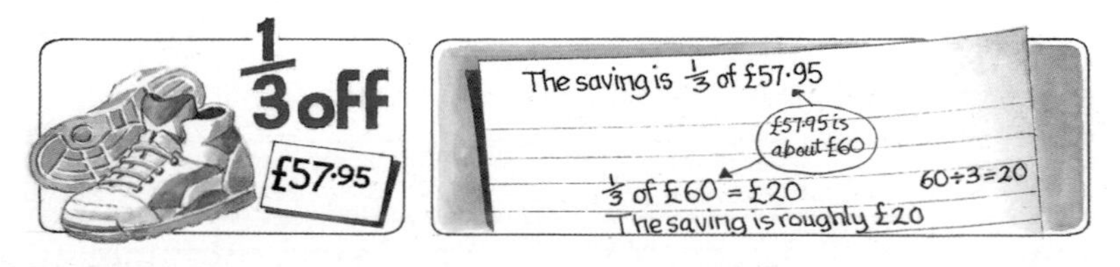

Estimate the saving on trainers priced at £89.45 and marked '$\frac{1}{3}$ off'.

Bridget earns £13 900. She receives a pay rise of 9.2%. She makes this rough estimate of how much extra she will earn.

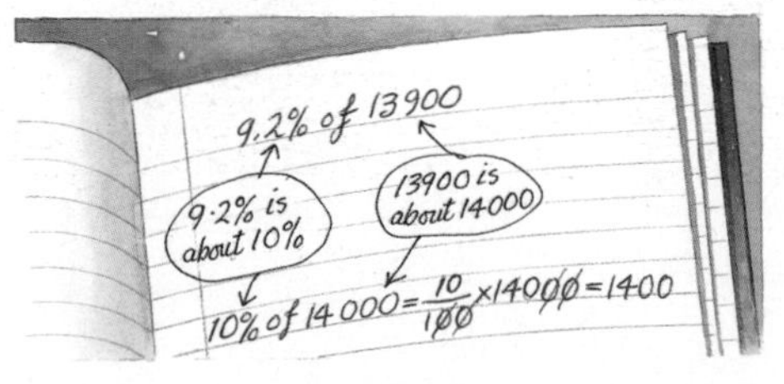

Use a calculator to work out Bridget's increase in salary. How do the results compare?

Exercise

1 Estimate the area of these shapes.

a) b) c)

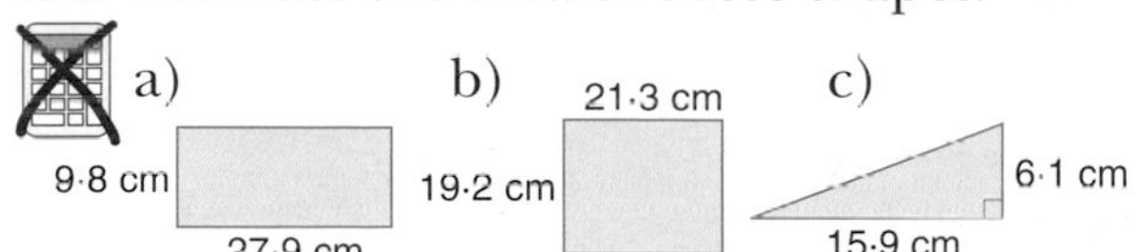

2 Five 42-seater coaches are almost full for a sight-seeing tour.

a) Estimate the total number of tourists in them.

b) Each tourist has paid £9.50 for the tour.

Estimate the total ticket income.

3 Estimate the answers to these calculations.

a) $\frac{1}{2}$ of 407 b) $\frac{1}{4}$ of 99 c) $\frac{1}{10}$ of 1793 d) $\frac{1}{3}$ of 595

e) $\frac{1}{6}$ of 1492 f) $\frac{3}{4}$ of 810 g) $\frac{2}{3}$ of 460 h) $\frac{3}{10}$ of 1187

4 This pie chart shows the distribution of a sports grant of £11 945.

Estimate the amount given to

a) athletics b) swimming

c) badminton d) golf.

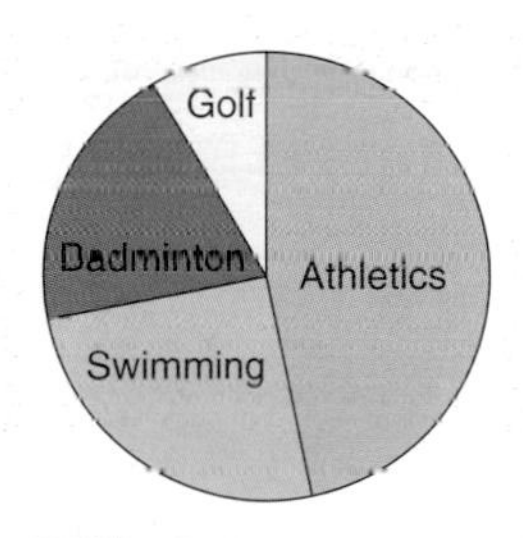

5 Estimate the answers to these calculations.

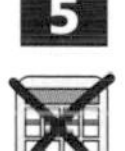

a) 51% of 607 b) 23% of 101

c) 48% of 395 d) 33% of 121

e) 49% of 362 f) 27% of 966

g) 73.5% of 197 h) 9.7% of 587

6 Estimate the perimeter and area of each of these shapes. (For rough calculations like this you can take π as 3.)

a) b) c)

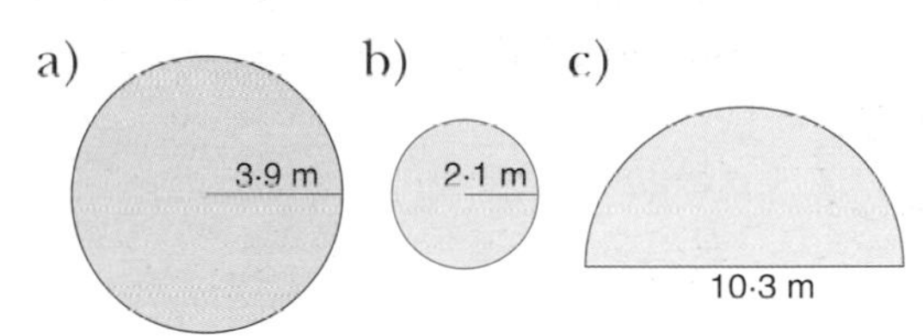

7 Estimate the sale price of

a) the shirt b) the jumper c) the trainers.

Errors

Jasmine's garden is a rectangle 50 m × 20 m.
Its area is 1000 square metres.

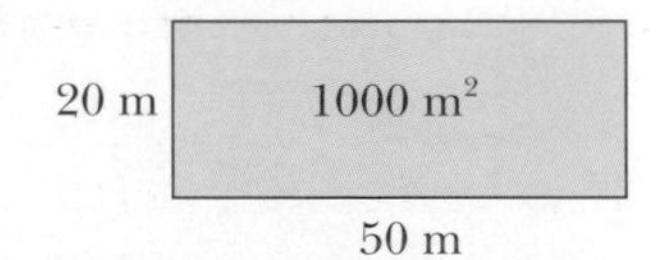

She tells a friend

> 'My garden is about 70 metres long and 30 metres wide. So its area is 2100 square metres.'

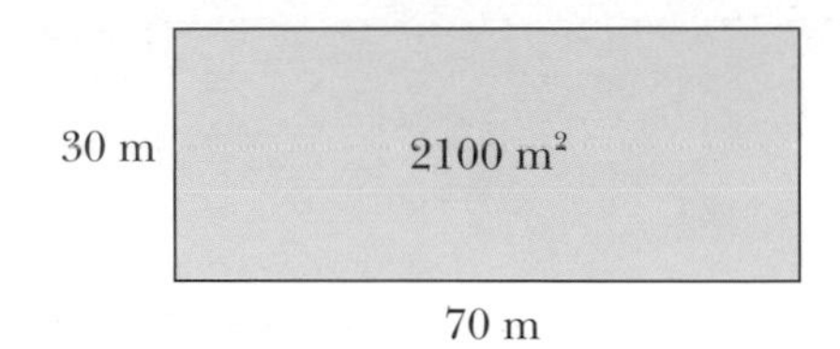

Look at the errors in what Jasmine says.

Length Error: 70 m – 50 m = 20 m

Percentage error: $\frac{20}{50} \times 100\% = 40\%$

Width Error: 30 m – 20 m = 10 m

Percentage error: $\frac{10}{20} = 50\%$

Area Error: $2100\ \text{m}^2 - 1000\ \text{m}^2 = 1100\ \text{m}^2$

Percentage error: $\frac{1100}{1000} = 110\%$

You always use true values when working out percentage error

You can see that the errors accumulate.

How is it that Jasmine's errors of 40 % and 50 %, for the length and widths, combine to give an error of 110 % for the area?

Some errors always occur when numbers are rounded.
Julian describes the top of his coffee table as 44 cm by 32 cm.
Both of these measurements are given to the nearest centimetre.

44 cm means between 43.5 cm and 44.5 cm
32 cm means between 31.5 cm and 32.5 cm

So the area of the top of Julian's coffee table lies between 31.5 × 43.5 and 32.5 × 44.5 cm^2. That is between 1370.25 and 1446.25 cm^2.

What can you say about the perimeter of Julian's coffee table?

Julian measures the table very carefully and finds it is 44.1 cm by 32.1 cm or 441 mm by 321 mm.

These measurements are to the nearest millimetre.

Using these new measurements what are the greatest and least values for
(i) the length (ii) the width (iii) the area?

Exercise

1 Robert is trying to impress his friends. He says

'I have had my job for 12 weeks and get £600 a week. That's £7200 already!'

Actually he had had the job for 10 weeks and earns £400 a week. Find the error and the percentage error in

a) the number of weeks

b) the weekly pay

c) the total amount of money Robert has earned.

2 Julie is moving to a new house. She measures the staircase and finds that there are 15 steps, i.e. 15 risers and 14 treads.

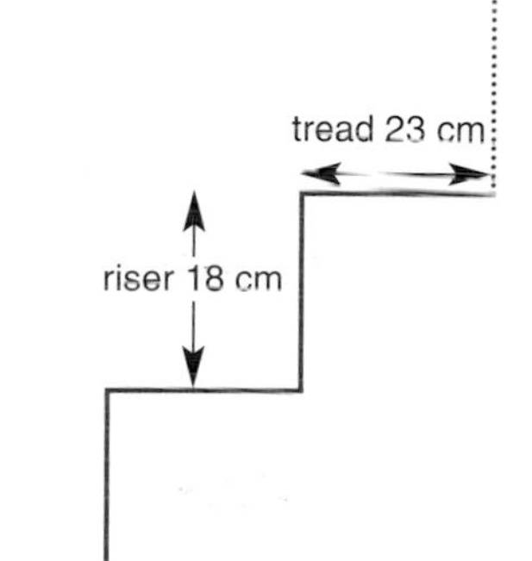

Each riser is 18 cm high to the nearet cm and each tread is 23 cm long to the nearest cm.

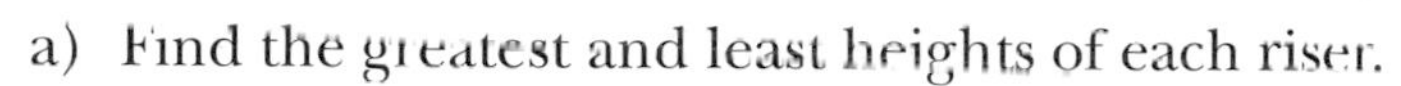

a) Find the greatest and least heights of each riser.

b) Find the greatest and least lengths of each tread.

c) Find the greatest and least amounts of carpet that Julie will need to cover the staircase.

3 Spiro knows his car contains 5 gallons of petrol, to the nearest gallon and that it will travel 40 miles per gallon, to the nearest 10 miles per gallon.

a) What is the greatest distance Spiro can hope to travel without running out of petrol?

b) What is the least distance Spiro can be certain to travel without running out of petrol?

4 A bird keeper describes her aviary as 15 metres by 10 metres by 3 metres, with each measurement to the nearest 1 metre.

State the smallest and largest possible volumes of the aviary.

Here are some well known approximate measures

Tip of your thumb to its first joint: 1 inch
Span of your hand when spread out: 8 inches
1 pace: 1 metre (or 1 yard)
Tips of fingers to chin (when your arm is outstretched): 1 metre (or 1 yard)

Measure these on yourself and estimate the percentage error when using each of these measures.

Finishing off

Now that you have finished this chapter you should be able to:

- ★ round to a given number of decimal places
- ★ round to a given number of significant figures
- ★ estimate costs
- ★ check that the answers on your calculator are sensible
- ★ make rough calculations
- ★ work out percentage errors.

Use the questions in the next exercise to check that you understand everything.

Mixed exercise

1 Use your calculator to find the square root of 28.

Write this value correct to

a) 1 decimal place b) 2 decimal places c) 3 decimal places.

2 Estimate each of these readings to 2 decimal places.

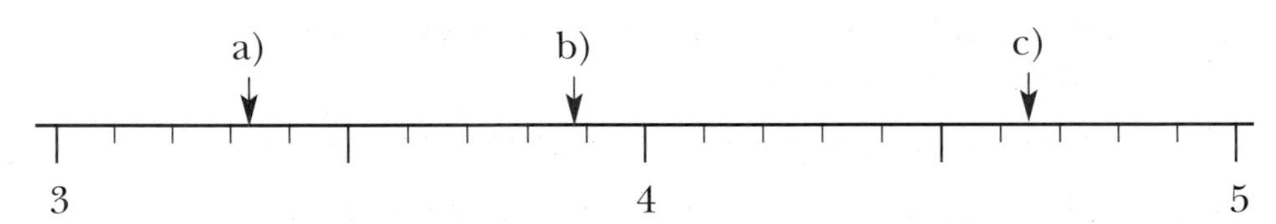

3 Work out the area of each shape giving your answer correct to 1 decimal place.

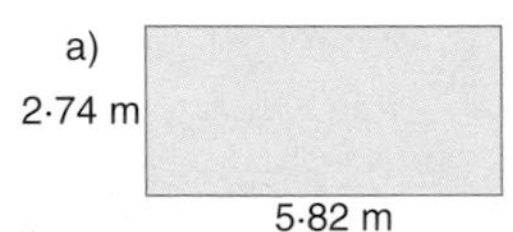

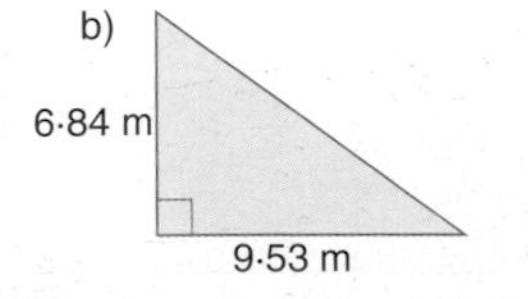

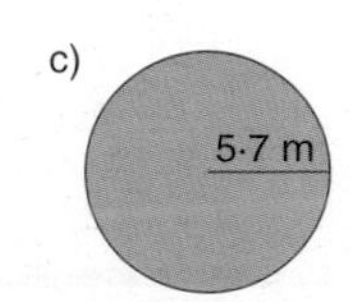

4 Write each of these to the number of significant figures (s.f.) shown.

a) 8397 to 1 s.f.

b) 764 729 to 3 s.f.

c) 14.7528 to 3 s.f.

d) 0.06527 to 2 s.f.

5 This bar chart shows the amount of money raised by each of the four houses at Westfield School.

a) Estimate, correct to 2 significant figures, the amount raised by each house.

b) Estimate the total amount raised correct to 2 significant figures.

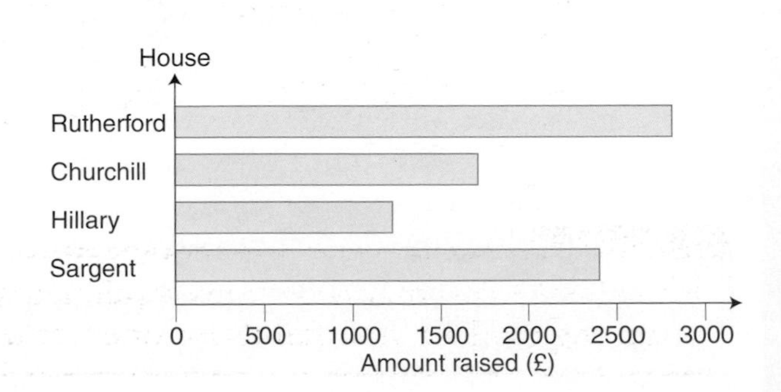

Mixed exercise

6 Estimate the cost of each person's bill from this takeaway.

a) Monika orders cod and chips twice and 2 colas.

b) Chris orders chicken and chips twice and 2 small pizzas.

c) Emma orders meat pie and chips, 2 large pizzas and 3 colas.

MENU

Cod	£1.99	Pizza (small)	£3.95
Chicken	£2.85	Pizza (large)	£4.95
Meat Pie	£1.10	Cola	£0.99
Chips	£0.99		

7 Kitchen tiles are 20 cm by 20 cm and cost £1.89 each. Lauren tiles an area 195 cm by 55 cm. Estimate the total cost of the tiles.

8 Estimate the answers to these calculations.

a) 9.4×30.2 b) $8.092 + 3.95$ c) $407.8 \div 4.1$

d) 19.7^2 e) $\sqrt{63.41}$ f) $26\,172 - 8395$

g) $\frac{1}{4}$ of 843 h) $\frac{2}{3}$ of 913 i) 74% of 982

9 Rio wants to returf his lawn. It is 8.1 m long and 5.8 m wide. The turf costs £7.95 per square metre. Estimate the cost of returfing the lawn.

10 The length of a rectangular room is 8 m, to the nearest metre. Its width is 6 m, to the nearest metre.

a) Find the smallest and largest possible values for its perimeter.

b) Find the smallest and the largest possible values for its area.

11 VAT at 17.5% increases the total cost of an item by about a sixth. Use this fact to estimate the cost (including VAT) of these.

a) b) c)

Estimate the cost of a wedding reception, a week's youth hostelling, or buying food for your family for one week.

Eleven

Algebra 1

Sequences

Look at this pattern.

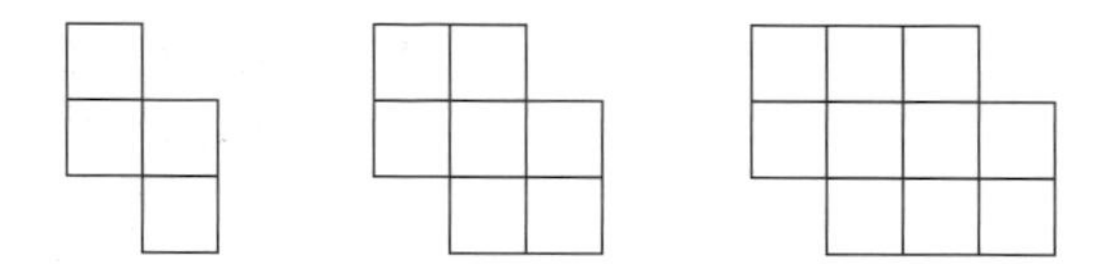

What is the next shape?

A pattern like this forms a **sequence**.

Now count the number of small squares in each shape.

This gives a **sequence of numbers**

4, 7, 10, 13, ...

4 is the first **term**
7 is the second term etc.

What are the next two terms in this sequence?

You can see that by *adding 3* you get from one term to the next.

Look at this sequence.

10, 8, 6, 4, ...

What are the next two terms in this sequence?

This time the rule is *subtract 2.*

Here are some more sequences and the rule used to form them.

2, 4, 8, 16, ...	The rule is *multiply by 2*
100, 10, 1, 0.1, ...	The rule is *divide by 10*
1, 3, 6, 10, ...	The differences are 2, 3, 4, ...
	They increase by 1 each time.

The rest of this chapter concentrates on sequences formed by adding or subtracting a fixed amount. It introduces the idea of using a letter (or symbol) to represent a number and develops some of the basic algebraic skills needed in later work.

1 Write down the next three terms of each sequence.

Write down the rule that you use.

a) 1, 3, 5, 7, …

b) 25, 22, 19, 16, …

c) 1, 10, 100, 1000, …

d) 7, 11, 15, 19, …

e) 50, 45, 40, 35, …

f) 64, 32, 16, 8, …

g) 1, 1.8, 2.6, 3.4, …

h) 15, 12, 9, 6, …

i) 1, 3, 9, 27, …

j) 1, −2, 3, −4, …

2 Ella is walking down the street.

a) She sees these house-number signs on the right.

EVENS 2 – 8	EVENS 10 – 16	EVENS 18 – 24

What are the next 3 signs in this sequence?

b) She sees these house-number signs on the left.

ODDS 1 – 15	ODDS 17 – 31	ODDS 33 – 47

What are the next 3 signs in this sequence?

c) Ella is visiting house number 71. Which sign is she looking for?

3 The Olympic Games are held every four years.

They were held in 1996.

Write down the years of the next three Olympic Games after this date.

4 Jake has 5 hospital appointments, one every three weeks beginning on 24 August.

Write down the sequence of dates for Jake's appointments.

5 Draw the next 2 shapes in this sequence.

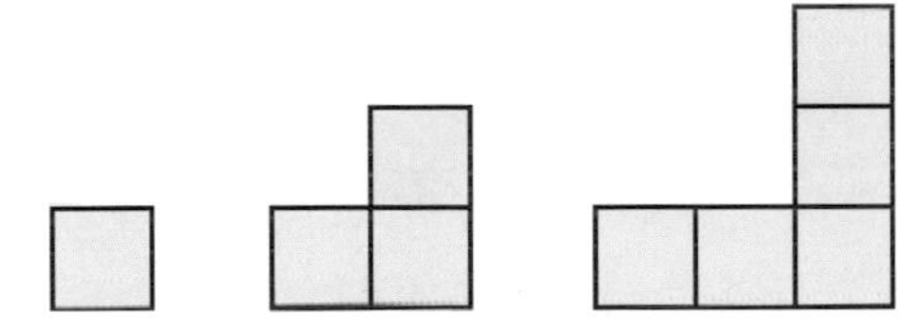

Write down the first 5 terms in the sequence of the numbers of squares.

6 Draw the next 2 shapes in this sequence.

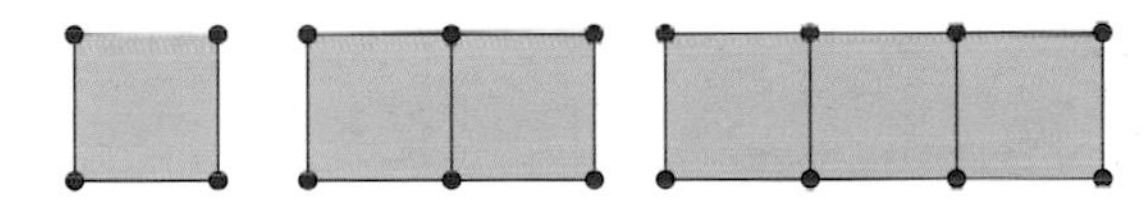

Write down the first 5 terms in the sequence of the numbers of

(i) squares

(ii) dots

7 Draw the next 3 patterns in this sequence.

Write down the first 6 terms in the sequence of the numbers of dots.

8 Write down the next 2 terms in each of these train time sequences.

a) 0900, 0915, 0930, …

b) 0800, 0930, 1100, …

c) 0700, 0745, 0830, …

Finding the nth term

Look at the sequence 6, 11, 16, 21, …

What is the hundredth term?

You need a quick way to work out the hundredth term unless you fancy writing them all out. The quick way is to see how the **term value** is connected to the **term number**. In this case

Term number	1	2	3	4	…	n	…
Term value	6	11	16	21	…	?	…

value of nth term

You can see that to get from one term value to the next you add 5.

This suggests a formula in which $+5n$ is involved.

5n means $5 \times n$

$5n$ gives	5	10	15	20	…	$5n$	…

You need to add 1 to each of these numbers to get the term value.

Value of nth term $= 5n + 1$

Check that this formula works for the first few terms.

When $n = 1$ value of first term $= 5 \times 1 + 1 = 6$ ✓

When $n = 2$ value of second term $= 5 \times 2 + 1 = 11$ ✓

Use the formula to work out the hundredth term.

What is the nth term of the sequence 15, 13, 11, 9, …?

Look at how the term value is connected to the term number.

Term number	1	2	3	4	…	n	…
Term value	15	13	11	9	…	?	…

value of nth term

You can see that to get from one term value to the next you subtract 2.

This suggests a formula in which $-2n$ is involved.

$-2n$ gives	-2	-4	-6	-8	…	$-2n$	…

You need to add 17 to each of these numbers to get the term value

Value of nth term $= -2n + 17$

$-2n + 17$ can also be written as $17 - 2n$

Check that this formula works for $n = 1$ and $n = 2$.

Exercise

1 Work out a formula for the value of the nth term of these sequences.

a) 5, 9, 13, 17, …
b) 2, 5, 8, …
c) 7, 9, 11, 13, …
d) 6, 12, 18, 24, …
e) 8, 6, 4, 2, …
f) 20, 17, 14, 11, …
g) 11, 12, 13, 14,
h) 1.5, 2, 2.5, 3, …
i) 100, 99, 98, 97, …
j) 16, 23, 30, 37, …
k) −8, −16, −24, −32, …
l) 4, 1, −2, −5, …

2 a) Work out the 25th term of the sequence 8, 13, 18, 23, …
b) Work out the 40th term of the sequence 7, 10, 13, 16, …
c) Work out the 75th term of the sequence 200, 198, 196, 194, …

3 Suzie works for a travel company. She has a file containing sheets like this one to help her work out the costs of holidays.

The first four numbers in the cost column are part of a sequence.

a) Work out a formula for the cost of a holiday of n nights.
b) Show that your formula works for
(i) $n = 7$ (ii) $n = 14$ (iii) $n = 21$

4 Here are the fares charged by Green Bus.

Journey length (miles)	1	2	3	4
Fare (pence)	30	50	70	90

The fares continue to increase by 20 pence for each extra mile.

a) Work out a formula for the fare for a journey of n miles.
b) Use it to find the fare for a journey of 20 miles.

5 a) Write down a formula for the nth even number.
b) Write down a formula for the nth odd number.

Simplifying expressions

Guy is building pens for the animals at a cattle market. The pieces of fencing he uses are each one metre long.

How many square metres of land (□ = 1 square metre) are needed for n pens?

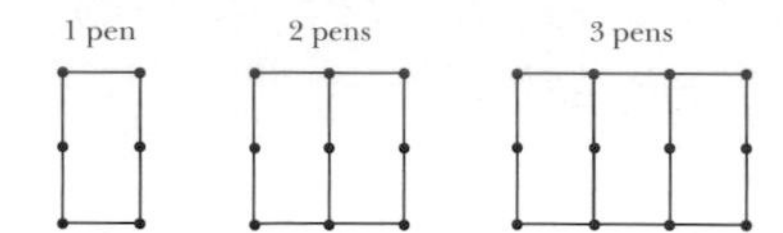

Guy sees that the number of squares in the top row matches the number of pens, n. He adds another n because there are two rows (alternatively he could double it).

Guy works out:

Number of square metres of land for n pens $= n + n$
$= 2n$

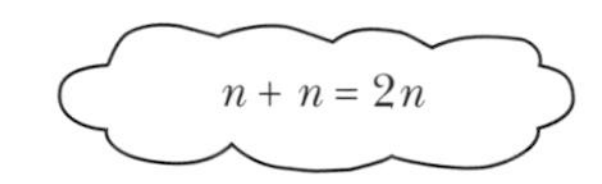

Here are two more examples of simplifying.

- $5n + n = 6n$

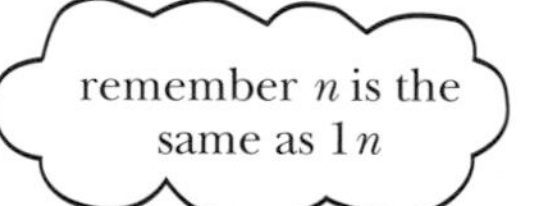

- $4n + 3n - 2n = 5n$

$4n + 3n = 7n$ and then $7n - 2n = 5n$

How many pieces of fencing are needed for n pens?

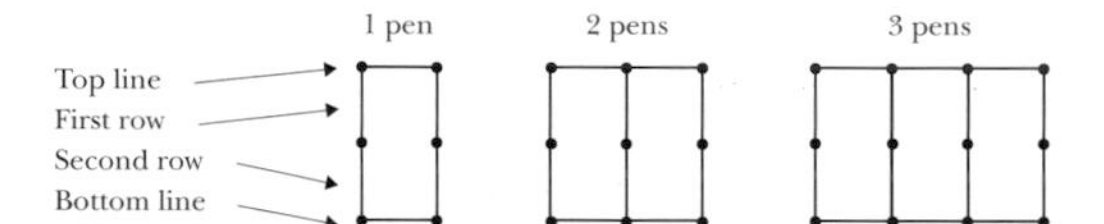

Guy counts the number of pieces of fencing for n pens like this.

The top and bottom lines both match the number of pens, giving n each.

The rows are both one more than the number of pens, giving $n + 1$ each.

Guy works out:

No. of pieces of fencing for n pens $= n + n + n + 1 + n + 1$
$= 4n + 2$

Collect n terms: $n + n + n + n = 4n$. Collect numbers: $+1 + 1 = 2$

Here is another example of simplifying an expression

$5x + y - 2x + 7y$
$= 3x + 8y$

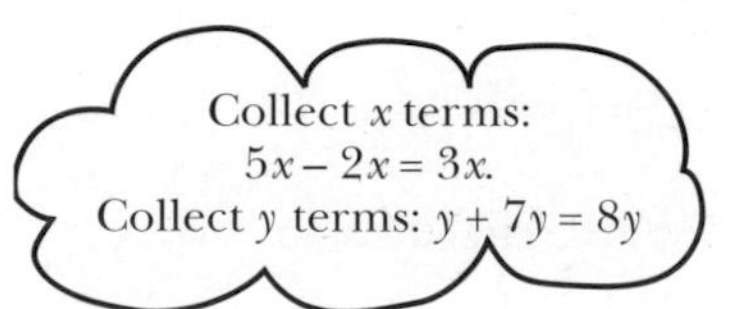

This expression, $3x + 8y$, is in its simplest form.

The x term and y term cannot be combined.

1 Marianne sells units of display space at an exhibition.

She arranges display screens for exhibitors like this.

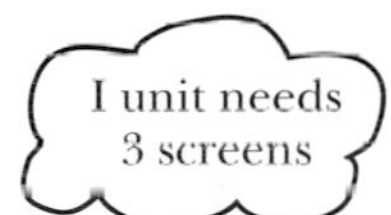

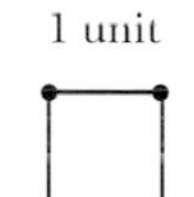

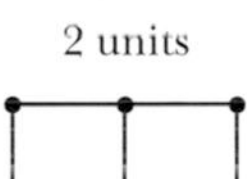

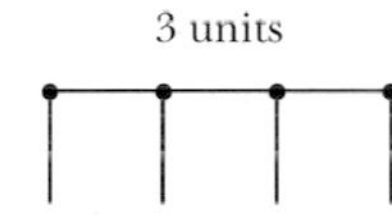

How many display screens are needed for n units?

2 The number of small squares in each pattern forms a sequence.

Work out a formula for the number of small squares in the nth pattern.

Write it in its simplest form.

a)

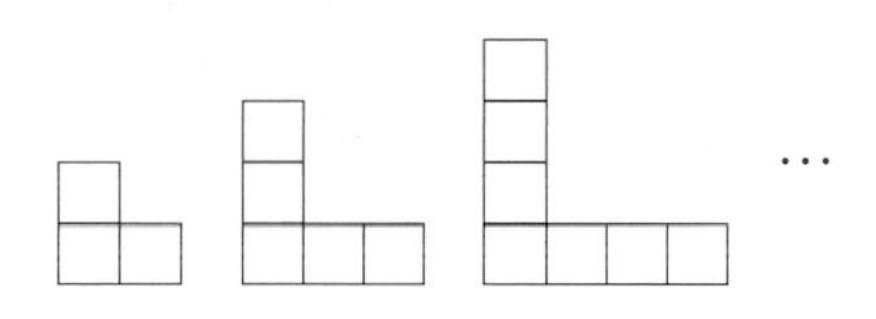

b)

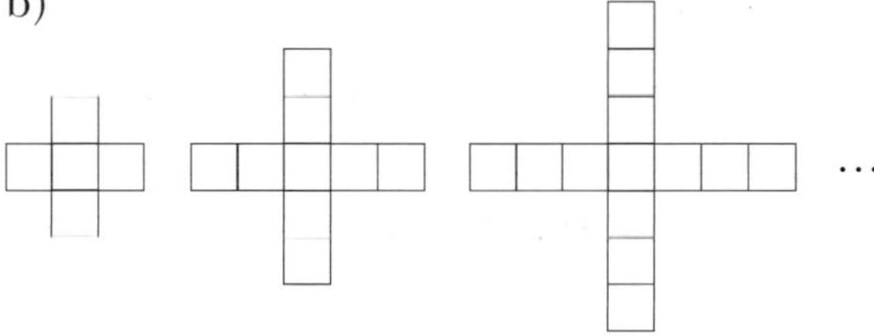

c)

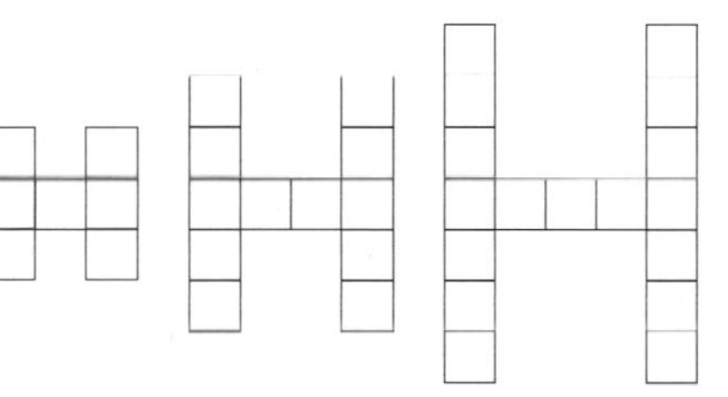

d)

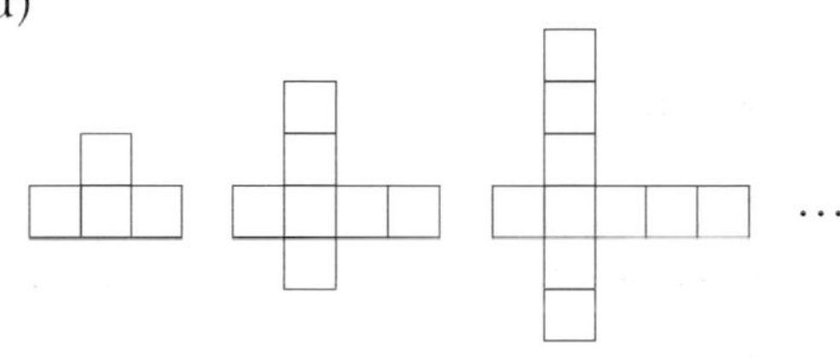

3 Write each of these as a single term.

a) $2n + 4n$

b) $3n + n$

c) $7n - 3n$

d) $6n + 3n - 7n$

e) $-2x + 3x + 5x$

f) $3x - 8x$

g) $y - 6y + 4y$

h) $-n + 2n + 3n - 4n$

4 Write each of these as simply as possible.

a) $2n + 3 + 4$

b) $6n - 1 + 2n$

c) $4n + 1 + 3n + 2$

d) $5n + 3 - 2n - 1$

e) $11 - 2a + 3a$

f) $2b + 1 + b - 4$

g) $4c + d + 3c$

h) $3e + 5f - 4e$

i) $7a + 2b + 4a + 3b$

j) $6p + q + 2p - 6q$

k) $2x - 3y + 5x + 4y$

l) $7r - 2s - s - 3r$

m) $3k - 4m - k + 4m + 1$

n) $e + 4f + 2 + 3e - 7f + 5$

Using brackets

On page 100 Guy counts the number of pieces of fencing needed to make n pens. He adds up 2 lots of n and 2 lots of $n + 1$ and writes

$$n + n + n + 1 + n + 1 = 4n + 2$$

As an alternative Guy can double each lot by multiplying by 2 and write

$$2n + 2(n + 1)$$

This must give the same answer, $4n + 2$, as before.

So $2(n + 1)$ must be the same as $2n + 2$

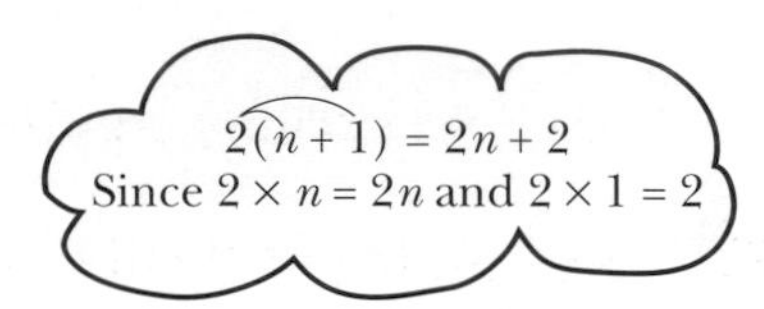

The term outside the bracket, 2, multiplies each term in the bracket.

The rules for simplifying expressions with negative numbers (see page 14)

$- \times + \rightarrow -$ **and** $- \times - \rightarrow +$

$+ \times - \rightarrow -$ $\quad\quad$ $+ \times + \rightarrow +$

are used when multiplying out (or expanding) brackets.

Here are some more examples:

- $5(n - 3) = 5n - 15$

$5 \times n = 5n$ and $5 \times -3 = -15$

- $4(3n + 1) = 12n + 4$

$4 \times 3n = 12n$ and $4 \times 1 = 4$

- $-2(4n - 3) = -8n + 6$

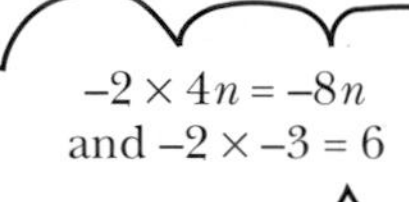

- $-(-5n + 2) = 5n - 2$

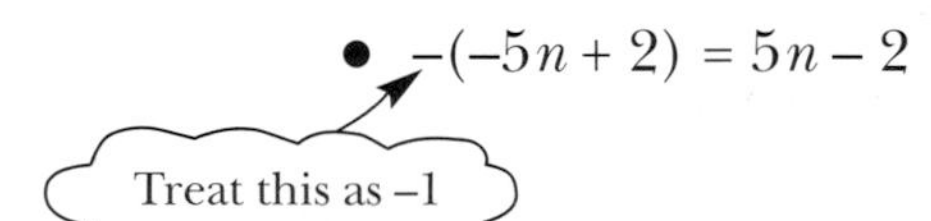

$-1 \times -5n = 5n$ and $-1 \times +2 = -2$

In longer expressions you may have to multiply out brackets and then simplify by collecting like terms. Here are some examples.

- $4(n + 2) + 1$
  $= 4n + 8 + 1$
  $= 4n + 9$

- $5a - 2(b + a)$
  $= 5a - 2b - 2a$
  $= 3a - 2b$

- $7(r + 3) - 3(2r - 1)$
  $= 7r + 21 - 6r + 3$
  $= r + 24$

- $4(5x - 2y) + 3(x + 4y)$
  $= 20x - 8y + 3x + 12y$
  $= 23x + 4y$

Exercise

1 Graham arranges bedding plants in garden boxes like this.

size 1 size 2 size 3

How many plants are there in a garden box of size n?

2 The number of small squares in each pattern forms a sequence. Work out a formula for the number of small squares in the nth pattern.

Write it in its simplest form.

a)

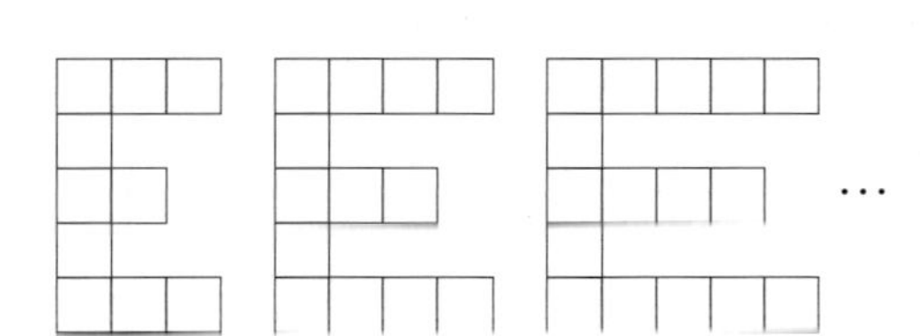

b)

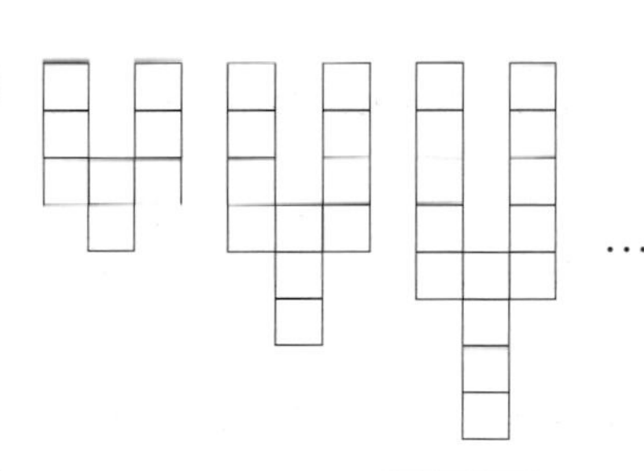

c)

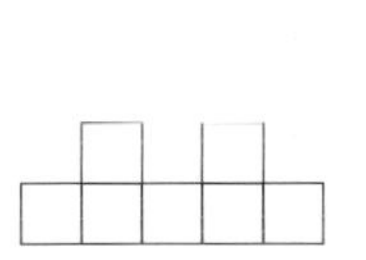

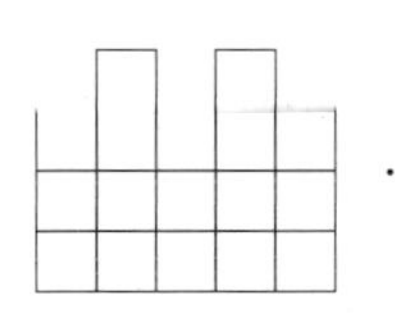

d)

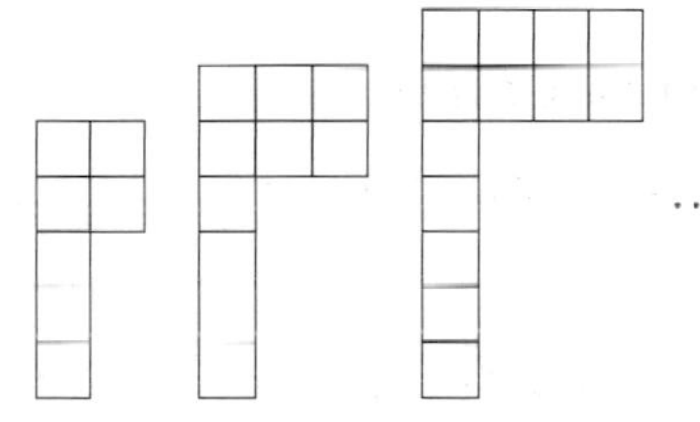

3 Multiply out these brackets.

a) $3(n + 4)$ b) $5(n - 2)$ c) $4(3 + n)$

d) $6(1 - n)$ e) $-2(x + 4)$ f) $4(2x + 1)$

g) $5(3 - 2y)$ h) $-3(7p - q)$ i) $9(5x + 4y)$

j) $-(a - 5)$ k) $-4(8h + 5)$ l) $5(x + 2y - 3z)$

4 Write each of these as simply as possible.

a) $2(n + 1) + n$ b) $3(n + 2) + 1$

c) $5(3n - 1) + 2n$ d) $13 - 2(6n - 1)$

e) $2x + 5(x - 4)$ f) $3(p + q) + 2p - q$

g) $2(5a + b) + 3(a + 4b)$ h) $4(3c + d) + 3(c - 2d)$

i) $7(2x - 5y) + 5(x + 2y)$ j) $8(r - 2s) - 3(s - 3r)$

k) $5(2a + 5b) - (9a - 7b)$ l) $4(f + 2) + 3(e - 6f) + 5$

m) $m + 2(5k - 4m) - 3(k + 4m)$ n) $4(x + 6y) - 7x - 8(4x + 3y)$

Finishing off

Now that you have finished this chapter you should be able to:

★ write down the next number in a sequence

★ find a formula for the nth term of a sequence (where the difference between terms is a fixed amount)

★ simplify expressions by collecting like terms

★ multiply a bracket by a single term.

Use the questions in the next exercise to check that you understand everything.

Mixed exercise

1 Write down the next two terms in each of these sequences and give the rule that you are using.

a) 3, 6, 9, 12, …

b) 2, 7, 12, 17, …

c) 50, 44, 38, 32, …

d) 2.3, 2.7, 3.1, 3.5, …

2 In football the World Cup is held every four years. It was held in 1998.

a) Write down the years of the next three World Cups after this date.

b) Is a World Cup due to be held in 2040?

3 Work out a formula for the nth term in each of these sequences.

a) 3, 7, 11, 15, …

b) 14, 17, 20, 23, …

c) 7, 6, 5, 4 …

d) 1.75, 2.5, 3.25, 4, …

4 At a farm show Mr Frost makes sheep pens with fences like this.

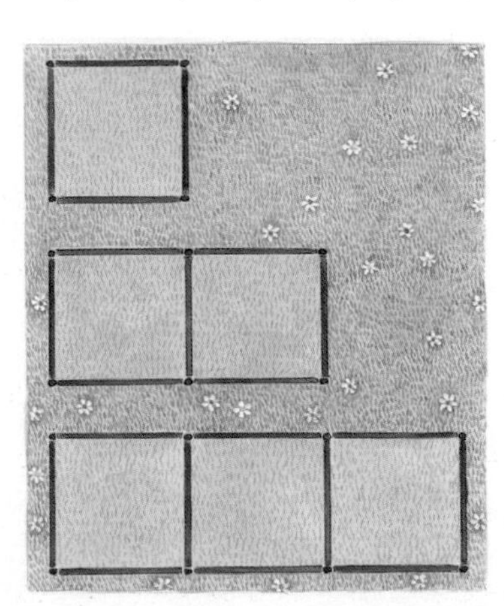

a) How many fences are used to make

(i) 1 pen?

(ii) 2 pens?

(iii) 3 pens?

b) Work out a formula for the number of fences needed to make n pens.

5 Geoff has a metal bar 100 cm long. He cuts off pieces of length 6 cm.

a) How much of the bar is left after Geoff has cut

(i) 1 piece? (ii) 2 pieces? (iii) 3 pieces?

b) Work out a formula for the length of bar left after n pieces are cut.

c) What value does the formula give when $n = 17$?
Explain what is happening.

6 Victoria is a landscape gardener. Her brochure shows these designs for garden ponds but she will build them as large as a customer wants.

Design 1 (8 slabs)   Design 2   Design 3

Work out a formula for the number of slabs required for design n.

7 Work out a formula for the number of small squares in the nth pattern.

a)

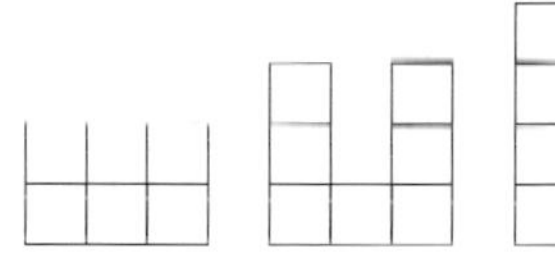

…

b)

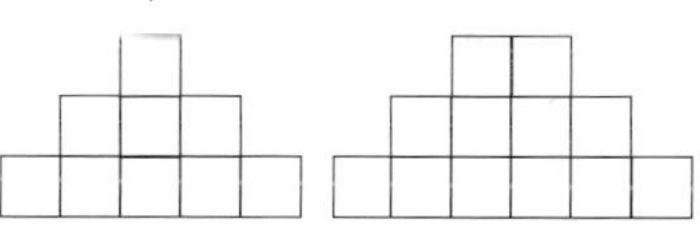

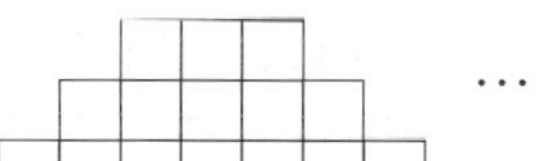

…

8 Simplify these expressions.

a) $4n + 5n$

b) $2 + 3n - 4$

c) $2n + 1 + 3n - 1$

d) $4n - 1 - n + 3$

e) $3a + 2b + 7a + 5b$

f) $6c + 4d - c + 2d$

g) $5r + 2 + s + 3r - 4s$

h) $2u - 5t + t + 7 - 3u + 1$

i) $4x - 3y + 2z + x - 5y$

j) $5a + 2b - 4c + a - 3b - c$

9 Multiply out the brackets and then simplify where possible.

a) $6(n + 3)$

b) $3(5n - 2)$

c) $2(3x + 4y)$

d) $5(x + 4y + 2z)$

e) $4(a + 2) + 3a$

f) $7 - 2(b + 2)$

g) $2(4c + d) - 5d$

h) $4p + 3(2q + p) - q$

i) $5(2e + f) + 3(4e + 3f)$

j) $3(x + 4y) - 2(3x - 2y)$

k) $4(3k + m) - 3(2k + 5m)$

l) $a - 2(a + 3b) + 5(7a + b)$

m) $5(4x - 3y - 1) + 3(3 + 2x + 5y)$

n) $6(s + 2t - 3u) - 5(2s - 4t + u)$

Twelve

Algebra 2

Using powers

Look at the sequence

$$1, 4, 9, 16, 25, \dots$$

What is the next term in the sequence?

What is the 25th term in the sequence?

One way of finding the next term is to see a pattern in the differences between the terms. This is fine if you just need the next term but it is a tedious way of finding the 25th term!

Another way of finding the next term is to recognise that this sequence contains the square numbers. The next term is 6×6 which is 36. The 25th term is 25×25 which is 625. In the same way the

$$n\text{th term} = n \times n = n^2$$

$n \times n = n^2$

Here are some more examples showing the simplest way of writing products.

You combine numbers and letters separately.

- $8 \times a = 8a$
- $3 \times a \times a = 3a^2$ ($a \times a = a^2$)
- $5a \times a^2 = 5a^3$ ($a \times a^2 = a^3$)
- $(3a)^2 = 3a \times 3a = 9a^2$ ($3 \times 3 = 9$ and $a \times a = a^2$)
- $(-6n) \times 3n = -18n^2$ ($-6 \times 3 = -18$ and $n \times n = n^2$)
- $3n \times (-4) \times (-5n) = 60n^2$

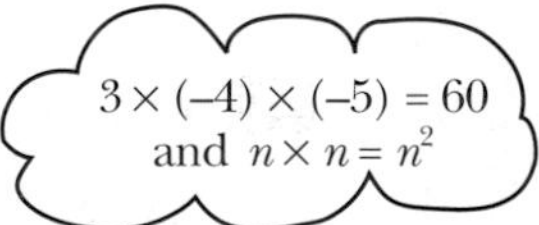

Here are some examples involving quotients and powers.

- $d^3 \div d = \dfrac{d \times d \times d}{d} = d \times d = d^2$
- $6n^2 \div 3n = \dfrac{6 \times n \times n}{3 \times n} = 2 \times n = 2n$
- $(u^2)^3 = u^2 \times u^2 \times u^2 = u^6$
- $u^{-1} = \dfrac{1}{u}$

The rules for letters follows the rules for numbers given pages 76 and 78.

$$\boldsymbol{a^m \times a^n = a^{m+n} \qquad a^m \div a^n = a^{m-n} \qquad (a^m)^n = a^{m \times n} \qquad a^{-1} = \frac{1}{a} \qquad a^0 = 1}$$

You must be particularly careful when finding the value of expressions involving powers.

When $a = 5$ how does $4a^2$ differ from $(4a)^2$?

$4a^2 = 4 \times a \times a$ so substituting $a = 5$ gives $4 \times 5 \times 5 = 100$

$(4a)^2 = 4a \times 4a$ so substituting $a = 5$ gives $20 \times 20 = 400$

Exercise

1 Write each of these as simply as possible.

a) $a \times a$ b) $a \times a \times a$ c) $5 \times a \times a$

d) $3p \times p$ e) $q \times 4q$ f) $2r \times 3r$

g) $t \times 2t$ h) $5c \times 6c$ i) $(4d)^2$

j) $e \times e^2$ k) $3f \times 4f^2$ l) $x^2 \times x^3$

m) $(6h)^2$ n) $5p \times (-3p)$ o) $2q^2 \times 7q$

p) $(-t) \times (-4t)$ q) $(-2a) \times 5a$ r) $2d^3 \times 5d$

s) $(2e)^2 \times (-e)^2$ t) $3f^2 \times 3f$ u) $(2a)^3$

v) $(a^3)^2$ w) $c^{-1} \times c$ x) $(t^4)^2 \times t^5$

2 Write each of these as simply as possible.

a) $a^4 \div a$ b) $a^3 \div a^2$ c) $6a \div 2$

d) $8x \div 4x$ e) $3x \div 12$ f) $15 \div 5y$

g) $20p^2 \div 10p$ h) $24q^3 \div 8q^2$ i) $16r \div 32r$

j) $2c^2 \div 8c$ k) $2d \div d^2$ l) $6e^3 \div 8e$

m) $(3h)^2 \div 6h^2$ n) $(-15k) \div 25k^2$ o) $8m \div 20m^3$

p) $(-a^2) \div (-a)^2$ q) $24b^2 \div 16b^3$ r) $(-15a) \div (-3a^2)$

s) $4p^3 \div 8p^2$ t) $4y^4 \div (2y)^2$ u) $9e^2 \div (-12e^3)$

v) $(r^2)^4 \div r^3$ w) $x^3 \div x^{-1}$ x) $a^{-3} \div a^{-2}$

3 Given $b = 5$ find the value of

a) $9b$ b) $3b + 7$ c) $2b^2$ d) $(2b)^2$

4 Given $t = -4$ find the value of

a) $3t$ b) $5t^2$ c) t^3 d) $(3t)^3$

5 Given $x = -2$ find the value of

a) $-x$ b) $\frac{12}{2x}$ c) $\frac{x+10}{4}$ d) $\sqrt{x^2+5}$

6 a) What is the nth term of the sequence 2, 4, 6, 8, …?

b) The squares of these terms are 4, 16, 36, 64, …

What is the nth term of this sequence?

c) The cubes of the terms of the sequence in a) are 8, 64, 216, 512, …

What is the nth term of this sequence?

Multiplying out brackets

On page 102 you saw that $2(n + 1) = 2n + 2$.

The term outside the bracket, 2, multiplies each term inside the bracket.

How is $n(3n + 5)$ multiplied out?

You use the same method as before. The term outside the bracket, n, multiplies each term inside the bracket.

$$n(3n + 5) = 3n^2 + 5n$$

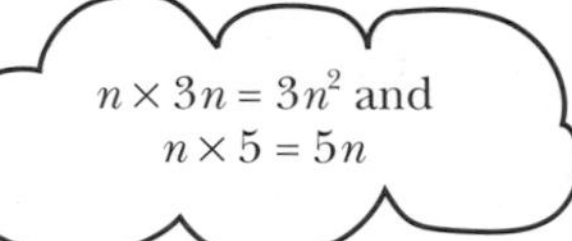

You can show this by looking at these rectangles.

$3n + 5$

n

$3n$ 5

n $n \times 3n = 3n^2$ $n \times 5 = 5n$

The whole area is $n(3n + 5)$.

The smaller areas added together are $3n^2 + 5n$.

These must be the same so $n(3n + 5) = 3n^2 + 5n$

Here are some more examples of multiplying out brackets.

- $n(4n + 1) = 4n^2 + n$
- $2n(n - 3) = 2n^2 - 6n$
- $x(x - 6) = x^2 - 6x$
- $-5x(3 + 2x^2) = -15x - 10x^3$

How is $(n + 1) \times (n + 2)$ multiplied out?

Look at these rectangles.

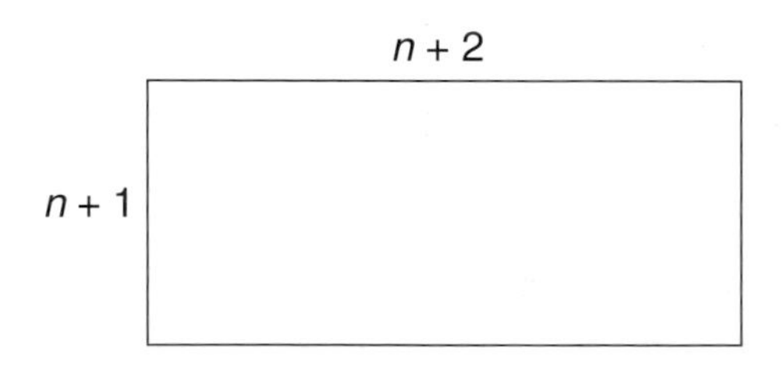

n 2

n $n \times n = n^2$ $n \times 2 = 2n$

1 $1 \times n = n$ $1 \times 2 = 2$

The whole area is $(n + 1)(n + 2)$

The smaller areas are $n^2 + 2n + n + 2$

The whole area is the same as the four smaller areas added together so

$$(n + 1)(n + 2) = n^2 + 2n + n + 2$$
$$= n^2 + 3n + 2$$

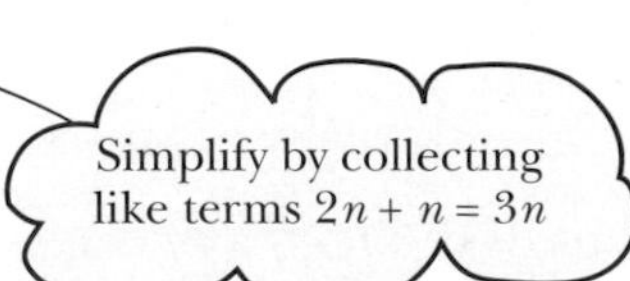

Here are two more examples

- $(n + 5)(n - 2) = n^2 - 2n + 5n - 10$
  $= n^2 + 3n - 10$
- $(n - 4)(2n - 3) = 2n^2 - 3n - 8n + 12$
  $= 2n^2 - 11n + 12$

Exercise

1 Simplify

a) $n(n+6)$ b) $n(n-2)$ c) $n(n+1)$
d) $n(n-1)$ e) $2n(n+5)$ f) $-4(2n+1)$
g) $x(2x-3)$ h) $-p(7p-1)$ i) $3a(5a+2)$
j) $-c(2c+5)$ k) $4k(3+5k)$ l) $5x(5x^2-1)$
m) $-a^2(a-4)$ n) $-4b(b^3+7)$ o) $3x^2(x^2-6)$

2 Write each of these as simply as possible.

a) $(n+2)(n+4)$ b) $(n+1)(n+6)$ c) $(n+5)(n-3)$
d) $(n-4)(n+7)$ e) $(n-8)(n-2)$ f) $(2n+3)(n+4)$
g) $(x+6)(3x+1)$ h) $(4x-5)(x+2)$ i) $(3x+7)(x-4)$
j) $(a+3)(5a-1)$ k) $(b-2)(4b-3)$ l) $(3c+4)(2c+7)$
m) $(8d+3)(2d-1)$ n) $(2t-9)(3t-4)$ o) $(5y-11)(4y+7)$
p) $(1+2y)(1+3y)$ q) $(4-s)(3+2s)$ r) $(7+u)(2-3u)$
s) $(3+4b)(b-2)$ t) $(4d-5)(4-3d)$ u) $(5+3f)(5-3f)$

3 Work out $(x+4)^2$ by multiplying out $(x+4)(x+4)$

4 Work out

a) $(x+5)^2$ b) $(x-6)^2$ c) $(3x+1)^2$ d) $(4x-3)^2$

5 Write each of these as simply as possible.

a) $n(n+2)+4n$ b) $12n^2+n(n-3)$
c) $(3x+5)^2+x^2$ d) $2p(p+2)+p(4p+1)$
e) $(2x-7)(x+3)+3(4x+5)$ f) $(4a+3)(4a-3)-16a^2$
g) $3c(c+4)-2(6c+1)$ h) $2n(n+1)-3n-2(n^2-5)$
i) $(7x-2)^2-6(x-3)$ j) $(2x+3)^2-(2x-3)^2$

6 For each of these sequences write down a formula for the nth term.

a) $1\times2, 2\times3, 3\times4$ … b) $4\times1, 5\times2, 6\times3$, …
c) $2\times3, 3\times4, 4\times5$, … d) $2\times0, 3\times1, 4\times2$, …
e) $1\times2\times3, 2\times3\times4, 3\times4\times5$, … f) $1\times3+2, 2\times4+3, 3\times5+4$, …

Factorising

On page 102 you saw that $4(3n + 1) = 12n + 4$ and that the method of multiplying out the bracket is that the term outside the bracket multiplies each term in the bracket.

Factorisation is the reverse process and involves finding the highest common factor (HCF) of the two terms.

- Factorise $12n + 4$

$12n + 4$	$12n$ and 4 have a HCF of 4.
$= 4 \times 3n + 4 \times 1$	Separate the HCF, 4, in each term.
$= 4(3n + 1)$	Take out 4 as a factor.
	The remaining terms in the bracket, $3n$ and 1, have no common factor other than 1 so the factorisation is complete.

Sometimes you may not spot the HCF straight away but don't worry you can still complete the factorisation successfully. The working for the last example may appear like this:

$12n + 4$	$= 2 \times 6n + 2 \times 2$	Using 2 (rather than 4) as a common factor
	$= 2(6n + 2)$	Take out 2.
		The terms in the bracket, 6n and 2, still have a common factor, 2, so the factorisation is incomplete.
	$= 2(2 \times 3n + 2 \times 1)$	Separate the 2 in each term.
	$4(3n + 1)$	Take out 2 and get the same answer as before.

On page 108 you saw that $n(3n + 5) = 3n^2 + 5n$ and $2n(n - 3) = 2n^2 - 6n$.

In these examples the HCF involves a letter.

- Factorise $3n^2 + 5n$

$3n^2 + 5n$	$3n^2$ and $5n$ have a HCF of n.
$= n \times 3n + n \times 5$	Separate the HCF, n, in each term.
$= n(3n + 5)$	Take out n as a factor. The factorisation is complete.

- Factorise $2n^2 - 6n$

$2n^2 - 6n$	$2n^2$ and $6n$ have a HCF of $2n$.
$= 2n \times n + 2n \times 3$	Separate the HCF, $2n$, in each term.
$= 2n(n - 3)$	Take out $2n$ as a factor. The factorisation is complete.

Exercise

1 Factorise

a) $6n + 8$ b) $5n - 15$ c) $20 + 8n$
d) $6 - 3n$ e) $2x + 2$ f) $12x + 30y$
g) $12 - 32y$ h) $36 - 27q$ i) $20s + 50t$
j) $48a - 80b$ k) $45 + 75b$ l) $96x + 54$
m) $42m - 66n$ n) $52p - 39q$ o) $84c + 28$
p) $22 + 77t$ q) $144x - 60y$ r) $34a - 51b$

2 Factorise

a) $4x + 8y + 12$ b) $5m - 10n + 20$
c) $12t - 6u + 3$ d) $2 + 4a + 10b$
e) $24p - 30q$ f) $36x + 72y - 132z$

3 Factorise

a) $n^2 + 5n$ b) $8n^2 - 3n$ c) $2n^2 + 6n$
d) $6a^2 - 4a$ e) $5y - 2y^2$ f) $15x + 3x^2$
g) $20x^2 + 8x$ h) $3p^2 - p$ i) $25x + 4x^2$
j) $28a^2 - 12a$ k) $4q^2 + 8$ l) $12y^2 - 21y$
m) $2t + 5t^3$ n) $6n - 10n^3$ o) $2x^2 - 4x^3$
p) $96u^4 + 80u^2$ q) $27d^3 + 24d^2$ r) $75t^3 + 125t^4$

4 Factorise

a) $x^3 + 5x^2 + 7x$ b) $4x - 6x^2 - 8x^3$
c) $3a^2 + 15a + 9$ d) $32p^4 + 24p^3 + 16p^2$

5 Write each of these as simply as possible and factorise your answer.

a) $3(x + 2) + 9$ b) $5(n - 3) + 10$
c) $4(x + 5) + x$ d) $3x + 10 - x + 4$
e) $5x - 7 + x + 9$ f) $x(x + 3) + 12x$
g) $2(x - 16) + 3(x + 4)$ h) $6(x + 3) - 2(x - 7)$
i) $(x + 4)^2 - 16$ j) $x^2 - (x - 5)^2$

Two unknowns

This pattern of wall tiles is 3 tiles high and 8 tiles wide.

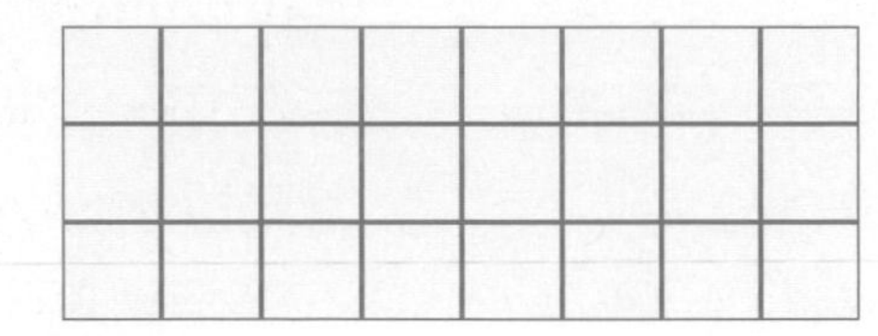

For a pattern 3 tiles high and 8 tiles wide 24 tiles are used.

For a pattern 3 tiles high and w tiles wide $3w$ tiles are used.

For a pattern h tiles high and w tiles wide hw tiles are used.

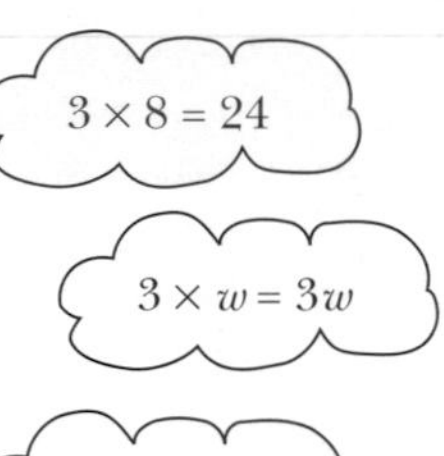

$h \times w$ is the same as $w \times h$. It is usual to write the letters in alphabetical order.

The following examples show how the work already covered in this chapter extends to 2 or more unknowns.

- $2p \times 4r = 8pr$

  $2 \times p \times 4 \times r = 8 \times p \times r$

- $3a^2b \times 5ab = 15a^3 \times b^2$

  $3 \times a^2 \times b \times 5 \times a \times b = 15 \times a^3 \times b^2$

- $mn^3 \div 2n = \dfrac{m \times n \times n \times n}{2 \times n} = \dfrac{mn^2}{2}$

- $3cd^2 \div c^2d = \dfrac{3 \times c \times d \times d}{c \times c \times d} = \dfrac{3d}{c}$

- $a(a + 5b) = a^2 + 5ab$

  $a \times a = a^2$ and $a \times 5b = 5ab$

- $2p(3q - 4p) = 6pq - 8p^2$

  $2p \times 3q = 6pq$ and $2p \times (-4p) = -8p^2$

- $(c + 4d)\ (2c + 3d) \quad = c \times 2c + c \times 3d + 4d \times 2c + 4d \times 3d$

  $= 2c^2 + 3cd + 8cd + 12d^2$

  remember $8dc = 8cd$

  $= 2c^2 + 11cd + 12d^2$

  collect like terms $3cd + 8cd = 11cd$

- Factorise $\quad x^2 + 4xy$

  $x^2 + 4xy = x(x + 4y)$

- Factorise $\quad 12ab - 15a^2$

  $12ab - 15a^2 = 3a(4b - 5a)$

Exercise

1 What is the cost (in pence) of:

a) 6 pens at 25 pence each?

b) 6 pens at x pence each?

c) n pens at x pence each?

2 How many pages are there in:

a) 3 magazines with 40 pages each?

b) m magazines with 40 pages each?

c) m magazines with p pages each?

3 Simplify

a) $2m \times 5n$ b) $4a \times 3b$ c) $5y \times x$

d) $-2r \times 3s$ e) $-5t \times -7u$ f) $3gh \times 6g$

g) $8a^2 \times 4ab$ h) $3x^2y \times y$ i) $2p^2q \times 7pq$

j) $-6r^3 \times 5rs^2$ k) $2c \times 3cd \times 4d^2$ l) $(-5ab)^2$

4 How much does one person get when:

a) 4 people share £200?

b) n people share £200?

c) n people share £x?

5 How much is one plant when:

a) 5 plants cost £10?

b) p plants cost £10?

c) p plants cost £c?

6 Simplify

a) $12a \div 4b$ b) $6x \div 15y$ c) $8xy \div 6y$

d) $2pq \div (-8q)$ e) $r^2s \div rs^2$ f) $10m^2 \div 15mn$

g) $20a^4b \div 5a^2$ h) $4t \div 20t^2u$ i) $12x^3y \div (-30xy)$

j) $45ab \div 18a^3b^2$ k) $c^2d^3e^4 \div c^4d^3e^2$ l) $(5tu)^2 \div 5tu^2$

7 Multiply out

a) $x(2 + y)$ b) $p(p - 4q)$ c) $3a(2b - a)$

d) $4r(r - 3s)$ e) $2m(3m + n)$ f) $5e(7f - 3e)$

g) $(x + 2y)(x + 3y)$ h) $(x - 5y)(x + 7y)$ i) $(2p - q)(5p - q)$

j) $(2c + d)(7c + 3d)$ k) $(5t - 3u)(4t + 7u)$ l) $(4x - 9y)(3x - 5y)$

8 Factorise

a) $4x + 3xy$ b) $5ab - 7b$ c) $6c + 8cd$

d) $5pq - 10q$ e) $6x^2 - xy$ f) $2g + 10gh$

g) $8r^2 - 4rs$ h) $ab^2 + a^2b$ i) $pq - 9p^2q$

Finishing off

Now that you have finished this chapter you should be able to:

- ★ multiply and divide simple algebraic terms involving powers
- ★ find the value of simple algebraic expressions
- ★ multiply out brackets of the type $3x(x+6)$
- ★ multiply out brackets of the type $(2x+1)(5x-4)$
- ★ simplify expressions by multiplying out brackets and collecting like terms
- ★ write down a formula e.g. for the nth term of a sequence
- ★ factorise expressions of the type $5x+20$
- ★ factorise expressions of the type $4x^2+12x$

Use the questions in the next exercise to check that you understand everything.

Mixed exercise

1 Write each of the following as simply as possible.

a) $3x \times 5a$ b) $(6b)^2$ c) $5cd \times 2d$

d) $3ef^3 \times 7ef$ e) $4g^2 \times (-9g)$ f) $2h \times (3h^3)^2$

2 Write each of the following as simply as possible.

a) $4p^2 \div 8p$ b) $20 \div 5q$ c) $6r^3 \div 21r$

d) $12st \div 8t^2$ e) $(5u)^2 \div 15u^3v$ f) $(-6wx^2) \div (-2w^3x)$

3 Given $x = 4$ find the value of

a) x^2 b) $3x^2$ c) $(6x)^2$ d) $2x^3$

4 Given $y = -3$ find the value of

a) $8 - 3y$ b) y^2 c) $2y^3$ d) $(3y)^3$

5 a) What is the nth term of the sequence 1, 3, 5, 7, …?

b) The squares of these numbers are 1, 9, 25, 49, …

What is the nth term of this sequence?

Give your answers in the form of $an^2 + bn + c$ where a, b and c are integers.

6 A football team get three points for a win and one for a draw.

How many points does a team get when they

a) win 10 games and draw 8 games?

b) win w games and draw d games?

7 Multiply out

a) $5(3 + x)$ b) $6p(p - 4)$ c) $a(3 - 7a^2)$

d) $-(3m + n)$ e) $-4t(2t - 3u)$ f) $3cd(7c - 5d)$

8 Multiply out

a) $(x + 2)(x + 8)$ b) $(a - 5)(a + 3)$ c) $(2p - 1)(3p - 5)$

d) $(3f + g)(4f + 3g)$ e) $(7 - y)(4 + 3y)$ f) $(4x + 5y)(4x - 5y)$

g) $(a - 3)^2$ h) $(3n + 5)^2$ i) $(8p - 3q)^2$

9 Write each of these as simply as possible.

a) $n(n + 6) + 3n$ b) $10a^2 + a(8 - a)$

c) $5x + 3x(x - 2)$ d) $y(y + 6) + y(3y + 1)$

e) $5a(b + 4) + 2a(b - 3)$ f) $2a(4a - 3) - 8(a^2 + 1)$

g) $(d - 2)(d + 4) + 4(d + 2)$ h) $(k + 6n)^2 + 3n(n - 4k)$

10 For each of these sequences write down a formula for the nth term.

a) $1 \times 3, 2 \times 4, 3 \times 5, \dots$ b) $3 \times 2 - 0, 4 \times 3 - 1, 5 \times 4 - 2, \dots$

c) $1 \times 2 + 3, 2 \times 3 + 4, 3 \times 4 + 5, \dots$ d) $1 \times 3 \times 5, 2 \times 4 \times 6, 3 \times 5 \times 7, \dots$

11 Factorise

a) $4n + 12$ b) $8t - 20$ c) $15x + 27y$

d) $6a - 2$ e) $3q^2 - 15$ f) $25x + 35y + 40$

12 Factorise

a) $4p - p^2$ b) $6x + 3x^2$ c) $24n^2 + 8n$

d) $6r^3 + 21rs$ e) $70f^2 - 28fg$ f) $112yz^2 - 48y^2z$

g) $24a + 12b - 15$ h) $x^3 + 3x^2 + 10x$ i) $35pr - 21qr + 56r^2$

13 What is the cost (in pence) of:

a) 4 oranges at 27 pence each?

b) n oranges at 27 pence each?

c) n oranges at p pence each?

14 How much is one book when:

a) 5 books cost £30?

b) x books cost £30?

c) x books cost £c?

Thirteen

Graphs

Co-ordinates

You will use graphs to display the relationship between two variables.

This diagram shows the axes meeting at the origin and explains how to write down co-ordinates.

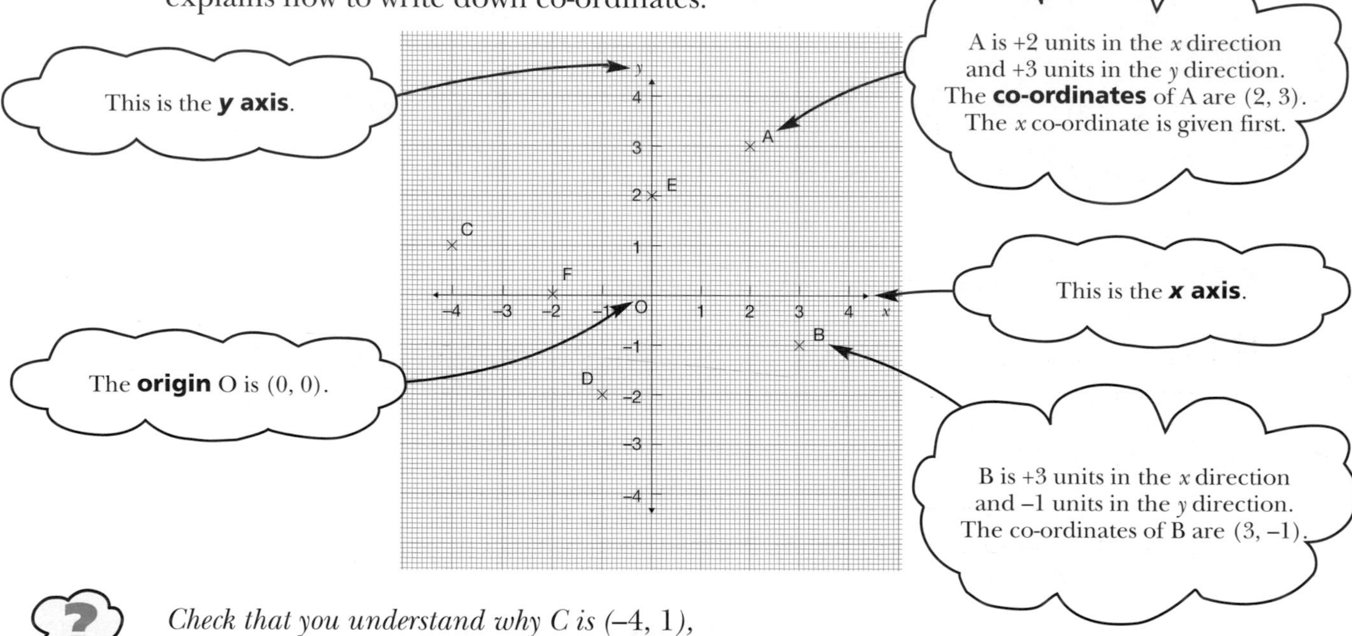

Check that you understand why C is (−4, 1), D is (−1, −2), E is (0, 2) and F is (−2, 0).

Equation of a graph

Look at the points A, B and C on this straight line graph.

A is (1, 0): $x = 1, y = 0$

B is (2, 1): $x = 2, y = 1$

C is (3, 2): $x = 3, y = 2$

You can see that there is a pattern.

This means that there is a relationship connecting x and y.

In each case the value of y is 1 less than the value of x.

$y = x - 1$

This is called the **equation of the graph**.

Do the points P and Q fit the same pattern?

1 Write down the co-ordinates of points A to J

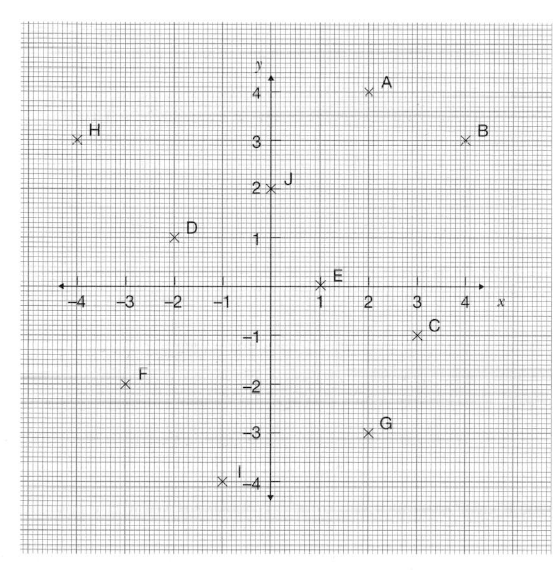

2 Draw a grid with –5 to 5 on each axis. Plot and label each of the following sets of points

a) A(1, 3), B(3, 2), C(4, 4) and D(2, 5). What shape is ABCD?

b) P(4, 1), Q(–2, 3), R(–3, 0) and S(3, –2). What shape is PQRS?

c) X(–4, –3), Y (–2, –5) and Z(–4, 0). What shape is XYZ?

3 This table gives the co-ordinates of the points on a straight line.

x	–2	–1	0	1	2	3
y	–3	–1	1	3	5	7

a) Choose suitable scales, plot the points and join them with a straight line.

b) As you read along the table, x increases by 1 each time.

Describe what happens to y as x increases.

c) Write down the equation of the line.

4 Draw a grid with –6 to 6 on each axis.

Look at these three sets of points.

A (–1, 0), (0, 1), (1, 2), (2, 3), (3, 4)

B (0, 5), (1, 4), (3, 2), (2, 3)

C (–1, –3), (0, 0), (1, 3), (2, 6)

a) Plot the points and join each set with a straight line;

b) Describe in words a formula to find y for a given value of x;

c) Write an equation for each line, of the form $y = \ldots$;

d) Write down the co-ordinates of two other points on each line and check that they satisfy the equation of the line.

Straight line graphs

Look at Carla's table of values for $y = 2x - 4$.

2x is twice x

x	0	1	2	3	4
$2x$	0	2	4	6	8
-4	-4	-4	-4	-4	-4
y	-4	-2	0	2	4

-4 stays the same whatever the value of x

These are the values of $y = 2x - 4$

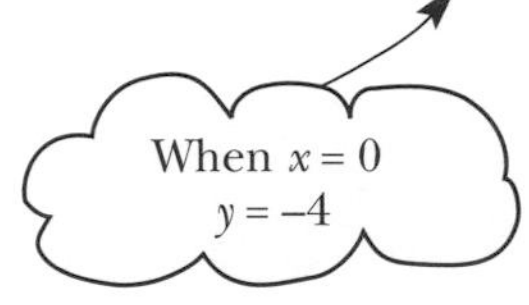

Carla draws the graph.

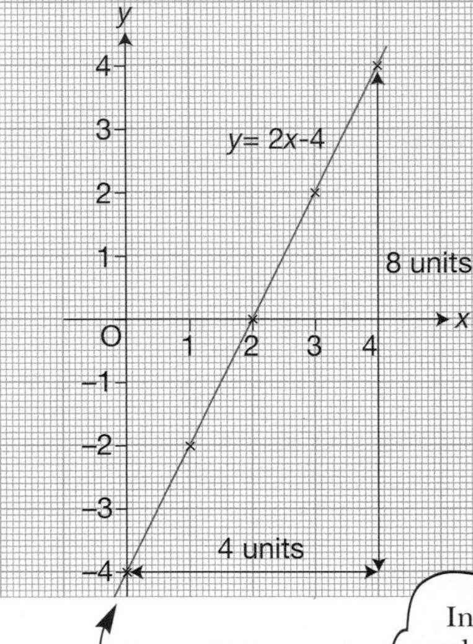

The steepness of a line is called its **gradient**.

In this graph the scales for x and y are the same. You can have different scales on the two axes.

Gradient $= \dfrac{\textbf{increase in } y}{\textbf{increase in } x}$

In the case of $y = 2x - 4$, when x increases by 4 then y increases by 8 so

gradient $= \dfrac{8}{4} = 2$.

When x increases by 3, what does y increase by?

Do these values still give a gradient of 2?

Look at the graph and the equation of the line.

The gradient is 2

$y = 2x - 4$

The ***y* intercept** is -4. It is the value of y when $x = 0$.

What are the gradient and y intercept of the line $y = mx + c$?

What can you say about the gradients of parallel lines?

Is $y = 2x + 1$ parallel to $y = 2x - 4$?

These diagrams show three diffferent gradients.

positive zero negative

1 Draw a grid from –8 to 8 on both axes.

Draw the graph of each of the following lines. Don't forget to label them. (Note: Not all x values will produce a y value that you can plot within the range given. If you are unsuccessful with one x value then think carefully before trying another.)

a) $y = x + 4$ b) $y = 2x$ c) $y = 4 - x$

d) $y = 3x - 8$ e) $x = 7$ f) $y = 5$

Write down the co-ordinates of the point where

g) $y = x + 4$ meets $y = 2x$

h) $y = 4 - x$ meets $y = 3x - 8$

i) $y = x + 4$ meets $y = 4 - x$

2 Find the gradient for each of these lines, stating whether it is positive or negative. Write down the y intercept. Then write down the equation of the line.

a)

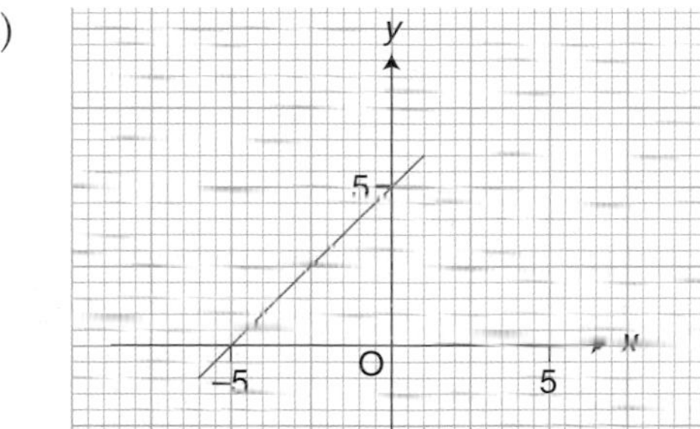

b)

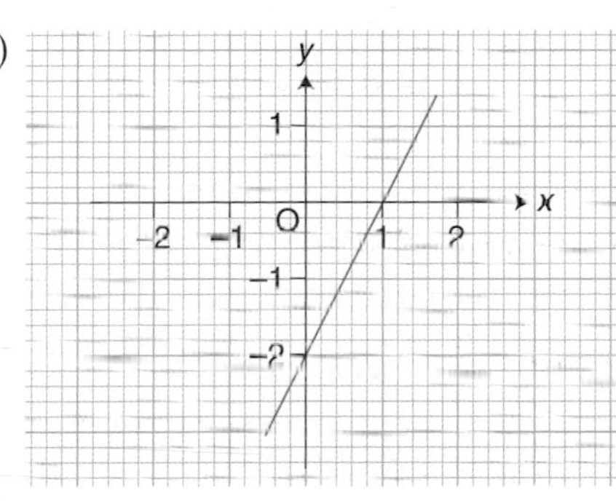

c)

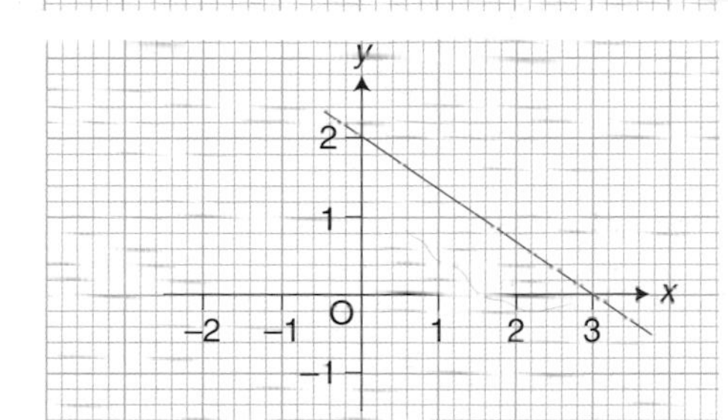

d)

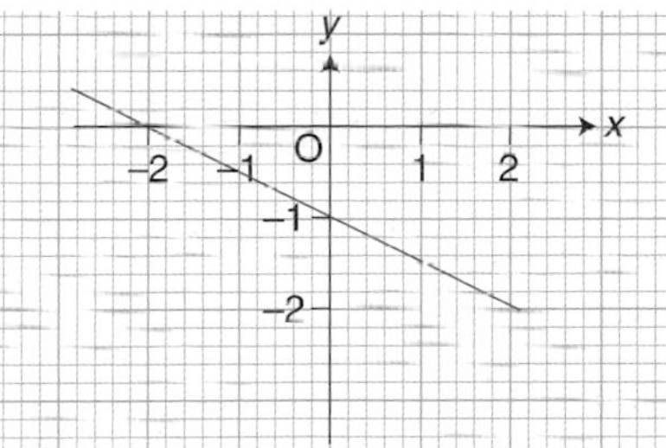

3 a) Draw a grid from –8 to 8 on both axes.

b) Draw the graph of each of the following lines. Don't forget to label them.

(i) $y = 11 - x$ (ii) $y = 2x + 8$ (iii) $y = 2x - 3$ (iv) $y = 3 - x$

c) Find the gradient of each line.

d) Which of these are parallel?

e) Explain how you can get the answers to d) from the equations in b).

Curved graphs

The graphs on the previous pages have all been straight lines. Sometimes you need to draw curves, as in the two examples on this page.

Example

Draw the curve $y = x^2 - 4x + 3$ for values of x between 0 and 4.

Solution

x	0	1	2	3	4
x^2	0	1	4	9	16
$-4x$	0	-4	-8	-12	-16
$+3$	$+3$	$+3$	$+3$	$+3$	$+3$
y	$+3$	0	-1	0	$+3$

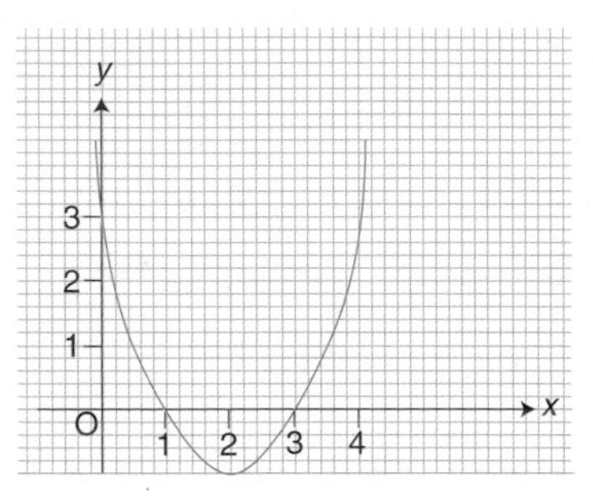

Why is it wrong to join the points with straight lines?

What are the values of x for which y = 2?

Example

Draw the curve $y = \frac{12}{x}$ for values of x from 1 to 6.

Solution

x	1	2	3	4	5	6
$y = \frac{12}{x}$	12	6	4	3	2.4	2

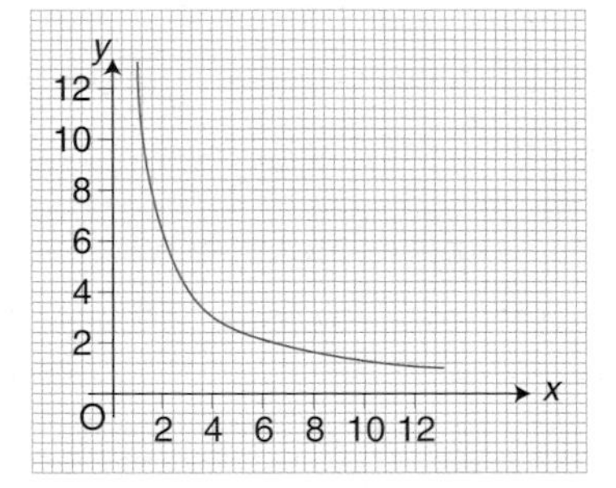

In this graph x cm and y cm could be the length and width of a rectangle of area 12 cm^2.

What is the length of the rectangle if the width is 2.6 cm*?*

What other meanings could x and y have?

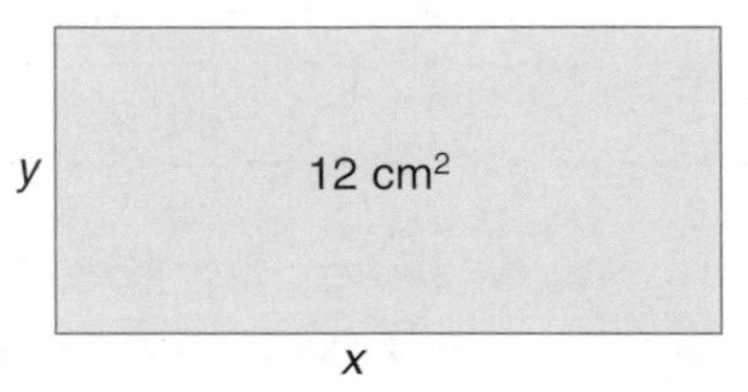

What happens to $y = \frac{12}{x}$ *when x is* 0*?*

$y = \frac{12}{x}$ is a function of x.

Using function notation we can write this as $f(x) = \frac{12}{x}$.

What is the value of $f(2)$*?*

1 A curve has equation $y = x^2 - 5$.

a) Copy and complete this table of values

x	−3	−2	−1	0	1	2	3
x^2	+9		+1	0			
−5	−5	−5		−5			
y	+4			−5			

b) Draw the graph. Use scales of 2 cm for 1 unit on each axis.

c) Find the value of y when $x = 1.2$.

d) Estimate the values of x for which $y = 2$.

2 A curve has equation $y = -4 + 5x - x^2$.

a) Copy and complete this table of values.

x	0	1	2	3	4	5
−4	−4					−4
$+5x$	0					25
$-x^2$	0					−25
y	−4					−4

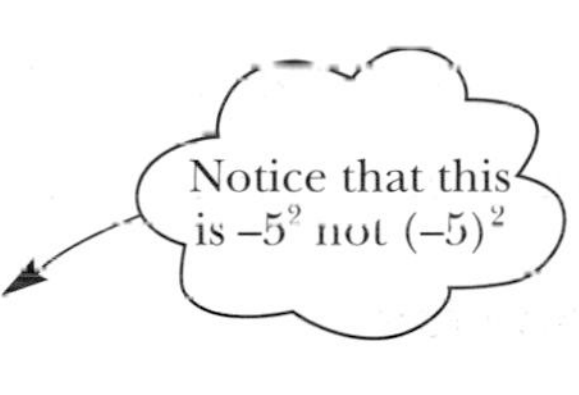

b) Choose suitable scales and draw the graph.

c) Estimate the greatest value of y (i.e. the highest point of the curve).

d) Estimate the values of x for which $y = 1$.

3 A curve has equation $y = x^3 - 2x$.

a) Copy and complete this table of values.

x	−2	−1	0	1	2
x^3	−8				8
$-2x$	+4				−4
y	−4				4

Notice that $(-2)^3$ is −8

b) Choose suitable scales and draw the graph.

c) Estimate the values of x where the curve crosses the x axis.

4 For $y = \frac{72}{x^2}$ work out the values of y when $x = 1, 1.5, 2, 3, 4, 5$ and 6.

a) Draw the curve $y = \frac{72}{x^2}$ for values of x from 1 to 6.

b) What is the curve like for values of x between −6 and −1?

c) What happens to the curve when $x = 0$?

Obtaining information

Catherine and her granny both live near a motorway, but at opposite ends of it. They agree to meet at a service station for lunch.

The red lines on the graph show Catherine's journey and the blue lines show Granny's.

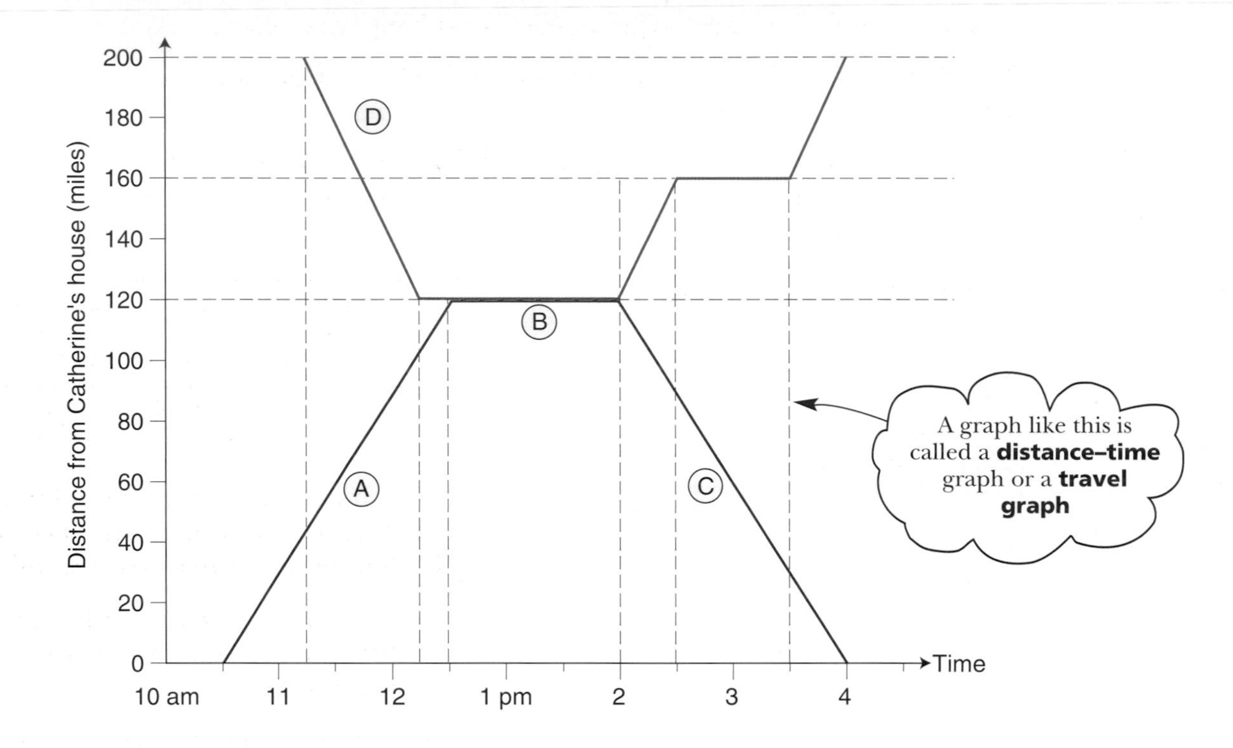

What is happening during part (B) of the graph?

Part (A) represents Catherine's journey to the service station.

She travels 120 miles in 2 hours.

The gradient of the line is $\frac{120 \text{ miles}}{2 \text{ hours}}$ = 60 miles per hour,

and this is her speed.

Notice that the units $\frac{\text{miles}}{\text{hours}}$ can be written as miles/hour, miles hour^{-1} or miles per hour (mph)

The gradient of a distance–time graph gives the speed.

What is Granny's speed on the way to the service station?

What is the meaning of a negative gradient on this graph?

The lines on this graph are all straight.

What does this mean?

Is it realistic?

Exercise

1 Use the graph on the opposite page to answer these questions.

a) When did Catherine arrive at the service station and how long did she stay?

b) How long did Granny have to wait for Catherine?

c) How long did Granny take to get home? What happened on the way?

d) Does Catherine drive more quickly, more slowly or at the same speed on the way home?

2 The graph shows the price of taxi fares in a city. There is a fixed charge plus a rate per mile travelled.

a) What does each small unit on the y axis represent?

b) What is the fixed charge?

c) Find the lengths of AC (in miles) and BC (in pounds) and use these to find the rate per mile.

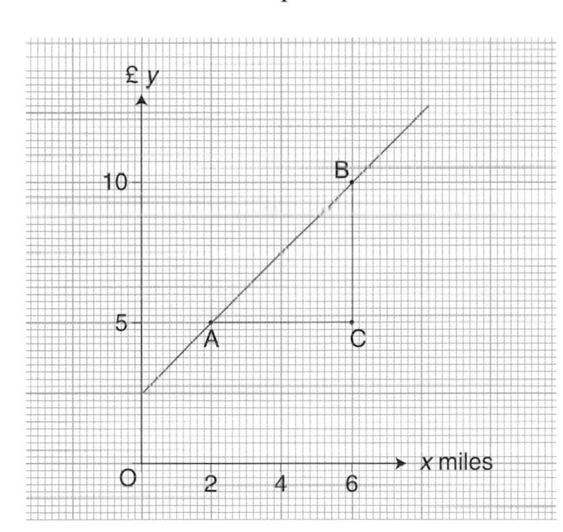

3 A train from Oxford to London stops at Didcot and Reading. Here is its timetable.

Oxford	Didcot	Reading	London
dep.	arr. dep.	arr. dep.	arr.
1200	1212 1215	1240 1245	1315

The first leg of the journey (Oxford to Didcot) is 10 miles. The second leg is 20 miles and the third is 35 miles.

a) Draw a travel graph of the journey using 1 cm for 5 minutes and 1 cm for 5 miles.

b) Find the average speed of the train for each part of the journey.

c) A non-stop London train travelling at 100 mph passes Didcot at 1230. Draw a line on your graph for this train. Find where and when it passes the first train.

4 The graph shows how the annual cost of running Mary's car depends on the number of miles she travels.

a) What is the y intercept on the graph?

b) What is this money spent on?

c) Find the gradient of the graph and write it in pence per mile.

d) What is this money spent on?

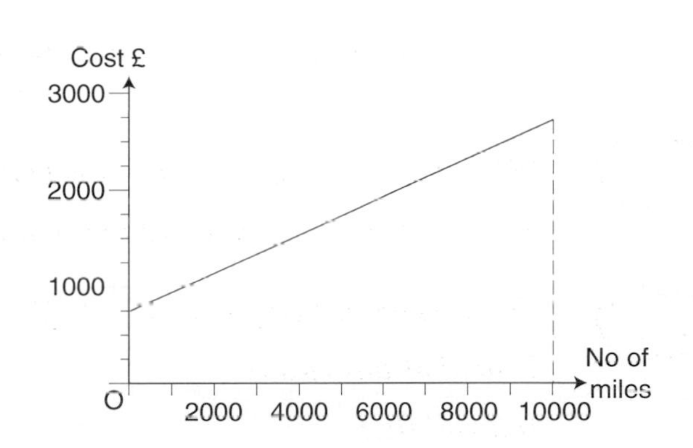

Finishing off

Now that you have finished this chapter you should be able to:

- ★ plot points (x, y) on a grid
- ★ find the co-ordinates of a point on a grid
- ★ construct a table of values and use it to draw a graph
- ★ find the gradient and y intercept of a straight line graph
- ★ recognise parallel lines
- ★ obtain information from a graph.

Use the questions in the next exercise to check that you understand everything.

Mixed exercise

1 Draw the x axis from $x = 0$ to $x = 5$ and the y axis from $y = -5$ to 10.

The following equations represent straight lines.

a) $y = 3x - 5$ b) $y = 9 - x$

(i) Draw the graph of each line from $x = 0$ and $x = 5$.

(ii) Write down the co-ordinates of the point where these lines intersect.

2 The following equations represent straight lines.

a) $y = \frac{1}{2}x + 4$ b) $y = 7 - 3x$

For each one

(i) construct a table of values from $x = -2$ to 4

(ii) draw the graph

(iii) calculate the gradient of the line

(iv) Write down the y intecept.

3 Alka throws a tennis ball up in the air.
Its height, h metres, above the ground at time t seconds is given by

$h = 20t - 5t^2$

a) Copy and complete this table of values of h.

t	0	1	2	3	4
$20t$	0				80
$-5t^2$	0	–5			–80
h					0

For $-5t^2$ work out t^2 first and then multiply by -5: when $t = 4$, $4^2 = 16$ and $-5 \times 16 = -80$

b) Choose suitable scales and draw the graph of h against t.

c) Use your graph to estimate the times at which the ball is 10 m above the ground. Why do you get two answers?

d) Why is it not sensible to take values of t greater than 4?

4 Which of the following lines are parallel to $y = 3x - 5$?

a) $y = 3x$ b) $y = 2x - 5$ c) $y = 1 + 3x$ d) $y = 4 - 3x$

5 The graph shows the cost of printing cards. There is a fixed cost for setting up the machine and a 'run-on cost' per 1000 cards.

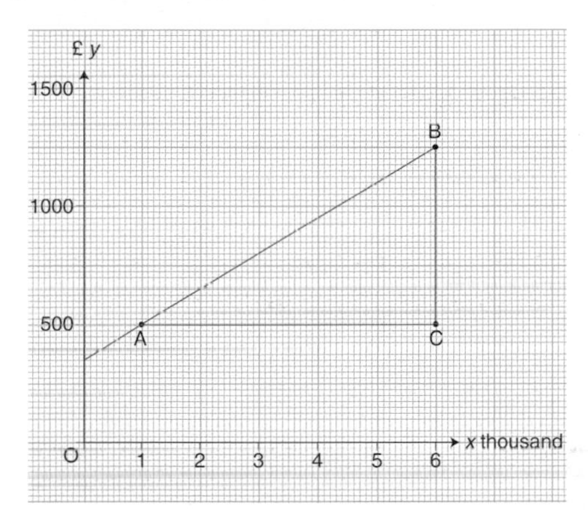

a) What does each small unit on the y axis represent?

b) What is the fixed setting-up charge?

c) Write down the co-ordinates of A and B and so find the lengths of AC (in thousand cards) and BC (in pounds).

d) Use your answers to c) to find the run-on cost per 1000 cards.

e) Work out the cost of printing 10 000 cards.

6 The Brown family from London are planning a holiday in Dumfries. The graph shows their estimated costs for travel to Dumfries and board for x days.

a) How much do they estimate for travel?

b) Find the gradient of the graph. What did they estimate for daily board?

c) Write down an equation for the total cost £C of a holiday of x days.

d) Use your equation to find the cost for 20 days.

7 Sol, Ali and Jo live in a block of flats.
Each person goes to the local shop and returns home 25 minutes later.

They each spend 5 minutes in the shop. The shop is 1200 m away.

Sol's speed to the shop is the same as his speed coming back.
Ali's speed to the shop is faster than his speed coming back.
Jo's speed to the shop is slower than her speed coming back.

Sketch a possible distance time graph for each person showing the journey to the shop and back.

Find out other situations where there is a fixed charge plus a rate depending on the amount a service is used. Draw a graph to illustrate one of these.

Fourteen

Equations

Finding unknowns

Andy and Beth play the game 'think of a number'.

Andy: 'Think of a number and multiply it by 2. What is your answer?'

Beth: 'Ten'.

How can Andy work out Beth's original number?

Andy does it like this.

x is the **unknown**; it is Beth's original number. She multiplies it by 2 to get 10.

$2x = 10$
divide BOTH SIDES by 2
$2x \div 2 = 10 \div 2$
$x = 5$

The **equation** $2x = 10$ has been **solved** to obtain the **solution** $x = 5$.

Andy: 'Think of a number and add 3. What is your answer?'

Beth: 'Seven'.

How can Andy work out Beth's original number?

Andy does it like this.

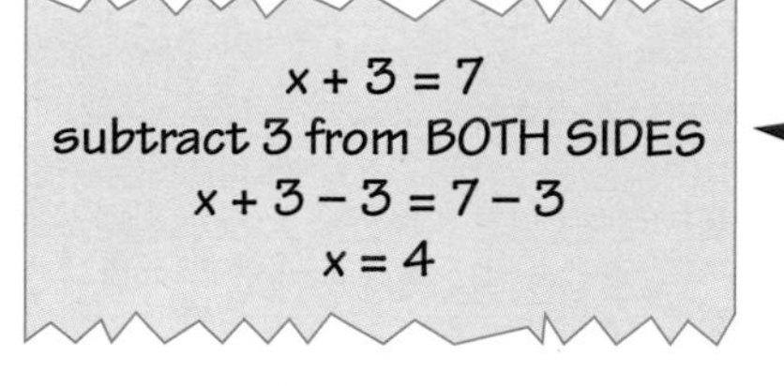

You must do the same to both sides to keep the equation balanced.

Andy: 'Think of a number, multiply it by 4 and subtract 5.
What is your answer?'

Beth: 'Nineteen'.

Remember that whatever you do to one side of the equation you must do to the other.

How can Andy work out Beth's original number?

Andy does it like this.

$4x - 5 = 19$ Add 5 to BOTH SIDES $4x - 5 + 5 = 19 + 5$ $4x = 24$	divide BOTH SIDES by 4 $4x \div 4 = 24 \div 4$ $x = 6$

Is Andy correct? Check by putting $x = 6$ in the left side of the equation.
You get $4 \times 6 - 5 = 24 - 5 = 19$. It is the same as the right side.
$x = 6$ satisfies the equation so Andy is correct.

Exercise

1 Andy: 'Think of a number and multiply it by 3. What is your answer?'

Beth: 'Twenty one'.

Write an equation and solve it to find Beth's original number.

2 Solve each of these equations, writing out the steps.

a) $2x = 8$ b) $5x = 30$ c) $3x = 27$ d) $4x = 32$

3 Andy: 'Think of a number and subtract 5. What is your answer?'

Beth: 'Nine'.

Write an equation and solve it to find Beth's original number.

4 Solve each of these equations, writing out the steps.

a) $x - 6 = 13$ b) $x + 7 = 10$ c) $x - 9 = 11$ d) $x + 5 = 17$

5 Andy: 'Think of a number, multiply it by 3 and add 2.
What is your answer?'

Beth: 'Twenty six'.

Write an equation and solve it to find Beth's original number.

6 Solve each of these equations, writing out the steps. Check your answer by substituting back into the original equation.

a) $2x + 5 = 17$ b) $5x - 4 = 31$ c) $3x + 7 = 19$

d) $8x - 1 = 15$ e) $7x + 3 = 45$ f) $2x - 5 = 13$

g) $3x + 7 = 19$ h) $4x - 1 = 5$ i) $3 + 2x = 25$

j) $6x + 11 = 5$ k) $18 = 7x + 4$ l) $3x - 12 = 0$

m) $5x + 9 = 37$ n) $15 - 2x = 7$ o) $7 = 4 - 5x$

7 Write an equation describing the situation below.

Solve it to find the value of x.

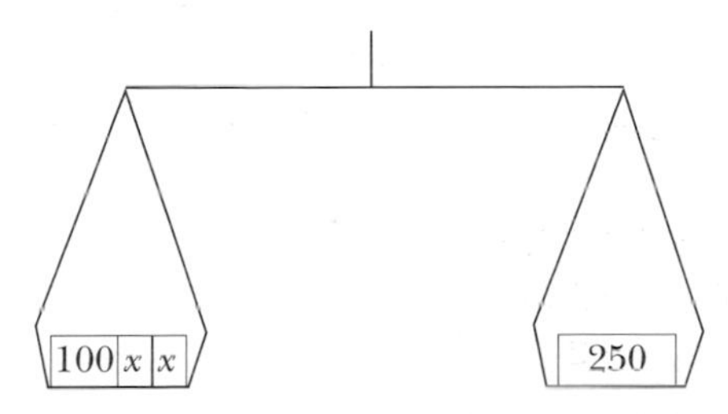

8 The perimeter of this rectangle is 84 units. Write an equation in x and solve it.

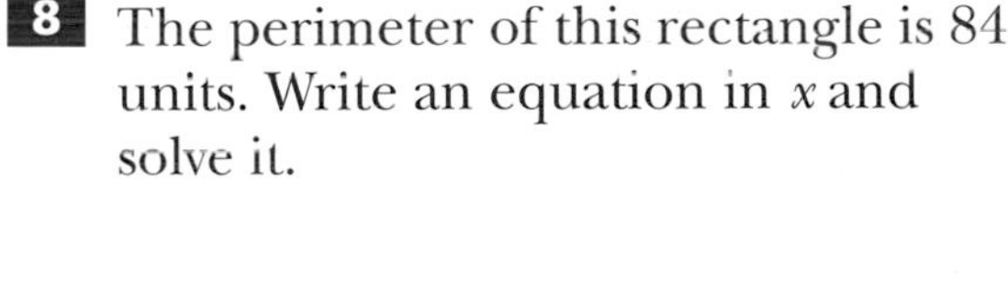

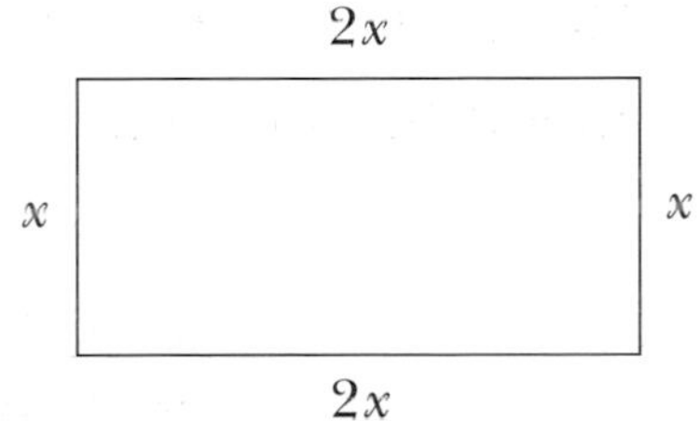

More equations

Sometimes you need to solve equations which contain brackets. You can do this by multiplying out the brackets and then continuing as before.

Example

Solve the equation $7(n + 2) = 35$

In this equation n is the unknown and we are going to find its value.

Solution

Multiply out the brackets $7n + 14 = 35$

Subtract 14 from both sides $7n = 35 - 14$

(Tidy up) $7n = 21$

Divide both sides by 7 $n = 21 \div 7$

$n = 3$

Check, by substitution, that $n = 3$ satisfies the original equation.

In the equations that you have solved so far the unknown has only appeared once. When it appears more than once you gather the terms containing the unknown on one side of the equation and the number terms on the other.

Example

Solve the equation $5x - 3 = 3x + 11$

Solution

Subtract $3x$ from both sides $5x - 3 - 3x = 3x + 11 - 3x$

(Tidy up) $2x - 3 = 11$

Add 3 to both sides $2x = 11 + 3$

(Tidy up) $2x = 14$

Divide both sides by 2 $x = 7$

Check your answer by substituting $x = 7$ into both sides of the original equation.

Left-hand side $= 5 \times 7 - 3 = 35 - 3 = 32$

Right-hand side $= 3 \times 7 + 11 = 21 + 11 = 32$

Both sides are equal so $x = 7$ is correct.

What happens when you try to solve $2(x + 3) = 2x + 6$?

What happens when you try to guess a solution by substituting $x = 1, 2, 3, 4, \ldots$?

$2(x + 3) = 2x + 6$ is true for all values of x. It is an **identity**.

Exercise

1 Solve these equations. Check your answers by substitution.

a) $3(n + 2) = 15$ b) $2(n - 5) = 16$ c) $4(x + 6) = 32$

d) $5(x + 4) = 10$ e) $8(y - 1) = 12$ f) $2(3y + 1) = 38$

g) $6(2a + 5) = 78$ h) $3(1 - 2x) = 21$ i) $4(1 - n) = 14$

2 Solve these equations. Check your answers by substitution.

a) $4x - 3 = x + 9$ b) $6x + 2 = x + 32$ c) $4n - 7 = 3n + 6$

d) $5n - 4 = 3n + 14$ e) $a - 3 = 2a + 7$ f) $3y + 15 = y - 3$

g) $4t + 5 = 11 + t$ h) $6x - 2 = 7 - 3x$ i) $13 - n = n + 5$

j) $d + 12 = 6d$ k) $1 - 2u = u + 13$ l) $7 + 2v = 6v - 11$

3 Solve these equations. Check your answers by substitution.

a) $2(x + 1) + 3 = 17$ b) $8x + 3 = 5(x + 3)$

c) $3(y - 4) = y + 8$ d) $11 - m = 2(m + 10)$

e) $5(a + 2) + 2a = 24$ f) $3(3n - 7) = 4(n + 1)$

g) $7(2p + 1) + 3(p - 2) = 69$ h) $5(7q - 3) - 2(5 - 2q) = 27$

4 Equations do not always involve simple numbers, but so long as you know the method for solving them, all you need is a calculator to help you.

a) $2.4c = 9$

b) $7y + 2 = 11$

c) $1.5 = 2.3 - 3x$

d) $7(1 - t) = 4$

e) $4.5n = 13 - 2.5n$

f) $6.2k + 3 = 5.5 + 4k$

5 The formula for converting a temperature $C°$ Celsius into $F°$ Fahrenheit is given by

$$F = 1.8C + 32$$

a) Find the value of F when $C = 25$

b) Find the value of C when $F = 53$

c) Put F equal to C in the formula and hence find the temperature at which Fahrenheit and Celsius give the same reading.

Investigation

One of these equations has no solution.

For the other equation every value of x is a solution.

(i) $3(x + 6) - 2x = x + 18$

(ii) $5(2 + x) + 3x = 4(2x + 3)$

Which equation is which and how can you tell?

Equations with fractions

Sometimes you need to solve equations with fractions in them, such as

$\frac{2}{3}x = 6$

In this case, the number on the bottom of the fraction is 3 so you multiply both sides of the equation by 3:

$$3 \times \frac{2}{3}x = 3 \times 6$$

(Tidy up) $2x = 18$

Divide both sides by 2 $x = 9$

Now the fraction has gone

What is $\frac{2}{3}$ of 9?

Example Solve $\frac{3}{4}(x + 2) = 12$

Solution

Multiply by 4 $4 \times \frac{3}{4}(x + 2) = 4 \times 12$

(Tidy up) $3(x + 2) = 48$

(Expand brackets) $3x + 6 = 48$

Subtract 6 $3x = 48 - 6$

(Tidy up) $3x = 42$

Divide by 3 $x = 14$

Substitute $x = 14$ into the original equation. Does the check work?

You use the same method to solve an equation when x is on the bottom line of a fraction. In this case you multiply both sides by x.

Example Solve $20 = \frac{360}{x}$

Solution

Multiply by x $x \times 20 = x \times \frac{360}{x}$

(Tidy up) $20x = 360$

Divide by 20 $x = 18$

This example could be about a pie chart with equal sectors of 20°.

The number of sectors is x. A more general formula is $A = \frac{360}{x}$ where A is the size of each sector.

Show how following the same steps gives the formula in the form

$x = \frac{360}{A}$

Exercise

1 Solve these equations.

a) $\frac{x}{2} = 5$ b) $\frac{x}{3} = 8$ c) $\frac{x}{100} = 100$ d) $-\frac{x}{4} = -1$

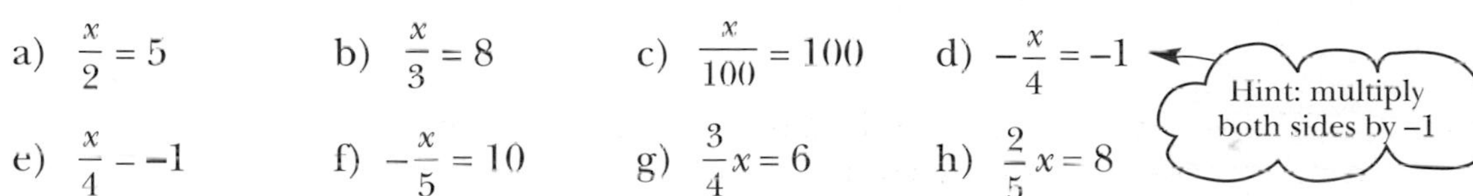

e) $\frac{x}{4} = -1$ f) $-\frac{x}{5} = 10$ g) $\frac{3}{4}x = 6$ h) $\frac{2}{5}x = 8$

2 a) Show that the equation $\frac{x}{12} + 2 = 5$ can be simplified to $\frac{x}{12} = 3$.

b) Solve the equation.

3 Solve these equations.

a) $\frac{x}{4} + 5 = 11$ b) $-\frac{2x}{3} + 8 = 6$

c) $\frac{5x}{4} - 5 = 10$ d) $\frac{1}{2}x + \frac{1}{2} = 6$

4 Solve these equations.

a) $\frac{1}{3}(x - 5) = 2$ b) $\frac{1}{3}(x + 5) = 1$

c) $\frac{1}{2}(x + 7) = 4$ d) $\frac{1}{5}(x - 1) = -2$

e) $\frac{2}{3}(x + 2) = 0$ f) $\frac{2}{3}(2x - 5) = 6$

g) $\frac{3}{4}(3x + 1) = 30$ h) $\frac{1}{3}(x - 1) + x = 13$

5 Solve these equations.

a) $\frac{120}{x} = 8$ b) $\frac{15}{x} = 3$ c) $\frac{4}{x} = 8$ d) $\frac{9}{x} = -1$

6 The formula for the electric current I (ampères) in a circuit is

$I = \frac{V}{R}$ where V is the voltage in volts and R is the resistance in ohms.

a) In the case when $V = 240$ and $I = 6$ find R.

b) In the case when $I = 8$ and $R = 25$ find V.

Investigation

The diagram shows a sector of a pie chart with angle $x°$ where x is a whole number.

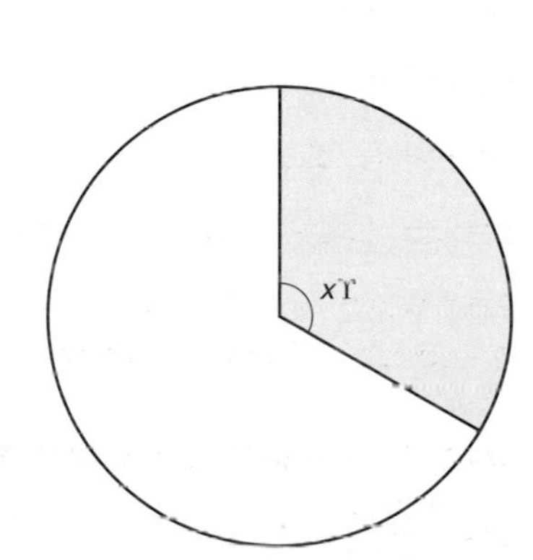

This sector represents $y\%$ of the total.

(i) Show that $x = 3.6y$.

(ii) When $y = 50$, $x = 3.6 \times 50 = 180$.
What other values can y and x have when y and x are both whole numbers?

Using equations to solve problems

Which of these telephone companies is cheaper if you make just a few calls a month?

Which is cheaper if you make a lot of calls each month?

How do you decide which is cheaper for you?

You can do this using trial and error, or you can gain a clearer picture of the situation by using algebra.

The charges each month depend on the number of minutes that your calls last. Call the number of minutes m.

The amount in pence charged by Venus is then $5m$.

The amount in pence charged by Connect is $1200 + 2m$.

These **expressions** are used to form an equation in m.

What does the 1200 *represent in the expression for Connect's charges?*

To find the value of m for which both companies charge the same amount, you form an equation in m and then solve it.

The companies charge the same amount when $5m = 1200 + 2m$

Subtract $2m$ from both sides $3m = 1200$

Divide both sides by 3 $m = 400$

The companies charge the same amount when $m = 400$.

Nick's calls last about 600 *minutes each month.*
Which company should he use?

Catherine's calls are usually under 300 *minutes each month.*
Which company should she use?

There are many situations in which forming and then solving an equation is helpful.

A chocolate bar is marked '25% extra free'. Its weight is 250 g.

How would you form an equation to find the weight of a normal bar?

1 Livewire Electricity charge 80 pence per unit. Sparks Electricity make a standing charge of £7.80 per quarter and then charge 65 pence per unit.

a) Write down an expression for the cost, in pence, to a consumer using x units per quarter with

(i) Livewire Electricity (ii) Sparks Electricity

b) Find the value of x for which both companies charge the same amount.

2 For each of the following situations

(i) form an equation in the unknown quantity given

(ii) solve the equation

(iii) check your answer.

a) Fanzia goes shopping with £100 in her purse.
She buys x CDs at £12 each and still has £16 left.

b) In one season, Totnes Wanderers Football Club scores 72 points.
They win w matches (3 points each), draw 6 (1 point each) and lose the rest (0 points).

c) The length of a field is 3 times its width of w metres.
The perimeter is 600m.

d) The largest angle of a triangle is 4 times the size of the smallest angle, $A°$. The third angle is 60°.

e) Halley's present age is y years. In 24 years time he will be 3 times as old as he is now.

3 Sara is finding out about monthly charges for using the Internet. She has written these notes.

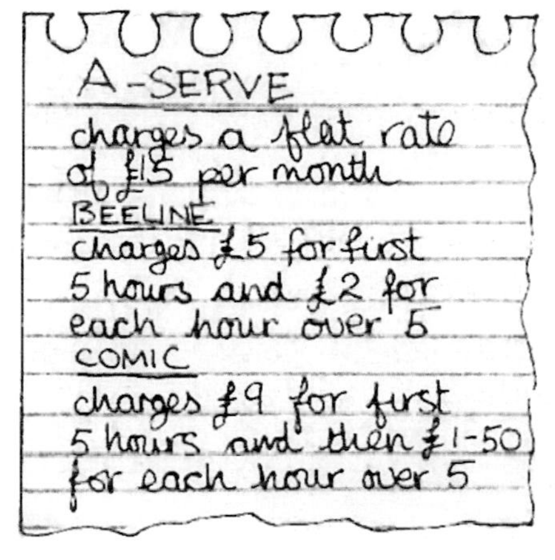

Sara intends to surf the Internet for more than 5 hours. Suppose x stands for the number of extra hours (above 5) that she spends.

a) Write down expressions for the amount Sara could expect to be charged by each company.
Which is cheapest when $x = 1$?
Which is cheapest when $x = 10$?

b) For a certain value of x, A-serve and Beeline must charge the same. Put those two expressions equal to each other, and solve the resulting equation to find this value of x.

c) Repeat b) to find a value of x for which Beeline and Comic charge the same.

d) Do the same for A-serve and Comic.

e) What advice would you give to Sara?

Using graphs to solve equations

In many real-life problems you get equations that are too complicated to solve by algebra, but you still need to know the answer. You can find it by drawing a graph.

This graph shows the fuel consumption of a car (in miles per gallon) being driven at different speeds (in m.p.h.). The equation of the curve is

$$y = \frac{1}{10\,000}(x^3 - 250x^2 + 15\,000x + 200\,000)$$

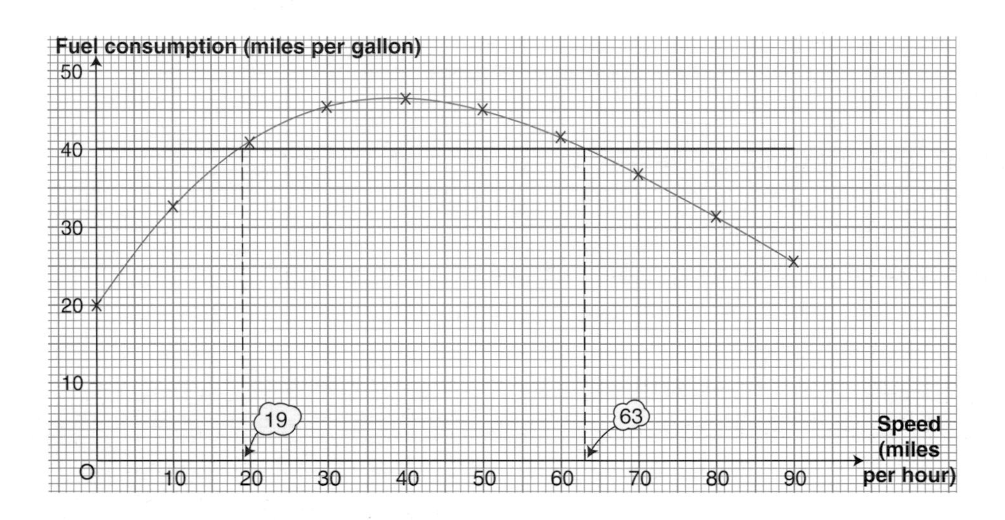

A car's fuel consumption is 40 miles per gallon. How fast is it travelling?

To find the answer you need to solve the equation

$$\frac{1}{10\,000}(x^3 - 250x^2 + 15\,000x + 200\,000) = 40$$

You cannot solve this equation by algebra but you can do it by looking at the graph. The red line is $y = 40$ and this crosses the curve when x is about 19 or 63.

A graph only gives an approximate answer, but you can find one as accurately as you like with a calculator using trial and improvement. This method is used here to find a more accurate answer for the speed near 63 m.p.h.

When $x = 63$ $y = 40.27\ldots$
When $x = 64$ $y = 39.8\ldots$
The answer is between 63 and 64. Try 63.5.

When $x = 63.5$ $y = 40.04$

The answer is between 63.5 and 64. It seems to be close to 63.5. Try 63.6.

When $x = 63.6$ $y = 40.002$ This is very close.

Is 63.6 closer than 63.7? How can you find the answer even more accurately?

Exercise

1 a) Draw the graph of $y = x^3 - 5x - 2$ taking values of x from -3 to $+3$.

b) Use your graph to solve the equations

(i) $x^3 - 5x - 2 = 0$

(ii) $x^3 - 5x - 2 = 4$

2 a) Draw the graph of $y = 4x^2 - x^3 - 2$ taking values of x from -1 to $+4$

b) Use your graph to solve the equations

(i) $4x^2 - x^3 - 2 = 0$

(ii) $4x^2 - x^3 - 2 = 2$

3 Use trial and improvement to solve $x^3 - 7x - 9 = 0$ to 3 decimal places.

4 Use trial and improvement to solve $x^2 - \frac{1}{x} - 1$ to 3 decimal places.

5 Jane is a scientist.

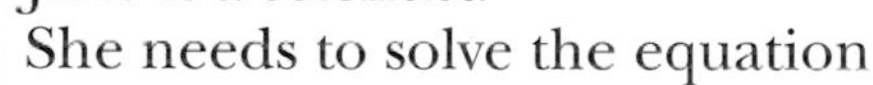

She needs to solve the equation

$$x^2 - 6x + \frac{12}{x} = 0$$

So she draws the curve

$$y = x^2 - 6x + \frac{12}{x}$$

It is shown here. She is only interested in positive (+) values of x.

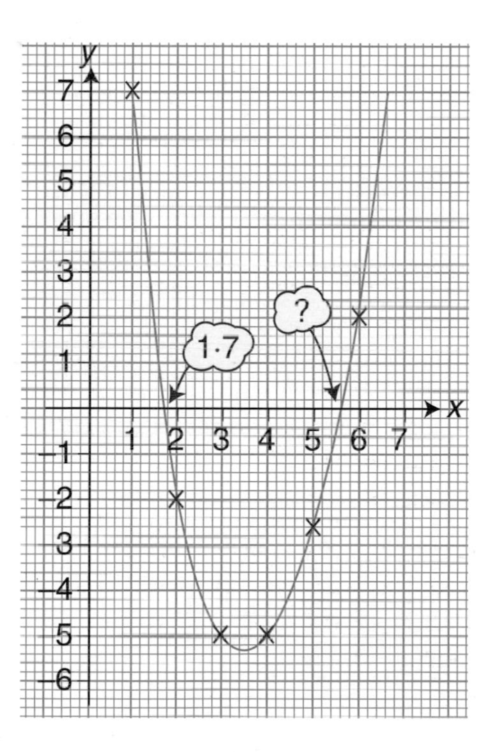

a) You can see from the graph that one of the values of x is near 1.7. Write down the other value, approximately.

b) Use trial and improvement to find the value of x near 1.7 to 2 decimal places.

c) Use trial and improvement to find the other value to 2 decimal places.

6 a) Make out a table of values for $y - \frac{20}{x} - 16 + x^2$ for $x = 1, 2, 3$ and 4.

b) Draw the graph.

c) Use your graph to find two values of x for which $\frac{20}{x} - 16 + x^2 = 0$.

d) Use trial and improvement to find your answers to c) to 2 decimal places.

Finishing off

Now that you have finished this chapter you should be able to:

- ★ solve simple equations using algebra
- ★ solve equations involving brackets and fractions
- ★ use equations to solve problems
- ★ solve equations graphically
- ★ use trial and improvement methods to solve equations.

Use the questions in the next exercise to check that you understand everything.

Mixed exercise

1 Solve these equations.

a) $5a = 15$
b) $15b = 21$
c) $14 = 35c$
d) $2.5d + 11 = 17$
e) $67 = 16e - 13$
f) $22 - 3f = 7$
g) $14 - 5g = 1.5$
h) $22.4 - 4h = 10$

2 Solve these equations by collecting the terms in the unknown together on one side.

a) $5z - 4 = 2z + 5$
b) $6y + 1 = 7y - 3$
c) $8x + 9 = 15 - 4x$
d) $3w + 4 = 39 - 2w$
e) $12 + 6v = 2v + 30$
f) $14 + 5u = 26 - 3u$
g) $62 - 9t = 2t - 4$
h) $999 - 50s = 49s + 900$

3 Multiply out the brackets and solve these equations.

a) $2(8 + 2x) = 44$
b) $3(10 - x) = 5x$
c) $22x = 7(x + 3)$
d) $25 + x = 4(x - 2)$
e) $3(x + 2) = 2(x + 1)$
f) $5(2x + 4) - 3(x - 3) = 57$

4 Ching asks Jo to think of a number, then to subtract 1, multiply by 4, and finally subtract 10.

a) Calling Jo's number n, find an expression for her answer in terms of n.

b) Jo says her answer is twice the number she first thought of.

Make an equation for n and solve it to find Jo's number.

5 Solve the following equations, which involve fractions.

a) $\frac{x}{3} = 6$

b) $\frac{x}{2} + 3 = 8$

c) $\frac{2x}{5} = 20$

d) $\frac{1}{4}(x + 3) = 2$

e) $\frac{1}{5}(x + 7) = 2$

f) $\frac{2}{3}(x + 8) = 6$

g) $\frac{42}{x} = 6$

h) $\frac{256}{x} = 32$

6 a) Draw the graph of $y = x^3 - 3x^2 + 3$ for values of x between –1 and 3.

b) Solve the equation

$x^3 - 3x^2 + 3 = 0$

c) Solve the equation

$x^3 - 3x^2 + 3 = 2$.

7 

a) Draw the graph of $y = \frac{12}{x} - x^2$ for values of x between 1 and 4.

b) Use your graph to solve the equation $\frac{12}{x} - x^2 = 0$.
Give your answer to 1 decimal place.

c) Use your calculator to find $\sqrt[3]{12}$.

d) Explain why your answers to parts b) and c) are the same.

8 Use trial and improvement methods to find the solution of the equations.

a) $x^3 - 2x^2 - 1 = 0$

b) $x - \frac{1}{x^2} - 2 = 0$

9 The Web electricity board has two ways of charging customers.
The Silver price is £12 per quarter plus 6p per unit of electricity.
The special Goldstar price is £36 per quarter plus 4.5p per unit.

Write down the total bill for x hundred units using

a) the Silver price

b) the Goldstar price.

c) Form an equation in x to find when the prices are equal. Solve it.

d) The Smiths use about 1400 units a quarter. Which price is cheaper for them?

Fifteen

Formulae

Substituting into a formula

Tom is looking up how long it takes to cook a chicken of weight 2.7 kg.

His cookery book says 45 minutes per kg plus 20 minutes.

This means multiply the weight by 45 and then add 20.

Tom works out the cooking time like this:

cooking time (minutes) is

$2.7 \times 45 + 20$

$= 121.5 + 20$

$= 141.5$

That's 2 hours $21\frac{1}{2}$ minutes

You can use algebra to write the formula. Calling the weight of the chicken w kg and the cooking time t minutes, the formula would be

$$t = w \times 45 + 20$$

Example

Work out the cooking time for a chicken of weight 3.2 kg.

Solution

Substituting $w = 3.2$ into the formula gives

$t = 3.2 \times 45 + 20$

$= 144 + 20$

$= 164$

How long is this in hours and minutes?

Exercise

1 Calculate s in each of these formulae by substituting the numbers given.

a) $s = 4u + 80$ (i) $u = 2$ (ii) $u = 5$ (iii) $u = 6.5$

b) $s = 4u + 8a$ (i) $u = 2,\ a = 5$ (ii) $u = -6.5,\ a = 10$

2 The area of a triangle is given by $A = \frac{1}{2}bh$. Find the value of A when

a) $b = 10,\ h = 5$ b) $b = 4,\ h = 4$

c) $b = 1,\ h = \frac{1}{2}$ d) $b = 0.4,\ h = 0.2$

3 You can estimate the depth, d m, of a well by dropping a stone into it and finding t, the number of seconds before it hits the bottom. You then use the formula

$$d = 5t^2$$

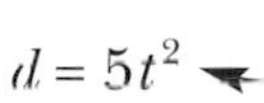

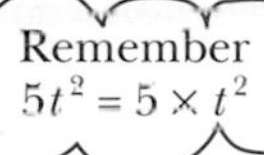

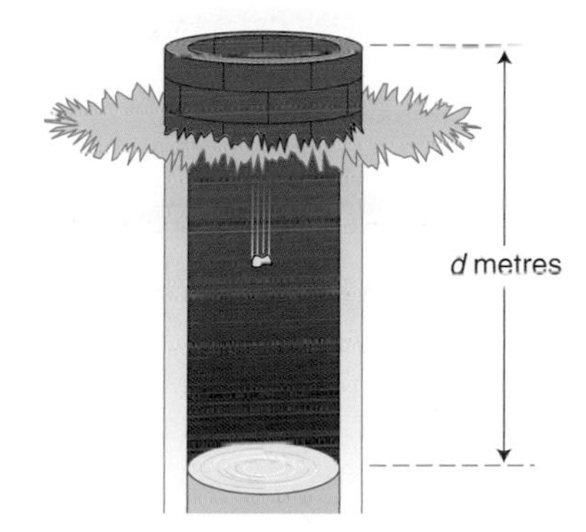

Find the value of d when

a) $t = 1$ b) $t = 2$

c) $t = 0.5$ d) $t = 1.5$

4 A printer says the cost £C of printing n greetings cards is given by

$$C = 25 + 0.05n$$

a) Find the cost of printing

(i) 100 cards (ii) 1000 cards and (iii) 10 000 cards.

b) Find the cost per card in pence in each of the cases in a).

c) Find the formula for the cost p pence per card when the number printed is n.

5 When a ball is thrown straight up in the air at 5 metres per second, its *upward* speed v metres per second after t seconds is given by the formula

$$v = 5 - 10t$$

Find v when t is:

a) 0 b) 0.2 c) 0.5 d) 1

Write down what is happening to the ball in each case.

Find three formulae which are commonly used in other subjects. They might be in words. If so, rewrite them using letters.

State clearly what the letters represent in all your formulae.

Making up formulae

Leanne is counting off the days before her summer holiday.

How many days does she have to wait if there are 14 *weeks to go before then?*

Suppose W stands for the number of weeks and D stands for the number of days.

You can write a formula to work out the value of D for different values of W:

$D = 7 \times W$ or $D = 7W$

When Leanne has 10 weeks to wait $W = 10$; the number of days Leanne has still to cross off is:

$$D = 7 \times 10$$
$$= 70$$

The formula above is written as $D = \ldots$ so D is called the **subject** of the formula.

To make W the subject of the formula $D = 7W$, first write $7W = D$ then divide both sides by 7, $W = \frac{D}{7}$.

Leanne is going to Norway for her holiday. The currency in Norway is the kroner. There are twelve kroners to the pound.

You can write a formula to work out the amount in pounds (P) that corresponds to an amount in kroners (k). The formula is

$$P = \frac{k}{12}$$

Check that this is correct by putting $k = 12$.

You can make k the subject of the formula by writing $\frac{k}{12} = P$ and multiplying both sides by 12. This gives $k = 12P$.

Exercise

1 Write the following as algebraic expressions.

a) Three times n

b) Half b times h

c) Six times a times a

d) Twenty divided by r^2

e) Three times h plus six times d

2 Today is Alex's birthday. Her age is y years.

a) Write down a formula for m, Alex's age in months.

b) Check your formula by letting $y = 10$.

c) Write down a formula to give y in terms of m.

3 A table is 2 m long. How would you find its length in centimetres?

a) Using C as the number of centimetres and M as the number of metres, write a formula of the form $C = \ldots \times M$.

b) Now write the formula with M as the subject, $M = \ldots$

4 The graph shows how many miles, m, a car can travel on l litres of petrol.

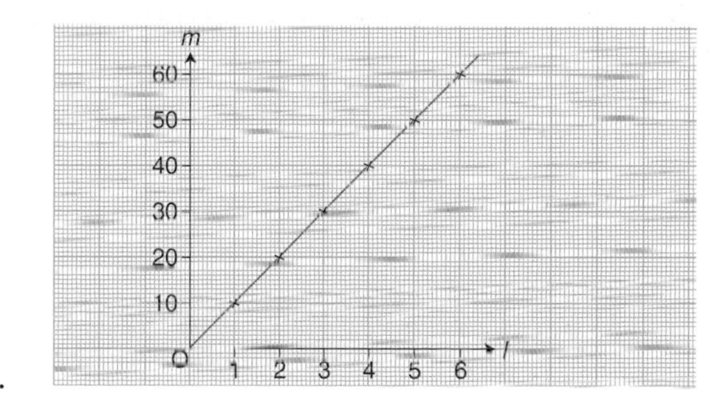

a) How many miles does the car do on 1 litre?

b) Write a formula giving m in terms of l.

c) Rewrite your formula with l as subject.

5 Write down formulae connecting

a) (i) the price in pence, P, of A apples costing 30p each

(ii) the number of apples you can buy for P pence.

b) (i) the number of centimetres, C, in I inches

(ii) the number of inches, I, in C centimetres.

6 The temperature at dawn is D °C.

The midday temperature, M, is 12 °C greater.

a) Write a formula giving M in terms of D.

b) Write a formula giving D in terms of M.

The density d (grams per cubic centimetre) of a substance may be found by finding the mass m (grams) of a volume V (cubic centimetres) of it.

$$d = \frac{m}{V}$$

a) Write this formula with (i) m (ii) V as the subject.
b) Conduct an experiment to find the density of various substances.

Changing the subject of a formula

Look at these two headlines.
They both tell you the same thing.

Pele is the subject of the first.
The winning goal is the subject of the second.

Look at this formula for finding the temperature, F, in degrees Fahrenheit from the temperature, C, in degrees Celsius.

$$F = 1.8C + 32$$

The Fahrenheit temperature F is on its own on the left. It is the subject of the formula. It is easy to find the value of F when you know the value of C.

Find the value of F when C is 20.

You probably found the formula very easy to use in this way. But what if you know the value of F and you want to find C?

Find the value of C when F is 77.

This would be easier with a formula for C in terms of F, in other words to have C as the subject of the formula. You can make C the subject by rearranging the formula as follows.

$$F = 1.8C + 32$$

Subtract 32 from both sides $\quad F - 32 = 1.8C$

Divide both sides by 1.8 $\quad \dfrac{F - 32}{1.8} = C$

or $\quad C = \dfrac{F - 32}{1.8}$

Rearranging the formula is a bit like solving an equation. You want to get C by itself on one side, but at each step you have to do the same thing to each side of the formula.

You can see that C is now the subject of the formula.

This can also be written as $C = \frac{5}{9}(F - 32)$. *Explain why.*

Some formulae have powers or roots.

- Rearrange $a^2 + b^2 = c^2$ to get $a = \ldots$.

  subtract b^2 $\quad a^2 = c^2 - b^2$

  square root $\quad a = \sqrt{c^2 - b^2}$

- Rearrange $x = \sqrt{y} - 2$ to get $y = \ldots$.

  add 2 $\quad x + 2 = \sqrt{y}$

  square $\quad (x + 2)^2 = y$

  so $y = (x + 2)^2$

1 Make x the subject of each of these.

a) $y = x + 4$ b) $y = x + a$ c) $y = 3 + x$ d) $y = c + x$

e) $y = x - 5$ f) $y = x - b$ g) $y = 6 - x$ h) $y = d - x$

2 Make x the subject in each of these.

a) $y = 2x$ b) $y = ax$ c) $y = \frac{x}{4}$ d) $y = \frac{x}{b}$

e) $y = \frac{3}{4}x$ f) $y = \frac{a}{b}x$ g) $y = \frac{4x}{5}$ h) $y = \frac{ax}{b}$

3 Make t the subject of each of these.

a) $x = 2t - 3$ b) $y = 3t + 4$ c) $p = 6 + 2t$

d) $c = 4 - t$ e) $z = 6 - 2t$ f) $s = 2t + a$

g) $x = 5t - c$ h) $n = 7t - 3x$ i) $v = u + at$

4 In each of these, make the given letter the subject.

a) $v = u + at,\ u$ b) $p = 2l + 2b,\ l$ c) $V = 4x - 9y,\ x$

5 In each of these, expand the bracket and then make x the subject.

a) $p = 2(x + y)$ b) $V = 12(r + x)$

c) $s = 4(2 - x)$ d) $y = 4(a - x)$

6 In each of these, make the given letter the subject.

a) $A = lw,\ l$ b) $V = lwh,\ h$ c) $V = IR,\ R$

d) $C = \pi d,\ d$ e) $C = 2\pi r,\ r$ f) $I = \frac{r}{100} \times P,\ P$

g) $I = \frac{PRT}{100},\ T$ h) $I = \frac{PRT}{100},\ R$

7 Rearrange the following to make x the subject.

a) $y = 6x^2$ b) $y = \frac{x^2}{3}$ c) $y = \sqrt{\frac{x}{5}}$ d) $y = \sqrt{2x}$

8 In each of these make the given letter the subject.

a) $E = mc^2,\ c$ b) $P = \frac{V^2}{R},\ V$ c) $v^2 = u^2 - 2gs,\ u$

d) $v = \sqrt{2gh},\ h$ e) $c = \sqrt{a + 2},\ a$ f) $d = 2\sqrt{\frac{A}{\pi}},\ A$

9 In each of the following make x the subject.

Hint: Collect the x terms together on one side and take out x as a common factor.

a) $7x = 5k + x$ b) $6x = t - 2x$ c) $ax + b = cx$

d) $3x - 2y = 2x + 7z$ e) $x = \frac{x}{n} + d$ f) $rx + sy = tx + u$

Finishing off

Now that you have finished this chapter you should be able to:

- ★ substitute numbers into a formula
- ★ change the subject of a formula
- ★ make up a formula.

Use the questions in the next exercise to check that you understand everything.

Mixed exercise

1 Here are the midday temperatures in four cities on 1st December.

Athens 15 °C   Buenos Aires 29 °C   Moscow –19 °C   Prague 0 °C

Change these temperatures from degrees Celsius (°C) to degrees Fahrenheit (°F) by using the formula $F = 1.8C + 32$.

Give your answers to the nearest degree.

2 The area, A, of a trapezium is given by $A = \frac{1}{2}(a + b) \times h$

Find the value of A when

a) $a = 6,\ b = 4,\ h = 2$

b) $a = 16,\ b = 4,\ h = 5$

c) $a = 8,\ b = 8,\ h = 8$

3 Work out the time Mac should cook a turkey weighing 6.3 kg by:

a) using the rule 15 *minutes per* 450 g *plus* 15 *minutes*;

b) using the formula $T = 33W + 15$ where T is the time in minutes for a turkey weighing W kg. Is your answer close enough to a)?

4 The surface area, A, and the volume, V, of a cone of radius r, height h and slant edge l are given by the formulae

$A = \pi r(r + l)$ and $V = \frac{1}{3}\pi r^2 h$

Find the value of A and V when

a) $r = 5,\ h = 12,\ l = 13$

b) $r = 8,\ h = 15,\ l = 17$

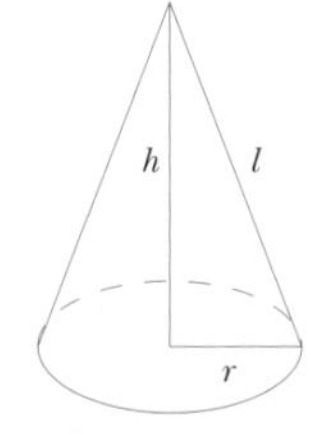

5 Write down the formula for

a) (i) the amount A litres of orange juice needed to fill g glasses, when each glass holds 0.25 litres

(ii) the number of glasses that can be filled from A litres of juice.

b) the dose d ml of medicine for a person weighing k kg when the instructions say '2 ml per 10 kg'.

c) the amount £C to be paid by each person going on a coach trip when N people want to go and the coach costs £150.

d) the price £T for a taxi for x miles, when it costs £1.80 standard charge plus 50p per mile.

6 In each of these, make the given letter the subject.

a) $y = x + 7$ x b) $y = 5 - x$ x c) $PV = k$ P

d) $P = \frac{F}{A}$ F e) $A = bh$ h f) $d = \frac{m}{V}$ V

g) $v = u + 10t$ t h) $P = 2(L + W)$ W i) $I = \frac{PRT}{100}$ P

7 Sunny Days Travel requires a deposit of 10% when a holiday is booked.

This can be written as a formula for the deposit £d for a holiday for n people priced at £p per person:

$$d = \frac{np}{10}$$

a) Calculate d when $n = 4$ and $p = 150$.

b) Jack paid a deposit of £85 for 2 people. What was the price per person?

c) Make p the subject of the formula.

8 In each of these make the given letter the subject.

a) $y = \frac{x^2}{2}$, x b) $s = \frac{1}{2}at^2$, t c) $V = \frac{4}{3}\pi r^3$, r

d) $a = \sqrt{\frac{b}{c}}$, c e) $y = 3\sqrt{x}$, x f) $T = 2\pi\sqrt{\frac{l}{g}}$, l

g) $ax + b = cx + d$, x h) $at = 2t + n$, t i) $h = \frac{h}{2n} + 3$, h

Sixteen

Inequalities

Using inequalities

These notices all describe restrictions on ages. These restrictions can be written as **inequalities** using special symbols.

< means *is less than*

> means *is greater than*

≤ means *is less than or equal to*

≥ means *is greater than or equal to*

Wilde (15) ***

RAIL CARD

Under 20?
60 or over?
Ask for details.

SINGLES CLUB

For people in their twenties and thirties meets Fridays at 8 pm

Let y stand for the age in years. Then each of the restrictions is written as an inequality and shown on a number line. (The negative number line is excluded because age is positive.)

a) $y \geq 15$ describes the age of a person who can go to the film *Wilde.*

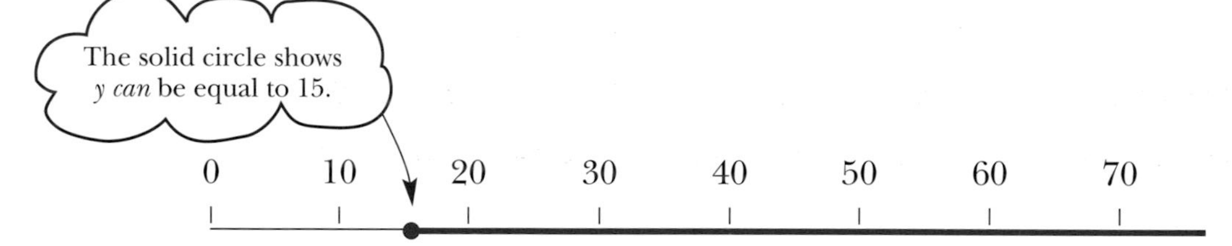

b) $y < 16$ describes the age of a person whose *Cinderella* ticket costs £2.50.

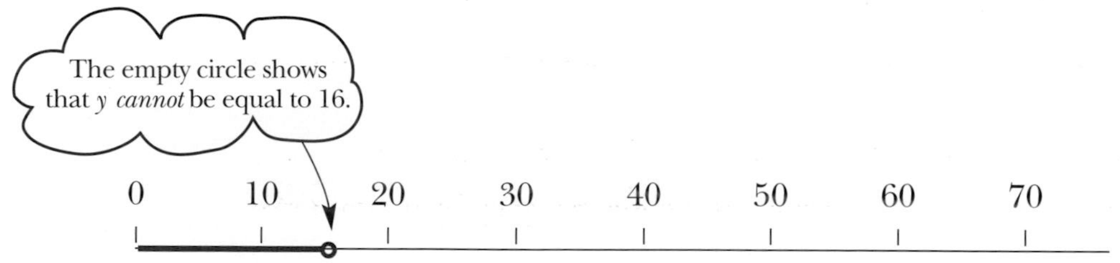

c) $20 \leq y < 40$ describes the age of a member of *Singles Club.*

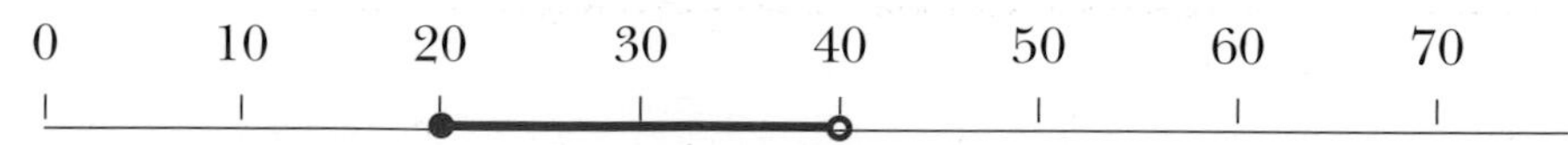

d) $y < 20$ or $y \geq 60$ describes the age of a person eligible for a railcard.

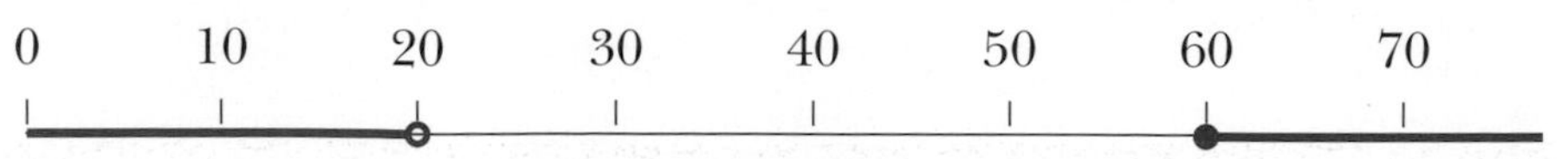

Exercise

1 Write these statements as inequalities.

a) x is greater than 3

b) x is greater than or equal to zero

c) x is less than 5

d) x is less than or equal to –2.

2 Look at each of these signs and write down the restriction shown using the inequality symbols. (Choose a suitable letter, in each case, to stand for the quantity that is being restricted.)

a

b)

c)

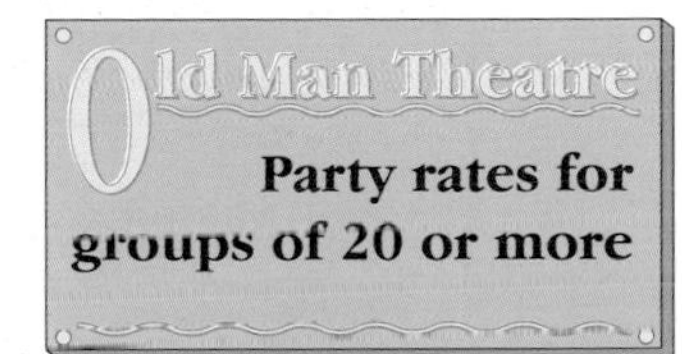

d)

e)

f)

3 Show each of these inequalities on a number line.

a) $x > 1$

b) $x \leq 3$

c) $x \geq 4$

d) $x < 0$

e) $x \geq -1$

f) $x < -2$

g) $x < 1$ or $x > 3$

h) $x < 4$ and $x > 1$

i) $x \leq 2$ and $x \geq -2$

j) $x > 3$ or $x < -1$

k) $x \geq -1$ or $x \leq -3$

l) $x < -1$ and $x > -2$

4 Write down the inequalities shown on these number lines.

a)

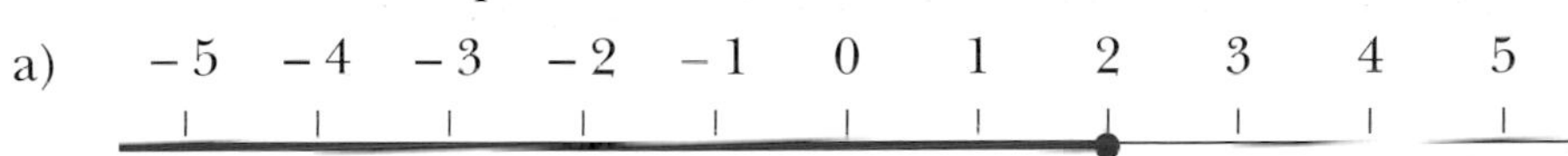

b)

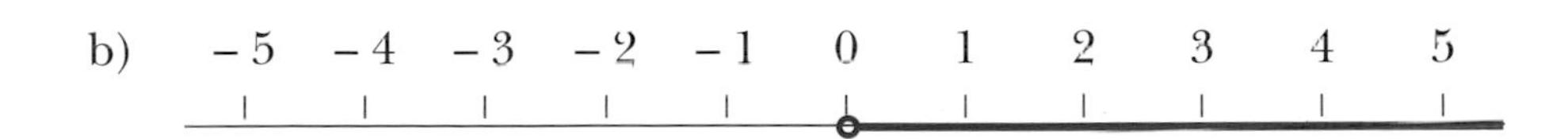

c)

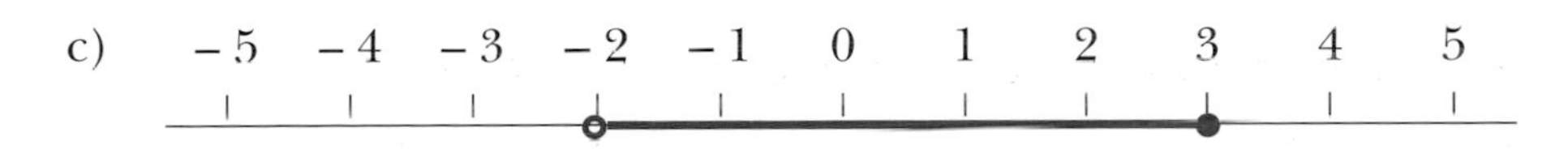

d)

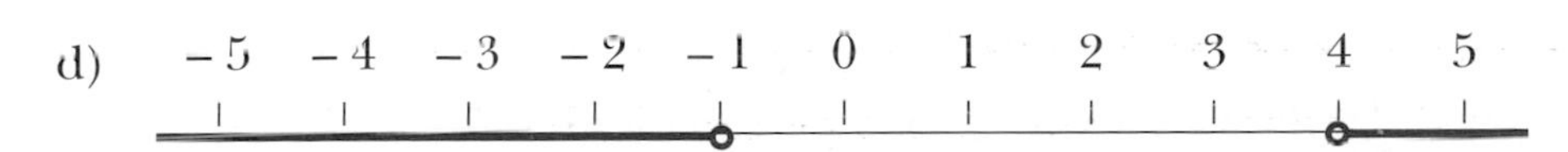

Solving inequalities

For what values of x is $3x - 5 > 16$?

You can solve this inequality in just the same way as you solve an equation.

An inequality

	$3x - 5 = 16$	$3x - 5 > 16$
Add 5 to both sides	$3x = 21$	$3x > 21$
Divide both sides by 3	$x = 7$	$x > 7$

The equation has just one solution

This is the solution of the inequality. It means that x can have any value greater than 7.

Check using a number just greater than 7

Check:	When $x = 7$	When $x = 8$
	$3x - 5 = 21 - 5 = 16$ ✓	$3x - 5 = 24 - 5 = 19$ ✓

The inequality holds because $19 > 16$

Check for yourself that when $x = 6$ *the inequality is not true.*

You can do what you like to an equation, so long as you do the same to both sides. You can add or subtract any number, and you can multiply or divide by any number.

Look at the inequality $30 > 20$. *You know that it is valid.*

Show that it is still valid if you add, subtract, multiply by or divide by 5.

Show that it is no longer valid when you multiply or divide by -5.

What happens to the inequality sign?

You solve inequalities just as you solve equations, except that if you multiply or divide both sides by a negative number you must reverse the inequality sign.

Example

Solve $-2x > 8$

Solution

$-2x > 8$

Divide both sides by -2 $\quad x < -4$

Dividing both sides by -2 reverses the inequality sign

Check: when $x = -5$, $-2x = (-2) \times (-5) = 10$ ✓

Another way to do this is to add $2x$ to both sides then subtract 8 from both sides. Try this to check you get the same answer.

Exercise

1 Solve these inequalities.

a) $x + 1 < 7$ b) $2x + 1 \leq 11$

c) $x - 3 \geq 5$ d) $3x - 2 > 10$

e) $x - 4 < -1$ f) $2x + 17 > 29$

g) $5x - 2 \leq 16$ h) $2x + 11 \leq 5$

i) $3 + 4x > 11$ j) $5 - x \geq 3$

k) $7 - 2x \leq 1$ l) $14 \leq 5 - 3x$

2 Find all the possible values of y when

a) $y < 20$ and y is a prime number

b) $20 \leq y \leq 40$ and y is a square number

c) $3 < y < 12$ and y is a factor of 12

d) $12 > y > 3$ and y is a multiple of 3.

3 Last Saturday Grandad went to the races. He placed a bet of £x on the first race and his horse came in first at 5 to 1. Grandad won £$5x$. After that he lost £200 of his winnings.

a) Write an expression for the winnings he had left.

b) This amount was still more than he had bet on the first horse.

Write this as an inequality and solve it for x.

4 Ben looks at the weight card his mother has kept from when he was a baby. He finds that he now weighs 52 kg more than he did when he was born. This is more than 14 times his birth weight.

Write this as an inequality for w, his birth weight, and solve it.

5 Solve these inequalities.

a) $2x + 5 > x + 16$ b) $4x - 3 \geq x + 3$ c) $8 < x + 1$

d) $2(x + 4) < 20$ e) $4 \leq x + 3 \leq 11$ f) $20 \geq 2x > 6$

g) $13 < 2x < 19$ h) $0 \leq x - 2 \leq 18$ i) $5 < 2x + 1 \leq 23$

j) $3(x - 2) > 2x + 5$ k) $10(x - 4) \geq 5(x + 8)$ l) $2 - x > 3 - 2x$

6 Madeleine has inherited some money from her aunt.

She puts it in one of these accounts.

It is earning 5.70% annual interest.

Write an inequality for the sum, £m, that Madeleine inherited.

RAINBOW Building Society SAVINGS RATES

Balance	Annual Interest
£1 – £4,999	5.35%
£5,000 – £9,999	5.70%
£10,000 – £24,999	6.20%
£25,000 – £49,999	6.50%
£50,000 – £99,999	6.70%
£100,000 +	6.80%

Inequalities and graphs

Look at this graph. It shows the **region** for which

$0 \leq x \leq 5$ and $0 \leq y \leq 3$.

All points in the shaded region have x and y values that satisfy these inequalities.

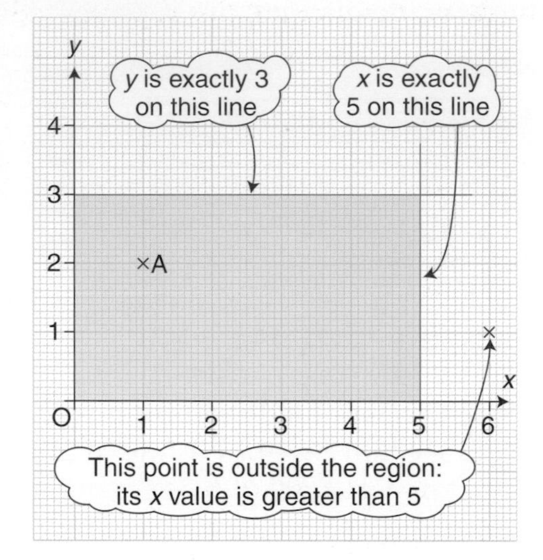

Check that this is true for point A.

This graph shows the region for which

$3 < x \leq 5$ and $y \geq 1$.

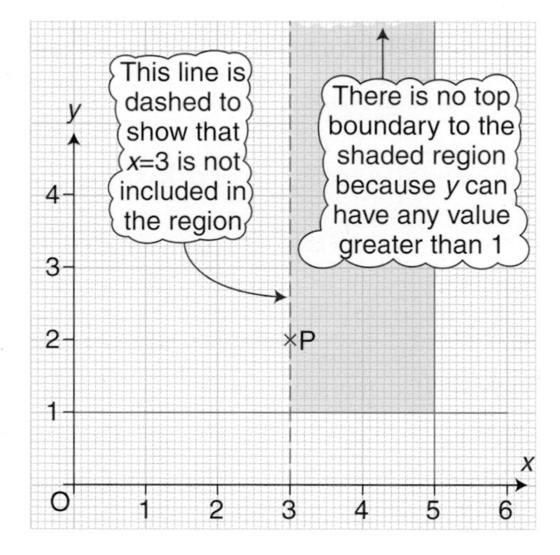

Is point P inside the region? Explain your answer.

Sketch a graph showing the region for which $3 < x \leq 5$ and $y \leq 0$.

So far you have met regions of a graph **bounded** by horizontal and vertical lines, but regions can be bounded by sloping lines.

Look at this graph of

$y = x + 1$

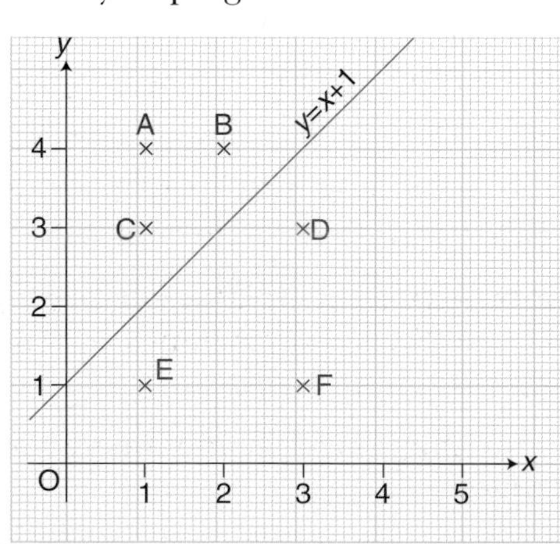

At each red point $y > x + 1$.

Check this for A, B and C.

At each blue point $y < x + 1$.

Check this for D, E and F.

All the points on one side of the line $y = x + 1$ satisfy $y > x + 1$.

All the points on the other side satisfy $y < x + 1$.

By testing a single point you can work out which side of the line satisfies which inequality.

Exercise

1 Write down the inequalities represented by the shaded regions.

a) b) c)

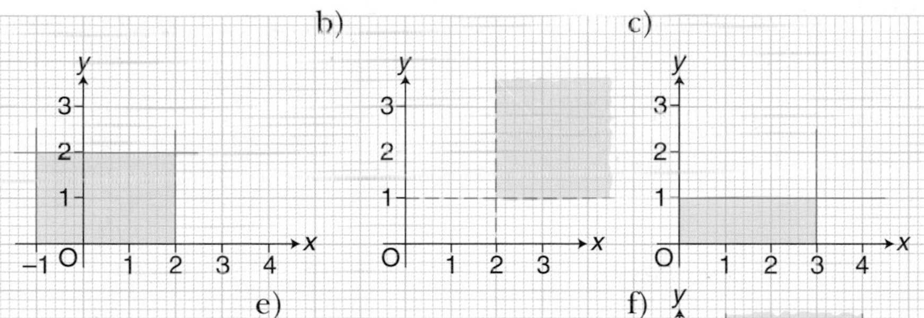

d) e) f)

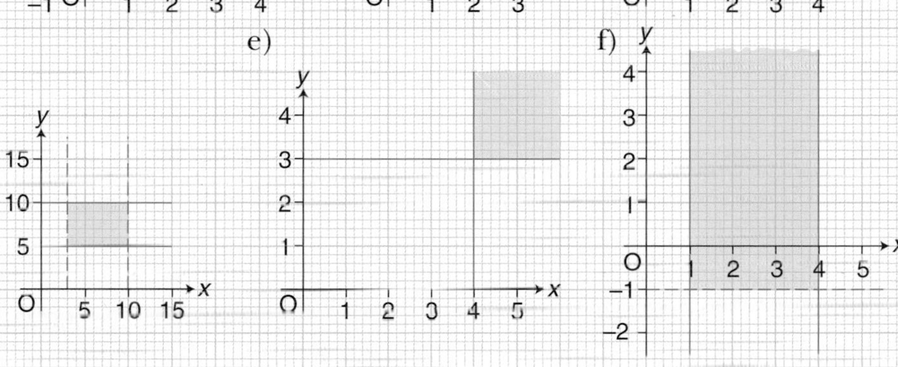

2 Sketch graphs and shade the regions that represent these inequalities.

a) $0 \leq x \leq 4$
$0 \leq y \leq 8$

b) $-1 \leq x \leq 3$
$2 < y < 4$

c) $x \geq 2$
$y > 1$

d) $x < 2$
$0 \leq y < 4$

e) $30 \leq x < 50$
$20 \leq y < 40$

f) $x \geq 1$
$y \leq 5$

3 In a hot air balloon race the winners were in the air for 2 hours and went a distance of 100 miles. They went furthest and were in the air for the longest time.

The shortest distance travelled by any of the balloons was 5 miles and the shortest time in the air was 20 minutes.

a) Write down suitable inequalities for the time taken and the distance travelled by the competitors in the race.

b) Illustrate these inequalities on a graph and shade the region in which all the balloons' times and distances could be plotted.

4 Sketch the graphs and shade the regions that represent these inequalities.

a) $y > x$ b) $x + y < 7$ c) $2x + y < 10$

Solution sets

Jackie has a £12 gift token to spend. She wants to use it to buy mugs and plates like these for her bedsit.

Find a combination of mugs and plates that costs exactly £12.

Find a combination that costs less than £12.

Jackie can buy any combination that costs up to £12.

You can write this as an inequality:

x is the number of mugs. $3x$ is their cost in pounds.

$3x + 2y \leq 12$

y is the number of plates. $2y$ is their cost in pounds.

The numbers x and y must be integers in this problem. Why?

You can show the solutions to this inequality on a graph. You start by drawing the line

$$3x + 2y = 12$$

Then use a test point to decide on which side of the line $3x + 2y \leq 12$.

The blue crosses show all the combinations that Jackie can buy. These points make up the **solution set**.

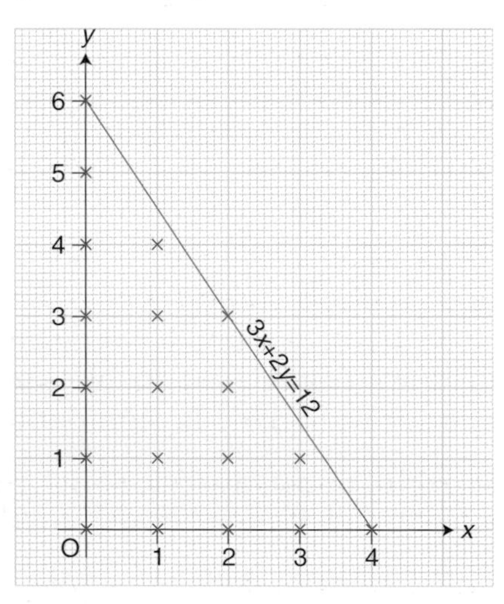

Why are none of the points to the right of the line?

Why is it not correct to shade the whole region to the left of the line in this case?

Points on the x and y axes are included in the solution. Why?

Exercise

1 For each of these, draw a graph and shade the region represented by the inequalities.

a) $x \geq 0,\ y \geq 0,\ x + y < 6$

b) $0 \leq x \leq 4,\ \ 2 \leq y \leq 6,\ \ x \geq y$

c) $x + y \geq 4,\ \ 0 \leq x \leq 8,\ \ 0 \leq y \leq 10$

d) $y < x + 3,\ \ y \geq 0,\ \ 1 < x < 5$

2 Kim is a gymnast. She has to train for at least 10 hours a week.

At the weekends (Saturday and Sunday), there is a 4 hour training session each day in the afternoon. The weekday sessions are 2 hours each evening.

a) Using w for the number of weekday sessions and s for the number of weekend sessions that Kim attends in a week, write an inequality for w and s.

b) Kim cannot attend more than 5 weekday sessions in a week. Write this as an inequality for w.

c) Write a similar inequality for s.

d) Show all the possible combinations of s and w on a graph.

3 Frances has saved £100 and uses it to buy CD-ROMS. She buys x of Type A at £8 each and y of Type B at £11 each.

a) Write down an inequality to show this.

b) She buys more of Type A than Type B. Write this as an inequality.

c) Frances buys at least 2 of Type B. Write this as an inequality.

d) Draw a suitable graph and mark on it the points in the solution set.

e) Can Frances spend exactly £100? Explain your answer.

4 Sean is organising a holiday for members of a social club. Members either book a single room (s) or share a double room (d). So far 13 people have booked a single room and 20 people will be in double rooms. The travel company have available a maximum of 15 double rooms and 28 single rooms.

a) Write an inequality for (i) d and (ii) s, and draw this region on a graph.

The maximum number of holiday makers is 50.

b) Why is $2d + s \leq 50$?

c) Draw $2d + s = 50$ on your graph.

d) Mark all possible combinations of d and s on your graph.

Finishing off

Now that you have finished this chapter you should:

★ understand the symbols $<, >, \leq, \geq$

★ be able to represent an inequality on a number line

★ be able to solve an inequality using algebra

★ be able to represent inequalities for x and y by shading areas or marking sets of points on a graph.

Use the questions in the next exercise to check that you understand everything.

Mixed exercise

1 Show each of these inequalities on a number line.

a) $x < 1$ b) $2 \leq x \leq 4$ c) $x < 0$ or $x > 2$

d) $x > 3$ and $x > 4$ e) $x \geq -1$ or $x \leq -4$ f) $x < -2$ and $x > -3$

2 Write the age ranges in these lonely hearts' advertisements as inequalities. Show each one on a number line.

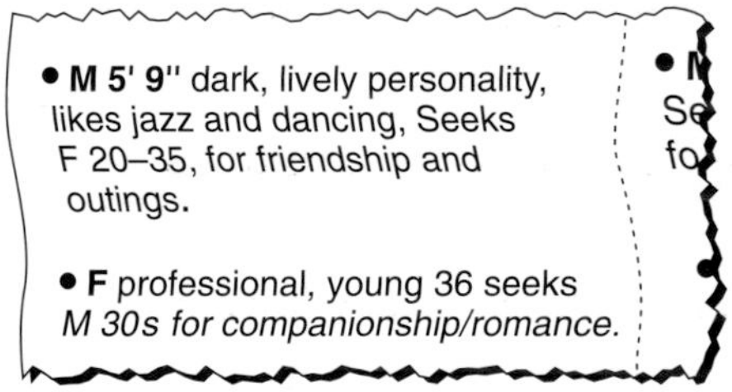
• **M 5' 9"** dark, lively personality, likes jazz and dancing, Seeks F 20–35, for friendship and outings.
• **F** professional, young 36 seeks *M 30s for companionship/romance.*

3 Given $9 < x < 18$ write down a value of y such that

a) y is a square number

b) y is a prime

c) y has a factor of 7

d) y is not an integer.

4 Write down the inequalities shown on these number lines.

a)

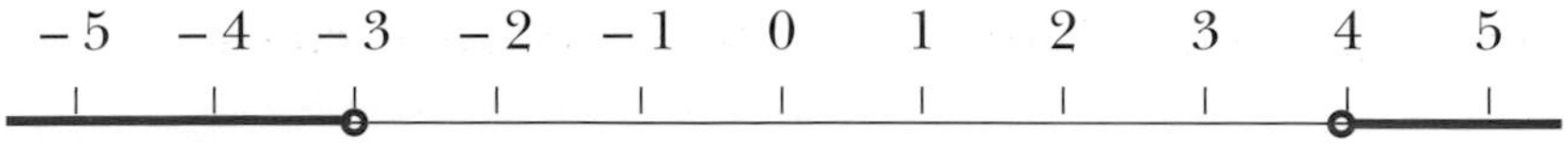
−5
−4
−3
−2
−1
0
1
2
3
4
5

b)

−5
−4
−3
−2
−1
0
1
2
3
4
5

5 Solve these inequalities.

a) $x + 3 \leq 8$ b) $2x - 5 \geq 21$

c) $2x + 1 > x + 9$ d) $5(x + 11) < 80$

e) $2(2x - 3) \leq 3x + 2$ f) $-3 < x + 1$

g) $4(x + 2) - 3(x + 1) > 12$

6 Write down the inequalities represented by the shaded areas.

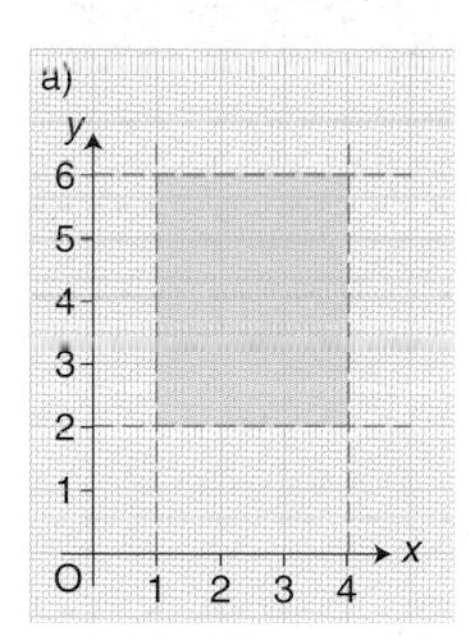

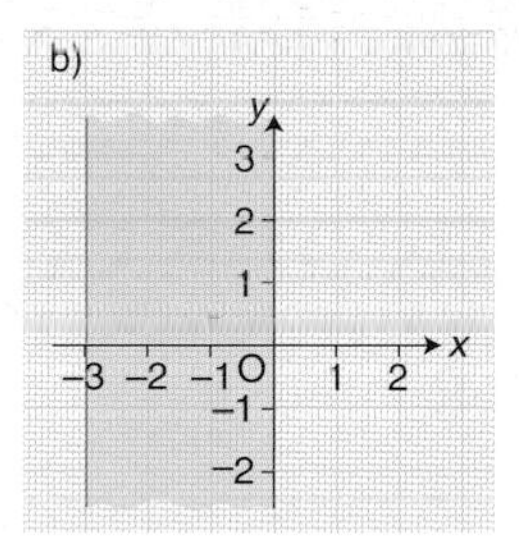

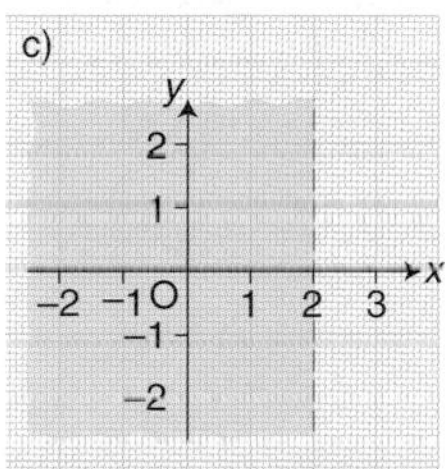

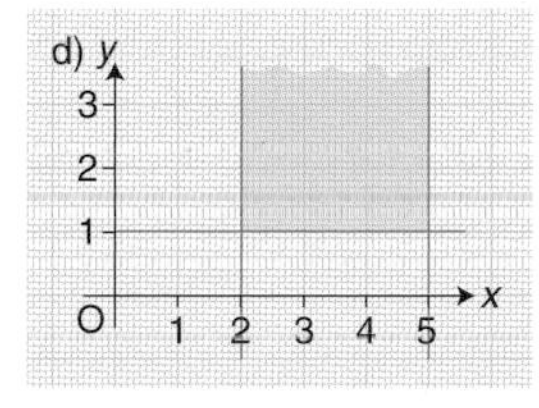

7 Draw suitable graphs and shade the regions that represent the following inequalities.

a) $1 \leq x \leq 6, -1 \leq y \leq 5$

b) $x + y \leq 10, x \geq 0, y \geq 0$

c) $y > x + 2, 0 \leq x \leq 5, y < 9$

d) $y \geq 0, x \geq 1, y \geq 2x$

e) $-2 \leq x \leq 2, -3 \leq x \leq 2, x + y \leq 0$

8 In a pub quiz, a team scores 10 points for a correct answer to a starter question. If they get the starter question right, they are asked a bonus question worth a further 5 points.

a) How many points does a team score when it gets 30 starter questions and 19 bonus questions right?

b) Write down an expression for the score when the team gets x starter questions and y bonus questions right.

c) Last season James's team's highest score was 430 points.

Write down an inequality that is true for each of his team's scores last season.

d) Explain why $x \geq 0$, $y \geq 0$ and $x \geq y$.

e) Draw a graph to illustrate the inequalities in c) and d) and show where you could mark the points in the solution set.

Seventeen

Simultaneous equations

Using simultaneous equations

Lisa and George enter a fishing contest.

Their catches are recorded as shown.

The roach are all of a similar size – they are from the same shoal. The same is true of the perch.

After the contest the fish are put back and they swim away.

Later on, Lisa and George try to work out from their results the weights of a typical roach and perch.

How can they work this out?

One approach is to write and solve a pair of **simultaneous equations**, as shown below.

Using r and p for the weights of a typical roach and perch, you can write

For Lisa's catch	$5r + 2p = 1600$	①
For George's catch	$3r + 2p = 1200$	②
Subtract ② from ①	$2r = 400$	
Divide by 2	$r = 200$	

① and ② are the simultaneous equations

Subtracting **eliminates** the term in p. You now have an equation just in r, to solve

Having worked out r you can now find p by substitution.

Substitute $r = 200$ in ①	$5 \times 200 + 2p = 1600$
	$1000 + 2p = 1600$
Subtract 1000	$2p = 600$
Divide by 2	$p = 300$

Check: substitute for r and p in the left-hand side of ②:

$3 \times 200 + 2 \times 300 = 1200$ ✓

The solution is $r = 200$, $p = 300$: a roach weighs about 200 g and a perch about 300 g.

Why is equation ② used for the check, rather than equation ①?

Exercise

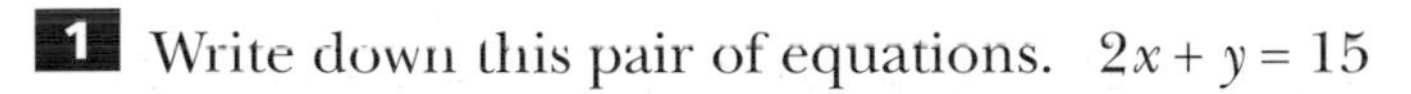

1 Write down this pair of equations. $2x + y = 15$ ①

$x + y = 8$ ②

Follow these steps to solve them.

a) Subtract ② from ① and so find x.

b) Substitute your answer for x in ① to find y.

c) Check that your answers fit equation ②.

2 Solve these pairs of simultaneous equations. (In some of them it is easier to subtract the first equation from the second.)

a) $2x + y = 17$, $x + y = 13$ b) $2x + y = 21$, $x + y = 11$ c) $x + y = 5$, $3x + y = 17$

d) $x + 4y = 10$, $2x + 4y = 12$ e) $2x + 3y = 19$, $x + 3y = 14$ f) $5x + 3y = 23$, $5x + y = 11$

In this case subtracting eliminates x

g) $7p + 5q = 40$, $7p + 2q = 37$ h) $6a + 5b = 160$, $11a + 5b = 210$ i) $4c + 19d = 49$, $11c + 19d = 35$

3 Lauren buys two cans of lemonade and a bag of crisps for 98 pence.

Abigail buys a can of lemonade and a bag of crisps for 63 pence.

Work out the cost of a) a can of lemonade and b) a bag of crisps.

4 Ella sends Sally out to buy 5 packs of white paper and 2 packs of blue paper. The bill comes to £15. The next day Ella sends Sally for 7 packs of white and 2 packs of blue. This time the bill is £19.

What is the price of each type of paper?

5 Solve the following equations. (Remember when you subtract something from itself the answer is always zero. For example, $-y - (-y) = 0$.)

a) $5x - y = 32$, $x - y = 4$ b) $4x - 3y = 11$, $2x - 3y = 5$ c) $7x - 2y = 41$, $3x - 2y = 5$

d) $3x - y = 3$, $x - y = 1$ e) $3x - 4y = 10$, $5x - 4y = 14$ f) $3x + 5y = 25$, $3x - 2y = 11$

When subtracting $5y - (-2y) = 5y + 2y = 7y$

g) $2x - 3y = 6$, $2x - 5y = 2$ h) $4x + 3y = 25$, $4x - y = 13$ i) $x - 2y = 11$, $x + 4y = 29$

6 Ally and Emile are in a quiz team. Ally scores 13 points. She gains points for making 5 correct answers but loses points for giving two wrong ones. Emile makes 9 correct answers and two wrong ones and scores 25 points.

a) How many points are gained for a correct answer?

b) How many points are lost for a wrong answer?

More simultaneous equations

Look at this pair of equations.

$$7p + 3q = 76 \quad ①$$
$$p - 3q = 4 \quad ②$$

What happens when you subtract ② from ①?

In this case, subtracting one equation from the other does not eliminate q. Instead you have to add the equations.

	$7p + 3q = 76$	①
	$p - 3q = 4$	②
Add	$8p = 80$	
Divide by 8	$p = 10$	
Substitute for p in ①	$70 + 3q = 76$	
Subtract 70	$3q = 6$	
Divide by 3	$q = 2$	

q is now eliminated

The solution is $p = 10$, $q = 2$.

Check that this solution fits equation ②.

In many cases, neither adding nor subtracting the equations eliminates an unknown.

For example, look at these.

$$3a + 2b = 19 \quad ①$$
$$5a + 6b = 45 \quad ②$$

Try adding and subtracting to eliminate an unknown.

In this case, you need to spot that multiplying ① by 3 gives you $6b$ in both equations.

① × 3	$9a + 6b = 57$
Copy ②	$5a + 6b = 45$
Subtract	$4a = 12$
Divide by 4	$a = 3$

Remember to multiply both sides by 3

Use this value of a to find b. Check your solution.

In some cases multiplying one equation by a whole number doesn't work.

Look at this example:

$$2x + 7y = 3 \quad ①$$
$$3x + 4y = 11 \quad ②$$

You can multiply ① by 3 and ② by 2 and then subtract. Alternatively you can start by multiplying ① by 4 and ② by 7.

Solve these equations and check that you get $x = 5$ and $y = -1$.

1 Solve these equations.

a) $8x + 3y = 86$
$4x - 3y = 34$

b) $x + y = 21$
$2x - y = 33$

c) $5x + 4y = 60$
$8x - 4y = 44$

d) $14y - 3x = 43$
$2y + 3x = 37$

e) $9a - 5b = 68$
$2a + 5b = 9$

f) $3p + 7q = 45$
$3p - 7q = 24$

2 Neha secretly writes down two numbers. She tells you that three times the first plus the second is 35 and that the first minus the second is 9.

Work out the two numbers that Neha has written down.

3 Solve these equations.

a) $5x + 3y = 23$
$2x - y = 7$

b) $4x + y = 3$
$6x + 5y = 8$

c) $3x - 2y = 20$
$8x - y = 75$

d) $2x + y = 12$
$5x - 4y = 17$

e) $2x - 3y = 7$
$4x - 7y = 13$

f) $2x + 3y = 15$
$4x + 7y = 34$

g) $5x + 2y = 47$
$7x - 8y = 28$

h) $6x - 5y = 50$
$x + 7y = 24$

i) $11x - 4y = 498$
$18x - 12y = 744$

4 Melissa is buying doughnuts for her family. Most of her family prefer jam ones. She has worked out these possible combinations.

9 jam + 1 ring → £1·74
7 jam + 4 ring → £1·74

Work out the price of each type of doughnut.

5 Henry buys two coffees and a tea for £2.95. Later on he buys three coffees and two teas for £4.80. How much is a) coffee and b) tea?

6 Write down this pair of equations.

$$3x + 5y = 29 \quad ①$$
$$5x - 2y = 7 \quad ②$$

Follow these steps to solve them.

a) Multiply ① by 2 and ② by 5.

b) Add your two new equations.

c) Find x and y and check your answers.

7 Solve each pair of equations.

a) $2x + 3y = 24$
$3x - 2y = 23$

b) $3x + 5y = 14$
$5x + 4y = 19$

c) $3x - 2y = 26$
$4x - 5y = 30$

d) $7x - 6y = 41$
$2x + 5y = 5$

e) $2x + 7y = 35$
$5x + 3y = 15$

f) $10x - 4y = 1$
$16x - 3y = 5$

g) $6x - 5y = 19$
$9x - 2y = 67$

h) $4x + 11y = 260$
$15x + 2y = 190$

Other methods of solution

Using graphs

Each Friday Sacha goes to Pastaland for a meal.

Last Friday Sacha took her two children. Each child had minipasta and Sacha had magnapasta. The total cost was £10.

This Friday Sacha and her two children are joined by Jason, Tracey and their two children. The four children each have minipasta while Sacha, Jason and Tracey each have magnapasta. The total cost is £24.

You can work out the cost of each pasta by letting £x be the cost of minipasta and £y be the cost of magnapasta and then forming an equation for each visit.

Last Friday $\quad 2x + y = 10 \quad$ ①

This Friday $\quad 4x + 3y = 24 \quad$ ②

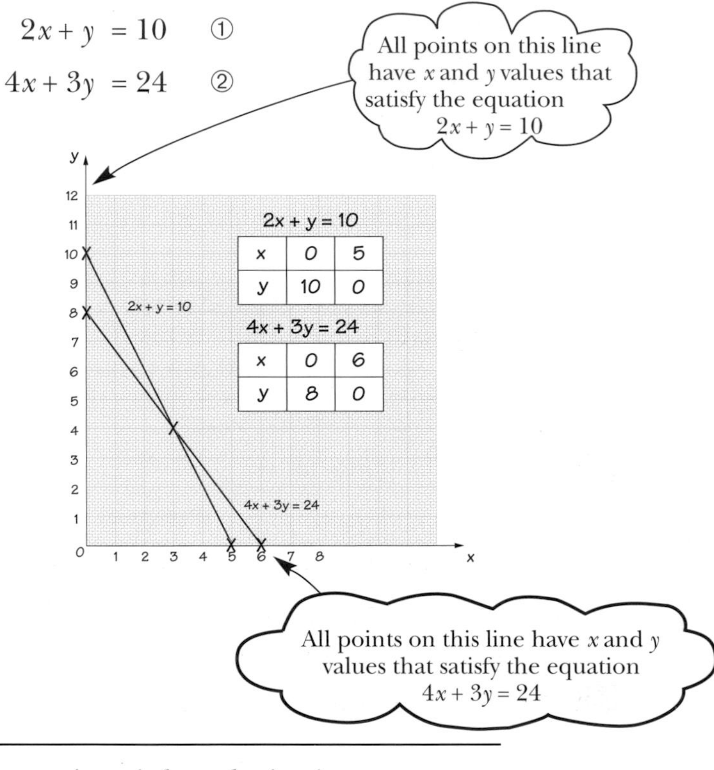

The graph of each of these equations is a straight line. You can see that the lines cross at the point (3, 4).

At this point the x and y values satisfy both equations, so the solution to the problem is $x = 3$, $y = 4$.

In other words the minipasta costs £3 and magnapasta costs £4.

Check that you get the same answer using algebra.

Which method do you find easier?

Using substitution

Another way to solve simultaneous equations is by substitution, as shown below.

$$y = 2x \quad ①$$

$$4x + 3y = 35 \quad ②$$

Substitute for y in ② $\quad 4x + 3 \times 2x = 35$

(Tidy up) $\quad 10x = 35$

Divide by 10 $\quad x = 3.5$

Substitute for x in ① $\quad y = 7$

Substitution is a good method when at least one of the equations is in the form $x = \ldots$ or $y = \ldots$

The solution is $x = 3.5$, $y = 7$.

Exercise

1 Solve each of these pairs of equations using a graph.

For a) to c) draw your x and y axes from 0 to 10.

a) $x + 2y = 16$
$2x + y = 14$

b) $x + y = 16$
$5x + 3y = 60$

c) $3x + 4y = 24$
$x + y = 7$

For d) to f) draw your x axis from 0 to 10 and your y axis from −10 to +10.

d) $x + y = 6$
$2x - y = 9$

e) $5x + y = 25$
$x - 2y = 16$

f) $x + 3y = 7$
$4x - y = 2$

2 Mike collects x 200 g jars and y 400 g jars to qualify for a free mug.

a) Write down an equation connecting x and y.

Draw a graph of this equation using axes for x from 0 to 12, and y from 0 to 10.

b) Mike has bought 8 jars of coffee altogether.

Write down a second equation connecting x and y.

Draw the graph of this equation on the same axes.

c) Use your graphs to find how many jars of each size Mike has bought.

COLLECT
12 VOUCHERS
and get this mug
FREE !
Aroma
Each 200 g jar of
Aroma coffee carries
1 VOUCHER
Each 400 g jar of
Aroma coffee carries
2 VOUCHERS

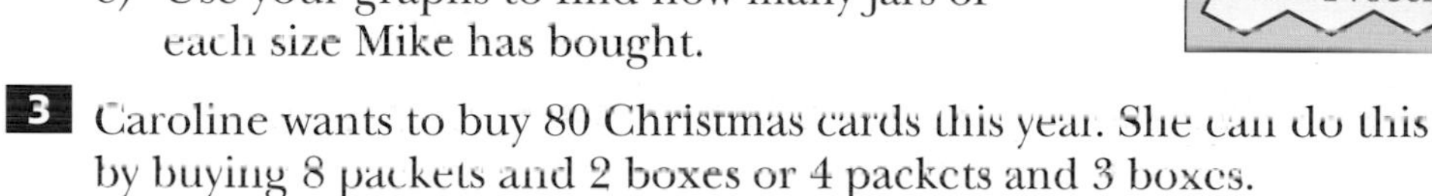

3 Caroline wants to buy 80 Christmas cards this year. She can do this by buying 8 packets and 2 boxes or 4 packets and 3 boxes.

Write down two equations to show this information.

Solve them by a graphical method to find how many cards are in each packet and each box.

4 Solve each of these pairs of equations by substitution.

a) $y = 2x + 1$
$3x + 2y = 23$

b) $3x + 2y = 16$
$y = 4x - 3$

c) $y = x + 2$
$y = 8 - 3x$

d) $2y + 3x = 7$
$y = 4 - x$

e) $y = 2x - 3$
$x - y = 2$

f) $y = 7 - x$
$2x + 3y = 15$

Investigation

What happens when you try to solve the following pairs of equations by an algebraic method?

a) $4x - 2y = 6$
$10x - 5y = 15$

b) $3x + y = 6$
$3x + y = 12$

Draw graphs for each pair of lines and use them to help you to explain your results.

Finishing off

Now that you have finished this chapter you should be able to:

★ solve simultaneous equations using algebra

★ solve simultaneous equations using graphs.

Use the questions in the next exercise to check that you understand everything.

Mixed exercise

1 Solve these equations. In some cases you can add or subtract the equations straight away, but in others you need to multiply first.

a) $x + y = 13$
$x - y = 3$

b) $x + 5y = 3$
$x - 4y = 12$

c) $7p + 4q = 19$
$2p + q = 5$

d) $5a - 2b = 7$
$8a - 4b = 10$

e) $4x - y = 300$
$3x - 2y = 0$

f) $2x + 3y = 16$
$3x - 2y = 11$

g) $3s + 2t = 26$
$2s - 3t = 26$

h) $2c + 7d = 19$
$5c - 2d = 28$

i) $2x + 11y = 52$
$4x - 11y = 5$

j) $3x + 20y = 25$
$4x + 10y = 25$

k) $4x - 3y = 50$
$5x - 2y = 80$

l) $2x + 5y = 72$
$11x - 2y = 101$

m) $2x - 3y = 34$
$11x - y = 1$

n) $5x + 7y = 6$
$7x - 3y = 2$

o) $9x + 10y = 57$
$3x + 2y = 15$

p) $4x - y = 36$
$8x + 2y = 88$

q) $3x + 5y = 22$
$2x + 3y = 14$

r) $x + y = 1$
$9x + 10y = 11$

s) $22x + 3y = 116$
$2x - 5y = 0$

t) $3x - 8y = -21$
$7x - 5y = -8$

u) $7p + 2q = 8$
$2p - 3q = 13$

2 The sum of two numbers is 21 and their difference is 5.
Write down two equations to represent this information, and solve them to find the two numbers. (The sum of x and y is $x + y$ and their difference is $x - y$.)

3 Use the information below to form two equations. Solve them to find the cost of 1 orange and the cost of 1 apple.

4 The Scotts and the Masons buy tickets for a pantomime.

The Scotts buy tickets for 2 adults and 3 children at a cost of £36.

The Masons buy tickets for 4 adults and 2 children at a cost of £48.

Work out the cost of an adult's ticket and a child's ticket.

5 David and Sian work part-time at Pizza Palace. After 11 p.m. they are paid at a higher rate.

Last week David was paid £149 for 34 hours' work at basic rate and 5 hours at the higher rate.

Sian was paid £229 for 50 hours at the basic rate and 9 hours at the higher rate.

What is the basic hourly rate and the rate after 11 p.m.?

6 Jess is a florist. She is preparing the flowers for a wedding.

For the bride's bouquet Jess uses 10 roses and 6 carnations. For each bridesmaid's posy she uses 4 roses and 5 carnations.

The flowers cost her £12 for the bouquet £6.10 for each posy.

What is the cost of

a) a rose b) a carnation

7 A company makes two types of medical instrument.

Instrument A takes 4 machine-hours and 7 operator-hours to make.

Instrument B takes 9 machine-hours and 8 operator-hours.

Find how many of each type of instrument the company makes in a week when it devotes 550 machine-hours and 730 operator-hours to making them.

8 Solve these pairs of equations using graphs.

a) $x + y = 18$
$4x + 3y = 60$

b) $2x + 5y = 40$
$2x - y = 16$

c) $x - y = 4$
$6x + 7y = 63$

9 Solve each of these pairs of equations by substitution.

a) $y = 3x + 2$
$3x + 2y = 13$

b) $y = 5x + 1$
$y = 7x - 2$

c) $2x - 3y = 9$
$y = 2x - 3$

Eighteen

Quadratics

Factorising quadratic expressions

Here are some examples of **quadratic** expressions in x.

$5x^2 - 12x - 8$ $\quad$ $2 - x + 3x^2$ $\quad$ $x^2 + 5x + 6$ $\quad$ $x^2 - 9$

You can see that the highest power of x is x^2.

In this chapter you learn how to solve quadratic equations. The first step in this is to **factorise** a quadratic expression. This is the opposite of multiplying two brackets together (or expanding them).

You can expand $(x + 2)\ (x + 3)$ as follows:

$$(x + 2)\ (x + 3)$$

$$= x^2 + 3x + 2x + 6$$

$$= x^2 + 5x + 6$$

Starting with $x^2 + 5x + 6$ how can you get back to $(x + 2)\ (x + 3)$?

Look at how $x^2 + 5x + 6$ was obtained from expanding $(x + 2)\ (x + 3)$

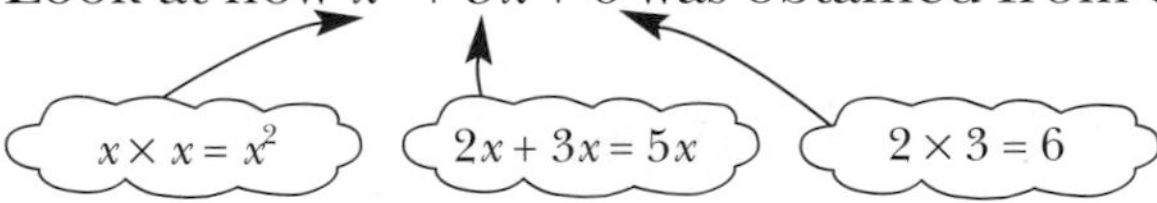

Since $x^2 = x \times x$ you begin factorising by writing

$$x^2 + 5x + 6 = (x \qquad)(x \qquad)$$

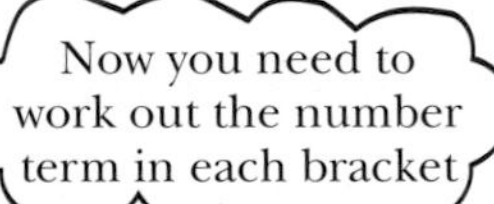

Look for two numbers that multiply to give 6.

They could be 1×6 or 2×3.

Your two numbers must also add up to 5.

They are 2 and 3.

$$x^2 + 5x + 6 = (x + 2)(x + 3)$$

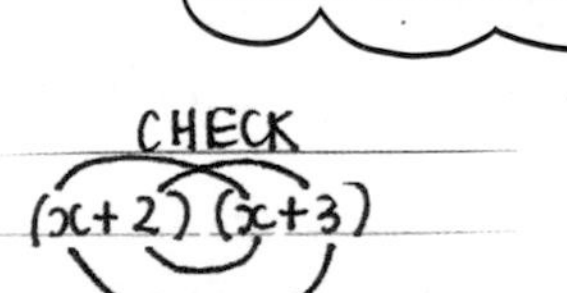

Example

Factorise $x^2 - 5x + 6$.

Solution

In this case the numbers are -2 and -3.

$$x^2 - 5x + 6 = (x - 2)(x - 3)$$

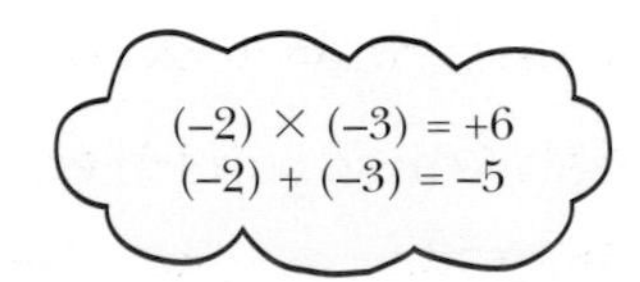

Check this is correct by multiplying out the brackets.

Exercise

1 a) Find two numbers with product 5 and sum 6.

b) Use these numbers to help you factorise $x^2 + 6x + 5$.

c) Check your answer by expanding your brackets.

2 a) Find two numbers with product 15 and sum 8.

b) Use these numbers to help you factorise $y^2 + 8y + 15$.

c) Check your answers by expanding your brackets.

3 Factorise these and check your answers.

a) $x^2 + 9x + 14$ b) $x^2 + 15x + 14$ c) $a^2 + 7a + 10$

d) $a^2 + 11a + 10$ e) $n^2 + 5n + 4$ f) $p^2 + 6p + 8$

g) $x^2 + 4x + 4$ h) $x^2 + 9x + 18$ i) $y^2 + 10y + 24$

j) $x^2 + 12x + 35$ k) $x^2 + 17x + 60$ l) $y^2 + 16y + 28$

4 a) Find two negative numbers with product +5 and sum –6.

b) Use these to help you factorise $x^2 - 6x + 5$.

c) Check your answer by expanding your brackets.

5 a) Find two numbers with product +15 and sum –8.

b) Use these numbers to help you factorise $y^2 - 8y + 15$.

c) Check your answer by expanding your brackets.

6 Factorise these and check your answers.

a) $x^2 - 3x + 2$ b) $x^2 - 4x + 3$ c) $x^2 - 6x + 5$

d) $x^2 - 12x + 11$ e) $x^2 - 9x + 14$ f) $x^2 - 6x + 8$

g) $p^2 - 9p + 18$ h) $a^2 - 15a + 44$ i) $x^2 - 45x + 44$

j) $t^2 - 12t + 20$ k) $x^2 - 10x + 25$ l) $y^2 - 13y + 36$

7 Factorise these and check your answers.

a) $x^2 - 8x + 15$ b) $a^2 + 13a + 22$ c) $x^2 - 10x + 21$

d) $p^2 - 10p + 16$ e) $t^2 - 8t + 12$ f) $y^2 + 7y + 12$

g) $a^2 + 2a + 1$ h) $x^2 - 6x + 9$ i) $x^2 - 25x + 100$

More quadratic factorisation

In quadratics you have factorised so far the number terms have been positive (+). When you factorised these you got two brackets with the same signs inside, such as $(x-3)(x-2)$ or $(x+5)(x+2)$.

In the examples on this page the number term is negative (–), and you have to be a bit more careful with the signs.

Look at $x^2 - x - 20$.

Note: $-x$ means $-1x$

You need to find two numbers that multiply to give –20 and add to give –1.

To multiply to get 20, you can have 20 and 1, 10 and 2 or 5 and 4.

Since it is –20, one of the numbers must be positive and one negative.

After trying the various pairs, you can see that the only one that works is –5 and +4.

The answer is $x^2 - x - 20 = (x-5)(x+4)$.

Check this by multiplying out the brackets.

Copy and complete this table. Look carefully at the pattern of signs in your table.

What signs should you expect in the brackets when you factorise $x^2 - x - 12$?

Brackets	Quadratic
$(x+3)(x+5)$	$x^2 + 8x + 15$
$(x-3)(x-5)$	
$(x-3)(x+5)$	
$(x+3)(x-5)$	

Some quadratic expressions are of the form $x^2 - a^2$ (the difference of two squares). For example, one way of factorising $x^2 - 9$ is to write it as $x^2 + 0x - 9$ and proceed as before.

$$x^2 - 9 = (x+3)\ (x-3)$$

Factorise $x^2 - 4$

These factorisations both follow the rule

$$x^2 - a^2 = (x+a)\ (x-a)$$

When you need to factorise a quadratic expression which has a common factor you take out the common factor first. Here are two examples.

(i) $x^2 + 4x$ has a common factor of x so

$$x^2 + 4x = x\ (x+4)$$

(ii) $2x^2 + 14x + 20$ has a common factor of 2 so

$$2x^2 + 14x + 20 = 2\ (x^2 + 7x + 10)$$
$$= 2(x+5)\ (x+2)$$

Exercise

1 a) Think of two numbers with product 6 and difference 5.

b) Use these to help you factorise $x^2 + 5x - 6$.

c) Factorise $x^2 - 5x - 6$.

Check your answers by expanding the brackets.

2 a) Think of two numbers with product 15 and difference 2.

b) Use these to help you factorise $y^2 + 2y - 15$.

c) Factorise $y^2 - 2y - 15$.

Check your answers by expanding the brackets.

3 Factorise these and check your answers.

a) $x^2 + 10x - 11$ b) $x^2 - 10x - 11$ c) $x^2 + 6x - 7$

d) $x^2 - 6x - 7$ e) $x^2 + 4x - 5$ f) $x^2 - 4x - 5$

g) $x^2 + 5x - 14$ h) $x^2 - 5x - 14$ i) $x^2 + 3x - 88$

4 Factorise these and check your answers.

a) $a^2 + 7a - 18$ b) $a^2 - 7a - 18$ c) $y^2 + 9y - 10$

d) $y^2 - 3y - 10$ e) $p^2 - 3p - 18$ f) $x^2 - x - 12$

g) $x^2 + x - 20$ h) $a^2 + 8a - 20$ i) $t^2 + 4t - 12$

5 Factorise these and check your answers,

a) $x^2 - 16$ b) $x^2 - 8x$ c) $y^2 - 49$

d) $t^2 - 4t$ e) $x^2 - 1$ f) $p^2 + p$

6 Here is a mixture of types to factorise. Check your answers carefully.

a) $x^2 + 7x + 6$ b) $x^2 + 6x + 8$ c) $r^2 - 5r + 4$

d) $x^2 - 8x - 9$ e) $y^2 + 3y - 4$ f) $x^2 + x - 12$

g) $t^2 - t - 12$ h) $x^2 - 11x + 18$ i) $p^2 + 4p - 12$

j) $y^2 - 81$ k) $b^2 - b - 20$ l) $a^2 - 11a$

7 Factorise these completely.

a) $4x^2 + 4x - 48$ b) $3a^2 - 9a + 6$ c) $3x^2 - 12$

d) $3x^2 + 6x + 3$ e) $10x^2 - 1000$ f) $5x^2 + 10x - 400$

Quadratic equations

Look at the equation

$$x^2 - 2x - 15 = 0$$

This is a **quadratic equation**.

Why is it called a quadratic equation?

To solve a quadratic equation, start by factorising the left-hand side.

$$x^2 - 2x - 15 = 0$$

$$(x - 5)(x + 3) = 0$$

This gives you two factors $(x - 5)$ and $(x + 3)$ that multiply to give 0. One of them must be 0.

Either $x - 5 = 0$ and so $x = 5$

or $x + 3 = 0$ and so $x = -3$.

The solution of the equation is $x = 5$ or -3.

Example

Solve $x^2 + 3 = 4x$

Solution

Before factorising you need to collect all the terms on one side of the equation.

$$x^2 + 3 = 4x$$

Subtract $4x$ $\quad x^2 - 4x + 3 = 0$

Factorise $\quad (x - 1)(x - 3) = 0$

Either $x - 1 = 0$ so $x = 1$

or $x - 3 = 0$ so $x = 3$

The solution is $x = 1$ or 3

You can get the same result by drawing a graph.

This is the graph of $y = x^2 - 4x + 3$

The points where the curve crosses the x axis (the line $y = 0$) give the solution of $x^2 - 4x + 3 = 0$.

You can see that $x = 1$ or $x = 3$.

Sometimes you will meet quadratic equations which cannot be factorised.

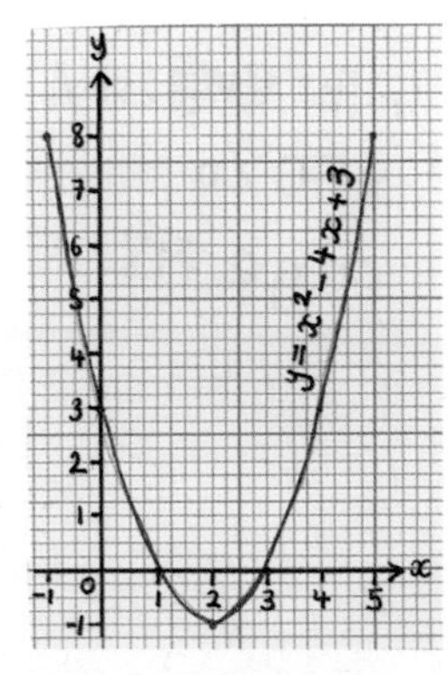

In such cases, you can still try to solve the equation by drawing a graph.

However, as you will find out from the Investigation on the opposite page, some quadratic equations have no solution.

1 Solve these equations.

a) $(x-2)(x-6)=0$ b) $(x-4)(x+3)=0$ c) $(x+7)(x-2)=0$

d) $(y+8)(y+5)=0$ e) $y(y-3)=0$ f) $(t-1)(t+6)=0$

2 Solve these equations.

a) $x^2-9x+14=0$ b) $a^2-7a+10=0$ c) $x^2+5x-14=0$

d) $n^2+5n+4=0$ e) $a^2+a-6=0$ f) $t^2-4t-12=0$

g) $x^2+5x-50=0$ h) $x^2-2x-63=0$ i) $x^2-12x+11=0$

j) $y^2+14y+24=0$ k) $a^2-a-30=0$ l) $t^2-6t+9=0$

m) $x^2-16=0$ n) $x^2-5x=0$ o) $n^2-7n-18=0$

3 Solve these equations.

a) $x^2+x=20$ b) $a^2+8a=20$ c) $r^2+4=5r$

d) $y^2-3y=4$ e) $x^2+18=-11x$ f) $x^2=x+12$

4 A curve has equation $y=x^2-5x+5$.

a) Copy and complete the following table.

x	0	1	2	3	4	5
x^2	0			9		
$-5x$	0			−15		
+5	+5	+5	+5	+5		
y	5			−1		

b) Draw the graph. Use a scale of 2 cm to 1 unit on each axis.

c) Use your graph to find an approximate solution of the equation $x^2-5x+5=0$.

5 a) Draw the graph of $y=1+7x-2x^2$ from $x=-1$ to $x=4$. Use a scale of 2 cm to 1 unit on the x axis and 1 cm to 1 unit on the y axis.

b) Use your graph to find an approximate solution of the equation $1+7x-2x^2=0$.

Investigation

Draw the graphs of

a) $y=x^2-6x+8$ b) $y=x^2-6x+9$ c) $y=x^2-6x+10$

Use them to solve the equations

$x^2-6x+8=0$, $x^2-6x+9=0$ and $x^2-6x+10=0$.

The equation $x^2-6x+k=0$ has no solution. What can you say about k?

Rational functions

Steve and Ray are doing their algebra homework.

Here is one of the questions.

Simplify $\dfrac{x^2 + 5x - 6}{x + 6}$

An algebra fraction like this is called a **rational function**

This is what Steve wrote.

Ray did it like this.

You must *never* cancel bits of an expression as Steve did. If you had $\frac{16}{68}$ you wouldn't cancel the 6s. You would look for a common factor.

You would write

$$\frac{16}{68} = \frac{4 \times 4}{4 \times 17} = \frac{4}{17}$$

You can cancel these 4s.

Why can't you cancel $\dfrac{x-1}{x+1}$?

Try several even-number values for x and check what happens.

Can you simplify $\dfrac{(x-1)(x+1)(x+2)}{(x+2)(x-3)(x+1)}$?

Exercise

1 Simplify these.

a) $\frac{x(x+1)}{(x+1)}$ b) $\frac{(x+1)x}{(x+1)}$ c) $\frac{x^2(x-2)}{(x-2)}$

d) $\frac{(x-1)x^3}{x(x-1)}$ e) $\frac{(x+4)x^4}{x^3(x+4)}$ f) $\frac{(2x-1)x^3}{x^4(2x-1)}$

2 a) Factorise (i) $3x^2 + x$ (ii) $4x^3 + x^2$ (iii) $5x^2 + x^3$

b) Use part a) to simplify (i) $\frac{3x^2 + x}{x}$ (ii) $\frac{4x^3 + x^2}{x^2}$ (iii) $\frac{5x^2 + x^3}{x^2}$

3 Simplify these.

a) $\frac{(x+1)(x-5)}{(x+1)}$ b) $\frac{(x+1)(x-2)}{(x+1)}$ c) $\frac{(x+3)(x-2)}{(x-2)}$

d) $\frac{(x-1)(x+1)}{(x-1)}$ e) $\frac{(x+4)(x-2)}{(x+4)}$ f) $\frac{(2x-1)(x-2)}{(2x-1)}$

4 Simplify these.

a) $\frac{(x+1)(x-5)}{(x+1)(x+5)}$ b) $\frac{(x+1)(x-2)}{(x+2)(x+1)}$ c) $\frac{(x+3)(x-2)}{(x-2)(x-3)}$

5 a) Factorise (i) $x^2 + 3x$ (ii) $x^2 - 1$ (iii) $x^2 - 9x$ (iv) $3x^2 + 2x$ (v) $x^3 + x^2$

b) Use part a) to simplify (i) $\frac{x^2 + 3x}{x+3}$ (ii) $\frac{x^2 - 1}{x+1}$

(iii) $\frac{x^2 - 9x}{x-9}$ (iv) $\frac{x^2}{3x^2 + 2x}$ (v) $\frac{x^3}{x^3 + x^2}$

6 a) Factorise (i) $x^2 + 7x + 10$ (ii) $x^2 - 5x + 6$ (iii) $x^2 + 2x - 3$

b) Use part a) to simplify

(i) $\frac{x^2 + 7x + 10}{x+2}$ (ii) $\frac{x^2 - 5x + 6}{x-3}$ (iii) $\frac{x^2 + 2x - 3}{x-1}$

Try simplifying these. You need to factorise the top and the bottom of the fraction.

a) $\frac{x^2 + 7x + 10}{x^2 - 25}$ b) $\frac{x^2 - x + 2}{x^2 - 4x + 4}$ c) $\frac{x^2 - 2x - 3}{x^2 - 1}$

d) $\frac{x^2 + x - 2}{x^2 + 3x + 2}$ e) $\frac{x^2 - 8x + 4}{x^2 + 3x - 10}$ f) $\frac{x^2 + 4x + 3}{x^2 - x - 2}$

Finishing off

Now that you have finished this chapter you should be able to:

- ★ factorise quadratic expressions
- ★ simplify rational expressions
- ★ solve quadratic equations

Use the questions in the next exercise to check that you understand everything.

Mixed exercise

1 Factorise each of these and check your answers.

a) $x^2 + 13x + 12$ b) $a^2 - 3a + 2$
c) $z^2 - 6z + 8$ d) $n^2 + 11n - 26$
e) $t^2 - 4t - 12$ f) $x^2 - 7x - 30$

2 Factorise each of these and check your answers.

a) $x^2 + 2x + 1$ b) $n^2 - 10n + 25$
c) $r^2 + 8r + 16$ d) $y^2 - 12y + 36$
e) $x^2 - 16$ f) $p^2 - 49$

3 Solve each of these equations and check your answers.

a) $2x = 0$ b) $3(x + 2) = 0$ c) $x(x - 2) = 0$
d) $a(a + 15) = 0$ e) $(t - 5)(t + 6) = 0$ f) $(a + 7)(a + 11) = 0$
g) $(x - 3)(x + 3) = 0$ h) $(2n - 3)(3n - 9) = 0$ i) $2(y - 5)(3y - 2) = 0$

4 Solve each of these equations and check your answers.

a) $x^2 - 5x + 6 = 0$ b) $b^2 + 7b + 12 = 0$ c) $x^2 + 18x + 81 = 0$
d) $x^2 + 3x - 18 = 0$ e) $d^2 - 4d - 21 = 0$ f) $x^2 - 10x - 24 = 0$
g) $x^2 + 2x = 0$ h) $n^2 - 4 = 0$ i) $3x^2 - 12 = 0$

5 Solve each of these equations and check your answers.

a) $x^2 - x = 56$ b) $x^2 + x = 90$ c) $y = y^2 - 12$
d) $t^2 + 6 = 5t$ e) $n^2 = 14 - 5n$ f) $x^2 = 3(6 - x)$
g) $x^2 = 20 + x$ h) $x^2 = 5x$ i) $2x^2 = 20x - 42$

6 Simplify the following.

a) $\dfrac{x(x + 6)}{(x + 6)}$ b) $\dfrac{(x + 2)(x - 7)}{(x + 2)}$ c) $\dfrac{10 + 5x}{2 + x}$

d) $\dfrac{3x^2 - 4x}{x}$ e) $\dfrac{x^2 + 7x + 12}{x + 3}$ f) $\dfrac{3x - 6}{x^2 - 4}$

Mixed exercise

7 Explain the difference between a quadratic expression and a quadratic equation.

8 a) Make out a table of values and draw the graph of $y = x^2 - x$ for values of x from -4 to 4.

b) Draw the line $y = 6$ on your graph and write down the values of x where the line meets the curve.

c) Solve the equation $x^2 - x = 6$ by factorisation.

d) Explain why your answers to b) and c) should be the same.

e) Use your graph to solve the equation $x^2 - x = 8$. Is it possible to solve this by factorising?

9 A rectangular garden is x metres by $x + 2$ metres. It has paths and flower beds one metre wide round the edge with a lawn in the middle.

a) Draw a diagram of the garden, showing all the measurements clearly.

b) Write down the dimensions of the lawn in terms of x and find an expression for its area.

c) The lawn requires 120 m^2 of turf. Write down an equation for x.

d) Solve your equation to find x. What are the dimensions of the garden?

10 For each of the following situations

(i) form an equation in the unknown quantity given

(ii) solve the equation

(iii) check your answer.

a) The length of a rectangular lawn is 3 m greater than its width, w m. Its area is 54 m^2.

b) The width of a box is 10 cm less than its length, l cm. Its height is 8 cm and its volume is 3000 cm^3.

c) A box has a square base of side x cm and height 2 cm. Its outside surface area (including its top and bottom) is 90 cm^2.

Design an exhibition space for showing paintings using 16 m of screens. Describe what you intend to exhibit and your reasons for using the screens as you have.

Draw a plan of your space showing any other furniture you might need.

Nineteen

Angles and shapes

Angles

Diana makes this scale drawing showing the slope of the roof of her solar-powered house.

She describes the face AC as sloping at an angle of 22 degrees (written 22°) to AB.

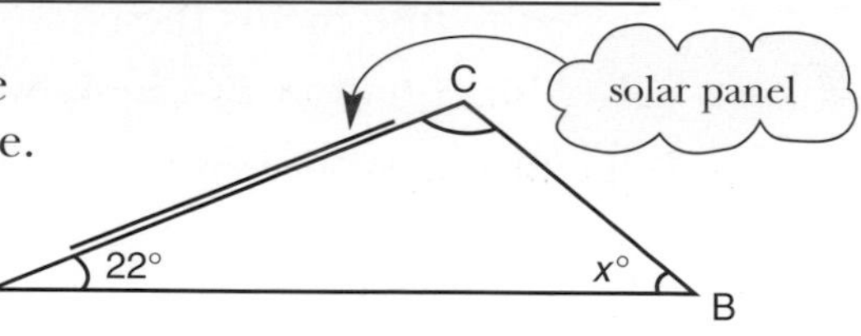

Angles are measured in **degrees** using a **protractor** or **angle measurer**.

By putting the centre of the protractor at A check that angle A is 22°.

∠A = 22°

By putting the centre of the protractor at B find the value of x.

Note: ∠B can be written more fully as ∠ABC or ∠CBA.

x = ∠B

Angles A and B on the diagram above are both **acute**. They are less than 90°.

An angle of 90° (a quarter turn) is called a **right angle**.

The lines are **at right angles** or **perpendicular**.

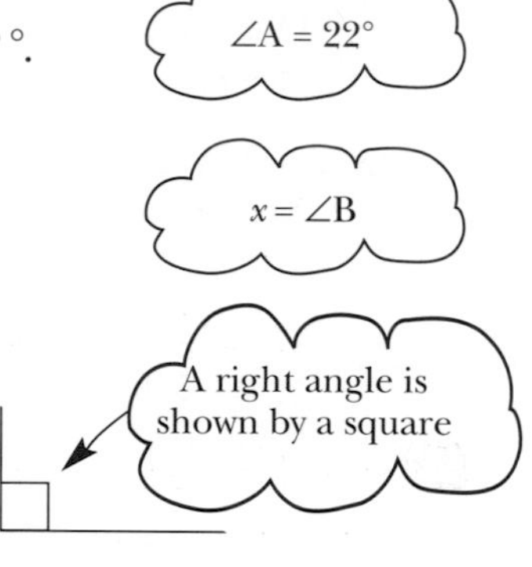

Angle C on the diagram above is **obtuse**. It is between 90° and 180°.

An angle between 180° and 360° is called a **reflex angle**.

Angles about a point or on a line

A whole turn is 360° so a half turn is 180°. This gives two important results.

Angles that fit round a point add up to 360°.

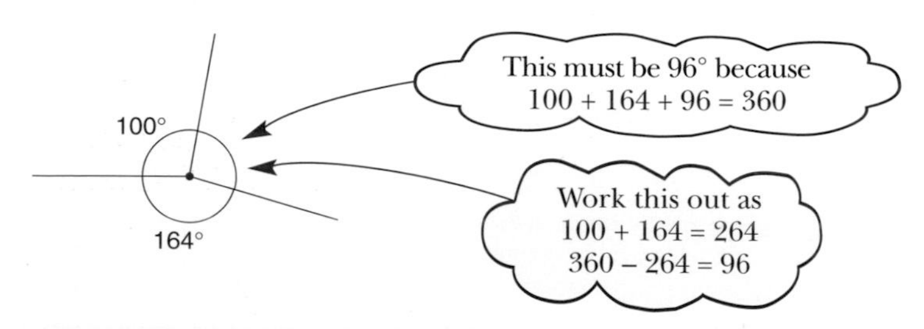

Angles that fit on a straight line add up to 180°.

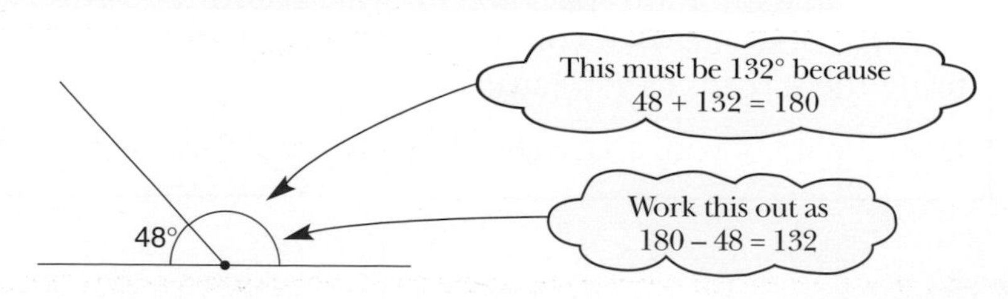

Exercise

1 Measure these angles as accurately as you can

a)

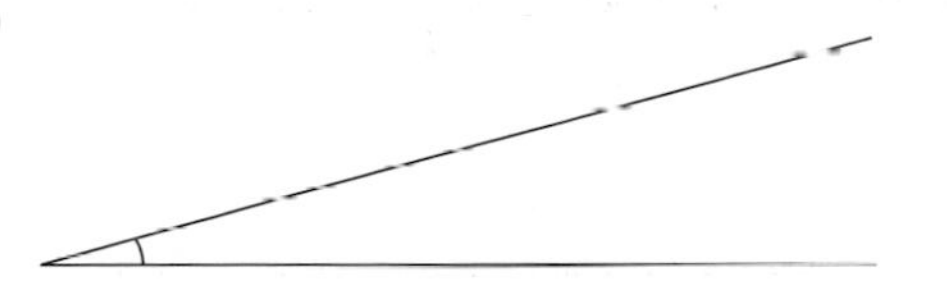

b)

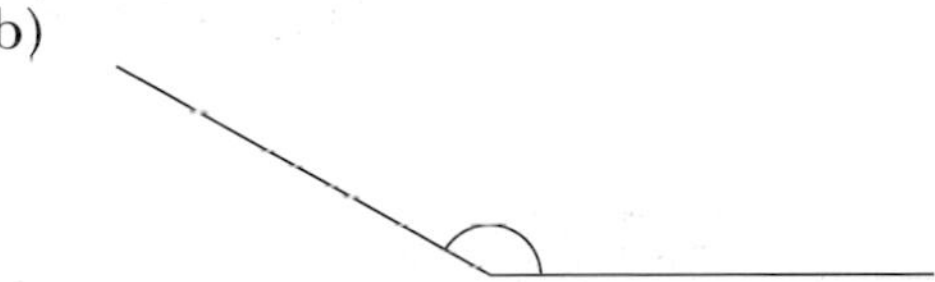

2 Work out the number of degrees in

a) 2 revolutions

b) $1\frac{1}{2}$ revolutions

c) 1.2 revolutions

3 What fraction of a complete turn is

a) 180°?

b) 90°?

c) 270°?

4 Label each of these angles as acute, obtuse or reflex.

a)

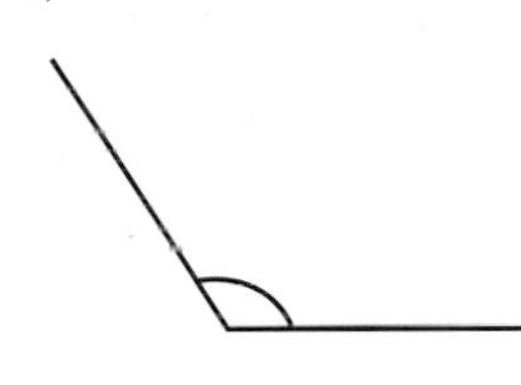

b)

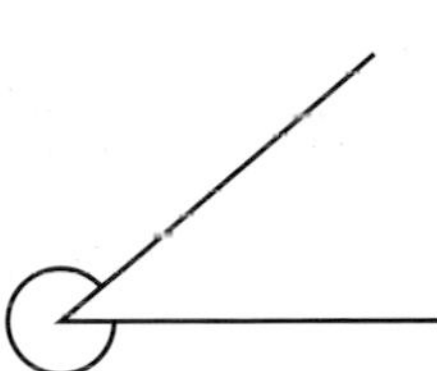

c)

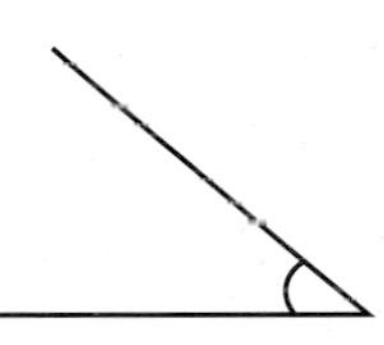

d)

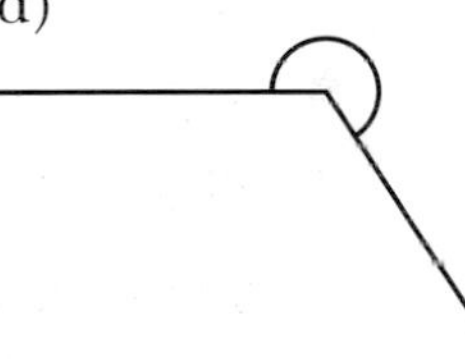

5 Find the angles marked with letters in these diagrams.
Give reasons for your answers.

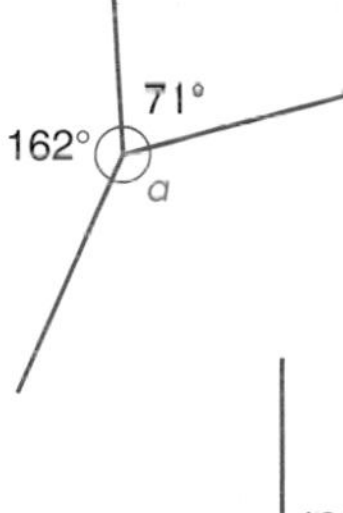

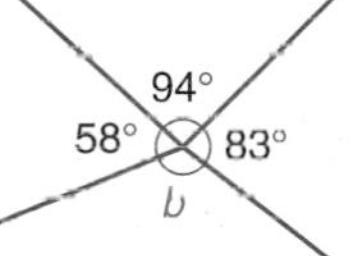

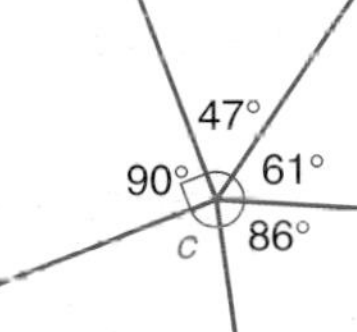

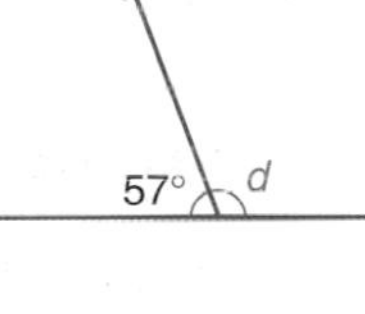

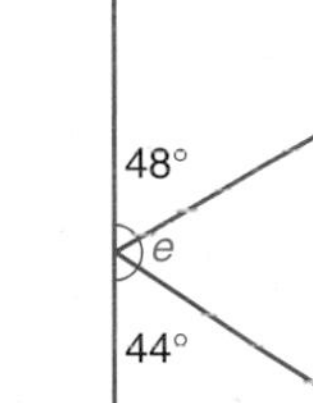

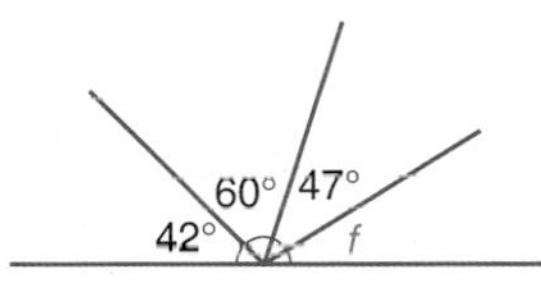

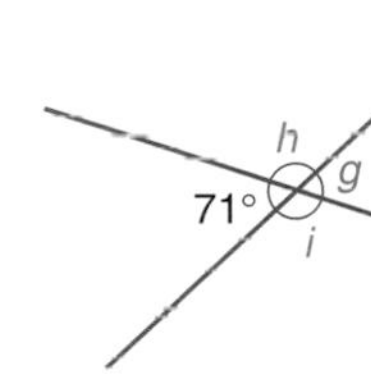

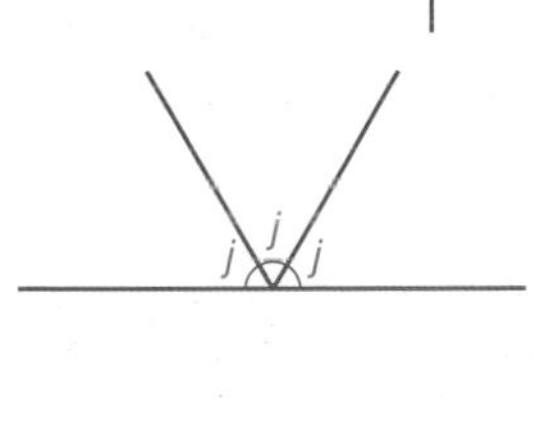

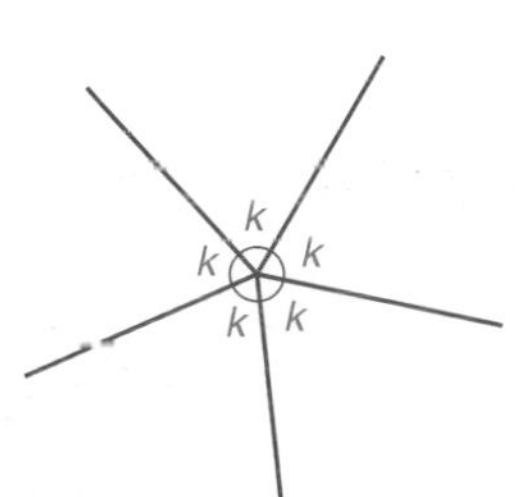

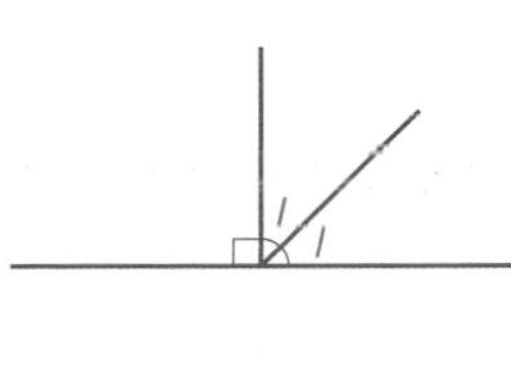

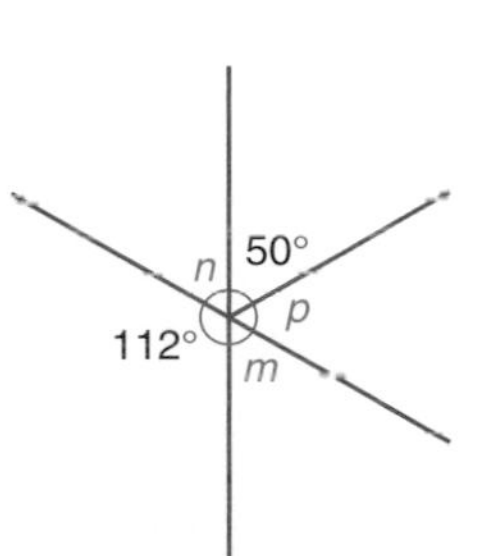

Parallel lines

Parallel lines are lines that go in the same direction and never meet.

A trellis like this is often used in gardens.

It is made of two sets of parallel lines.

Here is a larger diagram of part of the trellis with some of the angles marked with letters.

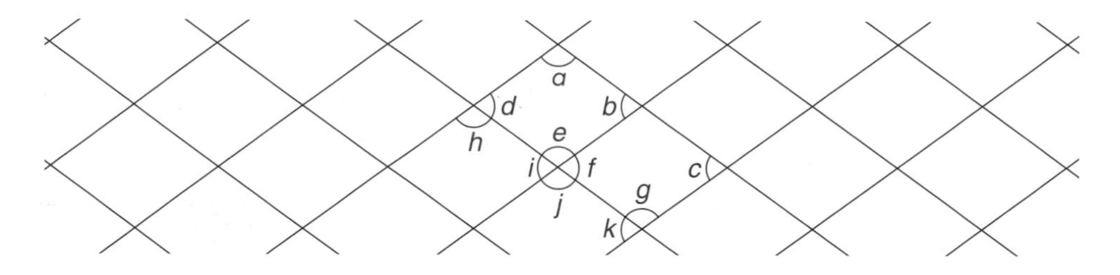

Which angles do you think are the same as a?

Which angles do you think are the same as b?

Find some rules for working out which angles are the same.

The rules

You may have found the rules below.

Where two lines cross, the angles opposite each other are equal. These are opposite angles.

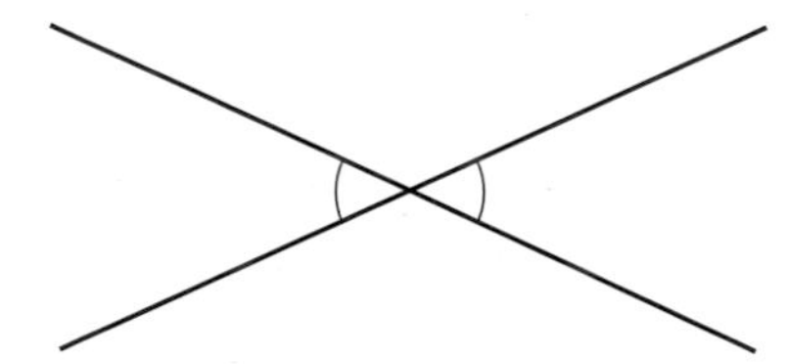

Where a line crosses two parallel lines, the two angles shown are equal. These are corresponding angles. Look for the shape like a letter *F*.

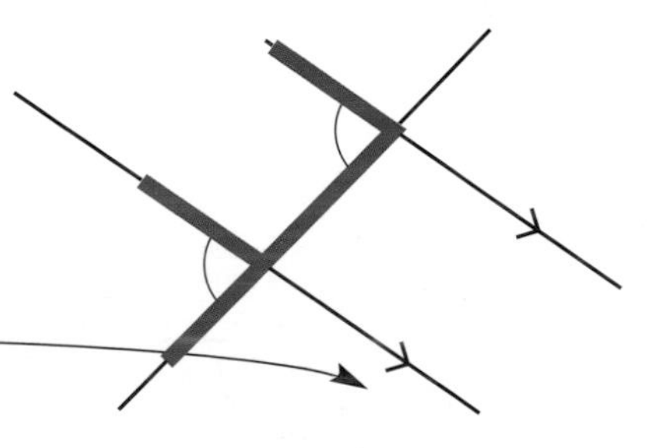

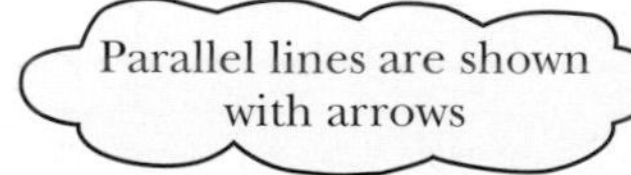

Where a line crosses two parallel lines, the two angles shown are equal. These are alternate angles. Look for the shape like a letter *Z*.

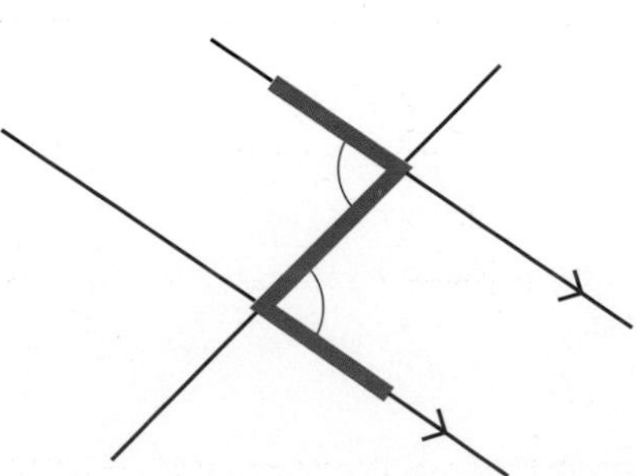

Exercise

1 Find the angles marked with letters in these diagrams.

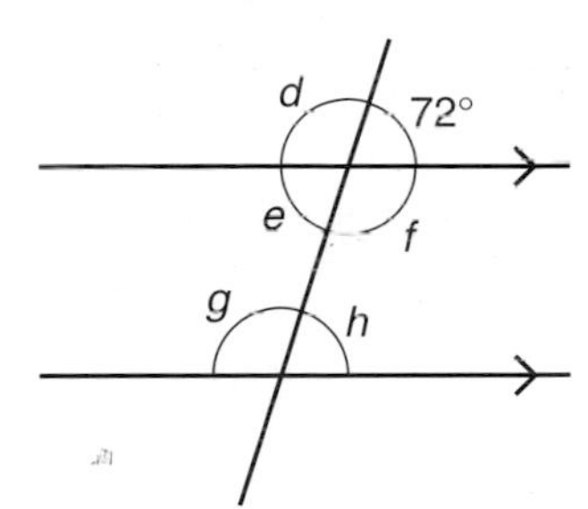

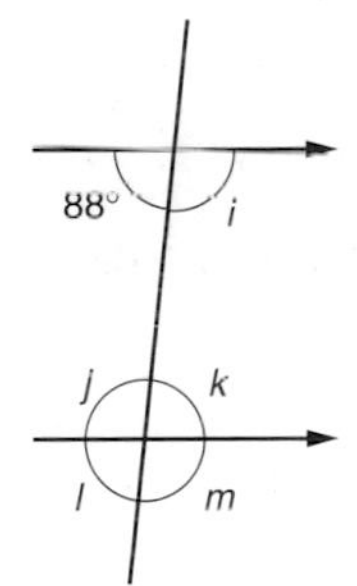

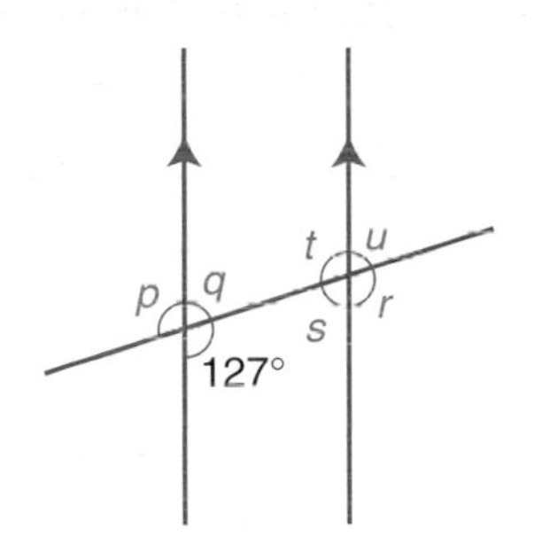

2 Look at this diagram.

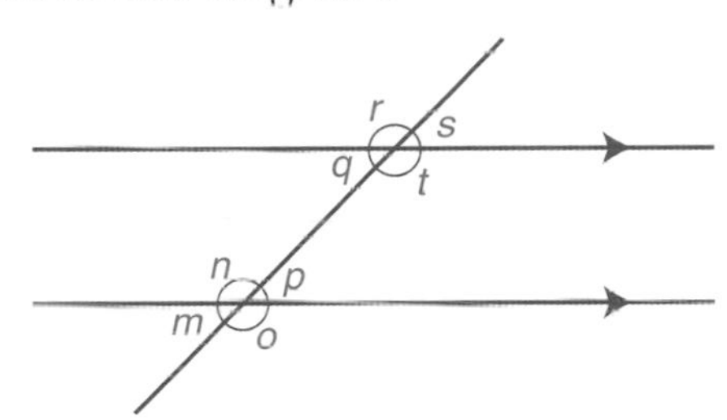

Write down as many pairs as you can of

a) opposite angles

b) corresponding angles

c) alternate angles.

3 The diagram shows a pair of parallel lines and a triangle.

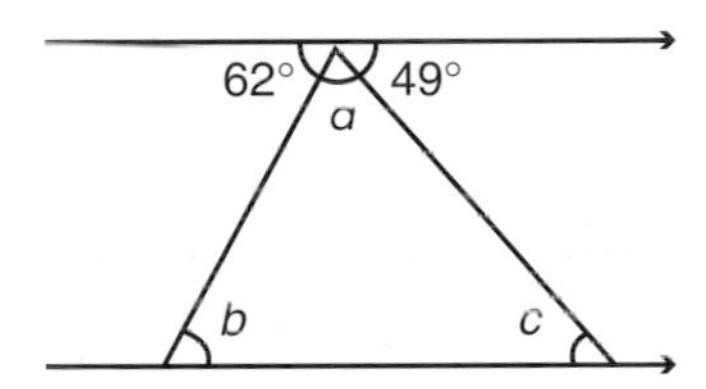

a) Work out a, b and c.

b) Angles a, b and c are the three angles in a triangle.

Add up these three angles.

4 Find the angles marked with letters in the diagrams below.
Give reasons for your answers.

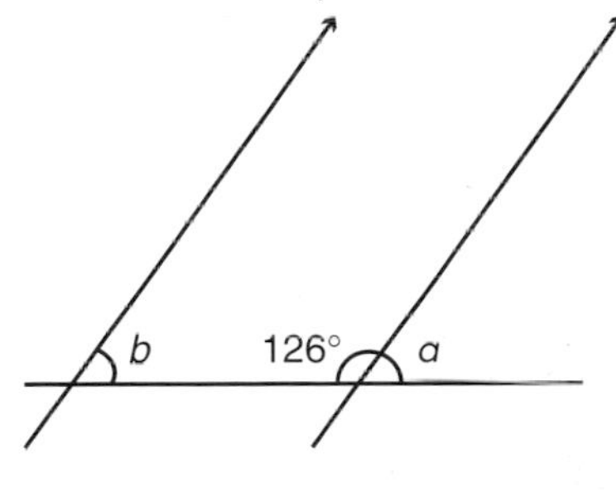

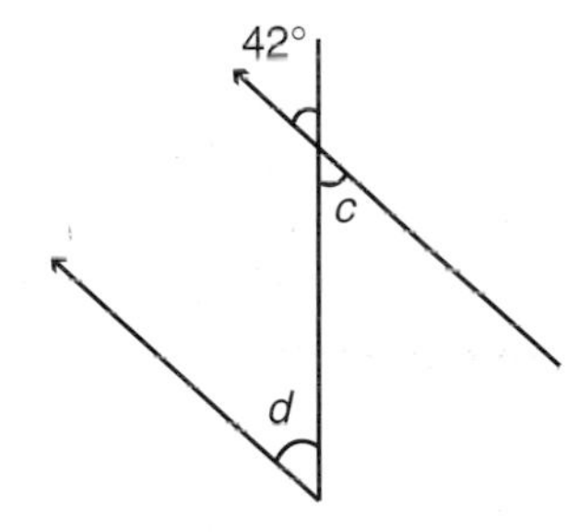

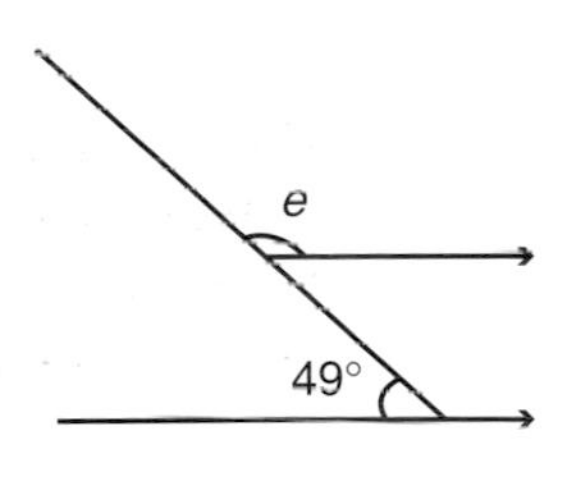

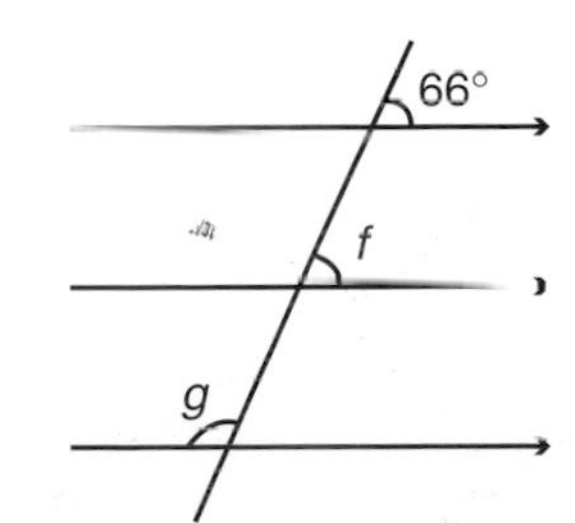

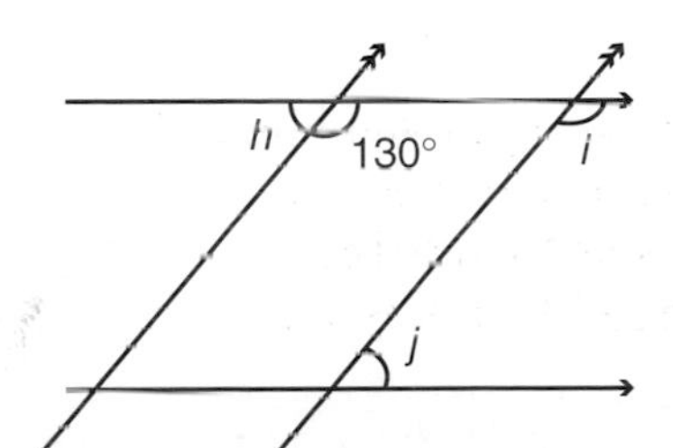

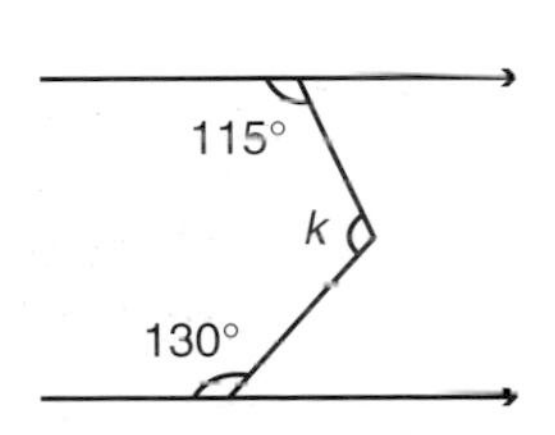

Triangles

A **triangle** is any three-sided shape.

Draw a triangle and measure the angles. What do the angles add up to?

Angles in a triangle add up to 180°.

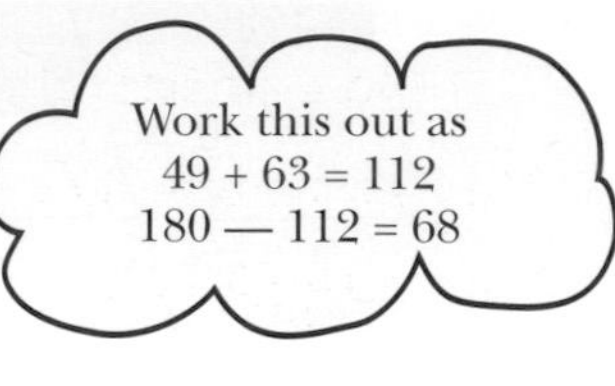

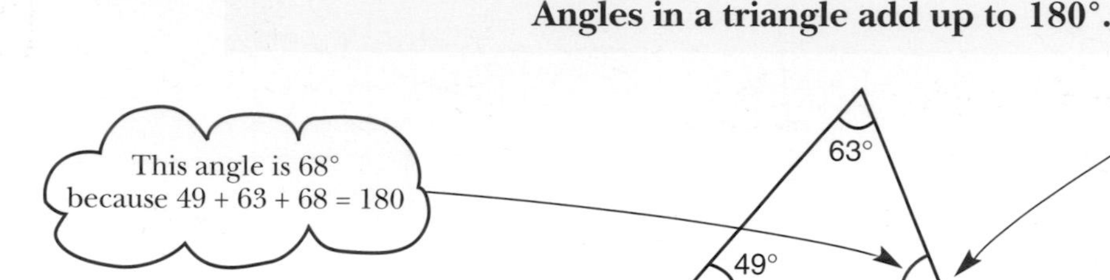

Types of triangle

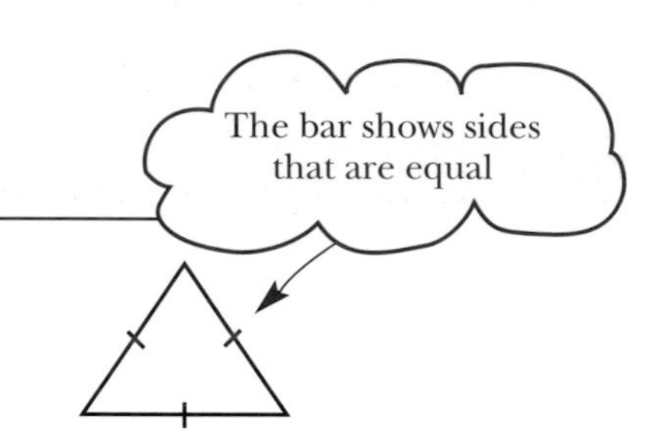

A triangle whose three sides are all the same length is called an **equilateral triangle**.

The three angles are also equal.

What is the size of an angle?

A triangle that has two sides the same length and the third side a different length is called an **isosceles triangle**.

The angles opposite the equal sides are equal.

In a triangle like the one shown, if the angle at the top is 84° then this means that the other two angles are each 48°.

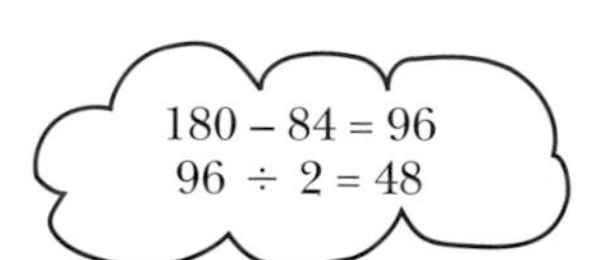

A triangle whose sides are all of different lengths is called a **scalene triangle**.

Triangles can also be described by what sort of angles they have.

An **acute-angled triangle** has three acute angles (all its angles are less than 90°).

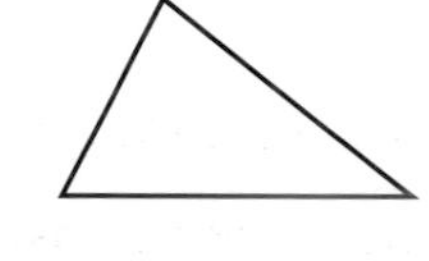

A **right-angled triangle** has one right angle (a 90° angle).

An **obtuse-angled triangle** has one obtuse angle (an angle greater than 90°).

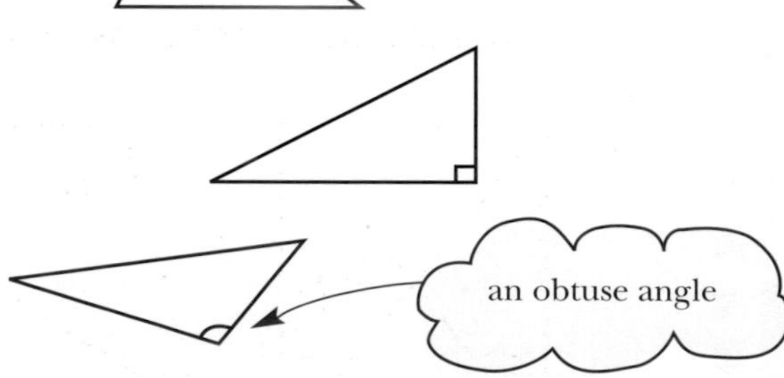

Draw a circle and mark on the diameter PQ. Choose a point R on the edge of the circle and join PR and RQ to form a triangle. Measure angle R.
What type of triangle is triangle PQR?

Exercise

1 Which of these triangles is

a) a right-angled scalene triangle?

b) an obtuse-angled isosceles triangle?

c) an equilateral triangle?

d) an acute-angled scalene triangle?

e) a right-angled isosceles triangle?

f) an obtuse-angled scalene triangle?

g) an acute-angled isosceles triangle?

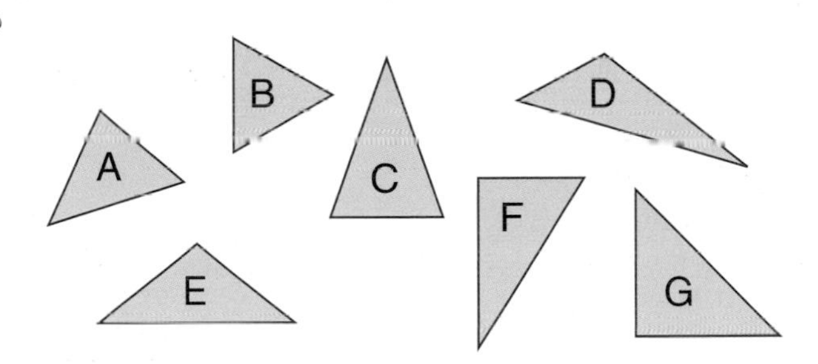

2 Find the angles marked with letters in these triangles.

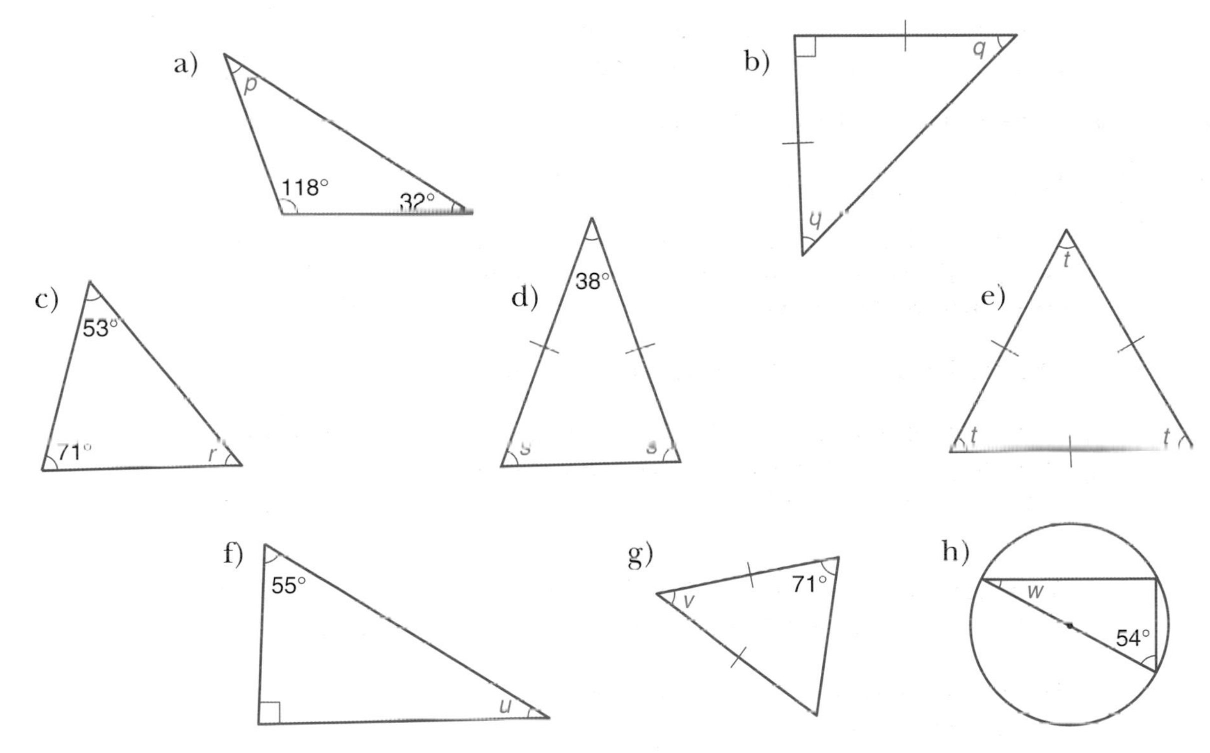

3 Find the angles marked with letters in the diagrams below.

Give reasons for your answers.

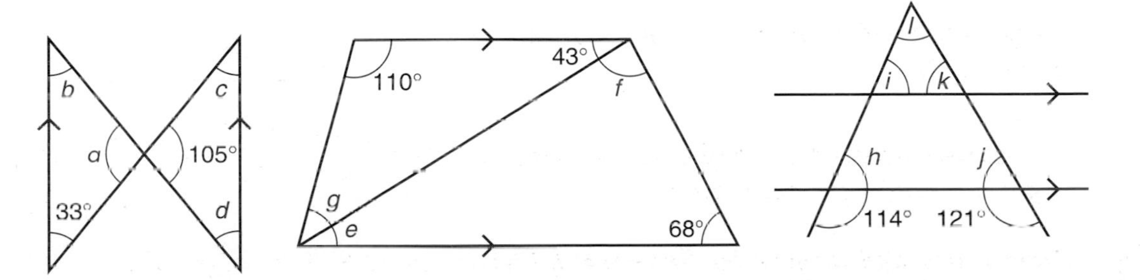

Quadrilaterals

A **quadrilateral** is any four-sided shape.

Any quadrilateral can be split up into two triangles by drawing in a diagonal, like this:

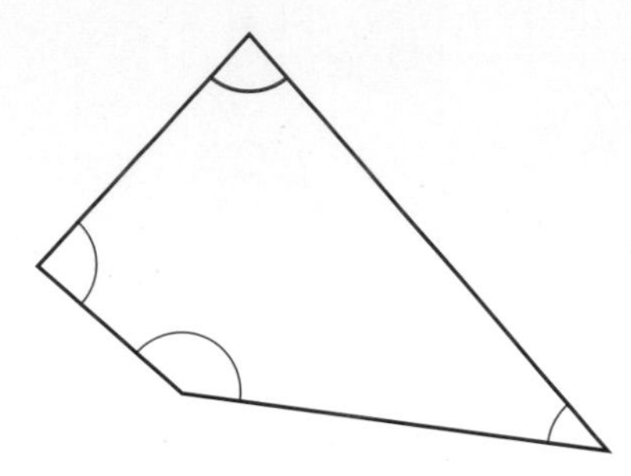

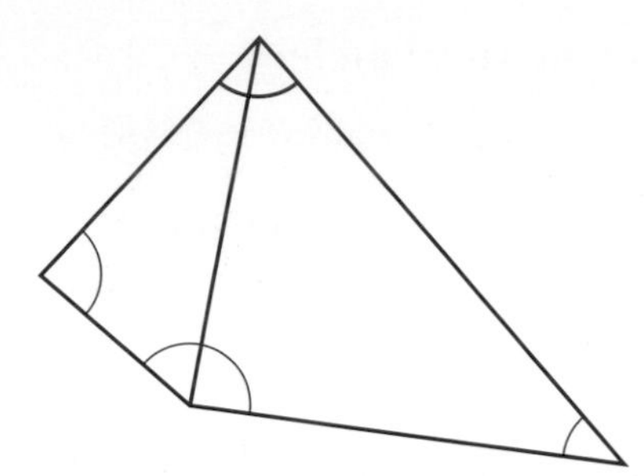

The angles in the quadrilateral must add up to the same as the angles in the two triangles.

So the angles in any quadrilateral must add up to $2 \times 180° = 360°$.

Angles in a quadrilateral add up to 360°.

Types of quadrilateral

A **square** has four equal sides. All of its angles are right angles.

A **rectangle** has two pairs of equal sides. All of its angles are right angles.

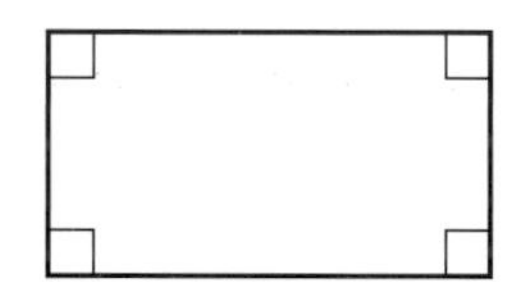

A **rhombus** has four equal sides. Its opposite sides are parallel.

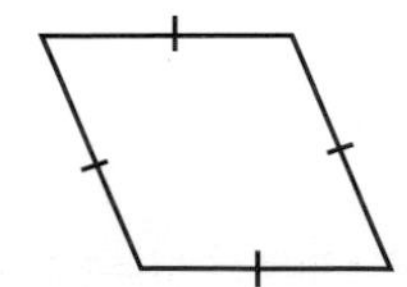

A **parallelogram** has its opposite sides equal and parallel.

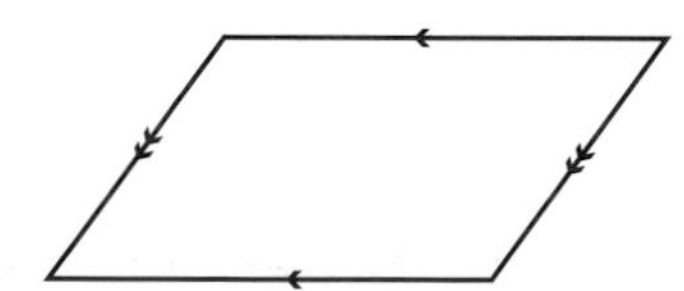

A **trapezium** has one pair of parallel sides.

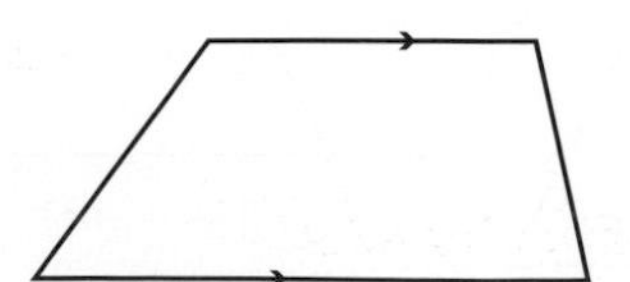

A **kite** has two pairs of equal sides. The equal sides are next to each other.

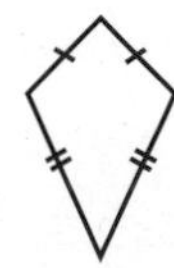

Exercise

1 Find the angles marked with letters in the quadrilaterals below.

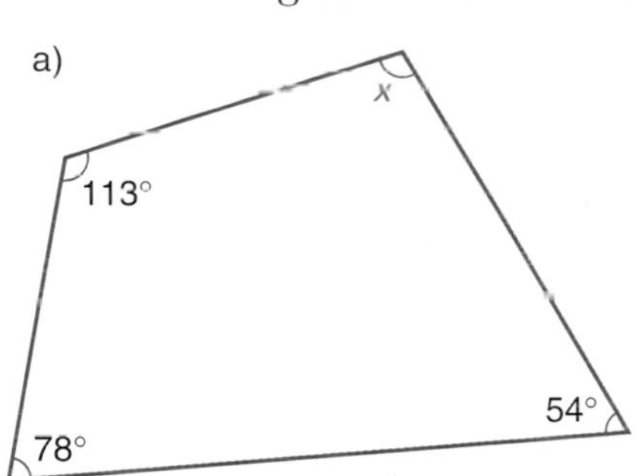

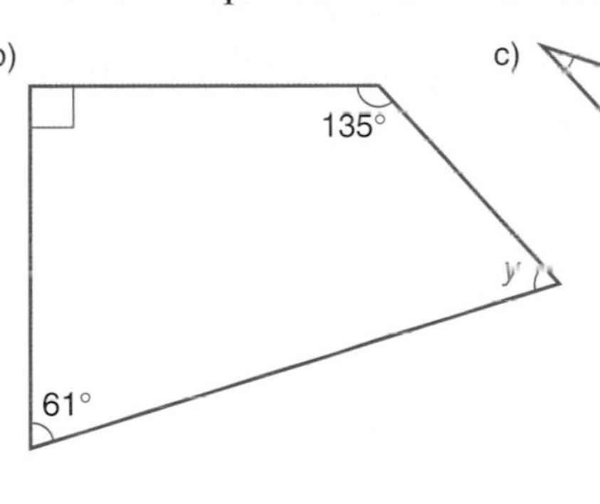

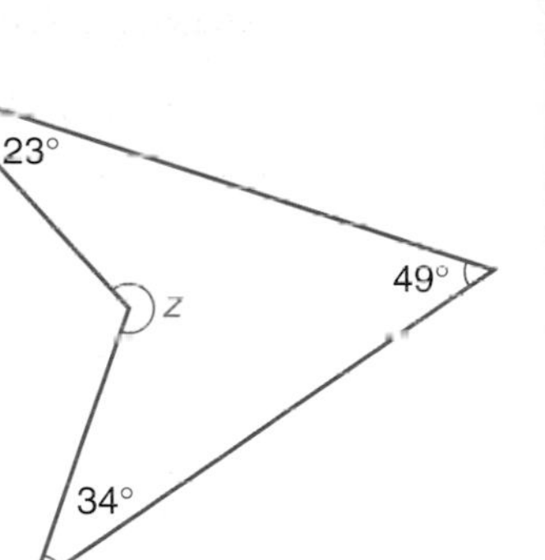

2 ABCD is a trapezium with AB parallel to DC. ∠A = 127° and ∠C = 76°.

Work out ∠B and ∠D. (Hint: draw a diagram first.)

3 JKLM is a rectangle. JK is the longer side. N is a point on JK such that KN = KL.

Given ∠MNL = 79° work out ∠JMN.

4 PQRS is a rhombus with the diagonal QS equal to PQ.

Work out ∠P.

5 ABCD is a rectangle. E is a point on AB.

a) What type of quadrilateral is BCDE?

b) Given ∠ADE = 50° work out ∠DEB.

c) If DE = EB work out ∠EBD.

6

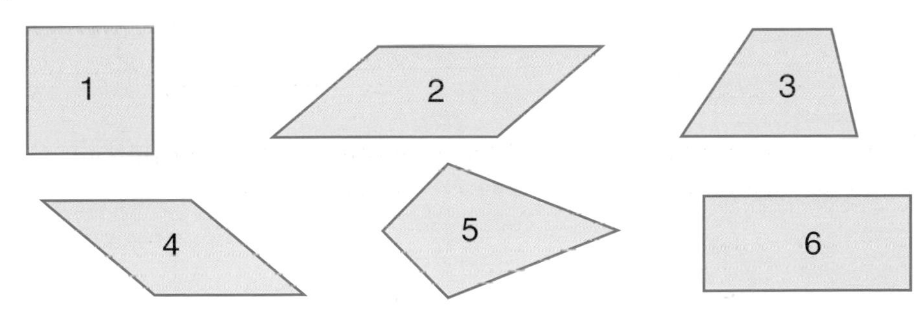

a) Write down the special name for each of these four-sided shapes.

b) For each shape say which statements in the list below are true.

A. The sides are equal.
B. Both pairs of opposite sides are parallel.
C. Only opposite angles are equal.
D. There is just one pair of parallel sides.
E. The four sides are made up of just two distinct lengths.
F. The diagonals make an angle of 90° where they meet.

Finishing off

Now that you have finished this chapter you should:

★ know that angles round a point add up to 360°

★ know that angles on a straight line add up to 180°

★ be able to find pairs of equal angles where two lines cross and where a line intersects parallel lines

★ know that the angles in a triangle add up to 180°

★ recognise equilateral, isosceles, scalene, acute-angled, right-angled and obtuse-angled triangles

★ know that the angle in a semi-circle is a right angle

★ know that the angles in a quadrilateral add up to 360°

★ recognise a square, a rectangle, a rhombus, a parallelogram, a trapezium and a kite.

Use the questions in the next exercise to check that you understand everything.

Mixed exercise

1 Find the angles marked with letters in these diagrams.
Give reasons for your answers.

a) 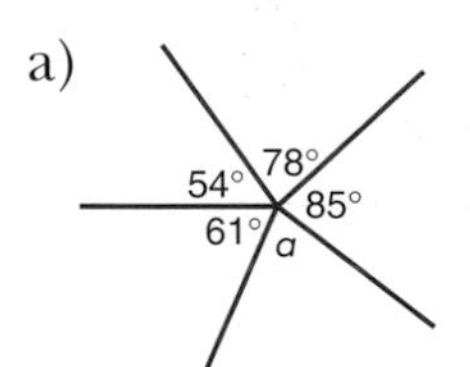

b) 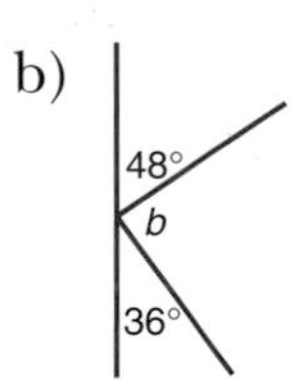

c) 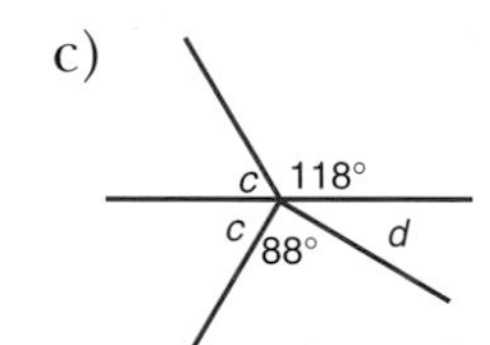

d) 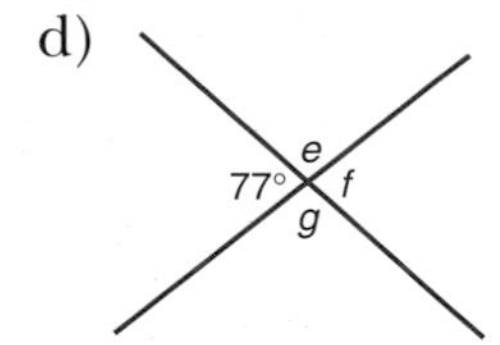

e) 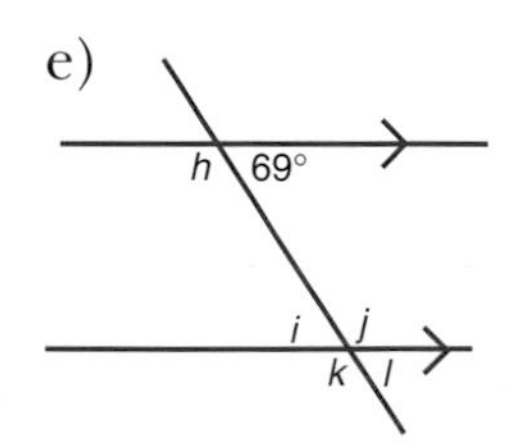

f) 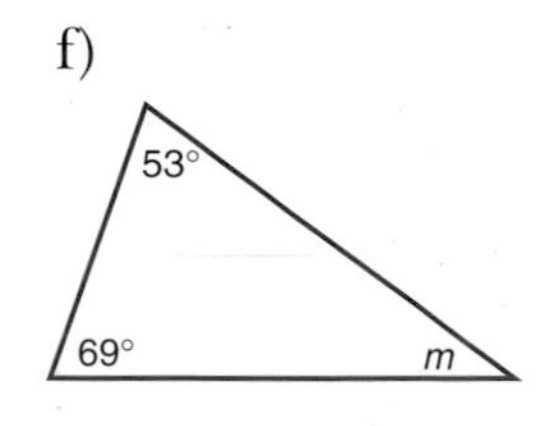

g) 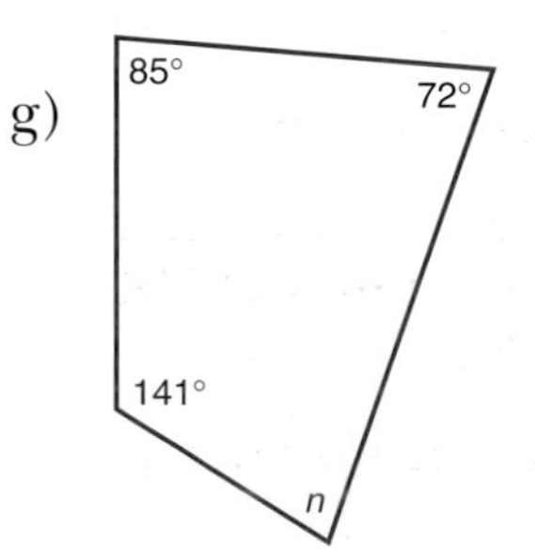

h) 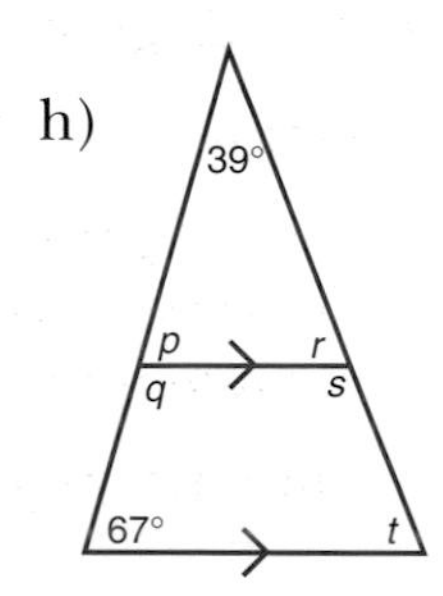

i) 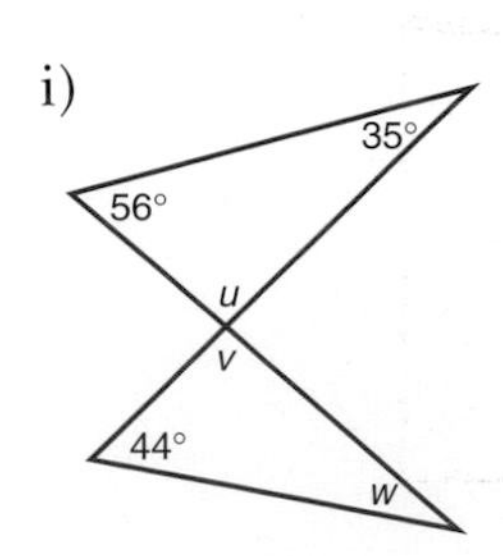

j) 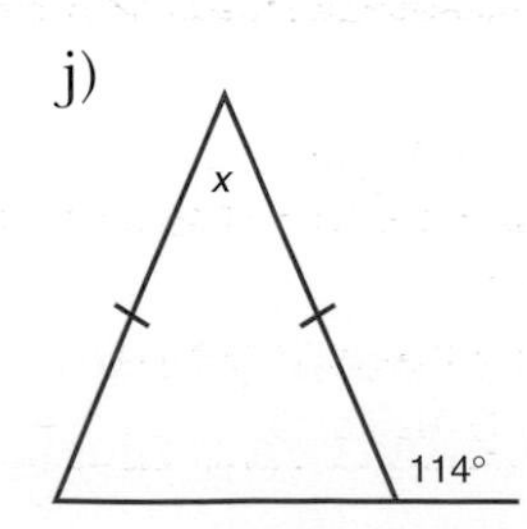

k) 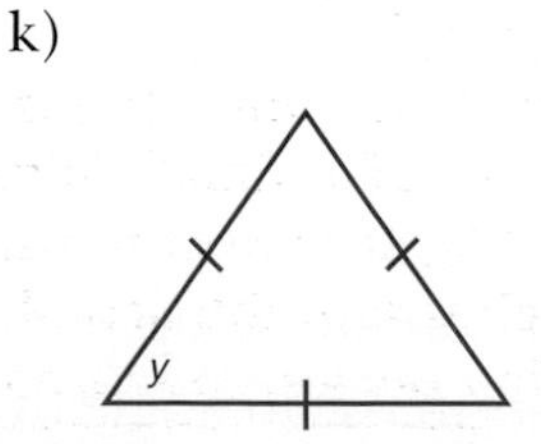

l) 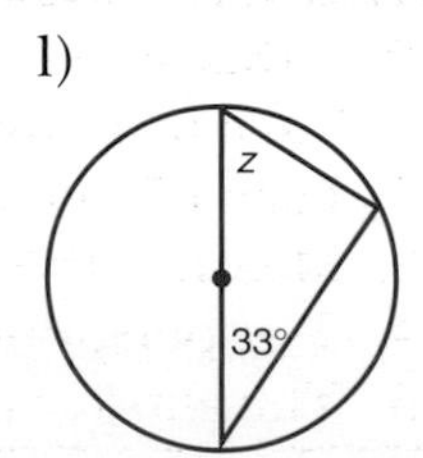

Mixed exercise

2 In triangle ABC, $\angle B = 90°$ and $\angle A$ is twice $\angle C$. Work out the size of $\angle A$.

3 ABC is an equilateral triangle. AC is extended to D and AC = CD. Join BD and work out $\angle CBD$.

4

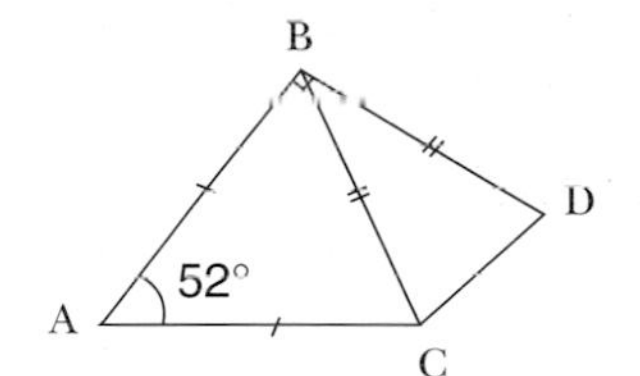

Triangles ABC and BCD are each isosceles. $\angle ABD = 90°$ and $\angle BAC = 52°$.

Work out $\angle ACD$.

5 In triangle PQR, $\angle Q = 80°$ and $\angle R = 55°$.

S is on PQ and T is on PR such that ST is parallel to QR.

Work out

a) $\angle QPR$ b) $\angle PST$ c) $\angle STR$

6 ACDF is a rectangle with B on AC and E on DF.

$\angle BEC = 100°$ and $ECD = \angle 35°$. FB is parallel to EC.

a) Write down the vertices of

(i) a right angled triangle

(ii) an obtuse-angled triangle

(iii) a trapezium

(iv) a parallelogram.

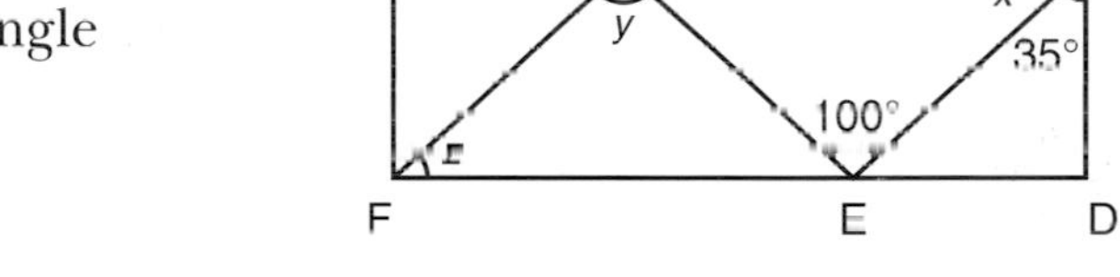

b) Work out the values of x, y and z.

7 The diagram shows a circle centre O.

OM = ON = OP (radii of circle).

The tangent QR meets the circle at P. QR is perpendicular to OP. MONR is a straight line and $\angle NPR = 35°$.

Find the angles marked with letters.

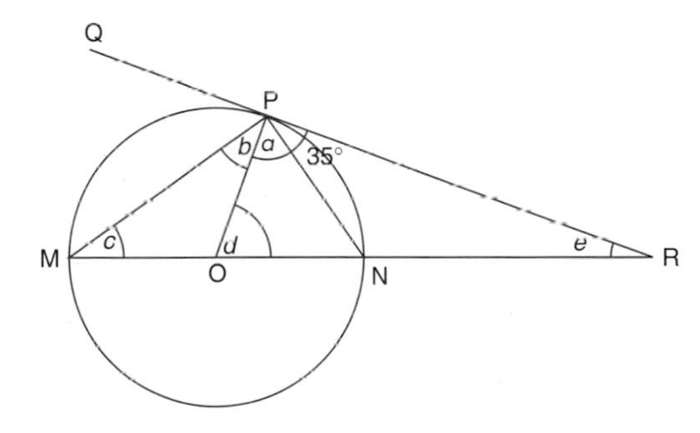

Play this game with a friend.
1. Take it in turns to draw two lines to make an angle.
2. Each of you writes down an estimate of the size of the angle in degrees.
3. Measure the angle. (Again take it in turns.)
4. The person who is closest to the actual size scores a point.
5. The winner is the first person to score five points.

Twenty

Polygons

Polygons

Shapes with three or more straight sides are called **polygons**.

You have already met two kinds of polygon: the triangle and the quadrilateral.

If all the sides of a polygon are the same length, and all its angles are the same, it is called a **regular polygon**.

Otherwise it is called an **irregular polygon**.

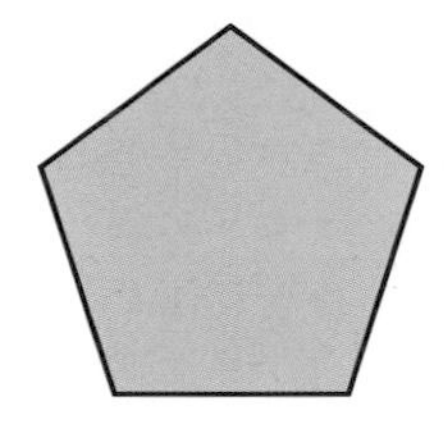

Regular pentagon (5 sides)

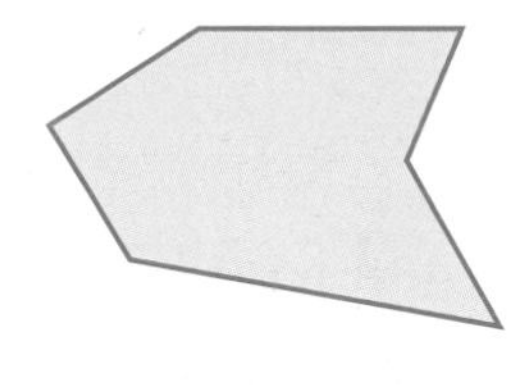

Irregular hexagon (6 sides)

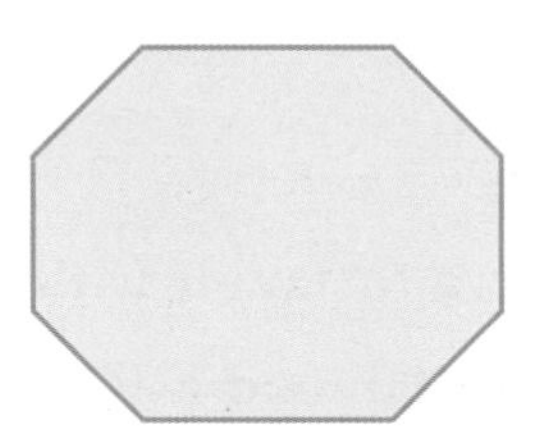

Irregular octagon (8 sides)

What is the special name for a regular quadrilateral?

Interior and exterior angles of a polygon

This diagram shows a pentagon.

The angles shown in red are called **interior angles**.

The angles shown in blue are called **exterior angles**.

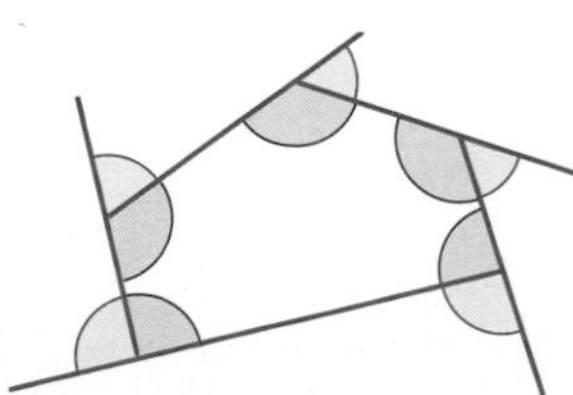

Imagine you are programming a robot to draw a polygon. The exterior angle is the angle through which the robot has to turn after drawing each side.

Through how many degrees would the robot have to turn, in total, to draw the whole polygon?

If a polygon is regular, all the exterior angles are the same size.

The sum of the exterior angles is 360°.

For a regular polygon with n sides, each exterior angle = $360° \div n$.

Exercise

1 Write down the name of each polygon below and say whether it is regular or irregular.

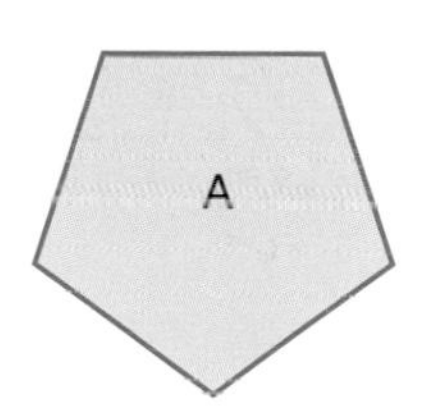

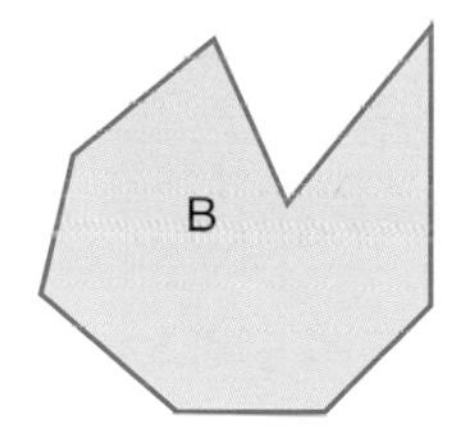

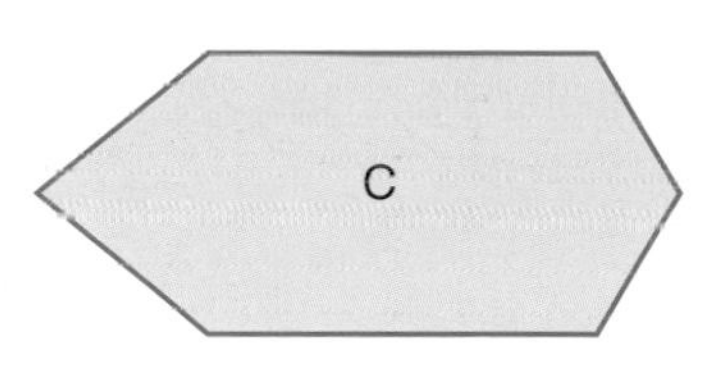

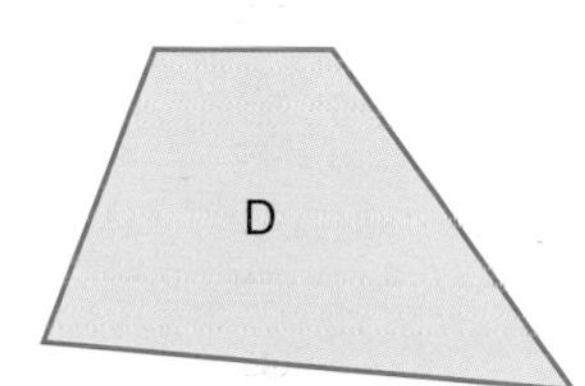

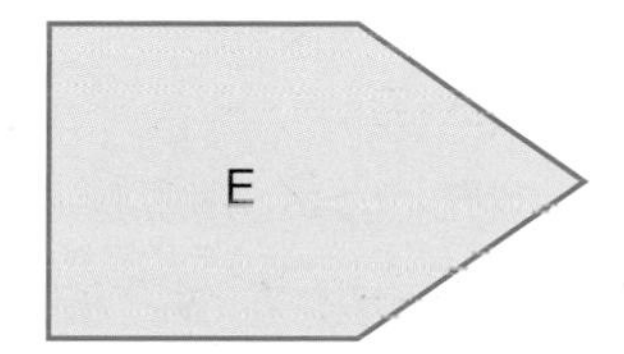

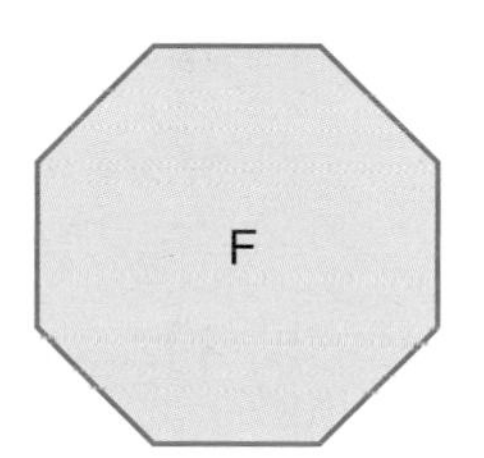

2 Find the exterior angle of each of the following polygons.

a) a regular (equilateral) triangle
b) a regular quadrilateral (a square)
c) a regular pentagon
d) a regular hexagon
e) a regular octagon
f) a regular decagon (10 sides)

3 **In this question you will discover a useful rule about the exterior angles of a triangle, which you should try to remember.**

a) For each of these triangles
   (i) find the sum of angles at P and Q;
   (ii) find the interior angle at R;
   (iii) find the exterior angle at R.

b) What have you found out about the exterior angle of a triangle and the sum of the two opposite angles?

c) Explain why this rule is always true.

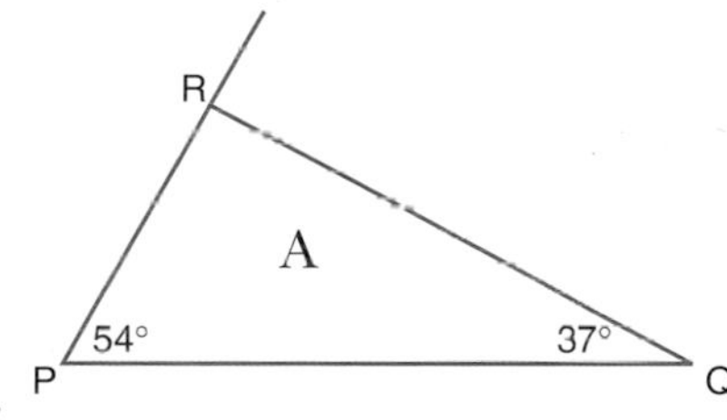

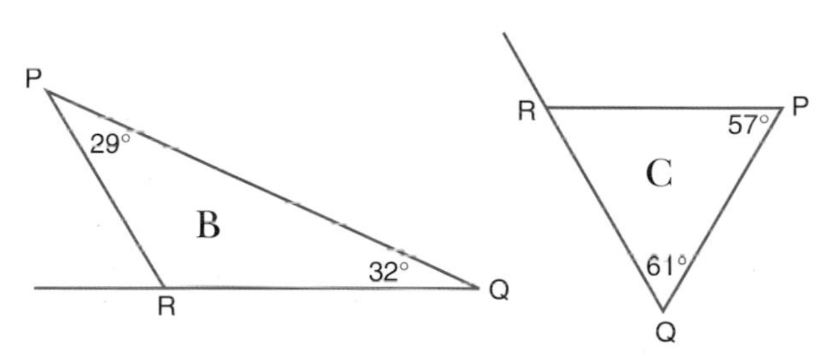

4 a) A regular polygon has an exterior angle of 15°.
How many sides does it have?

b) Explain why no regular polygon has an exterior angle of 25°.

Angle sum of a polygon

Any polygon can be split into a number of triangles by drawing diagonals.

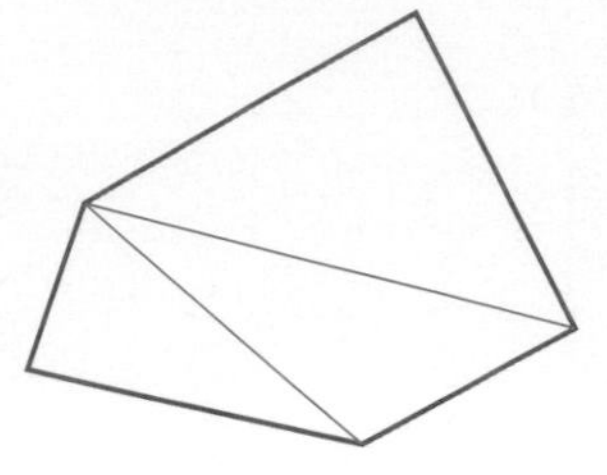

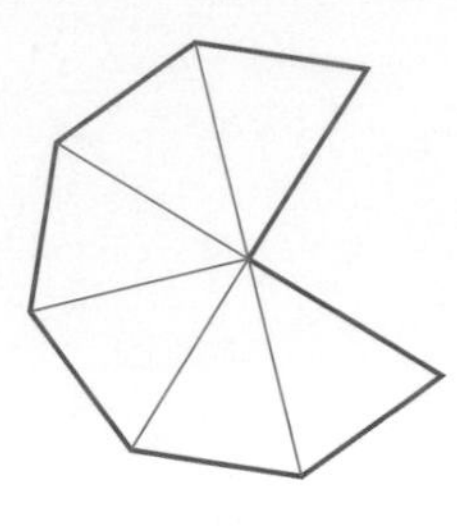

Into how many triangles could you split a 10-sided polygon?

What about a 15-sided polygon?

What rule can you use to work out the number of triangles?

The angles in each triangle add up to 180°, so you can find the sum of the interior angles of the polygon by multiplying the number of triangles by 180°.

Make sure you can see how this works.

You can write this as a formula, using n for the number of sides of the polygon:

$$\textbf{Angle sum of a polygon} = 180^\circ \times (n - 2)$$

Using this formula, the angle sum of a pentagon

$= 180^\circ \times (5 - 2)$

$= 180^\circ \times 3$

$= 540^\circ$

Check that this formula works for a triangle ($n = 3$) and a quadrilateral ($n = 4$).

Angles of a regular polygon

If a polygon is regular, all the interior angles are the same.

You can find each interior angle by dividing the angle sum of the polygon by the number of angles in the polygon.

The interior angle of a regular pentagon is

$540^\circ \div 5 = 108^\circ$

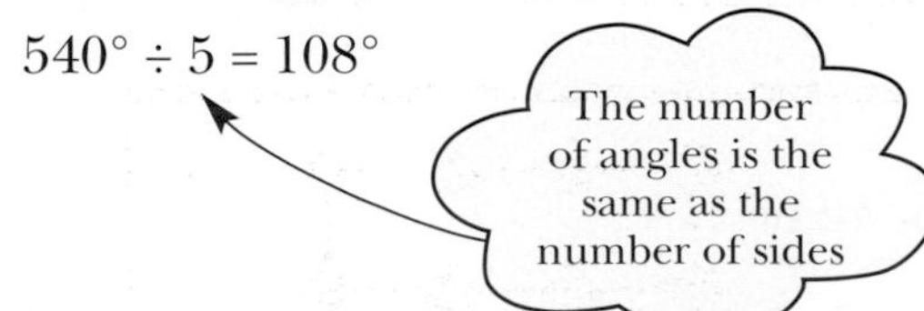

What is the relationship between the interior angle and the exterior angle of a polygon?

Exercise

1 Find the sum of the interior angles of each of these polygons.

a)

b)

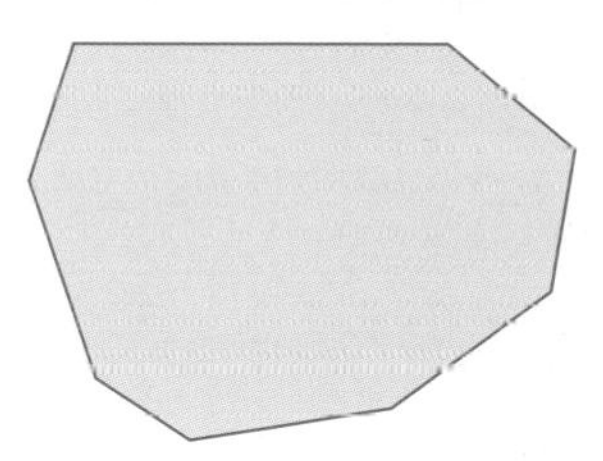

c)

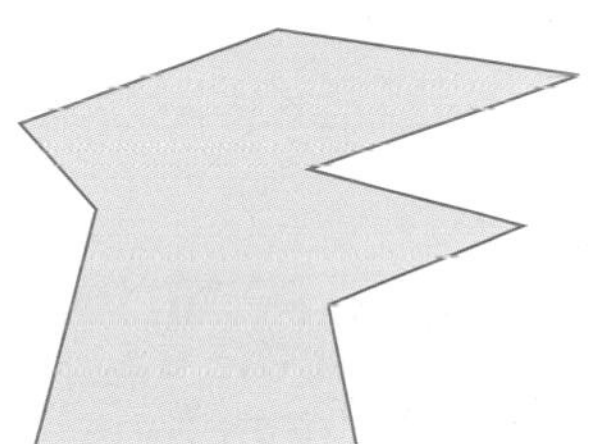

d)

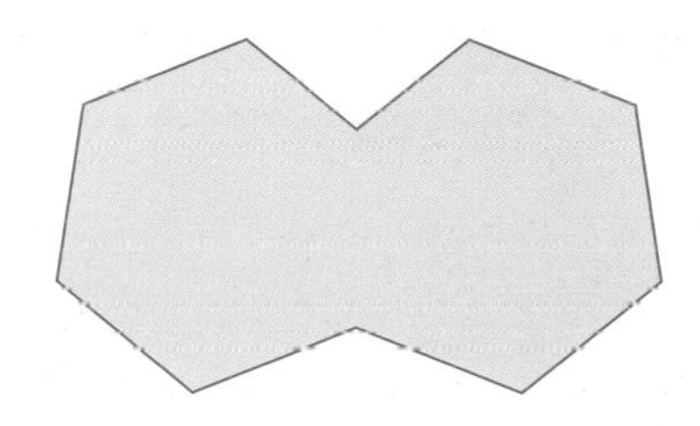

2 Use your answers to question 1 to work out the interior angle of each of the following regular polygons.

a) a regular hexagon
b) a regular octagon
c) a regular nonagon (9 sides)
d) a regular decagon (10 sides)

3 Find the exterior angle of each of the regular polygons in question 2. (Use the rule that for a regular n-sided polygon the exterior angle is $360° \div n$.)

4 a) Use your answers to question 3 to find the interior angle of each regular polygon.

b) Check your answers with those for question 2.

5 You have now used two methods for finding the interior angle of a regular polygon: by finding the angle sum first and by finding the exterior angle first.

Which do you find easier, and why?

6 a) A regular polygon has an interior angle of 160°.

How many sides does it have?

b) Explain why no regular polygon has an interior angle of 130°.

The diagram shows part of a tiling pattern made up of regular octagons and squares.
Draw another tiling pattern that uses a regular polygon and one other shape.

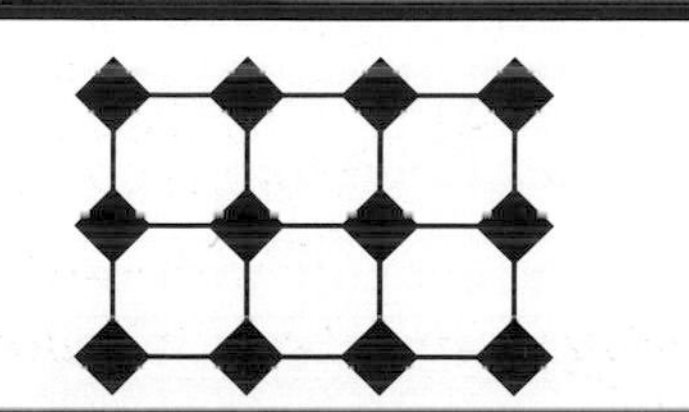

Congruent shapes

Trace triangle A.

Try to fit your tracing over each of the other triangles. You may need to turn it over for some of them. Which triangles fit the tracing exactly?

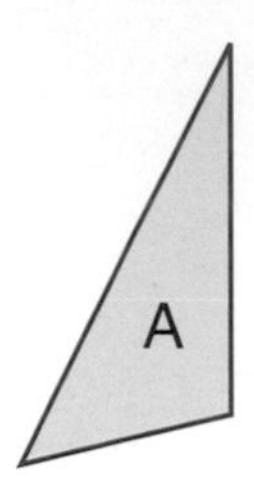

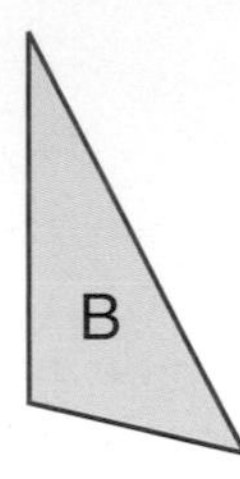

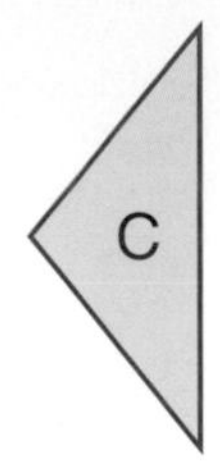

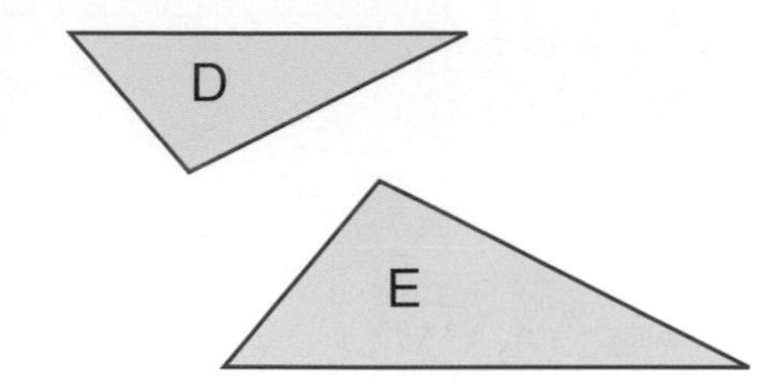

Shapes which are exactly the same shape and size are called **congruent**. Congruent shapes may be turned round or flipped over, but the tracing of a shape will always fit exactly over a shape which is congruent to it.

When congruent shapes are identified by their vertices you must be careful to ensure that the vertices correspond.

These two triangles are congruent.

$\angle A = \angle Q$, $\angle B = \angle P$, $\angle C = \angle R$, $AB = QP$, $BC = PR$ and $AC = QR$ so we say that:

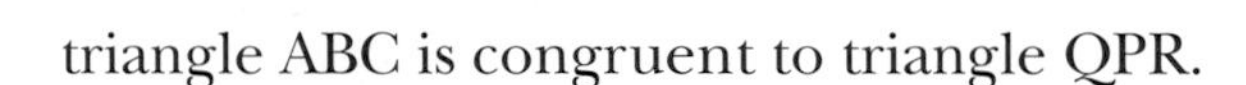

triangle ABC is congruent to triangle QPR.

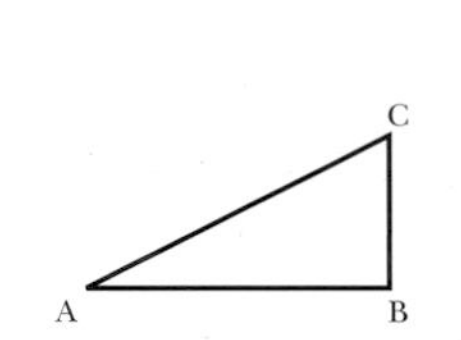

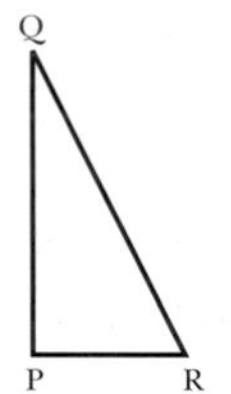

Tessellations

Look at these designs for floor tilings:

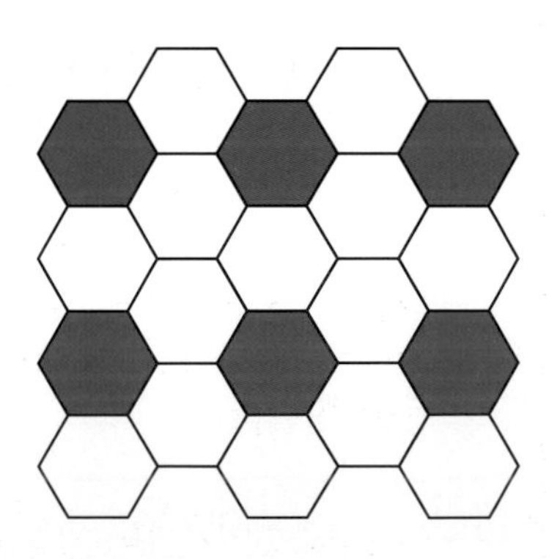

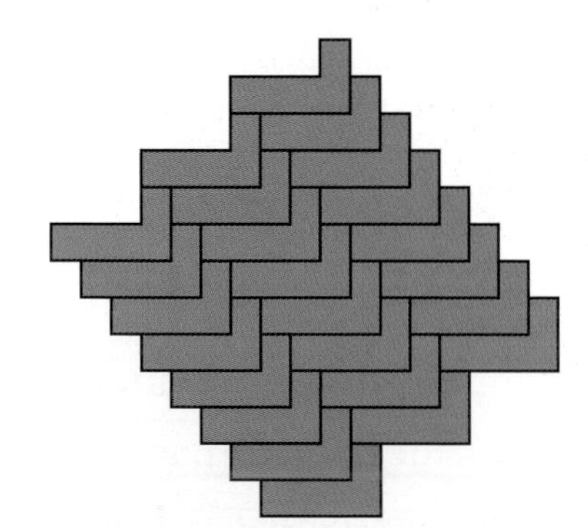

These patterns are examples of **tessellations**.

A tessellation is a repeating pattern without any gaps, made up of congruent shapes fitted together.

Tessellations can be made with very complicated shapes, as well as simple ones like those above. The Dutch artist M.C. Escher used fascinating tessellations in his work.

Exercise

1 Which of these shapes are congruent to shape A?

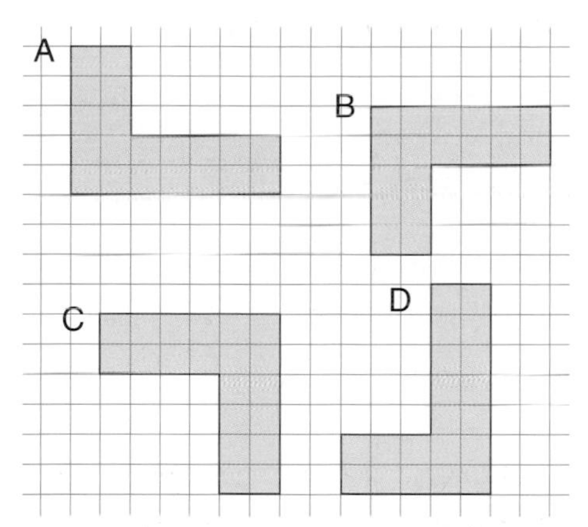

2 Draw a kite and add the line which divides it into two congruent triangles.

3 Draw a rectangle ABCD. Draw the line AC. Complete the following:

triangle ABC is congruent to …

4 Draw a trapezium ABCD with AB and DC parallel but not equal.

Under what circumstances is it possible to divide the trapezium into two congruent shapes?

5 Use squared paper to draw tessellations using the following shapes.

a)

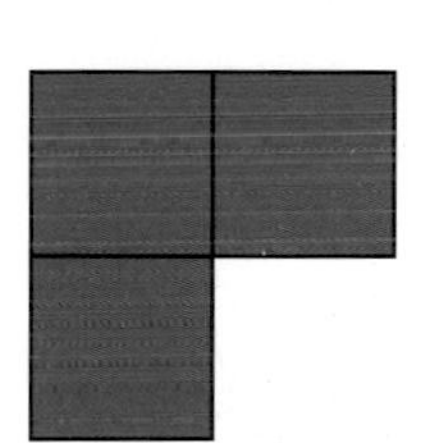

b)

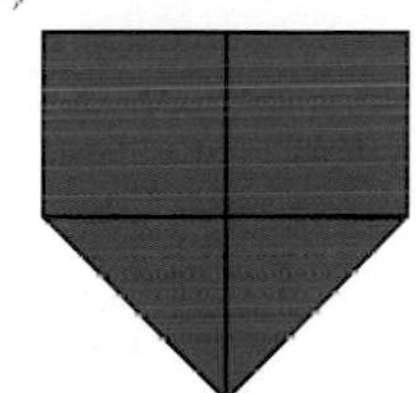

c)

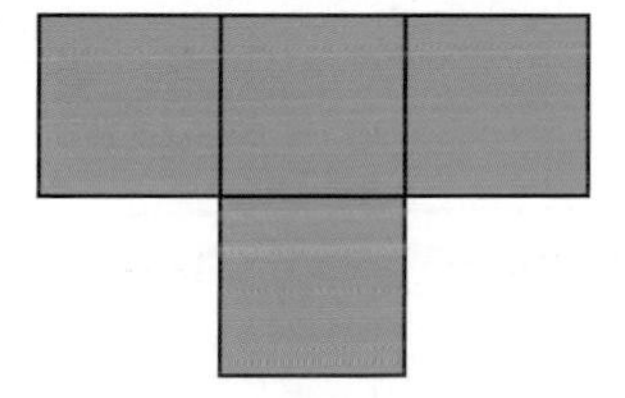

1. Start with a simple shape that tessellates such as a square or a hexagon.

2. Draw a shape on to one side. Use tracing paper to remove the same shape from the opposite side.

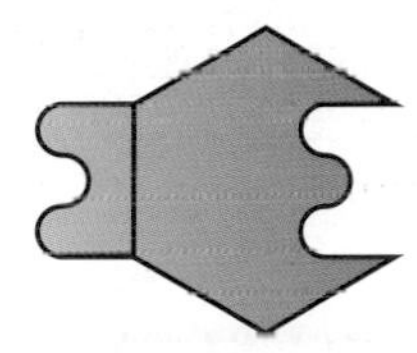

3. Do the same on any other sides that you want to.

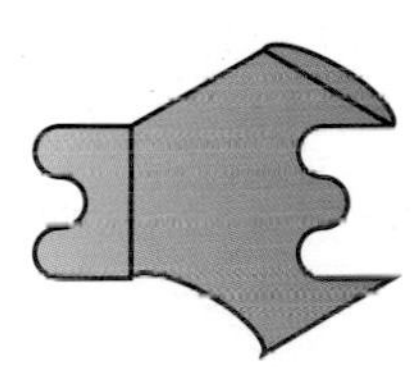

4. Cut out your design to make a template. Use your imagination to add any details you like.

5. Use your template to draw your tessellation. The shapes should fit together perfectly.

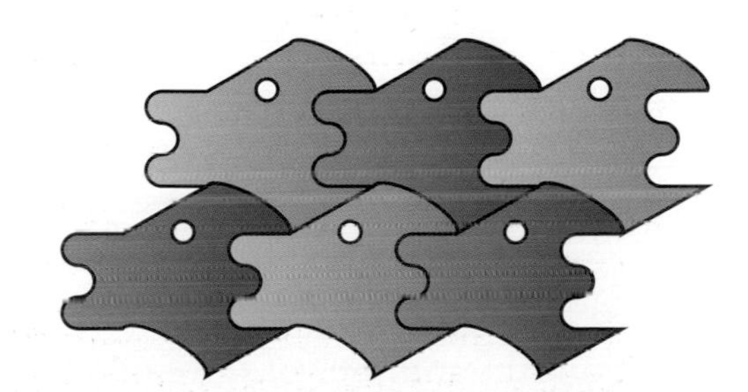

Finishing off

Now that you have finished this chapter you should:

- ★ know the meaning of the words *polygon, pentagon, hexagon* and *octagon*
- ★ know that difference between a regular and an irregular shape
- ★ be able to find the sum of the interior angles of any polygon
- ★ be able to find the interior and exterior angles of any regular polygon
- ★ be able to find the number of sides of a regular polygon if you know the interior angle
- ★ recognise congruent shapes
- ★ be able to draw tessellations with simple shapes.

Use the questions in the next exercise to check that you understand everything.

Mixed exercise

1 Here is a regular pentagon which has been split into 5 congruent triangles.

a) Work out the size of each **blue** angle.

b) Use your answer to a) to work out the size of each **red** angle.

c) Use your answer to b) to work out the interior angle of the pentagon.

2 Find the sum of the interior angles of each of these polygons.

a)

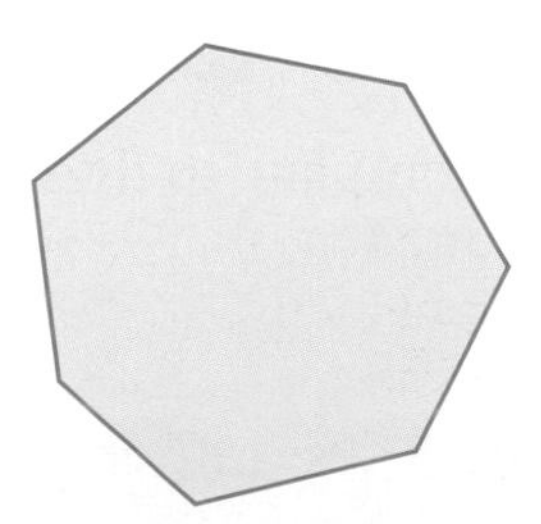

b)

3 Find the exterior angle and the interior angle of each of the following polygons.

a) regular decagon (10 sides)   b) regular dodecagon (12 sides)

4 Work out the number of sides of a polygon with an interior angle of

a) 120°   b) 156°   c) 135°

5 This photograph shows a walkway in a Spanish holiday resort. It is based on hexagons that tessellate.

These hexagons are made up of three shapes A, B and C as shown.

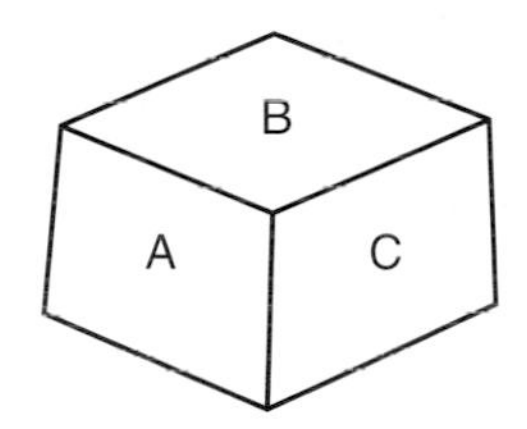

a) Use a single word to describe, as accurately as possible,

(i) shape A
(ii) shape B

b) Which shape is congruent to shape C?

6 Draw a regular hexagon ABCDEF. The diagonals AD, BE and CF meet at O.

a) Write down the vertices of a shape congruent to

(i) rhombus ABOF
(ii) trapezium ABCF
(iii) triangle ABE

b) Work out (i) $\angle AOB$ (ii) $\angle BAO$

c) What type of triangle is triangle ABO?

7 Sophie wants to make a tessellation which has 6 regular polygons fitting round a point.

a) What must the interior angle of her polygon be?

b) How many sides does the polygon have?

c) Draw Sophie's tessellation.

8 a) Work out the interior angle of a regular 9-sided polygon.

b) Is it possible to make a tessellation of these polygons?

9 a) Explain why it is not possible to make a tessellation of regular octagons.

b) What angle is left over when you have fitted as many octagons as possible together?

c) What shape would fit in the gap?

Draw a repeating pattern using octagons and this other shape.

Twenty one

Perimeter and area

Rectangles

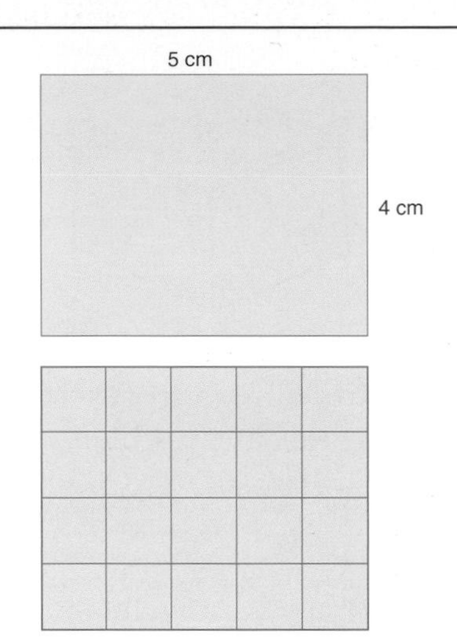

Here is a rectangle.

The **perimeter** of a rectangle is the distance round the outside of it.

Perimeter = 5 cm + 4 cm + 5 cm + 4 cm
= 18 cm

To find the **area** of a rectangle you can split it into lots of small squares each 1 cm long and 1 cm wide. Each square has an area of 1 square centimetre (1 cm^2).

The total number of squares is $5 \times 4 = 20$ so the area of the rectangle is 20 cm^2.

For any rectangle

Area = Length × Width

Write down a formula for the perimeter of a square of side x cm.

Shapes made from rectangles

This shape is not a rectangle, but you can split it up into rectangles to work out its area.

Here is one way of splitting it.

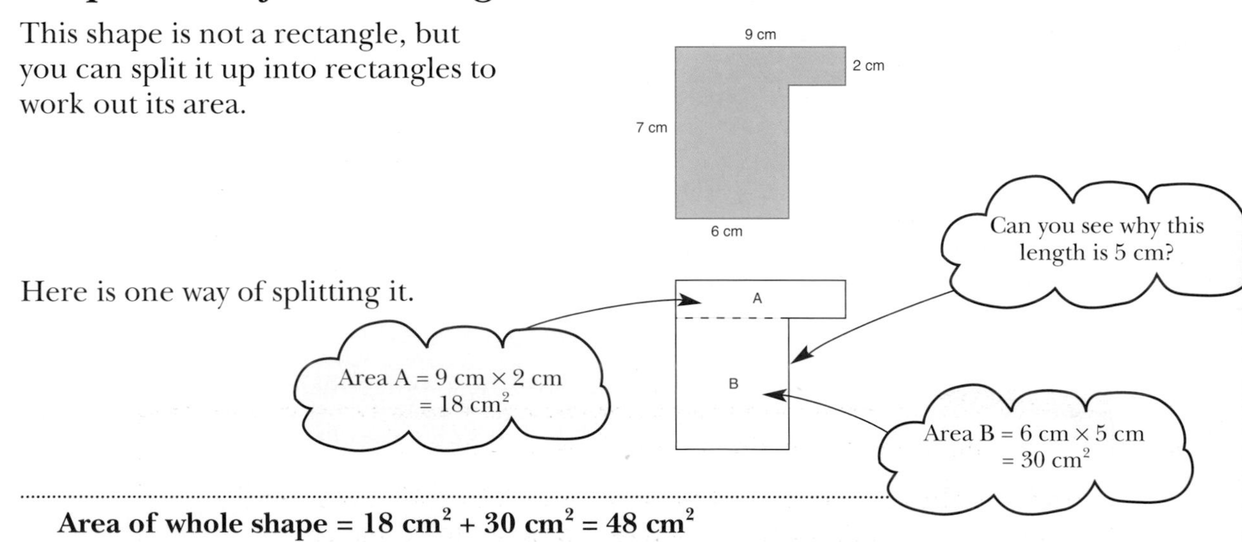

Area of whole shape = 18 cm^2 + 30 cm^2 = 48 cm^2

How else can the rectangle be split up to find the area?

Find a way to work out the area using subtraction rather than addition.

1 Find the areas and perimeters of the following rectangles.

a) Length 9 cm, width 7 cm b) Length 12 cm, width 5 cm

c) Length 16 m, width 11 m d) Length 7.5 m, width 4 m

2 A jewellery designer wants to make some pendants using letters of the alphabet.

Find the area of metal she needs to make each of the letters below.

a)

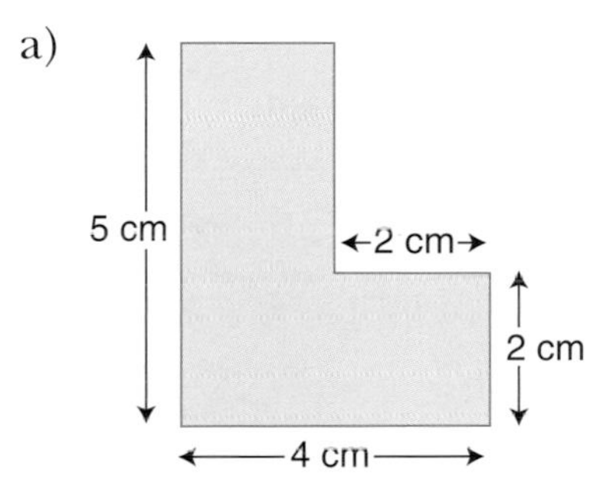

b)

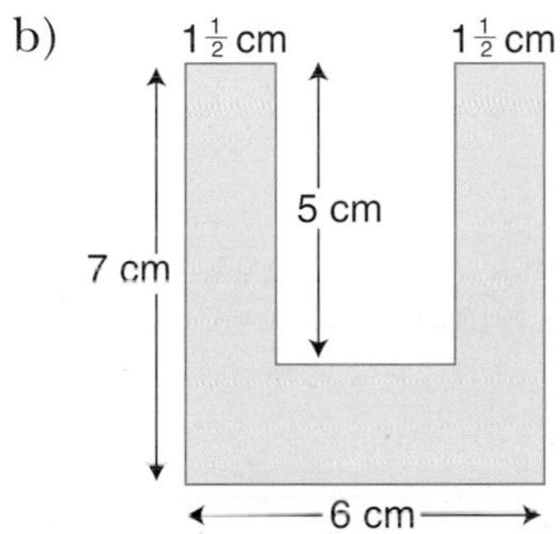

c)

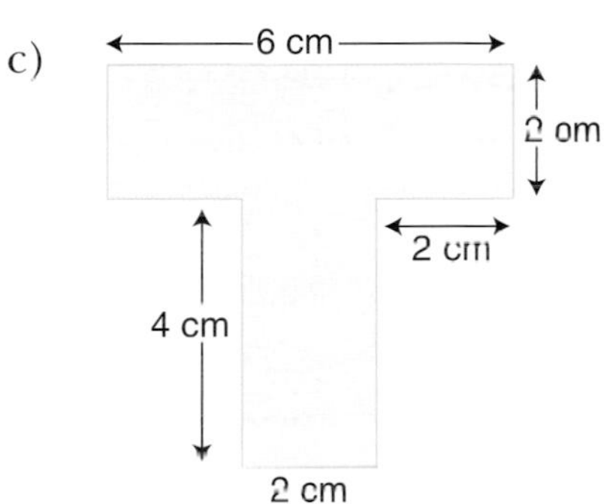

d)

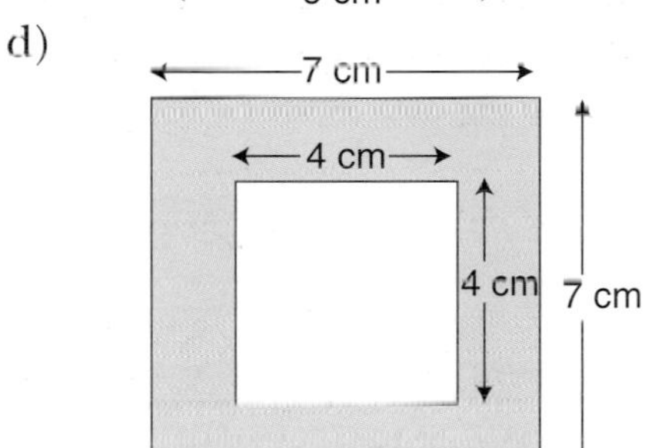

e)

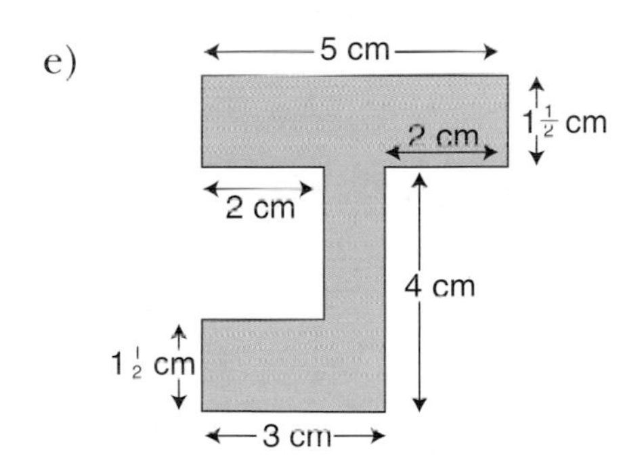

f)

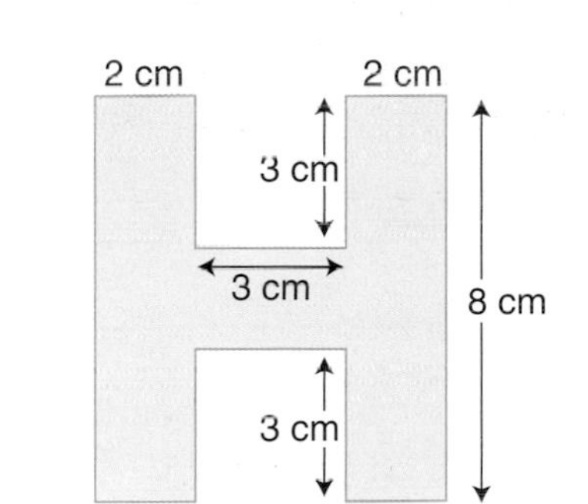

You are planning to repaint a room in your house.

Work out the area of each of the walls. Don't include doors and windows or any other areas which do not need to be painted.

Find out the cost of a tin of paint and what area it covers, and work out how much it would cost to paint the room.

Other areas

Triangles

Look at these two triangles. Each triangle is half the rectangle.

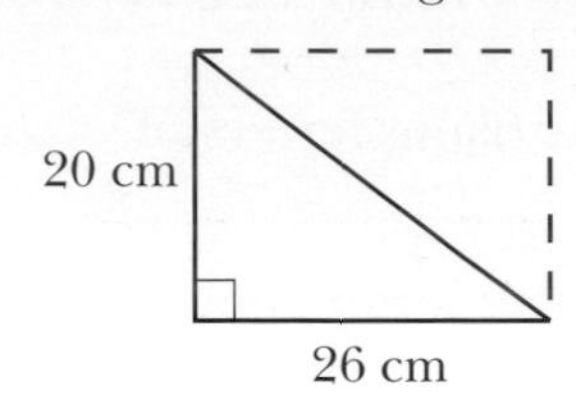

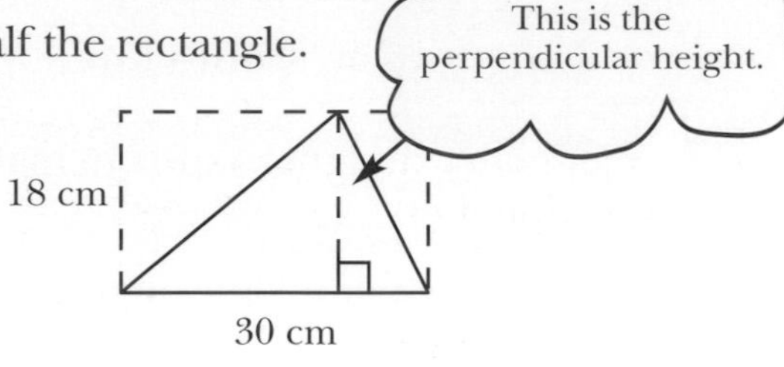

Area = $\frac{1}{2} \times 26 \text{ cm} \times 20 \text{ cm}$

= 260 cm^2

Area = $\frac{1}{2} \times 30 \text{ cm} \times 18 \text{ cm}$

= 270 cm^2

Area of a triangle = $\frac{1}{2}$ × base × perpendicular height

Parallelograms

Area of a parallelogram = $b \times h$

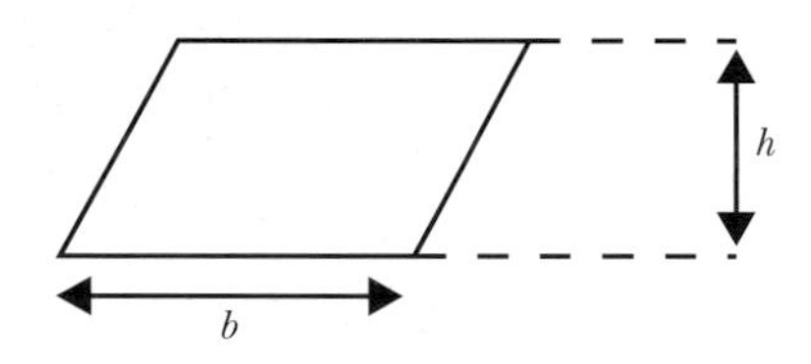

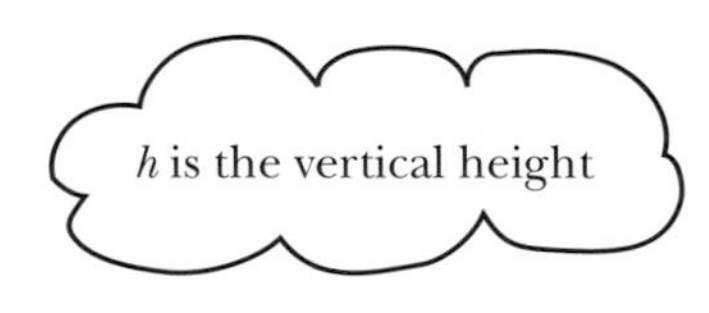

Explain why the rule works.

This diagram may help.

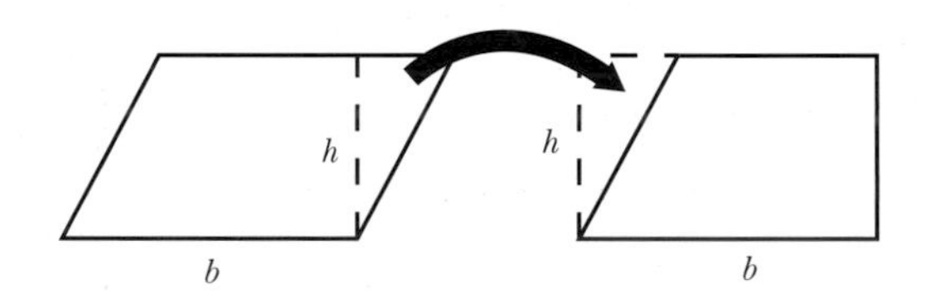

Trapezia

Area of a trapezium = $\frac{1}{2}(a + b) \times h$

Note: the plural of trapezium is trapezia.

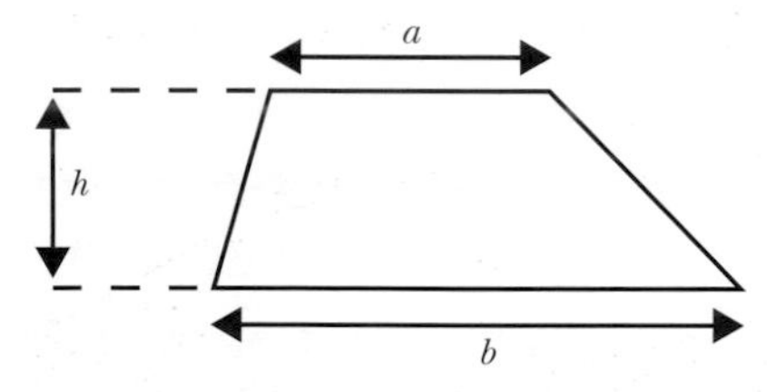

Explain why the rule works.
This diagram may help.
It shows two trapezia fitted together to make a parallelogram.

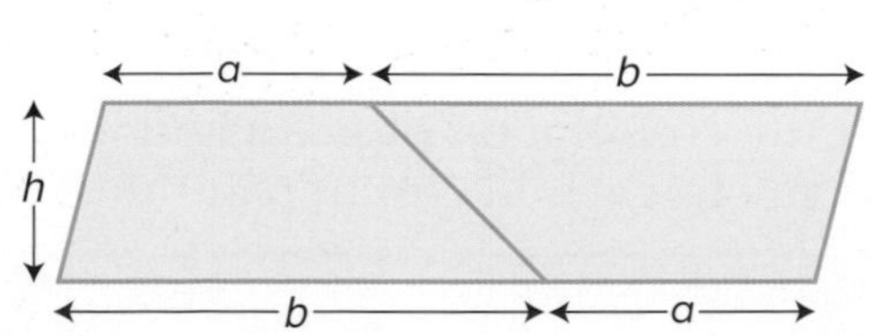

Exercise

1 Find the areas of the shapes below.

a) 6 cm, 5 cm

b)

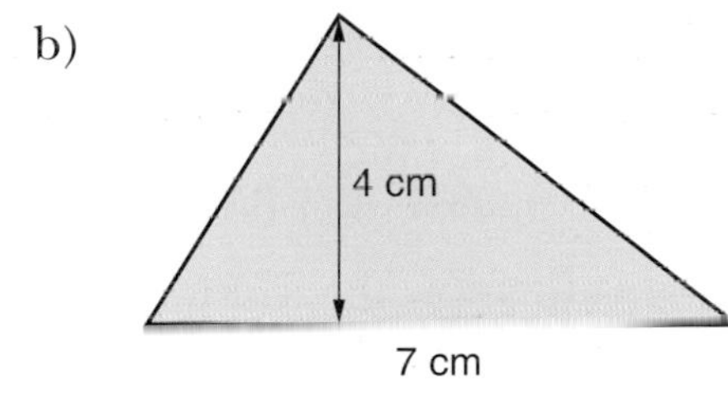

c)

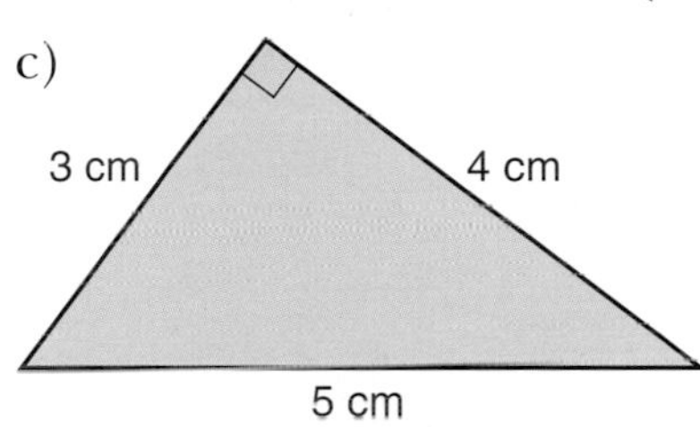

d)

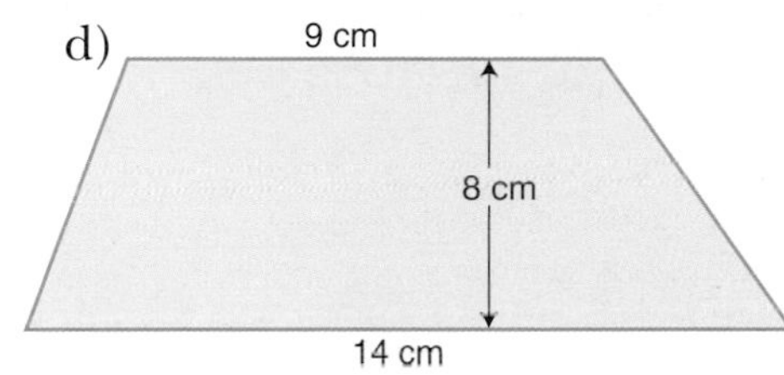

e)

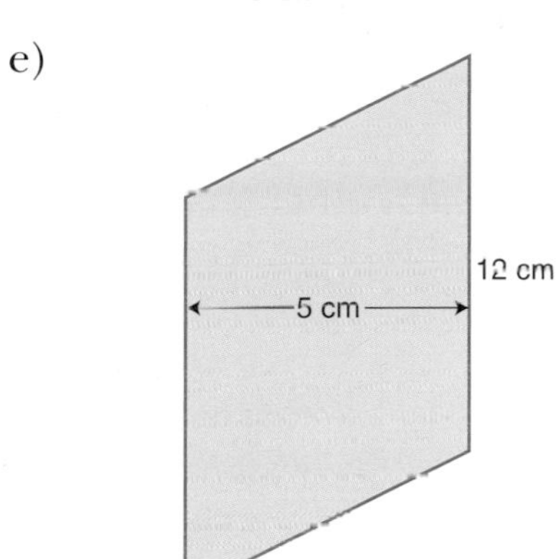

f)

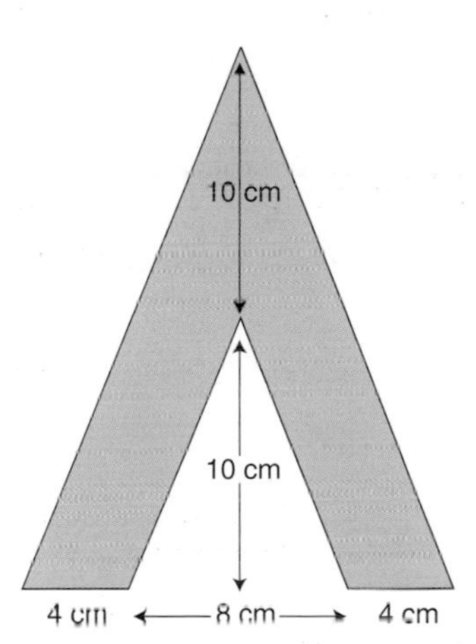

2 This diagram shows the sail of a boat.

The area of the sail is 24 m^2.

The sail is 2.5 m wide at its widest point.

How high is the sail?

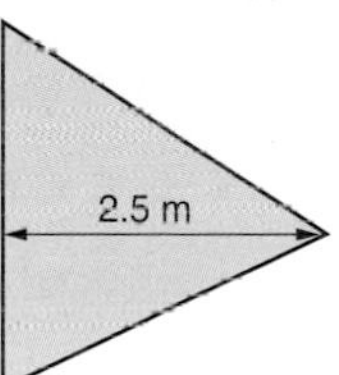

3 You are making the roof for a dolls' house out of two trapezia and two triangles, as shown. Work out the area of wood you need to make the roof.

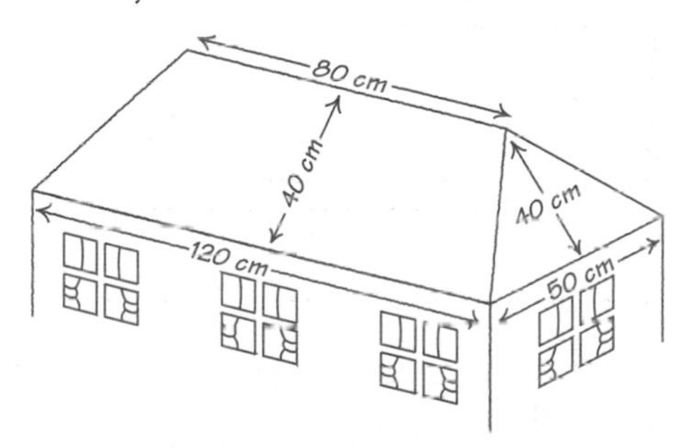

4 This map is divided into 1 km by 1 km squares. By counting squares, half squares and so on, estimate the area of each island.

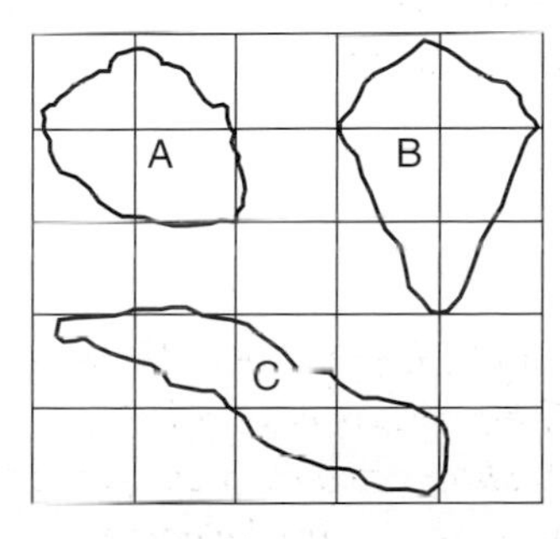

Using similarity

Justin wants to buy a larger television. His TV has a 34 cm screen and he would like a 51 cm one. Television screen sizes are given by the length of the diagonal, which is usually 34 cm, 48 cm, 51 cm, 59 cm or 66 cm.

First Justin wants to know the width of a 51 cm screen so that he can get an idea of how much space it will take up. He measures his 34 cm screen and finds that it is 29 cm wide.

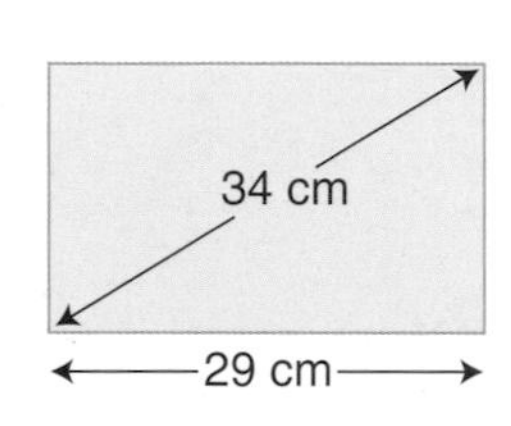

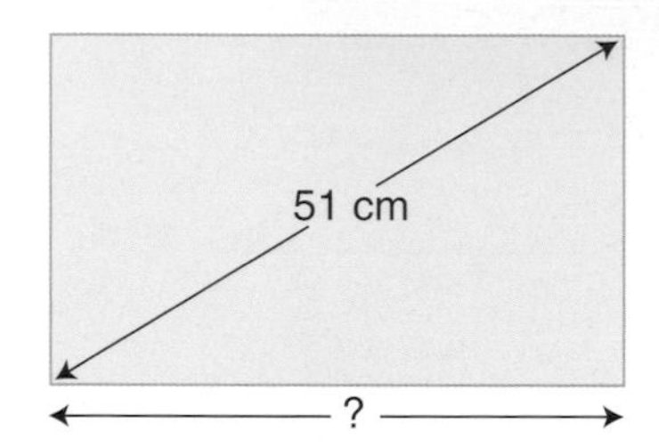

Justin's 34 cm screen and the larger 51 cm screen have similar shapes. **Similar shapes** are shapes that are enlargements of each other.

Are all rectangles similar?

Are all squares similar?

Are all circles similar?

The screens are similar so all the dimensions have been multiplied by the same scale factor.

Justin works out the scale factor like this:

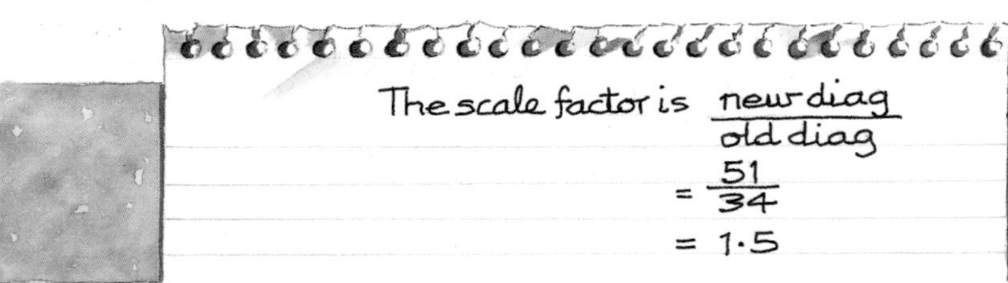

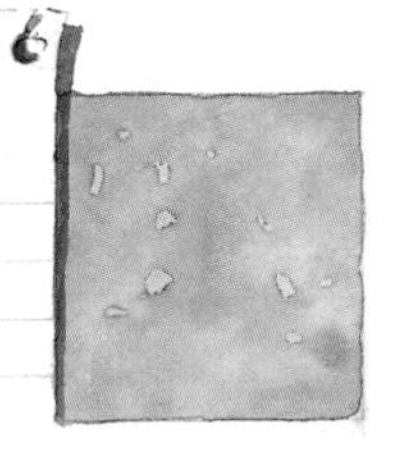

Check that multiplying 34 *by* 1.5 *gives* 51.

Justin works out the width of the new screen like this:

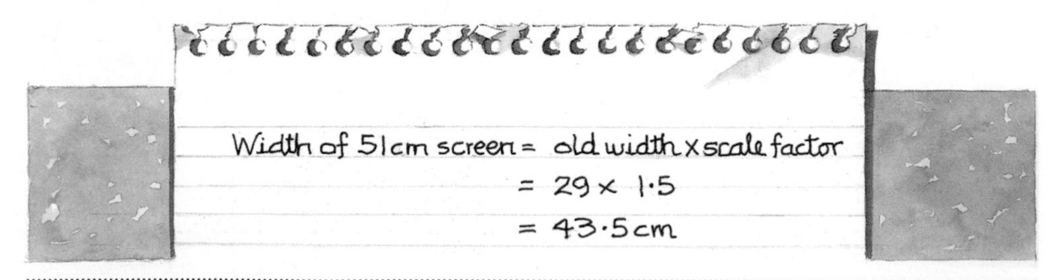

The new screen is 43.5 cm wide.

Exercise

1 Find the scale factor of these enlargements.

a) A photograph 15 cm long is enlarged to 20 cm long
b) A drawing 8 cm wide is enlarged to 18 cm wide
c) A diagram 24 cm long is reduced to 9 cm long.

2 a) Which of the photographs B–F are similar to photograph A?
Explain your answers.

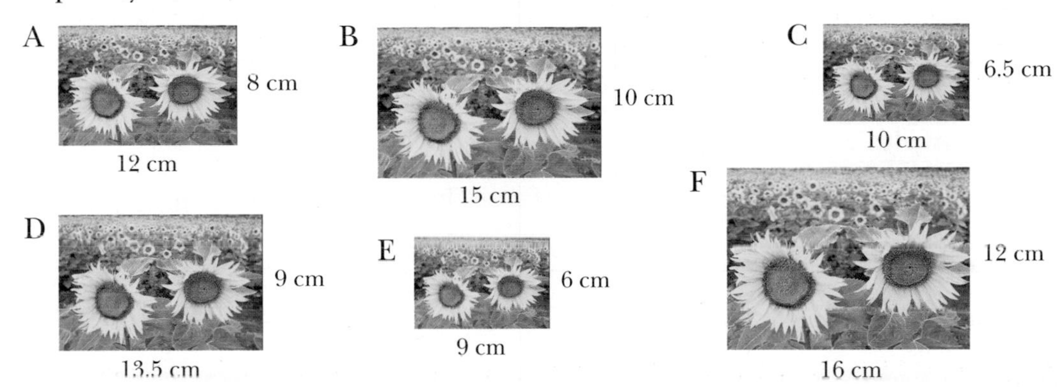

b) I need an enlargement of photograph A, 24 cm high.

(i) How wide will the enlarged photograph be?
(ii) Find, in its simplest form, the ratio
perimeter of photograph A : perimeter of enlargement.

3 In each of these questions there is a pair of similar shapes.
In similar shapes corresponding angles are equal. In each case draw the shapes side by side and the same way up and then work out the missing lengths.

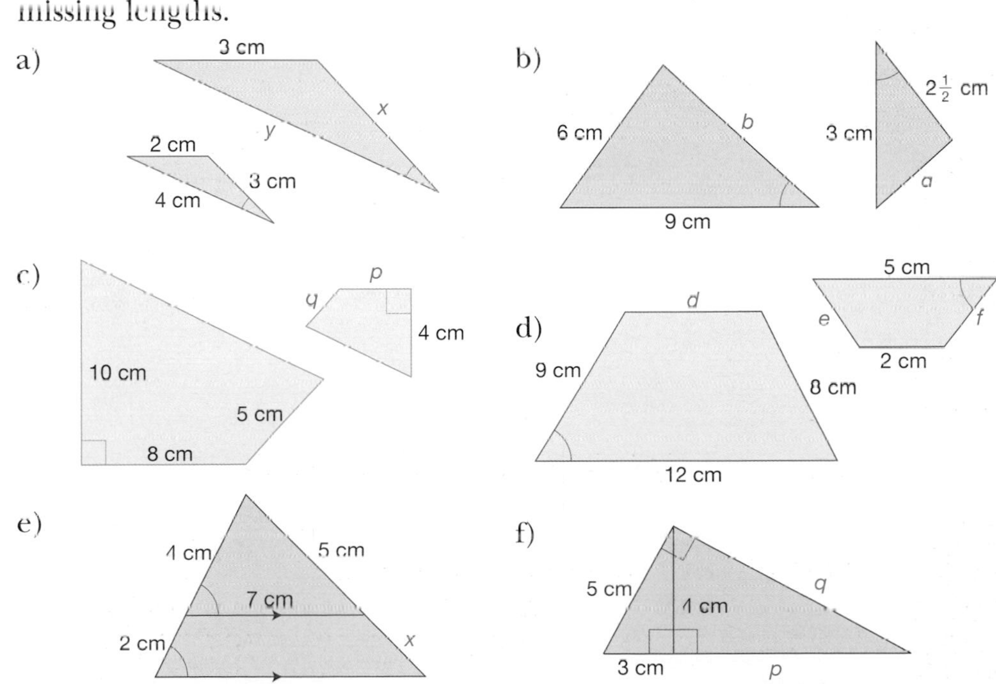

Circumference and area of a circle

The **circumference** of a circle is the distance round the edge.

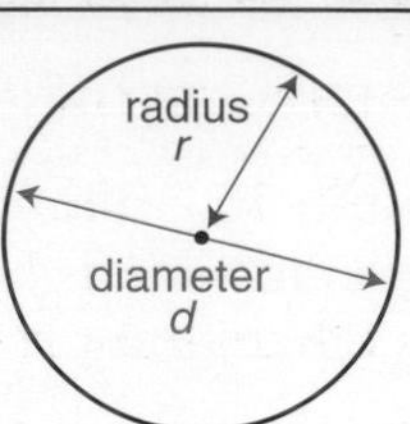

You can find the circumference, C, and area, A, of a circle using these formulae

You may have a key for π on your calculator. If not, use the value 3.14 for π

$$C = \pi d \qquad A = \pi r^2$$
$$\text{or } C = 2\pi r$$

Square the radius first, then multiply by π

Example

Find the circumference and area of a circle with diameter 6 cm.

Solution

Circumference $= \pi d$
$= \pi \times 6$
$= 18.85$

Area $= \pi r^2$
$= \pi \times 3^2$
$= 28.27$

The radius r is half the diameter

The circumference is 18.85 cm, and the area is 28.27 cm^2.

Example

Find the diameter of a circle with circumference 15 cm.

Solution

Circumference $= \pi d$

$15 = \pi \times d$

$d = 15 \div \pi = 4.77$

To work backwards you need to do the opposite of multiplying by π. You have to divide by π

The diameter is 4.77 cm.

Example

Find the radius of a circle with area 60 cm^2.

Solution

Area $= \pi r^2$

$60 = \pi \times r^2$

$r^2 = 60 \div \pi = 19.1$

$r = \sqrt{19.1} = 4.37$

When you work backwards you must do everything in reverse order. Divide by π first

Finding the square root is the reverse of squaring

The radius is 4.37 cm.

Exercise

1 Find the circumference of each of these circles.

a) Diameter = 5 cm b) Diameter = 8 cm c) Radius = 13 cm

2 Find the area of each of these circles.

a) Radius = 2 cm b) Radius = 7 cm c) Diameter = 11 cm

3 Find the diameter of the circle whose circumference is:

a) 25 cm b) 12 cm c) 34 cm

4 Find the radius of the circle whose area is:

a) 30 cm^2 b) 56 cm^2 c) 112 cm^2

5 A label is to be wrapped round a tin which has radius 4 cm.

Find the length of the label.

6 The distance round the edge of a circular flower bed is 24 metres.

a) Find the radius of the flower bed.

b) Find the area of the flower bed.

7 Find the area of each of the following shapes. The dotted lines are there to help you.

a)

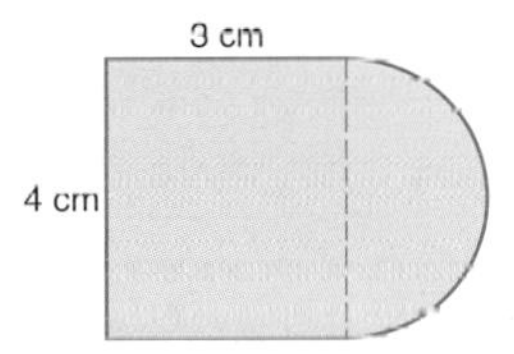

b)

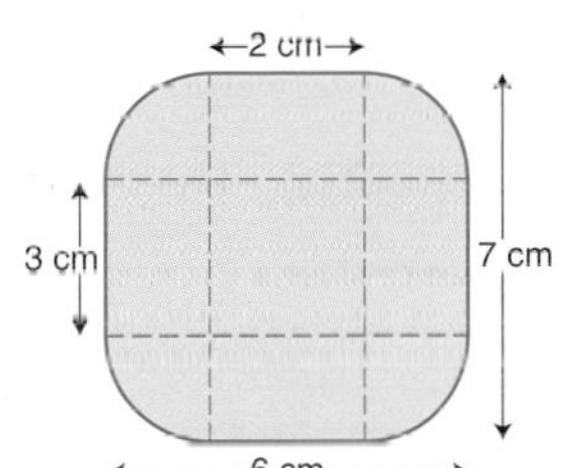

c)

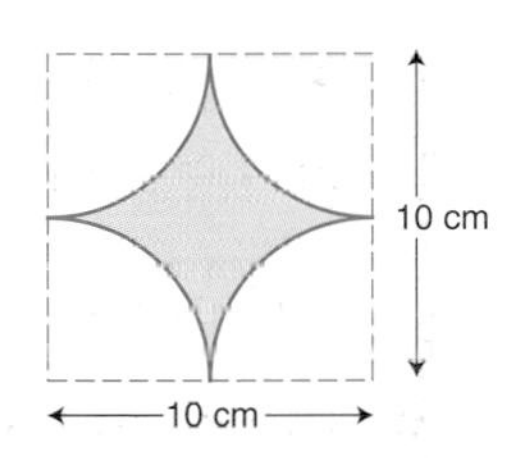

8 The diagram shows a large circular grassed area, radius 40 m, in a park.
Six seats, labelled A, B, C, D, E and F, are equally spaced around the circumference.

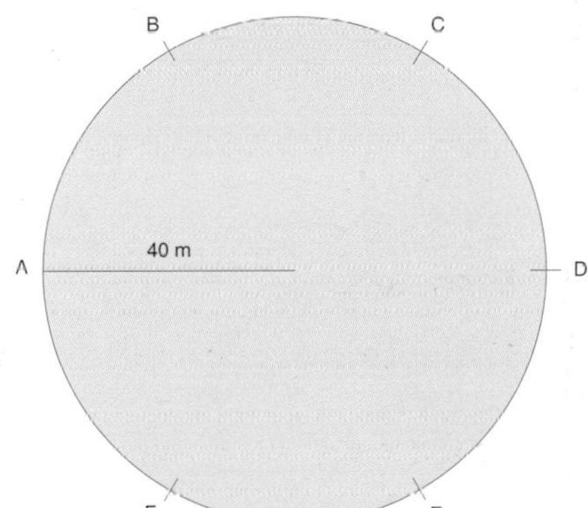

Dan and Abi walk from A to B.

a) Dan walks around the edge of the grassed area. (This curved path AB is called an **arc**.) What is the length of the arc AB?

b) Abi walks directly from A to B. (The straight line AB is called a **chord**.) What is the length of the chord AB?

c) How much further than Abi does Dan walk?

9 A circle has a radius of 9 metres.

Work out a) the circumference b) the area.

Give your answers as multiples of π.

Finishing off

Now that you have finished this chapter you should be able to:

- ★ find the perimeter of any shape with straight sides
- ★ find the area of a rectangle, a triangle, a parallelogram, a trapezium or of a shape made up from these
- ★ find an unknown side given similar shapes
- ★ find the circumference and area of a circle
- ★ work out an unknown length when the area or perimeter is known.

Use the questions in the next exercise to check that you understand everything.

Mixed exercise

1 This is the floor plan of a room.

a) A decorative border is to be put round the top of the walls.

How many metres are needed?

b) The room is to have a fitted carpet. Carpet costs £16.50 per square metre.

How much does the carpet for the room cost?

6 m
4 m
1 m
2·5 m
2 m

2 Find the areas of these shapes.

a)

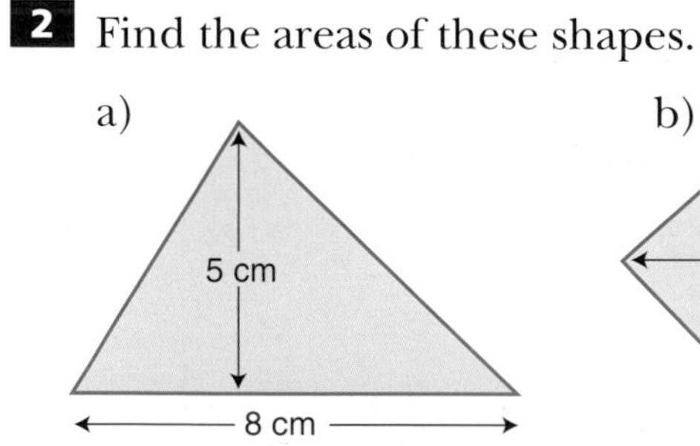

b)

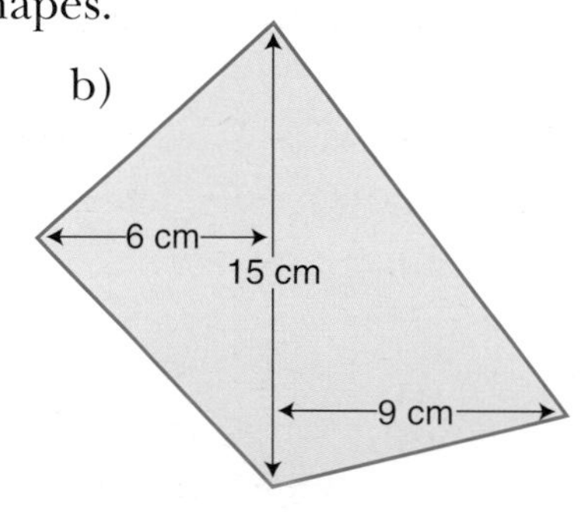

c)

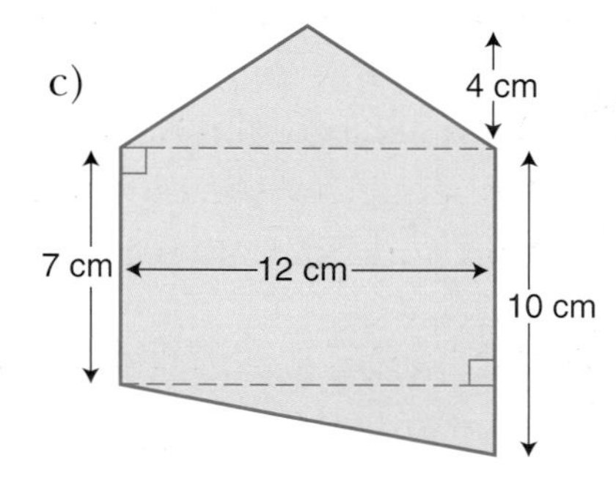

d)

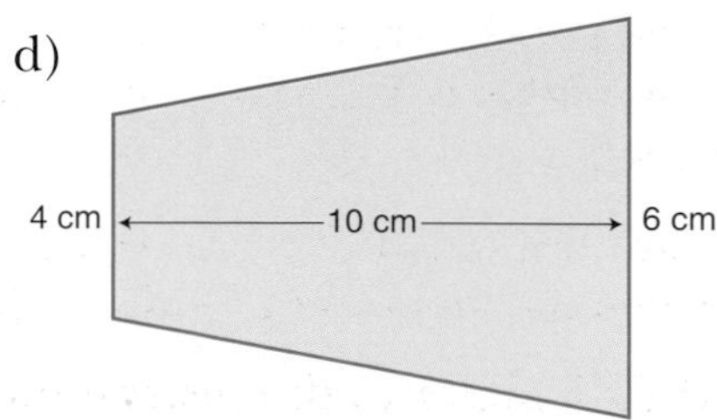

e)

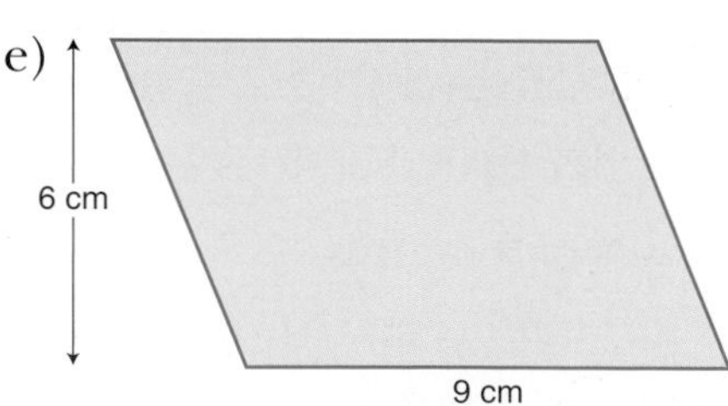

Mixed exercise

3 These three triangles are similar. Find the missing lengths.

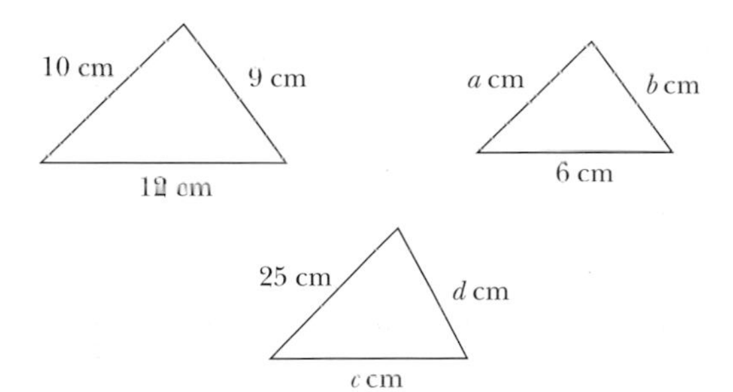

4 Oliver's house is built on a rectangular plot 25 m long and 10 m wide. On a plan the plot is 2 cm wide.

a) Work out the scale of the plan.

b) How long is the plot on the plan?

c) Work out the area of the plot

(i) on the plan (ii) in real life.

d) Explain how the answers in c) and the answer in a) are connected.

5 The different 'A' sizes of paper, such as A3, A4 and A5, are all similar shapes. As the numbers go up, the paper sizes get smaller. The height of a particular size of paper is always the same as the width of the previous size. A4 paper is 210 mm wide and 297 mm high.

Work out the dimensions of all paper sizes from A0 (the biggest) to A6. Write your answers in a table like this.

Paper size	Height (mm)	Width (mm)
A0		
A1		
A2		
A3		
A4	297	210
A5	210	
A6		

6 Mitchell's bicycle wheel has a diameter of 70 cm.

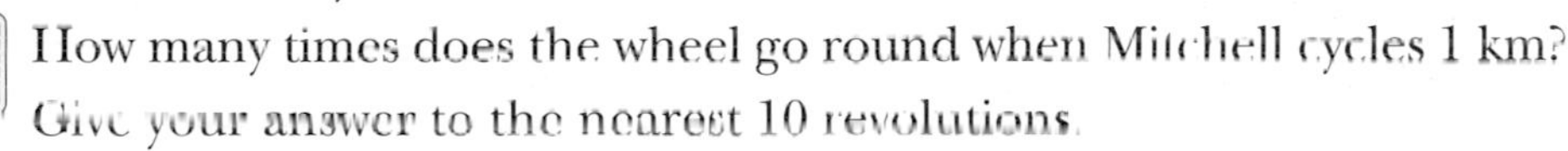

How many times does the wheel go round when Mitchell cycles 1 km?

Give your answer to the nearest 10 revolutions.

7 The circumference of the Earth is about 40 000 km.

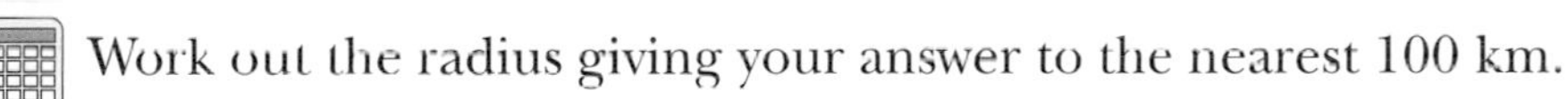

Work out the radius giving your answer to the nearest 100 km.

8 A cake band, which goes round the outside of a cake, is 65 cm long.

Work out the diameter of the largest circular cake that the band will fit.

Give your answer as a whole number of centimetres.

9 Work out the area of this washer.

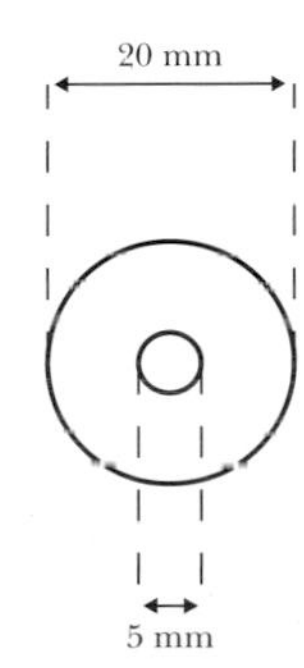

10 Last Friday Richard and Louise each ate half of a 12 inch (diameter) circular pizza.

This Friday they want different toppings on their pizzas so they have a pizza each.

What size of pizza will they order if they want the same amount of food as before?

Give the diameter to the nearest inch.

Twenty two

Three dimensions

Cubes and cuboids

The diagram shows a small box of cereal.

How much room is there in the box?

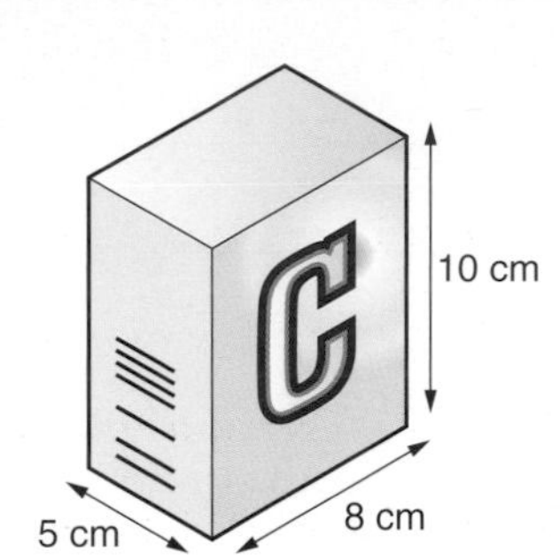

The amount of space taken up by the box is called **volume**.

To find the volume of the box, you need to know how many cubic centimetres it takes up.

Imagine that you have lots of cubes each 1 cm long, 1 cm wide and 1 cm high. Each cube has a volume of 1 cubic centimetre (1 cm^3).

The bottom of the box is a rectangle. It is 8 cm long and 5 cm wide. Each layer has 40 cubes.

$8 \times 5 = 40$

The box is 10 cm high. Each layer has 40 cubes and there are 10 layers.

$40 \times 10 = 400$

There are 400 cubes in total

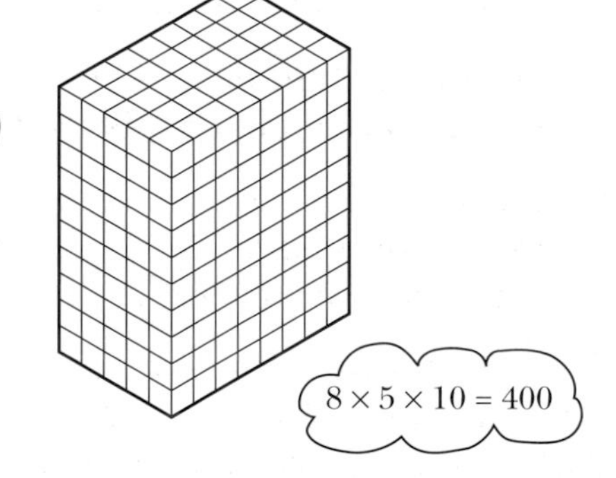

There are 400 cubes so the box has a volume of 400 cm^3.

Volume = Length × Width × Height

A larger box has a volume of 500 cm^3. *The length is* 8 cm *and the width is* 5 cm *as before. What will the height be?*

Faces, edges and vertices

The cereal box is made up of 6 rectangles. These are called **faces**.

A box with 6 rectangular faces is called a **cuboid**.

A box whose 6 faces are all squares is called a **cube**.

How many **edges** *are there on a cereal box?*

How many **vertices** *(corners) are there?*

You will find that a cuboid has 12 edges and 8 vertices.

Exercise

1 Find the volume of each of the following cuboids.

a)

6 cm

9 cm

5 cm

b)

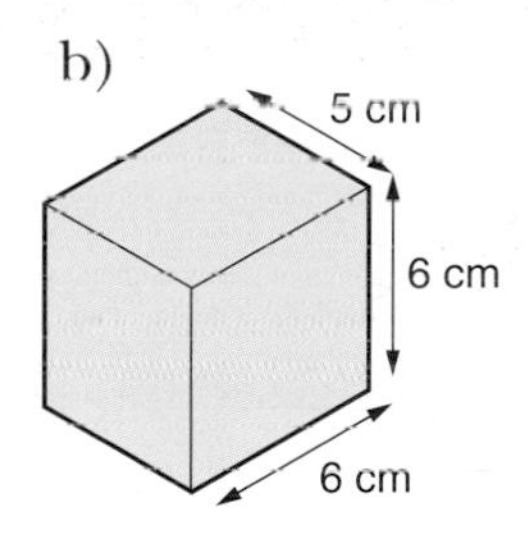

c)

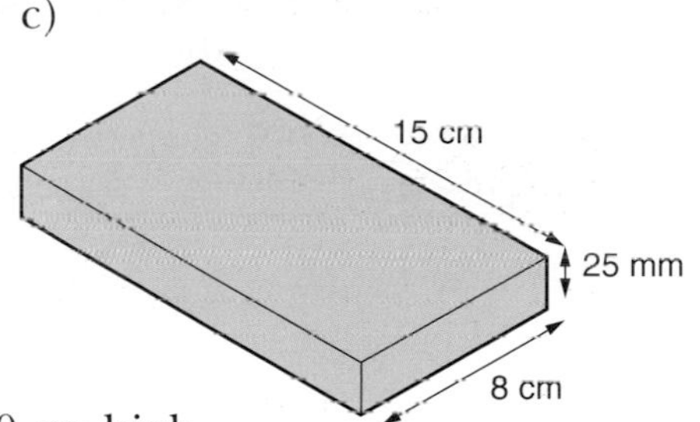

2 A container is 150 cm long, 120 cm wide and 130 cm high.

a) Work out its volume in cm^3.

b) Change each length into metres and work out the volume in m^3.

c) How do you convert between cubic centimetres and cubic metres?

3 A fish tank is 60 cm long and 40 cm wide.
It is filled with water to a depth of 30 cm.

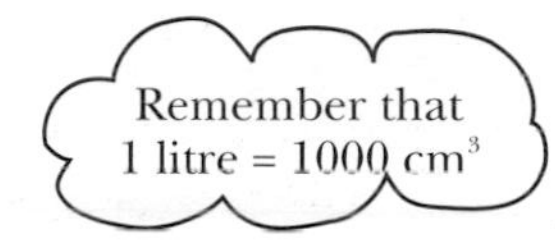

How many litres of water are there in the tank?

4 Abdul is designing a rectangular container to hold 1 litre of fruit juice.

a) His first design has a square base of side 5 cm. Work out the minimum height of this container.

b) His second design has a base 9 cm by 6 cm. Work out the minimum height of this container. Give your answer as a whole number of centimetres.

5 An open topped box is made by cutting out a rectangle 15 cm by 12 cm, and cutting a square 3 cm by 3 cm off each corner.

Find the volume of the box.

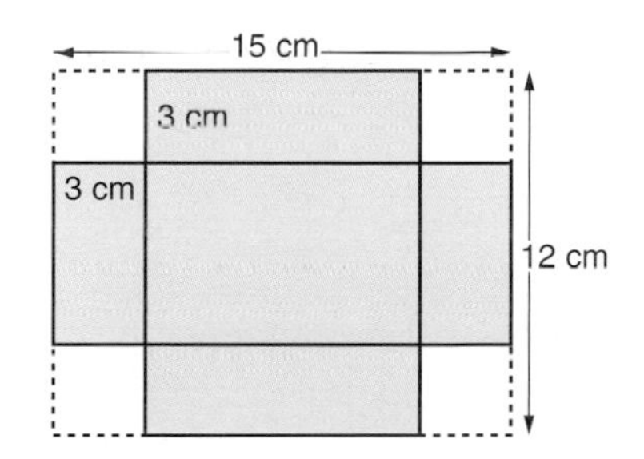

6 A block of wood is 30 cm long, 8 cm wide and 2 cm thick.

a) Calculate the volume.

The density of the wood is 0.7 g/cm^3.

b) Calculate the mass of the block.

7 A box is 50 cm long, 30 cm wide and 40 cm high. A packet is 10 cm long, 6 cm wide and 13 cm high.

What is the maximum number of packets that will fit into the box.

8 A swimming pool has 3750 m^3 of water in it.
It is 3 metres deep, and the length of the pool is twice its width.

Find the length of the pool.

Volume of a prism

When slices are cut (as shown) from this wedge of cheese, each piece is the same size and shape.

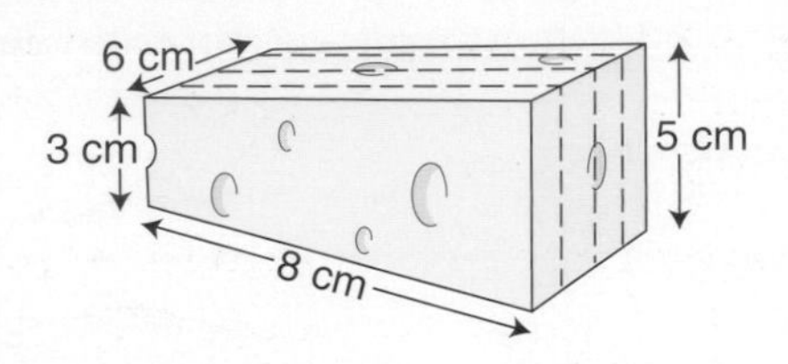

The wedge of cheese is an example of a **prism**.

A prism is a solid which has the same cross section all the way along its length.

Think of 3 other familiar objects whose shapes are prisms.

If you want to find the volume of a prism, the first step is to find the area of the cross section. In the case of the wedge of cheese, this is a trapezium.

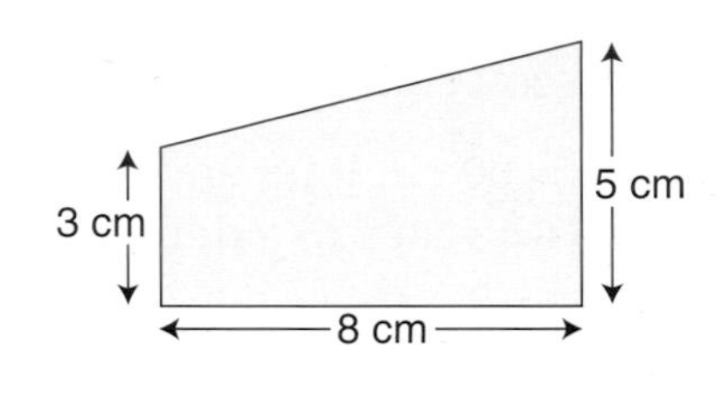

$$\begin{aligned}\text{Area (in cm}^2) &= \frac{1}{2}(a+b) \times h\\ &= \frac{1}{2} \times (3+5) \times 8\\ &= \frac{1}{2} \times 8 \times 8\\ &= 32\end{aligned}$$

A slice of cheese 1 cm thick has a volume of $32 \text{ cm}^2 \times 1 \text{ cm} = 32 \text{ cm}^3$.

Two slices, each 1 cm thick, have a combined volume of

$2 \times 32 \text{ cm}^3 = 64 \text{ cm}^3$

The whole wedge of cheese is 6 cm long, so its volume is the same as the volume of 6 slices of cheese, each 1 cm thick.

The volume of the wedge of cheese = $32 \text{ cm}^2 \times 6 \text{ cm} = 192 \text{ cm}^3$.

The volume of any prism can be found in the same way, by multiplying the area of the cross section by the length of the prism.

Volume of a prism = area of cross section × length

This Swiss roll is a prism with circular cross section. This shape is usually called a **cylinder**.

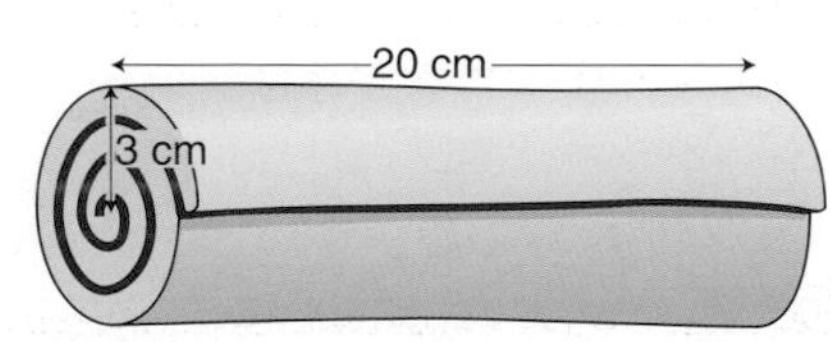

Using the above rule,

Area of circular cross section (in cm^2) = $\pi r^2 = \pi \times 3^2 = 9\pi$

Volume = area of circular cross section × length

$= 9\pi \text{ cm}^2 \times 20 \text{ cm}$

$= 565 \text{ cm}^3$ (to the nearest cm^3)

It is often easiest to leave π in your working until the end. Sometimes an answer is given as a multiple of π.

Exercise

1 A child's toy consists of five plastic prisms (shown below) with different cross sections, and a box with matching holes through which to post them.

Each prism is 5 cm long. Find the volume of each prism.

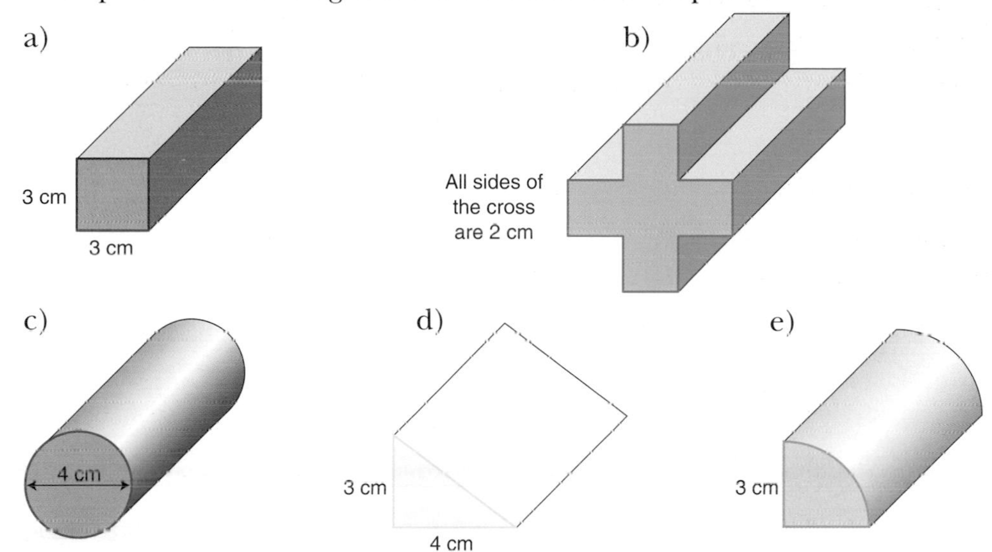

2 The cross section of a swimming pool is shown in this diagram.

The pool is 10 m wide, and it is filled to the brim.

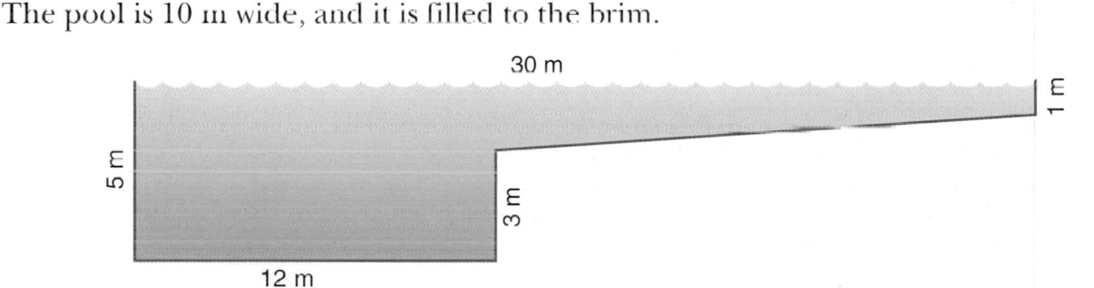

Find the volume of water in the swimming pool.

3 Metal discs for pet collars have a diameter of 4 cm. Each disc is 0.1 cm thick. How many discs can be made from 1000 cm^3 of metal?

4 A solid cylinder is 10 cm long and has a diameter of 6 cm. Work out the volume giving your answer as a multiple of π.

Measure the diameter and thickness of a 1p coin and a 2p coin and work out the volume of each coin. Does a 2p coin contain twice as much metal as a 1p coin?

Compare the amounts of metals in other coins, and make a table of your results.

More about prisms

Claire designs food packaging. She is working on a new design for a packet of ground coffee.

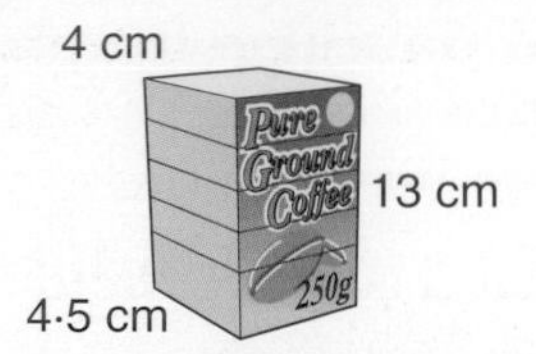

The existing packet is a cuboid as shown. The volume of this packet is

$$13 \text{ cm} \times 4.5 \text{ cm} \times 4 \text{ cm} = 234 \text{ cm}^3$$

The new design must have the same volume so that it holds the same amount of coffee.

Claire's first idea is a triangular prism like this:

Claire needs to work out how long the packet must be to have a volume of 234 cm^3.

Here is Claire's calculation.

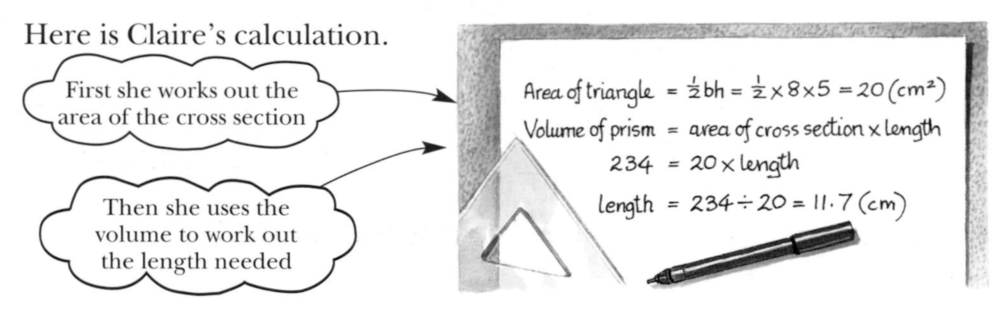

The packet needs to be 11.7 cm long.

Claire's next idea is a packet in the shape of a cylinder. She wants a cylinder 12 cm high. She needs to work out the radius of the cylinder.

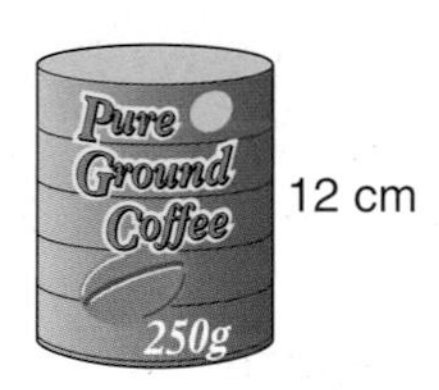

Here is Claire's calculation.

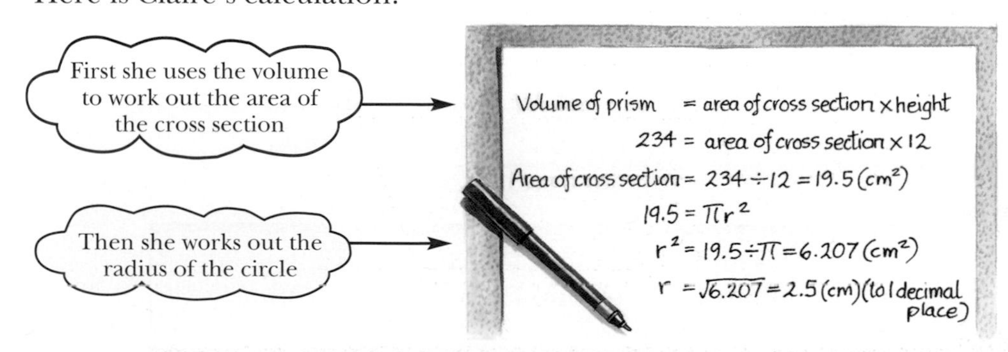

The radius of the cylinder should be 2.5 cm.

What are the advantages and disadvantages of Claire's designs?

Exercise

1 The volume of each prism is marked on it. Find the unknown length in each case.

2 Find the surface area of the four prisms involving a, c, d and e in question 1.

3 A drinks company is trying out different sizes for cans of fizzy drink.

a) One size is to hold 250 ml. The radius of the base of the can is 3 cm. Find the height of the can.

b) Another size is to hold 400 ml. The height of the can is 12 cm. Find the radius of the base.

4 A skip has a capacity of 6 m^3. It is in the shape of a prism with the cross section shown. How wide is the skip?

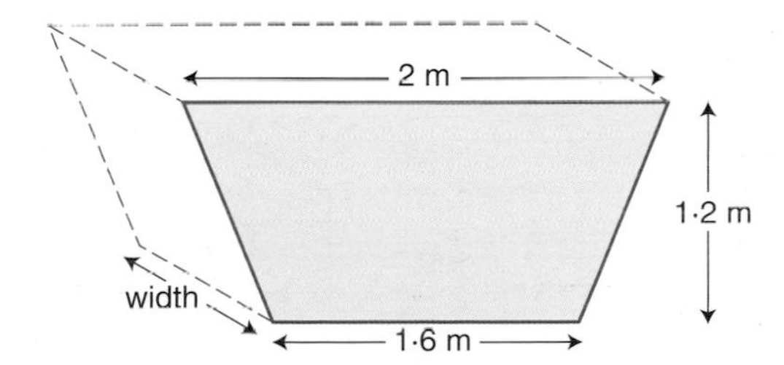

Design two different cartons, one a cuboid and the other a cylinder, to hold 500 ml of orange juice. Which shape do you think is better?

Drawing solid objects

All solid objects have three dimensions: length, width and height.

A drawing has only two dimensions: length and width.

This makes solid objects difficult to draw.

There are several ways of representing solid objects on paper. Here are three different ways of drawing the solid shape called a triangular prism.

Three-dimensional drawings

These are drawings made to look like three-dimensional objects. They may use perspective.

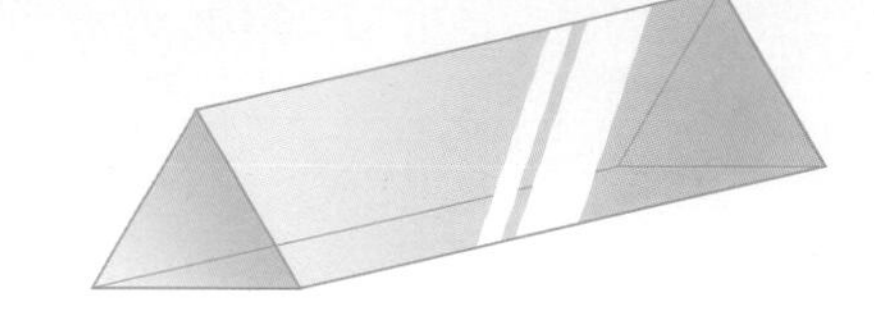

Nets

Nets show what the shape would look like if it could be opened up and made flat.

Views

These show what the shape looks like from the top, the front and the side.

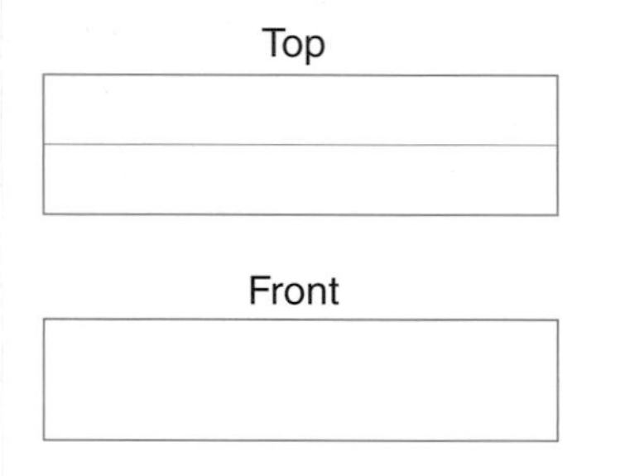

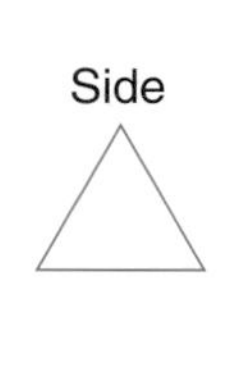

Which drawing gives the best idea of what the prism really looks like?

Solid objects made up from cubes can also be drawn using **isometric paper**. Make sure that you use the paper the right way round and only put in edges that can be seen. Here are two examples.

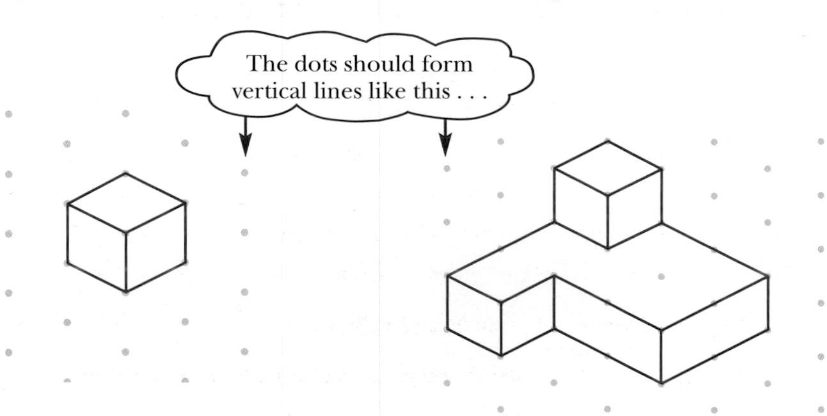

This is a cube of side 1 cm.

This is a drawing of an arrangement of 8 cubes – 7 cubes on the bottom layer and 1 on the top.

Exercise

1 Which of the following diagrams are nets for a cube?

A
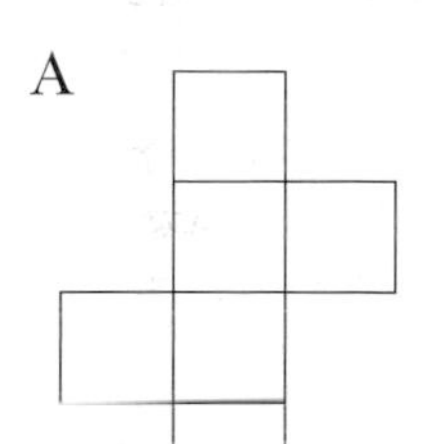
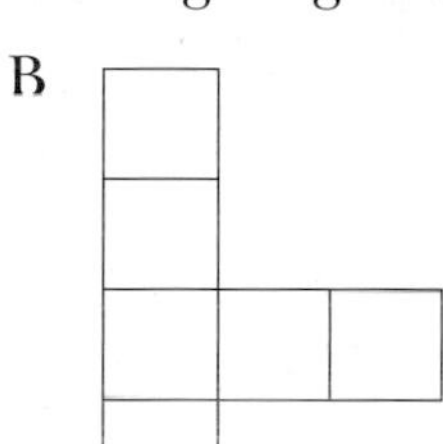
B
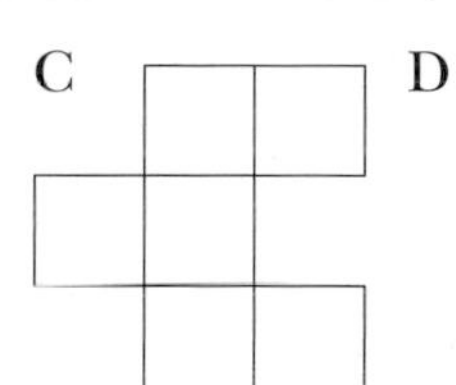
C
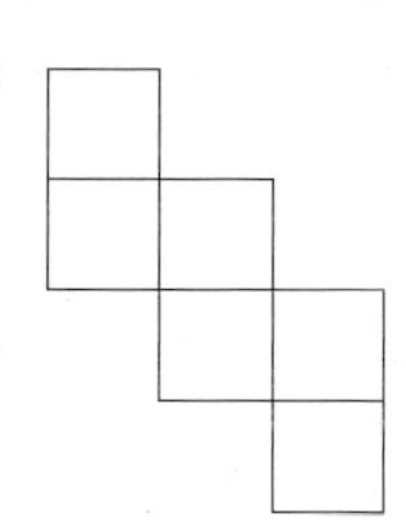
D

Copy your nets on to squared paper and cut them out to check your answers.

2 The diagram shows a cuboid 4 cm long, 3 cm wide and 2 cm high.

a) Draw a full size net for the cuboid on a piece of centimetre squared paper.

b) Work out the surface area of the cuboid.

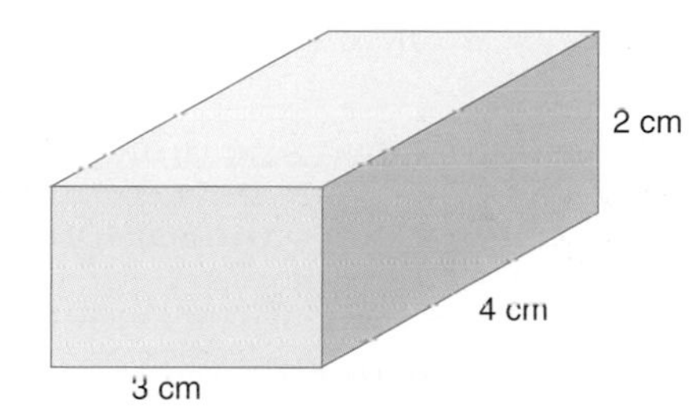

3 Here are drawings of a) a house and b) a tent.

a)
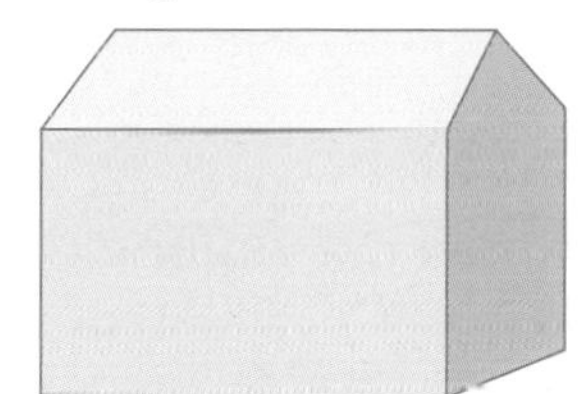
b)
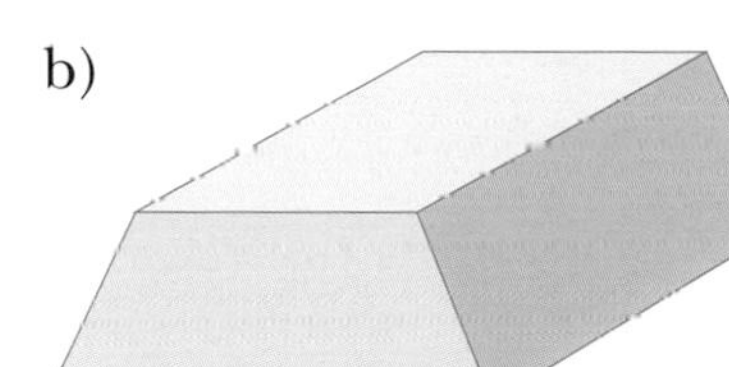

In each case draw diagrams to show

i) the front view ii) the side view iii) the top view

4 This diagram shows a net for a solid called a **tetrahedron**.

The tetrahedron is formed by folding along AB, BC and CA so that D, E and F meet at a point.

a) How many (i) faces, (ii) edges, (iii) vertices does a tetrahedron have?

b) Draw another net for this tetrahedron, which has a different shape from the net above.

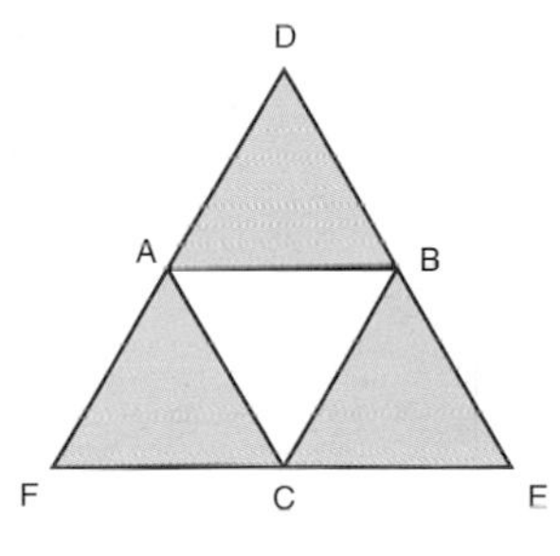

5 Draw the following shapes on isometric paper.

(Check that the paper is the right way round before you start.)

a) A cube of side 3 cm

b) A cuboid 2 cm long, 1 cm wide and 4 cm high

c) A cuboid 4 cm long, 4 cm wide and 3 cm high

Using dimensions

You have now met quite a lot of different formulae for lengths, areas and volumes. There are many more! Some of them are quite similar and it can be difficult to remember which is which. It is possible to tell just by looking at a formula whether it is for a length, an area or a volume. The way you can tell is by deciding how many **dimensions** there are in it. This means the number of length measurements that have been multiplied together. Numbers, like 2 or π or $\frac{1}{2}$, don't count, because they aren't measurements of length.

$2\pi r$ $\quad \pi r^2 h$ $\quad lwh$ $\quad 2l+2w$ $\quad \frac{1}{2}(a+b)h$ $\quad \pi r^2$

- Lengths have one dimension
- Areas have two dimensions
- Volumes have three dimensions.

Example 1

lwh

l, w and h are all measurements of length. The three measurements are multiplied together

The formula has **three** dimensions. So this formula is a **volume**.

Which solid has the formula Volume = lwh?

Example 2

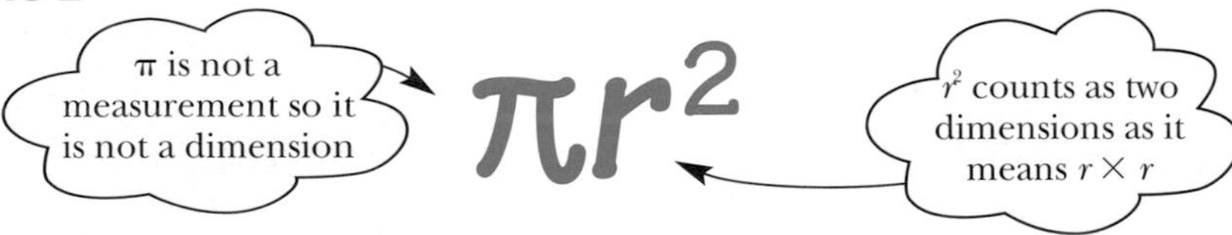

The formula has **two** dimensions. So this formula is an **area**.

Which shape has the formula Area = πr^2?

Example 3

In some formulae there are two or more parts added together. Each part must have the same number of dimensions. The number of dimensions of the whole formula is the same as the number of dimensions of each part.

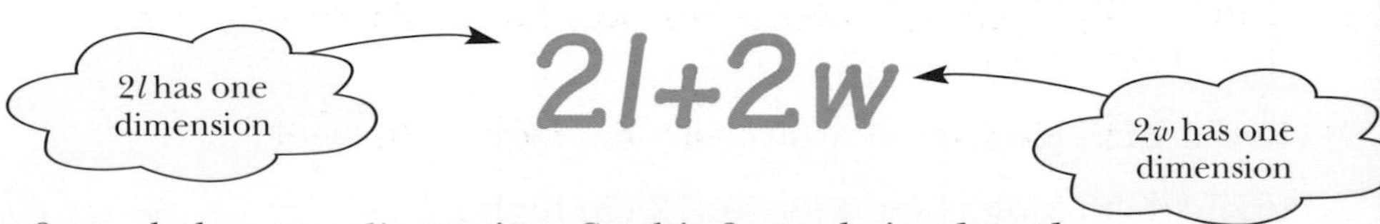

The formula has **one** dimension. So this formula is a **length**.

Which shape has the formula Perimeter = $2l + 2w$?

Exercise

1 Each of these formulae is either a length (perimeter or circumference), area or volume of one of the shapes shown below. For each formula:

a) say how many dimensions it has and what they are (Example: $3a^2b$ has three dimensions, a, a and b).

b) say whether the formula is a length, area or volume.

c) say which of the shapes A – F it belongs to.

(i) x^2y (ii) $2(x + y)$ (iii) πy^2 (iv) xy

(v) $\frac{1}{2}x(x + y)$ (vi) πx^2y (vii) $\frac{1}{2}xy$ (viii) $2\pi y$

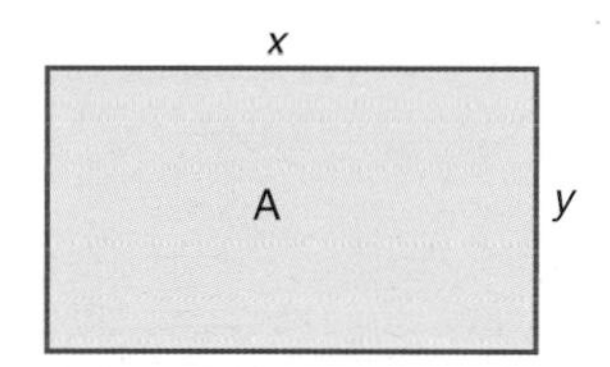

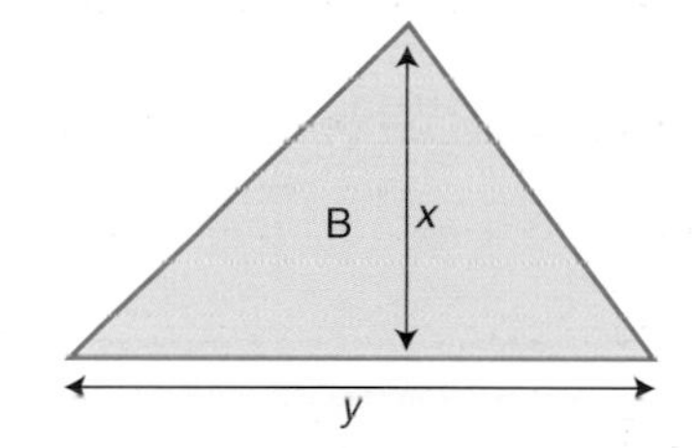

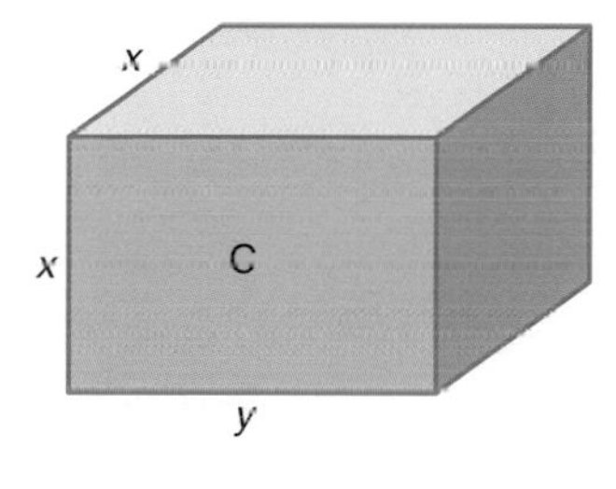

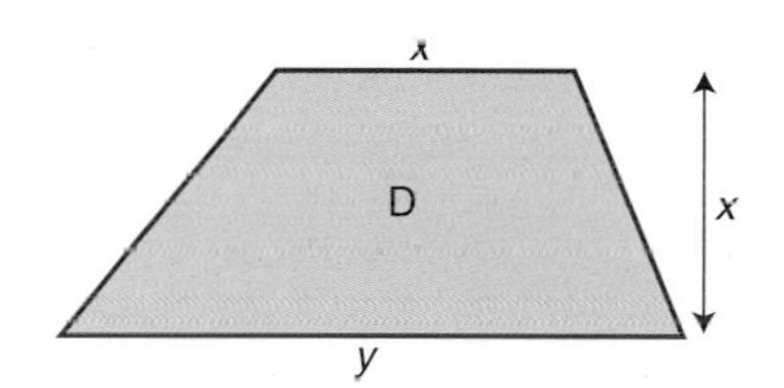

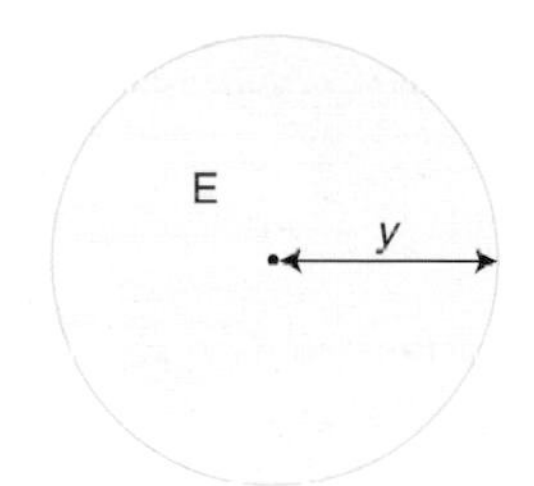

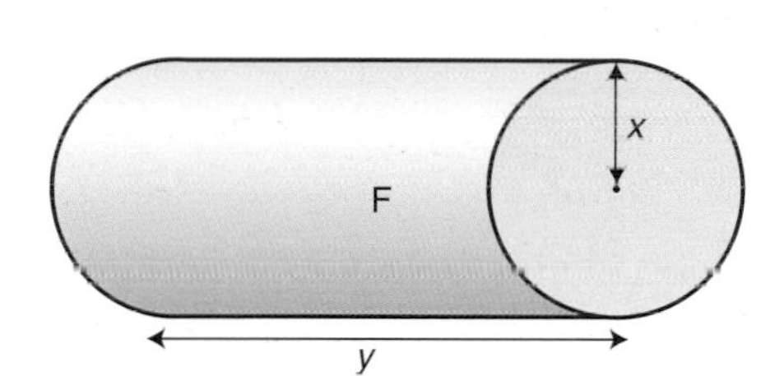

2 Some of the formulae below are real formulae for length, area or volume. Some could not be real formulae because they have the wrong number of dimensions. For each formula, write down whether it is a length, an area, a volume or not a real formula. r, h, a and b are all measurements of length. Any other letters used are not.

a) $\frac{1}{3}\pi r^2h$ b) $2(ah + bh + ab)$ c) $4a + 2b$ d) $2\pi rh + 2\pi r^2$

e) πr^2a^2 f) $\pi r + 2r$ g) $2r^2 + a^2b$ h) $\frac{4}{3}\pi r^3$

Finishing off

Now that you have finished this chapter you should be able to:

★ find the volume of a cuboid or a solid made up of cuboids

★ count the faces, edges and vertices of a solid

★ recognise and use different ways of drawing solids

★ find the volume of a prism

★ find an unknown length when the volume is known

★ distinguish between formulae for length, area and volume by using dimensions.

Use the questions in the next exercise to check that you understand everything.

Mixed exercise

1 Find the volume of each of these prisms.

a)

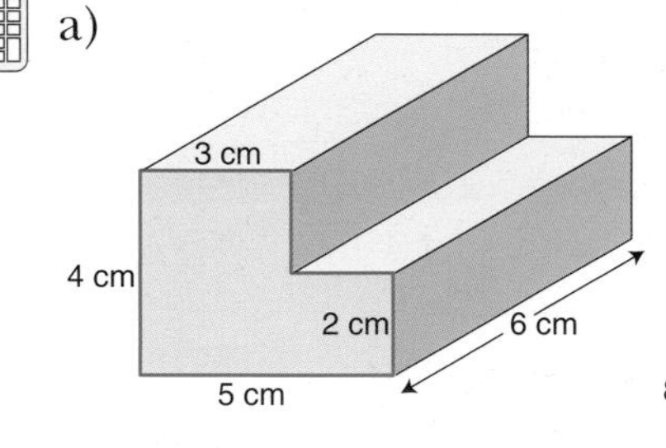

b)

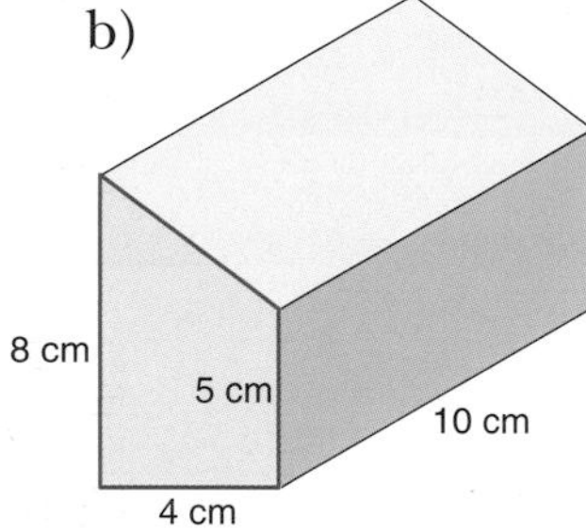

c)

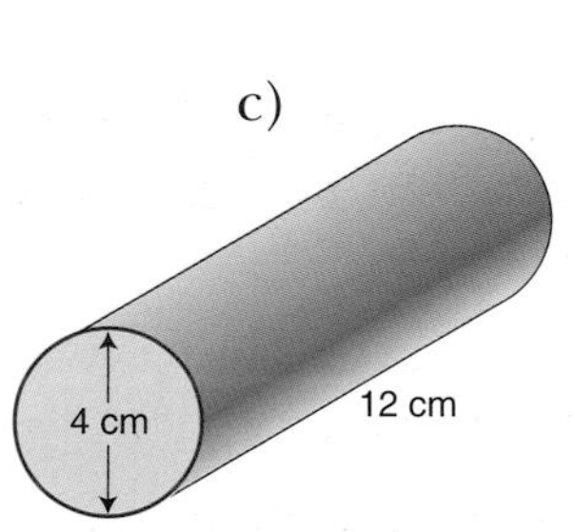

2 A baby's bottle is approximately the shape of a cylinder. The diameter of its base is 6 cm. The bottle can be filled to a maximum depth of 9 cm.

a) What is the maximum amount of milk that the bottle can hold?

b) Baby Sam drinks 150 ml of milk at each feed (1 ml is the same as 1 cm^3).

To what depth should his bottle be filled?

3  A measuring cylinder is marked in 20 ml increments. The marks are 1 cm apart. Find the radius of the cylinder.

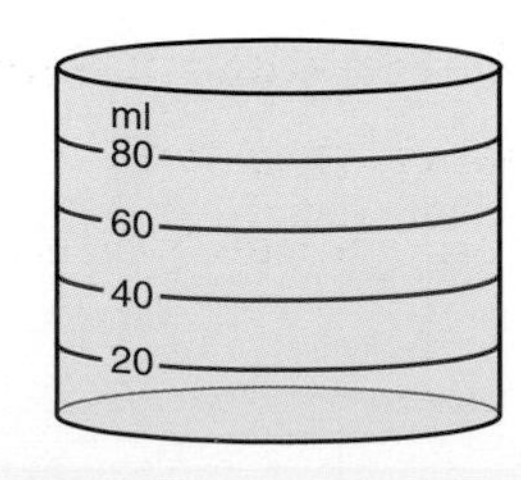

4 This greenhouse has a capacity of 300 cubic feet.

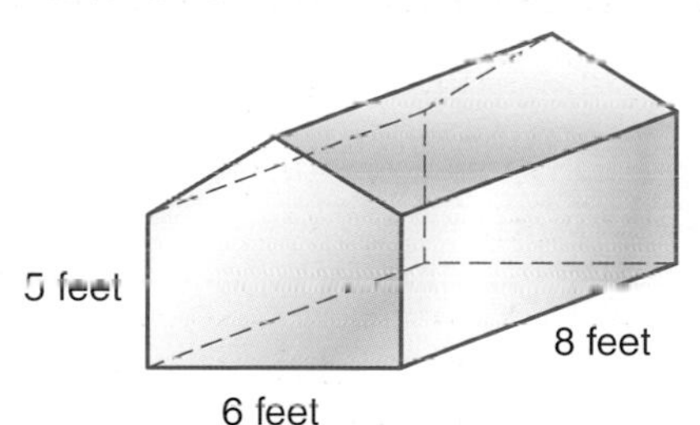

Find the total height of the greenhouse.

5 This solid shape is made up of centimetre cubes.

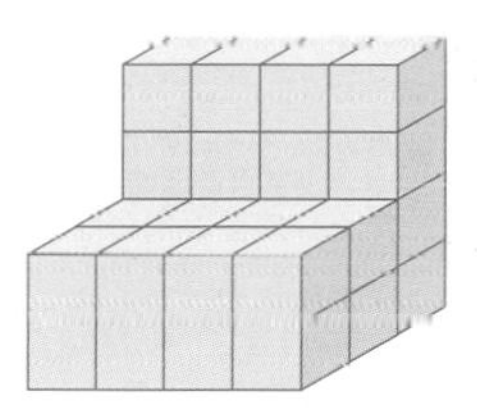

a) Draw it on isometric paper.

b) Work out its volume.

6 This diagram on centimetre squared paper shows the net for a pyramid. The pyramid is formed by folding along AB, BC, CD and DA so that P, Q, R and S meet at a point.

a) Work out the surface area of the pyramid.

b) How many (i) faces, (ii) edges, (iii) vertices does a pyramid have?

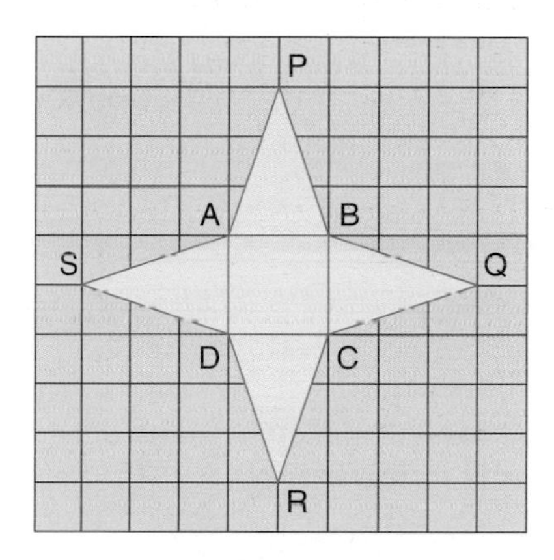

7 This sweet box is in the shape of a prism. The cross-section is a right-angled triangle.

a) Sketch a net for the prism and on it mark the length of each line.

b) Work out the surface area of the prism.

c) Work out the volume of the prism.

8 For each formula, say whether it is a formula for length, area, volume or none of these. The letters p, q, r and h are all measurements of length.

a) $3pq^2$ b) πrh c) pr^2h d) $\pi(p+q)$

e) p^2r+qh f) πr^3 g) $r(p+q)$

9 A solid metal cylinder has a radius of 6 mm and a length of 25 mm.
The density of the metal is 7.8 g/cm^3.
Work out the mass of the cylinder.

Twenty three

Transformations

Reflection symmetry

The McDonald's logo has **reflection symmetry**. If you put a mirror on the dotted line shown, the logo looks exactly the same in the mirror as it does without the mirror.

The logo has **one line of reflection symmetry**.

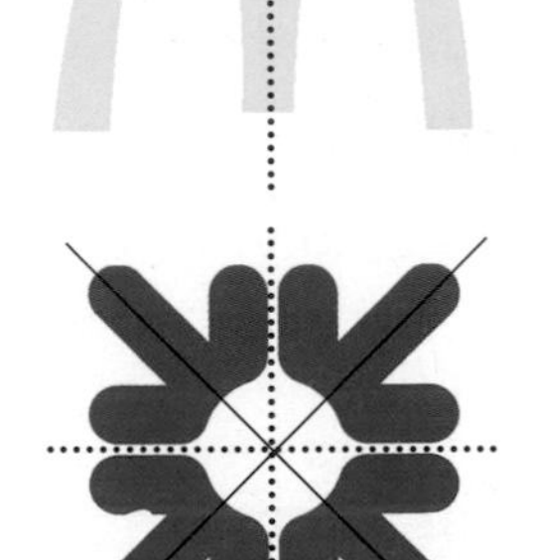

The Royal Bank of Scotland logo has four lines of reflection symmetry.

These are shown with dotted lines.

Find a logo with two lines of symmetry.

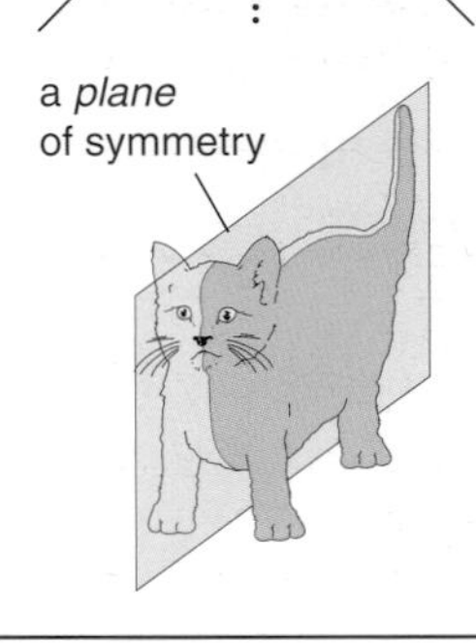

This picture shows reflection symmetry in three dimensions. Instead of a line you have a **plane of reflection symmetry**. The cat is the same on each side.

What other three dimensional objects have a plane of symmetry?

Rotational symmetry

The train station logo has no lines of reflection symmetry. However, it has another kind of symmetry, called **rotational symmetry**. When it is rotated through half a turn it looks exactly the same. So it has two positions in which it looks the same.

It has **rotational symmetry of order 2**.

The NatWest logo has rotational symmetry of order 3.

The Royal Bank of Scotland logo has rotational symmetry as well as the reflection symmetry which is shown above. What is its order of rotational symmetry?

Exercise

1 Copy these shapes and draw in all the lines of reflection symmetry, if any. Write underneath the order of rotational symmetry, if any.

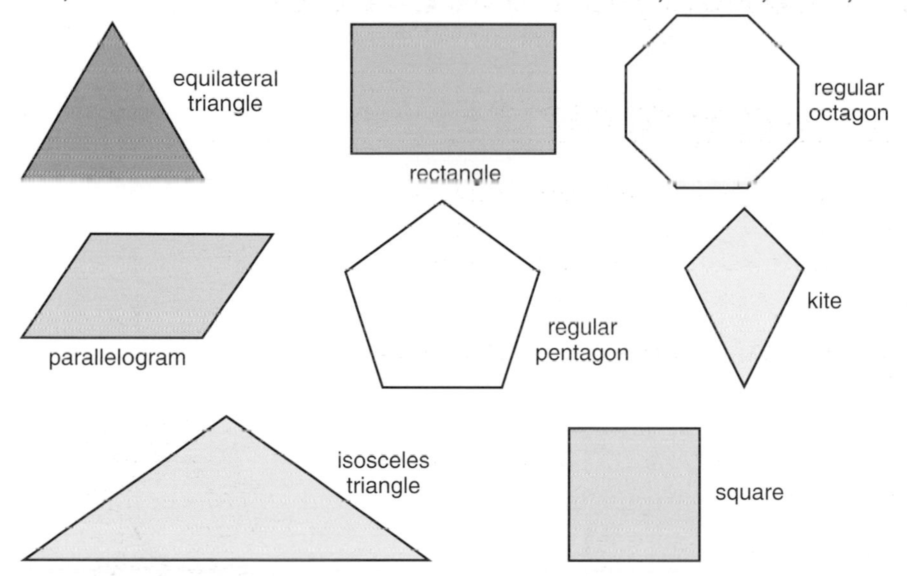

2 Copy each of the patterns below and complete them so that the dotted lines are lines of reflection symmetry.

a)

b)

c)

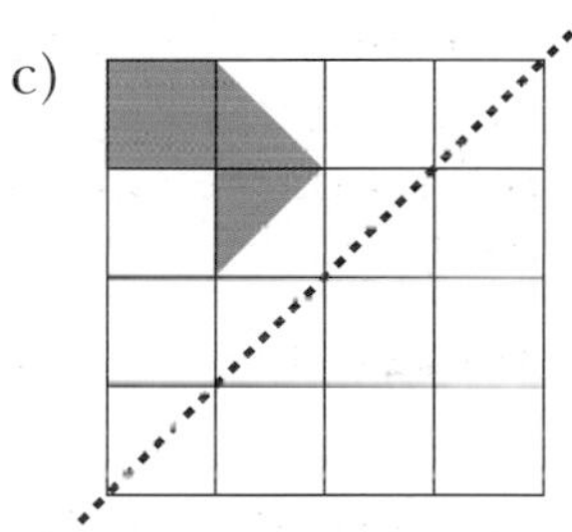

3 Copy each of the diagrams below and complete them so that they have rotational symmetry of the order stated.

a)

Rotational symmetry of order 2

b)

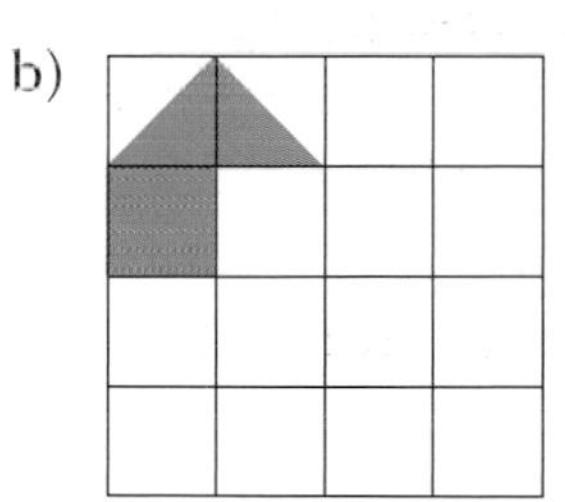

Rotational symmetry of order 4

c)

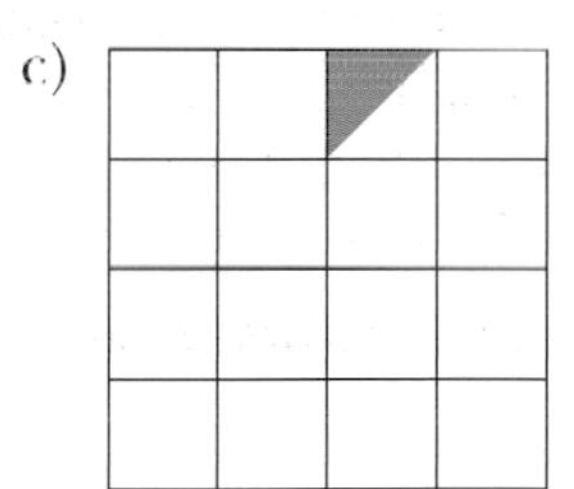

Rotational symmetry of order 8

Find some common signs and symbols (such as road signs) and describe their symmetries.

Translations

A translation is described by saying how many squares the shape has moved in each direction.

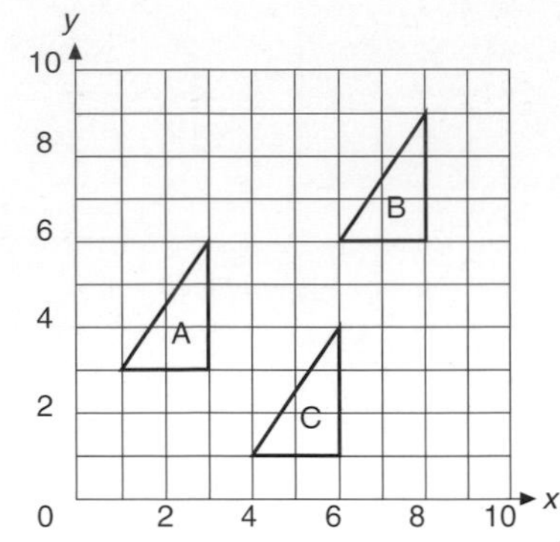

The translation A to B is 5 units in the x-direction and 3 units in the y-direction.

The translation B to C is –2 units in the x-direction and –5 units in the y-direction.

Describe the translation A to C.

A quicker way of describing a translation is to use a **column vector**.

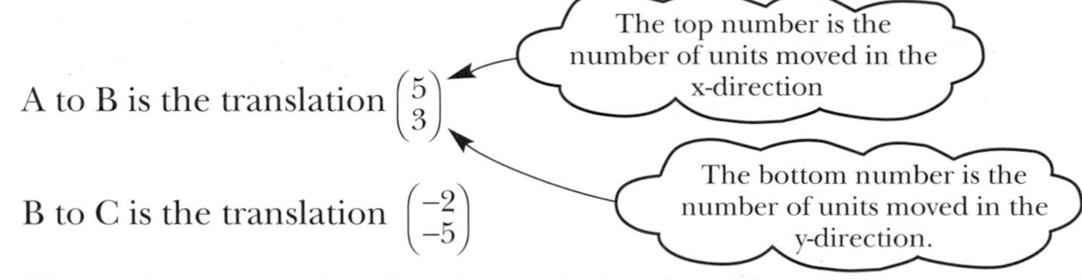

A to B is the translation $\begin{pmatrix} 5 \\ 3 \end{pmatrix}$

B to C is the translation $\begin{pmatrix} -2 \\ -5 \end{pmatrix}$

What column vector describes the translation A to C?

Reflections

The reflection of a shape must be the same distance from the mirror line as the original shape. When you draw a reflection of a shape it is best to reflect the corners one at a time. Here are some examples.

A to B is a reflection in the y axis.

C to D is a reflection in the x axis.

A to E is a reflection in $y = x$

B to C is a reflection in $y = -x$

E to F is a reflection in $y = -1$.

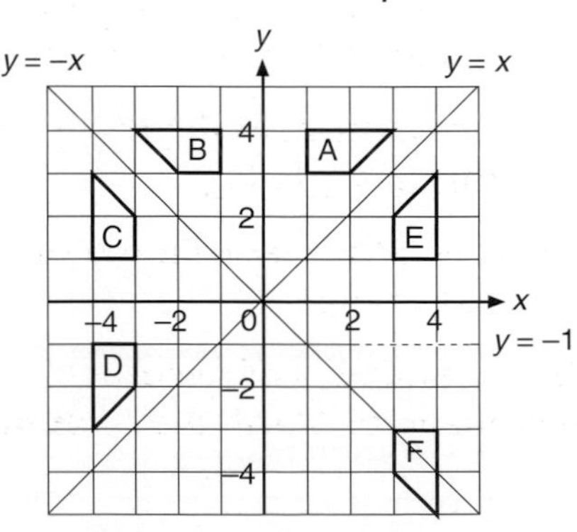

Describe the transformation D to E.

Describe the transformation A to D.

Exercise

1 Write down the column vectors for the following translations.

a) A to B b) B to A c) B to C

d) D to C e) C to A f) B to D

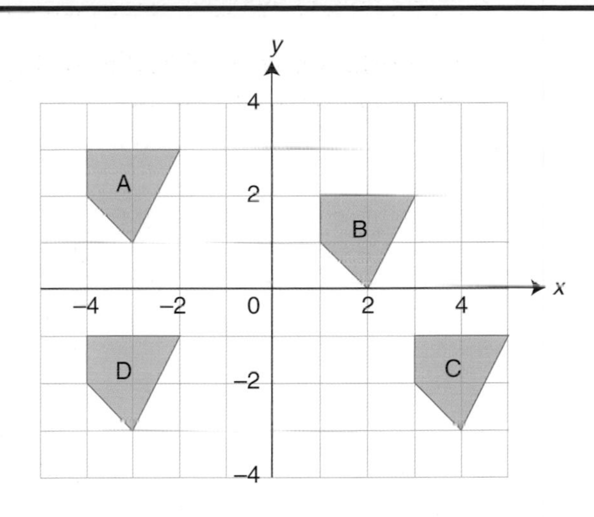

2 Look back at your answers to question 1.

a) What is the rule connecting a vector with its reverse (e.g. A→B and B→A)?

b) Compare the vector A→C with the vectors A→B and B→C. What do you notice?

3 Describe the reflection which takes

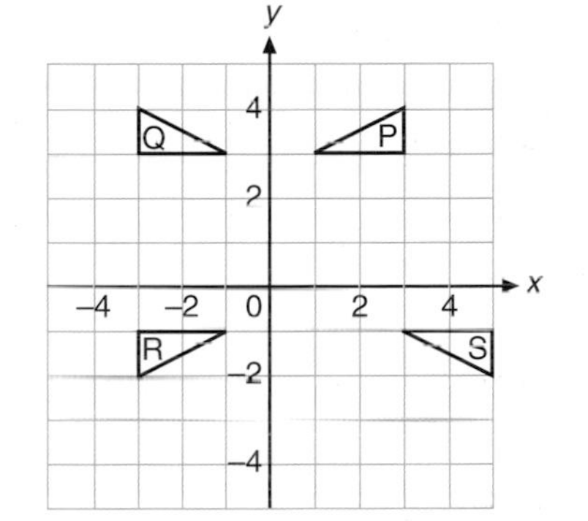

a) P to Q

b) Q to R

c) R to S

S is reflected in the x axis to form T.

d) Write down the vertices of triangle T.

e) What translation takes P to T?

4 a) Write down the co-ordinates of each corner of triangle A.

b) Copy the diagram and reflect triangle A in the line $y = x$. Label this triangle B.

c) Write down the co-ordinates of each corner of triangle B.

What is the rule connecting the co-ordinates of A with the co-ordinates of B?

d) Reflect triangle A in the line $y = -x$. Label this triangle C.

e) Write down the co-ordinates of each corner of triangle C.

What is the rule connecting the co-ordinates of A with the co-ordinates of C?

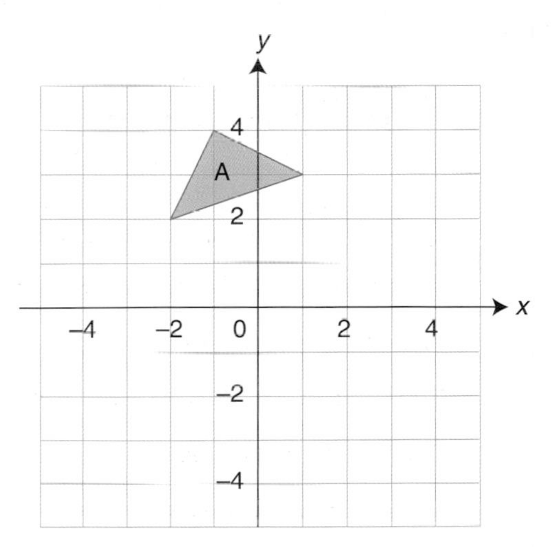

Rotations

Here are two examples of rotations.

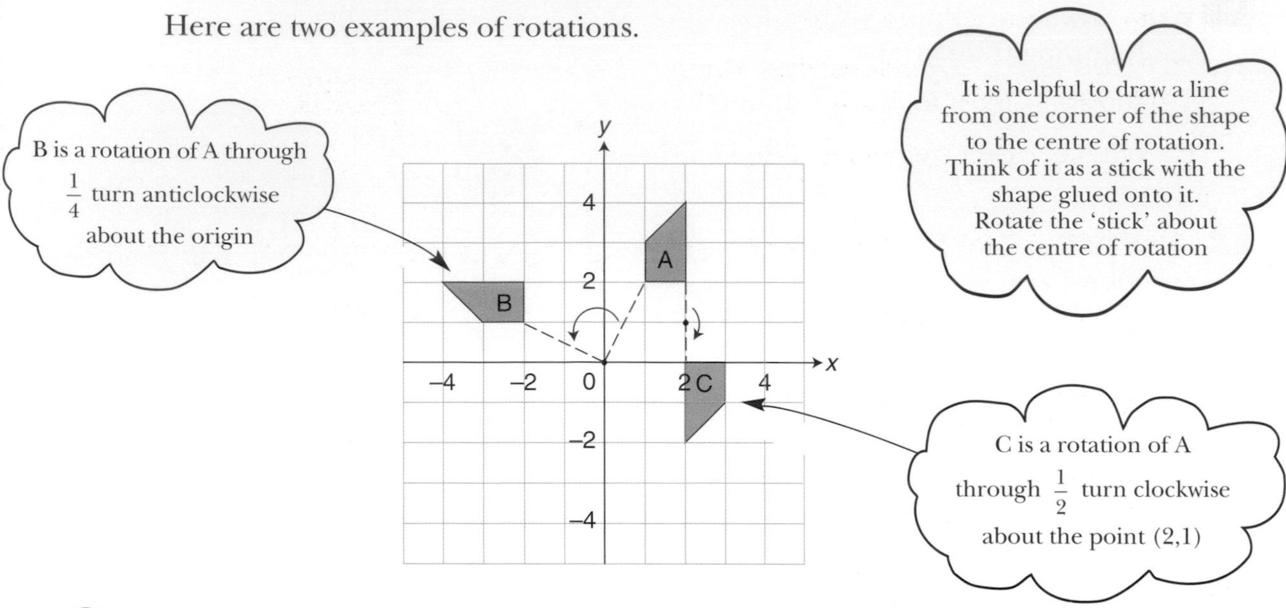

Why isn't it necessary to use the term clockwise or anticlockwise to describe the rotation from A to C?

Recognising reflections, rotations and translations

Sometimes it is easy to recognise a reflection, rotation or translation and to spot the mirror line or centre of rotation. In cases where you are not sure, try joining up pairs of corresponding points.

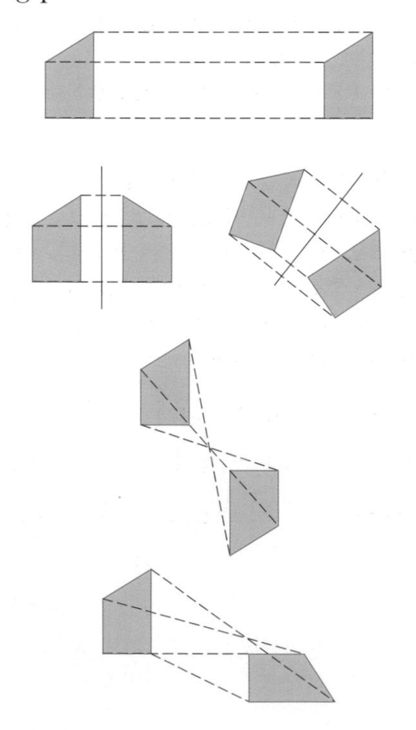

For a translation, the lines will be parallel and the shapes will be the same way round.

For a reflection, the lines will be parallel and the shape will be 'flipped over'. The mirror line goes down the middle.

For a rotation through $\frac{1}{2}$ turn, the lines all meet at a point.

What is special about this point?

If the lines are not parallel and do not meet at a point, you need to check for $\frac{1}{4}$ or $\frac{3}{4}$ turn. (In some cases it is hard to find the centre of rotation, but not in the examples in this book.)

Are the object and image congruent under

a) *a translation* b) *a reflection* c) *a rotation?*

Exercise

1 Copy this diagram.

For each of these rotations, draw the rotation and label it clearly.

a) $\frac{1}{4}$ turn clockwise about (–4, 1)

b) $\frac{3}{4}$ turn clockwise about (1, 2)

c) $\frac{1}{2}$ turn about (1, 0)

d) $\frac{1}{4}$ turn anticlockwise about (–1, 1)

e) $\frac{1}{2}$ turn about (1, –1)

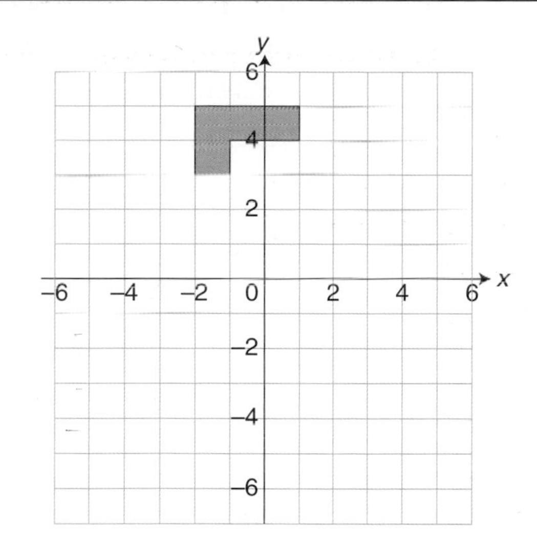

2 Describe each of these rotations fully.

a) A to B b) A to C

c) D to B d) C to E

e) A to F f) E to D

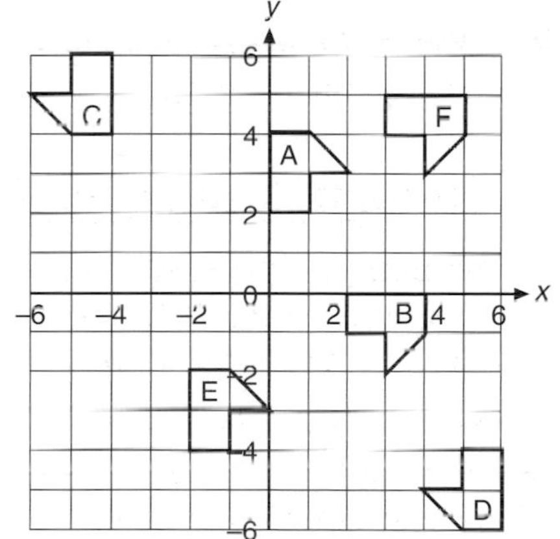

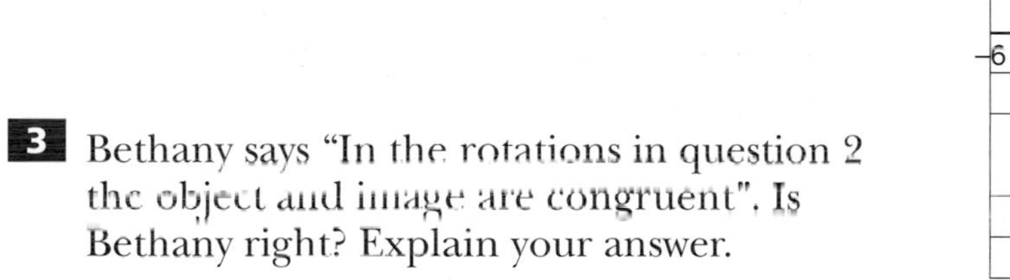

3 Bethany says “In the rotations in question 2 the object and image are congruent”. Is Bethany right? Explain your answer.

4 For each of the following transformations, state whether it is a reflection, a rotation or a translation. For a reflection, give the equation of the mirror line, for a rotation, give the angle turned through, the direction and the co-ordinates of the centre of rotation, and for a translation, give the translation vector.

a) A to B b) B to C c) C to D

d) D to E e) E to F f) F to G

g) G to H h) H to I i) I to A

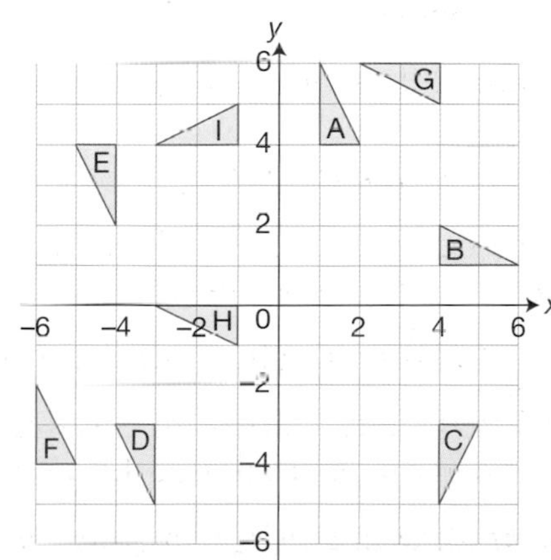

Enlargements

Shape A has been **enlarged** by scale factor 2 to give shape B. All the lines are twice as long.

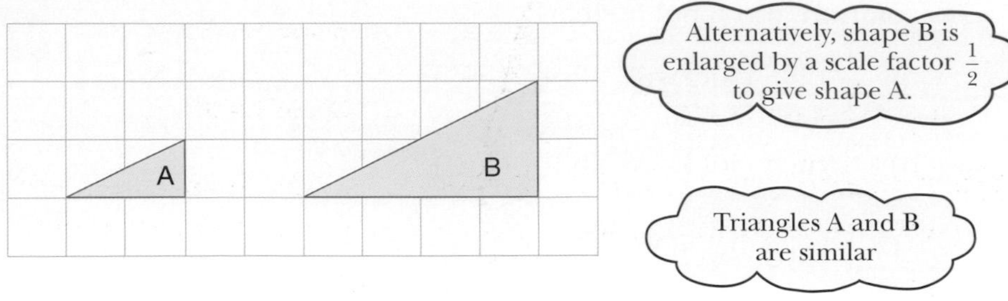

How is the value of the perimeter of shape A affected by enlargement?

How are the sizes of the angles of shape A affected by enlargement?

This example shows how to draw an enlargement of scale factor 2 using a **centre of enlargement** (C).

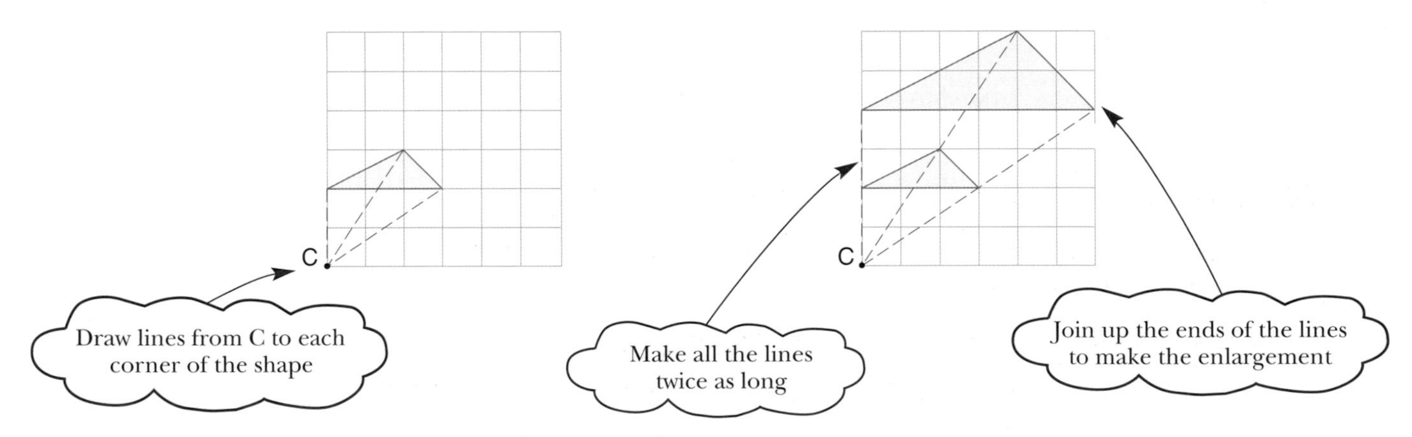

You can use the same method in reverse if you are given a shape and its enlargement and you want to find the centre of enlargement and the scale factor.

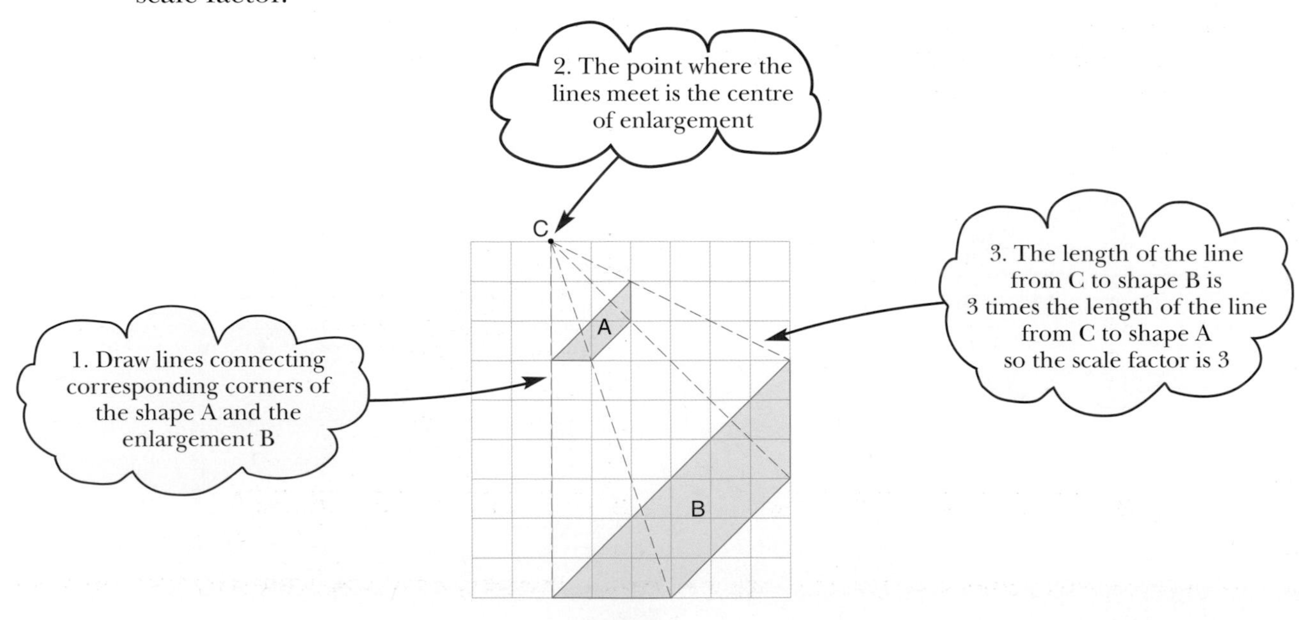

Exercise

1 A drawing is 18 cm wide.

Find the width of each of these photocopied enlargements.

a) Scale factor 3 b) Scale factor $\frac{1}{4}$

c) Scale factor $\frac{3}{2}$ d) Scale factor $\frac{2}{3}$

2 For each of these, copy the diagram and draw an enlargement with scale factor 2, using the point C as the centre of enlargement.

a)

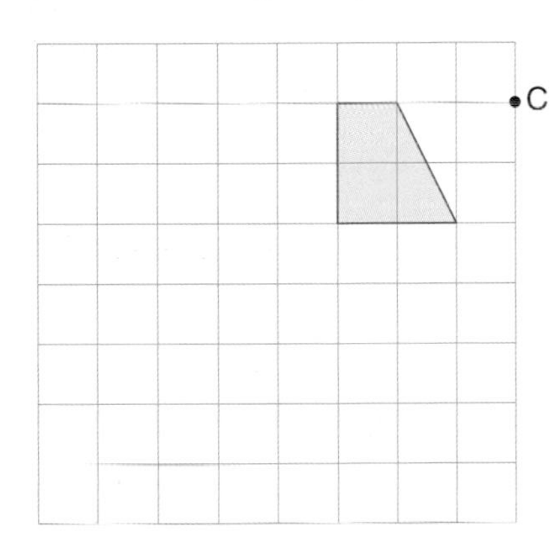

b)

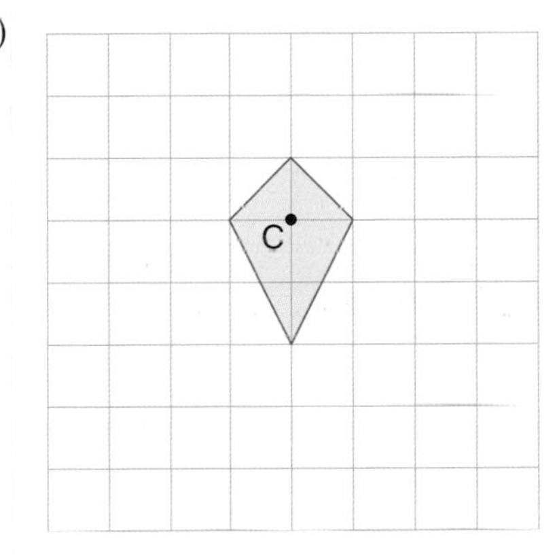

3 Draw a grid from –10 to 10 on each axis.

Plot the points (0, 1), (0, 3) and (3, 3). Join these points to form a triangle and label it T.

Draw and label the enlargement of triangle T by

a) a scale factor of 3 using the origin as the centre of enlargement

b) a scale factor of 2 using (–3, 4) as the centre of enlargement.

c) a scale factor of $\frac{1}{2}$ using (–6, 1) as the centre of enlargement.

4 In each of the diagrams below, shape A has been enlarged to make shape B. In each case find the scale factor and the co-ordinates of the centre of enlargement.

a)

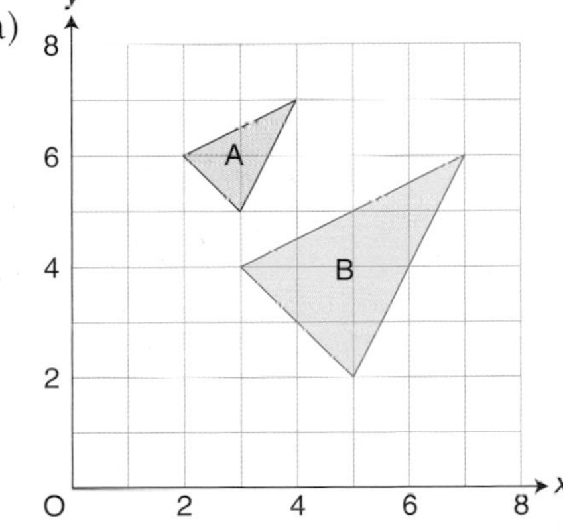

b)

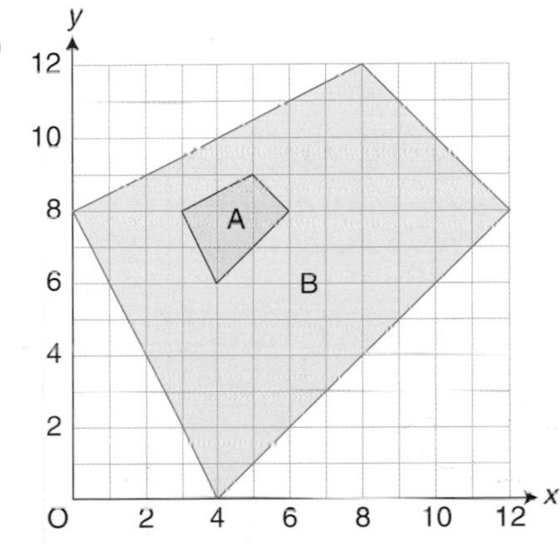

Combining transformations

This diagram shows part of a repeating pattern. It is made out of transformations of triangle A.

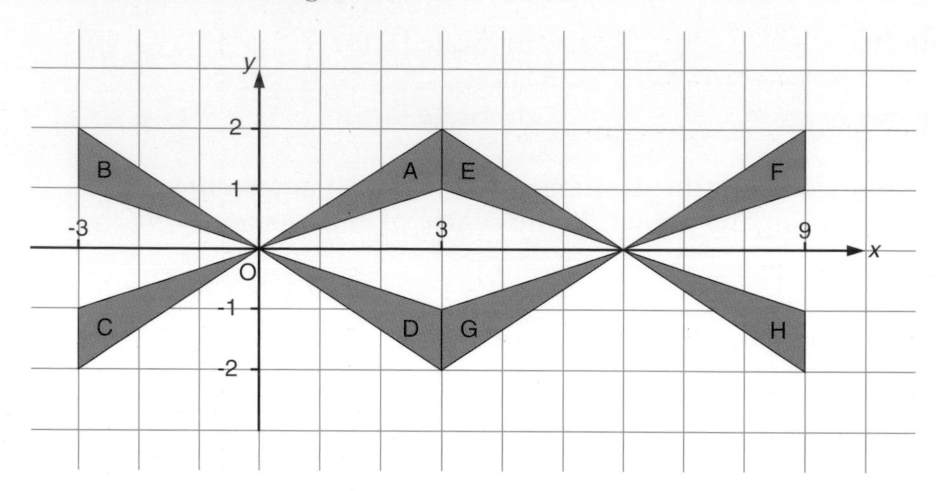

You can map A onto each of B, C, D, E, F and G by a single transformation. In each case it is either a reflection or a rotation or a translation. How would you describe these six transformations?

When you come to H, you find that there is no single transformation. You need two. One way of doing this is

> First rotate A centre (0, 0) through 180° then do a reflection in the vertical line $x = 3$.

How many more ways can you find to map A → H in two steps? Remember to be careful to say which of the two comes first and which second.

Here is a real-life example of combining transformations.

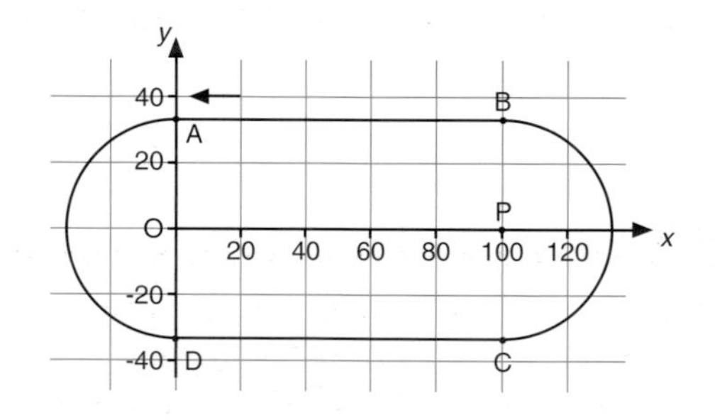

The diagram shows a 400 m running track. An athlete starts at point A (0, 32) and runs round the track in an anticlockwise direction.

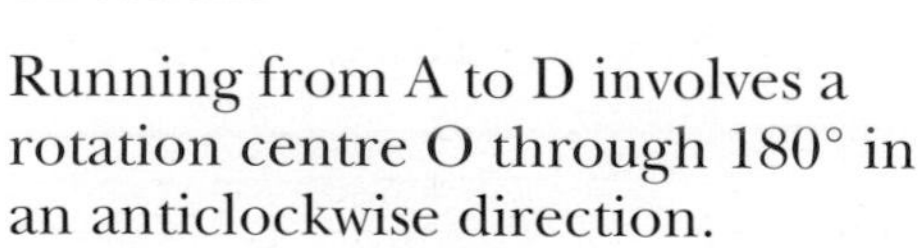

Running from A to D involves a rotation centre O through 180° in an anticlockwise direction.

Running from D to C involves a translation $\begin{pmatrix} 100 \\ 0 \end{pmatrix}$.

What transformations represent the athlete's run from C back to A?

Which of the four transformations cancel each other out?

Exercise

Questions 1–4 of this exercise refer to the triangles in this diagram.

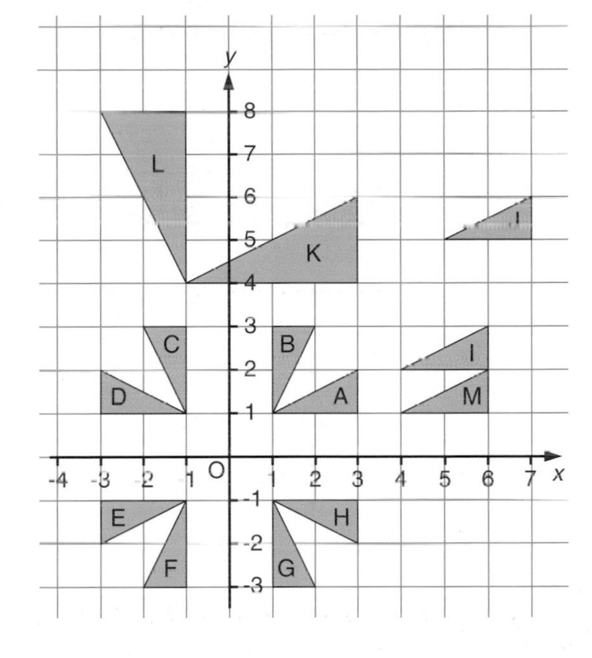

1 Describe the transformations which map

a) (i) A → D (ii) D → E (iii) A → E

b) (i) A → D (ii) D → M (iii) A → M

c) Is it possible for 2 reflections to be the same as
   (i) a translation?
   (ii) a rotation?
   (iii) another reflection?

2 Describe the transformations which map

a) (i) A → I (ii) I → J (iii) A → J

b) Is it possible for 2 translations to be the same as
   (i) another translation?
   (ii) a rotation through 90°?
   (iii) a reflection?

3 Describe the transformations which map

a) (i) A → K (ii) K → L

b) (i) A → C (ii) C → L

c) Is there a single transformation which maps A → L?

4 Give examples, using the triangles in the diagram, to show that

a) two rotations are equivalent to another rotation.

b) the same reflection carried out twice cancels itself out.

The diagram shows a settee in one corner of a room (position A) which has to be turned round and moved to the opposite corner (position B). The settee has legs at its four corners. It is too heavy to lift or slide and so has to be rotated about one of its legs.

Copy the diagram and shade, in different colours, the moves needed to get the settee to its new position. Mark in the centre of each rotation.

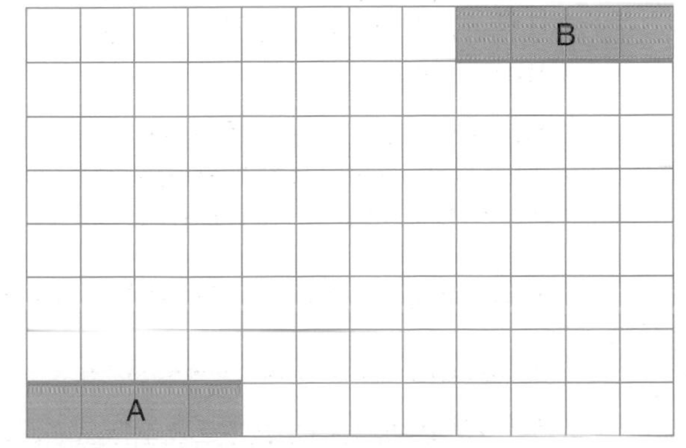

Finishing off

Now that you have finished this chapter you should be able to:

★ say whether a shape has reflection symmetry

★ say whether a shape has rotational symmetry

★ draw a translation, a reflection a rotation and an enlargement

★ recognise and describe fully a translation, a reflection, a rotation and an enlargement

★ investigate situations involving two transformations

Use the questions in the next exercise to check that you understand everything.

Mixed exercise

1 For each of the following signs and logos

(i) write down how many lines of reflection symmetry it has;

(ii) write down whether it has rotational symmetry, and if it has, write down the order of rotational symmetry.

a) 

b) 

c) 

d) 

2 A triangle A has co-ordinates (0, 0), (1, 0), (3, 3).

a) Draw triangle A, using the same scale for both x and y axes. Take values between –3 and +3 for both x and y.

b) Triangles B, C and D are formed by rotating A anticlockwise about O through 90°, 180° and 270°, respectively. Draw, B, C and D.

c) Triangles E, F, G and H are formed by reflecting A, B, C and D in the x axis. Draw E, F, G and H.

d) Describe single transformations which map

(i) A → G (ii) A → H (iii) A → F.

3 Look at each of these statements and explain whether they are true or false.

a) Aaron says 'In a reflection the value of the perimeter is preserved.'

b) Elizabeth says 'In an enlargement length and angles are preserved.'

4 Describe the following transformations fully.

a) A to B b) A to C

c) C to D d) A to E

e) B to F f) D to E

g) C to F h) D to F

Describe a combination of transformations that goes

i) from A to D

j) from F to A

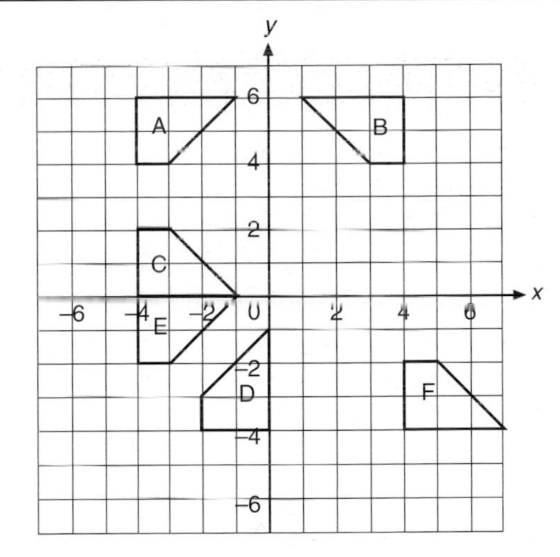

5 For each of these, copy the diagram and draw an enlargement with the given scale factor and centre of enlargement (C).

a)

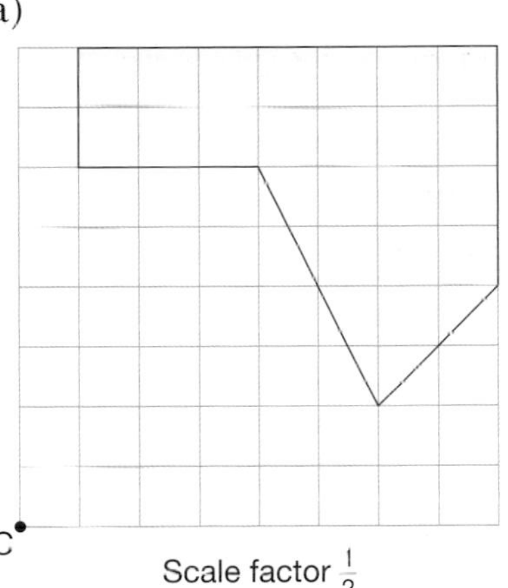

Scale factor $\frac{1}{2}$

b)

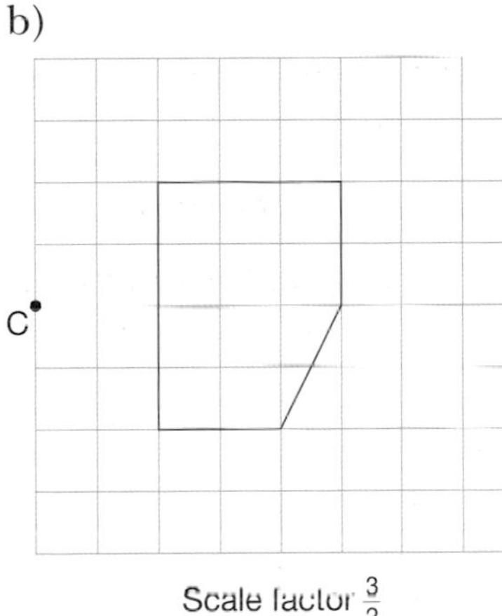

Scale factor $\frac{3}{2}$

c)

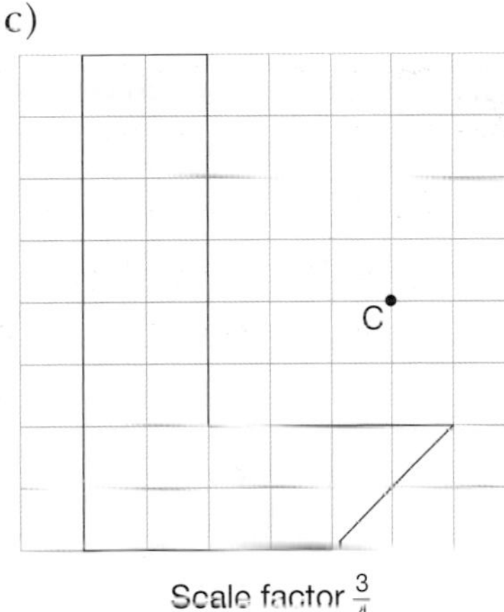

Scale factor $\frac{3}{4}$

6 Answer this question on graph paper. Use the same scale for the x axis (values −4 to 12) and the y axis (values 0 to 12).

A triangle L has co-ordinates (4, 3), (6, 3), (6, 4).
Another triangle P has co-ordinates (0, 6), (4, 6), (4, 8).

a) Find the centre of the enlargement which maps L onto P, and the scale factor.

b) Triangle P is given a translation $\begin{pmatrix} 4 \\ 0 \end{pmatrix}$ to form triangle Q, and a translation $\begin{pmatrix} 8 \\ 0 \end{pmatrix}$ to form R. Find the centres of the enlargements which map L onto Q and R.

c) Triangle L is now enlarged from the same three centres but with scale factor 3. Do the enlarged triangles touch each other, like P, Q and R do?

Twenty four

Drawings and loci

Scale drawings

A chapter on bones in a biology textbook contains these pictures.

Human tooth

Scale 5:1

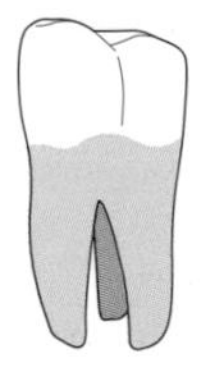

The tooth is small in real life, so an enlargement has been drawn. The scale of the drawing is 5:1. This means that 5 cm on the drawing represents 1 cm in real life. The drawing is 5 times as long as the actual tooth.

Human skull

Scale 1:10

Read this as 'one to ten'

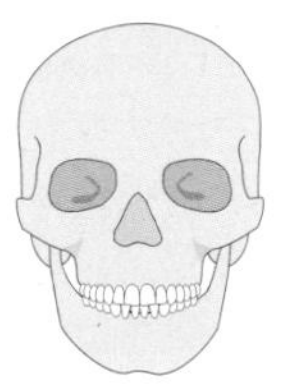

The skull is too big in real life to draw full size, so the drawing has been reduced. The scale is 1:10. This means that 1 cm on the drawing represents 10 cm in real life. The actual skull is 10 times as high as the drawing.

When you see a scale like 5:1 or 1:10, remember that the first number refers to the drawing and the second number refers to real life.

When making a scale drawing, you do calculations like these:

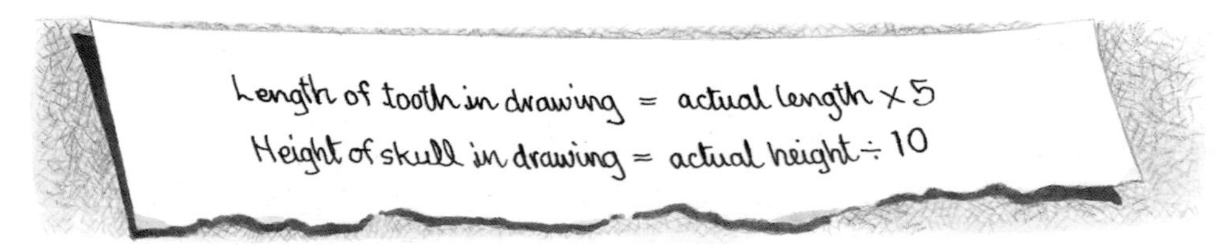

When reading or using a scale drawing, you do calculations like these:

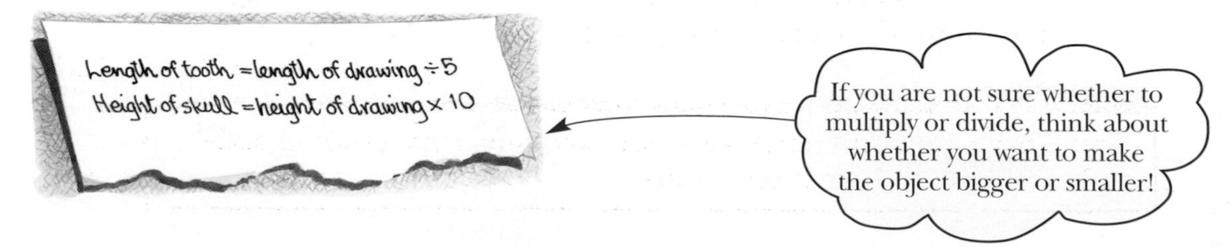

Exercise

1 a) A swimming pool 50 m by 20 m, is drawn to a scale of 1:250. How big is it in the drawing?

b) A ladybird, length 8 mm and height 5 mm, is drawn to a scale of 5:1. How big is it in the drawing?

2 This plan of Laura's bedroom is not drawn to scale.

A is the bed (180 cm by 80 cm)

B is the wardrobe (90 cm by 50 cm)

C is the dressing table (100 cm by 40 cm)

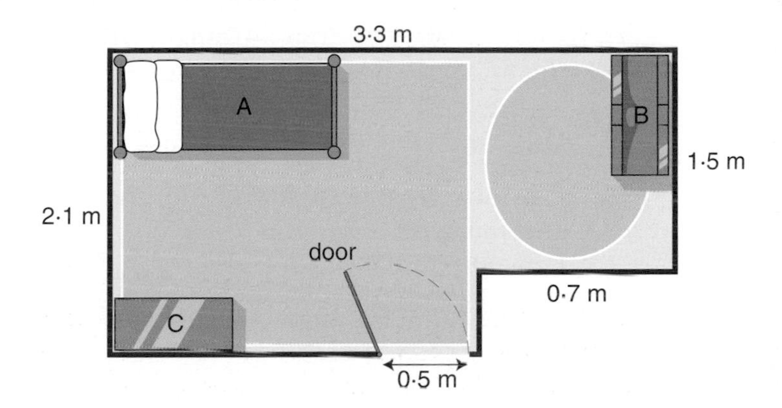

a) Draw an accurate plan of Laura's bedroom using a scale of 1:20.

b) Measure the diagonal distance across the drawing from the corner with the dressing table to the corner with the wardrobe.

Work out what this distance would be in real life.

c) What is the perimeter of Laura's bedroom (i) in real life (ii) on your plan?

d) Use a compass to draw the arc through which the end of the door travels when it is opened or closed.

Measure how close this arc goes to the edge of the bed.

What would this distance be in real life?

Is this enough space to walk through?

3 a) This diagram of a red blood cell has a diameter of 3 cm.

What is the diameter of the real red blood cell?

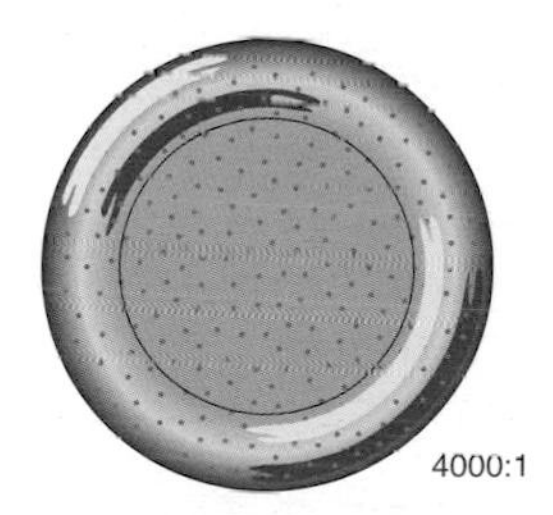

4000:1

b) A real human hair has a width of 0.1 mm.

If you drew it to the same scale as the red blood cell, what would be the width of the hair in the picture?

Find 3 maps with different scales. For each one, write down its scale and say how long 1 km is on the map.

Drawing triangles

To draw a triangle accurately, you need to know at least three facts about the length of its sides and the size of its angles. You need a ruler, a compass and a protractor.

Drawing a triangle given one side and two angles

Follow these steps to draw a triangle ABC with side AB = 6 cm, angle BAC = 62° and angle ABC = 47°.

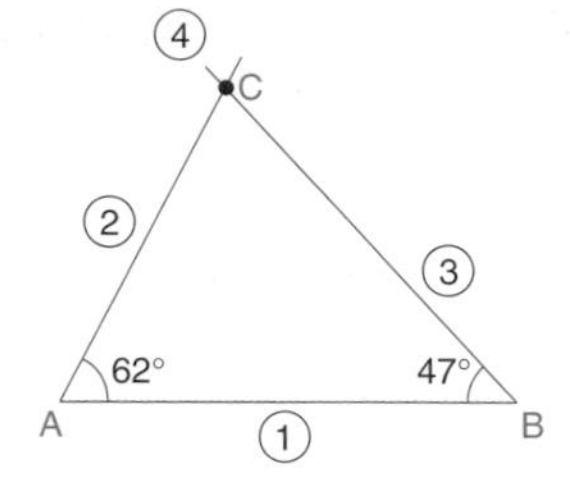

① Draw and label the side AB (6 cm long).

② Put your protractor at end A of the line and mark off angle BAC.
Draw the line.

③ Do the same for angle ABC. Make sure the lines cross.

④ The point where the lines cross is point C.

Drawing a triangle given the lengths of all three sides

Follow these steps to draw a triangle ABC with AB = 8 cm, AC = 6 cm and BC = 5 cm.

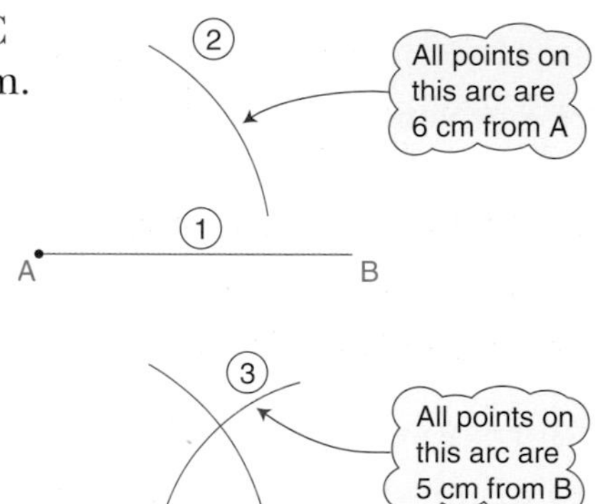

① Draw a line 8 cm long and label it AB. (You could use any of the sides to start the triangle.)

② Open your compass to a length of 6 cm. Put the point of the compass on A. Draw an arc (part of a circle).

③ Now open the compass to a length of 5 cm. Put the point of the compass on B. Draw another arc to cross the first arc.

④ The point where the arcs cross is the third corner, C, of the triangle. Join this point to both A and B, and the triangle is complete.

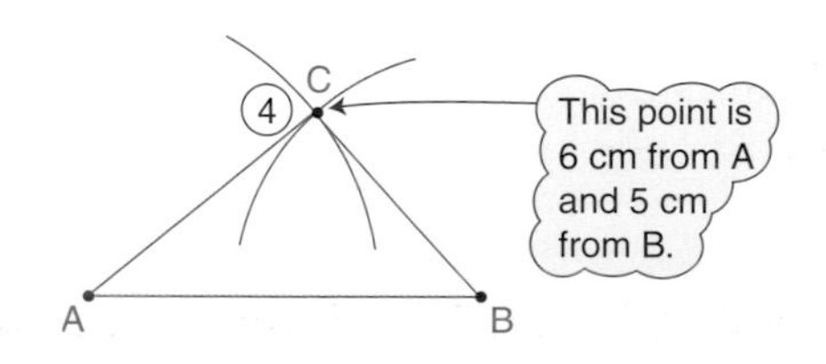

Drawing a triangle given two sides and an angle

There are two distinct cases:

(i) The angle is between the two sides.
You are guided through this in question 4 on the opposite page.

(ii) The angle is not between the two sides.
You are guided through this in question 5 on the opposite page.

Exercise

1 Make a rough sketch and then an accurate drawing of each of these triangles. On each accurate drawing, measure and label the lengths of the sides that were not given.

a) Triangle ABC with AB = 7 cm, angle ABC = 46° and angle BAC = 62°.

b) Triangle PQR with PQ = 5 cm, angle PQR = 108° and angle PRQ = 31°. (Hint: work out angle QPR first.)

2 Make accurate drawings of triangles with sides as follows. On each drawing, measure and mark on the sizes of all the angles.

a) 3 cm, 5 cm, 6 cm

b) 4.5 cm, 8 cm, 9 cm

c) 4 cm, 5.2 cm, 6.8 cm

d) 7.1 cm, 3.2 cm, 8.4 cm

3 a) Try to draw a triangle with sides 7 cm, 4 cm and 2 cm.

b) Explain why it is not possible to draw this triangle.

4 Follow these steps to make an accurate drawing of triangle ABC, with AB = 7 cm, AC = 5 cm and angle CAB = 54°.

1 Draw the line AB 7 cm long.

2 Put the protractor at A and mark off the angle CAB. Draw a line.

3 Mark point C on this line 5 cm from A.

4 Join B to C to complete the triangle.

Measure

a) angle ABC

b) angle ACB

c) the side BC

5 Follow these steps to make an accurate drawing of triangle ABC, with AB = 4 cm, AC = 7 cm and angle ABC = 110°.

1 Draw the line AB 4 cm long. (The first line you draw must be the one with the given angle at one end of it.)

2 Use a protractor to mark off an angle of 110° at B, and draw a line.

3 Open your compass to a length of 7 cm. Put the point of the compass at A. Draw an arc so that it crosses the line you have just drawn from B. Mark point C.

4 Join C to A to complete the triangle.

Measure the side BC.

6 Make a rough sketch and then an accurate drawing of each of these triangles.

a) Triangle FGH with FG = 5 cm, FH = 6 cm and angle FGH = 54°.

Measure and mark on your diagram the length GH and the angle GFH.

b) Triangle XYZ with XY = 4 cm, XZ = 5.6 cm and angle YXZ = 135°. Measure and mark on your diagram the length of side YZ.

Using bearings

An aeroplane is flying from London to Manchester.

The pilot needs to know exactly what direction to take.

She can do this by measuring the angle on the map between the direction she needs to fly in and a line going north from London.

The angle is measured clockwise from the North line.

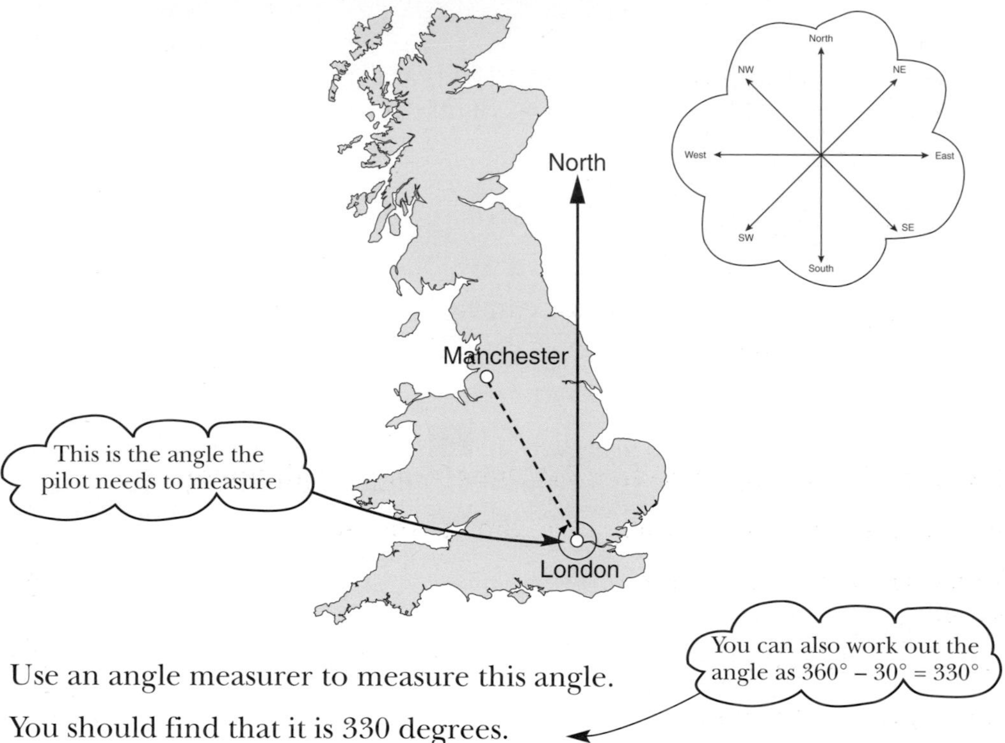

You can also work out the angle as $360° - 30° = 330°$

Use an angle measurer to measure this angle.

You should find that it is 330 degrees.

This angle is called the **bearing** of Manchester from London.

Remember that bearings are always measured clockwise from North.

Bearings are always written with three figures.

A bearing of 62° is written as 062°.

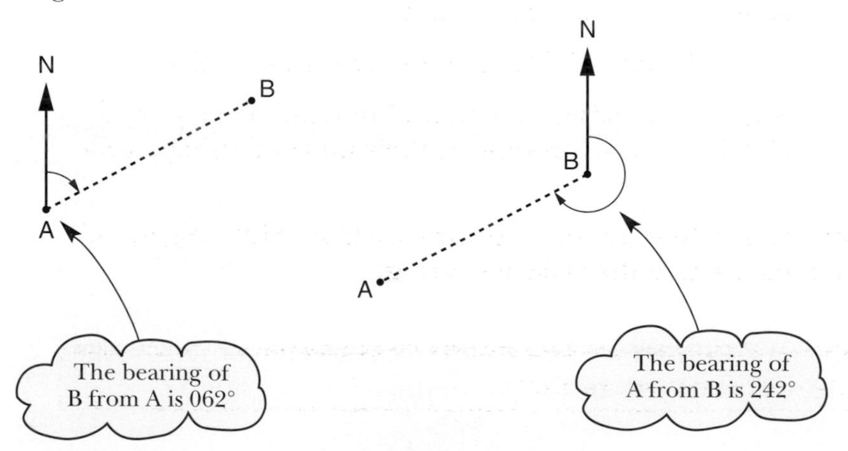

Exercise

1 A, B, C and D are 4 ships.

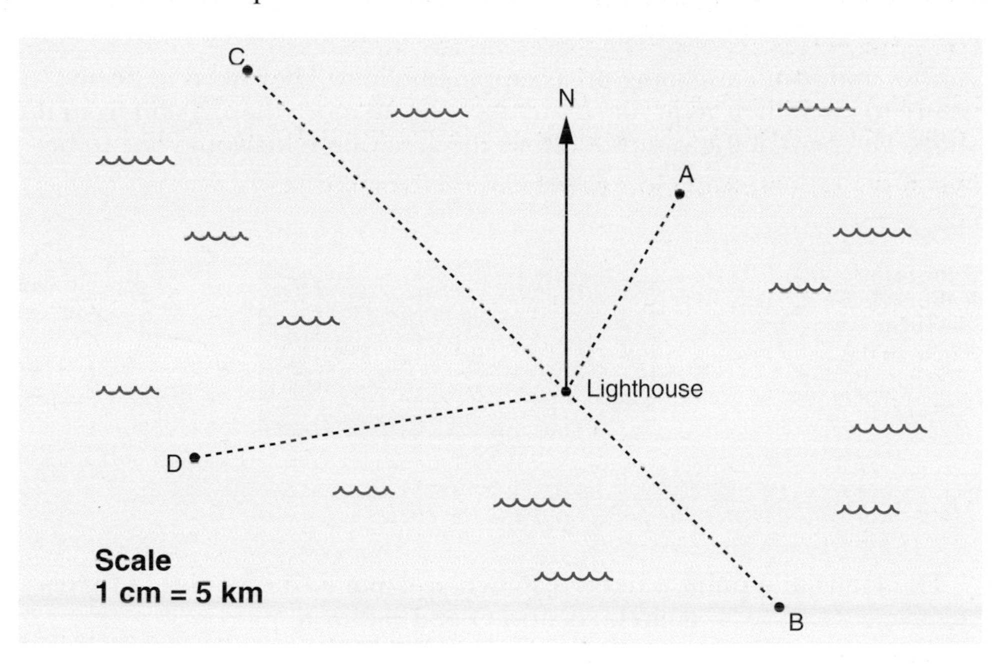

a) Measure the bearing of each of the ships from the lighthouse.

b) Find the distance of each ship from the lighthouse.

2 This diagram shows three towns, A, B and C.

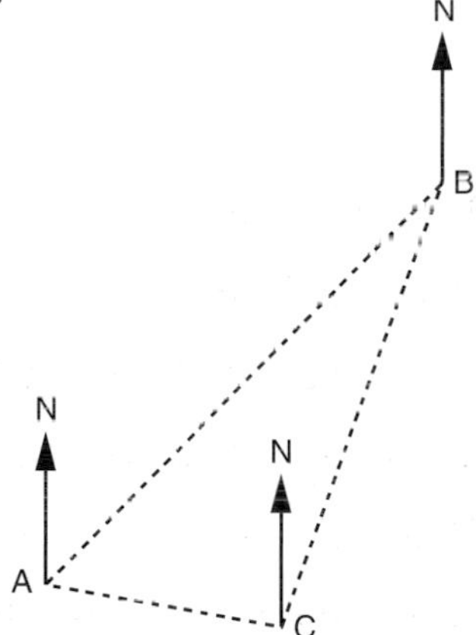

Measure the bearing of

a) A from B b) B from A

c) A from C d) C from A

e) B from C f) C from B

g) What do you notice about each pair of bearings?

3 Andrew and Ranjit are on a hike in the hills.

They find that a nearby hill, Mill Crag, is on a bearing of 158°.

They walk for 4 km north-east and then find that Mill Crag is on a bearing of 205°. Make a scale drawing of their journey, using a scale of 1:50 000.

Use your drawing to find out how far they are from Mill Crag at each of the points where they take a bearing.

Plan a walk through the open country.

Simple loci

The **locus** of a point means all the possible positions for that point. (Note: The plural of locus is loci.)

Ainsley and Sarah are going on a camping holiday. They need to decide where to pitch their tent. Ainsley wants to be no more than 200 m from the shop. The map of the campsite shows the area where he would like to be.

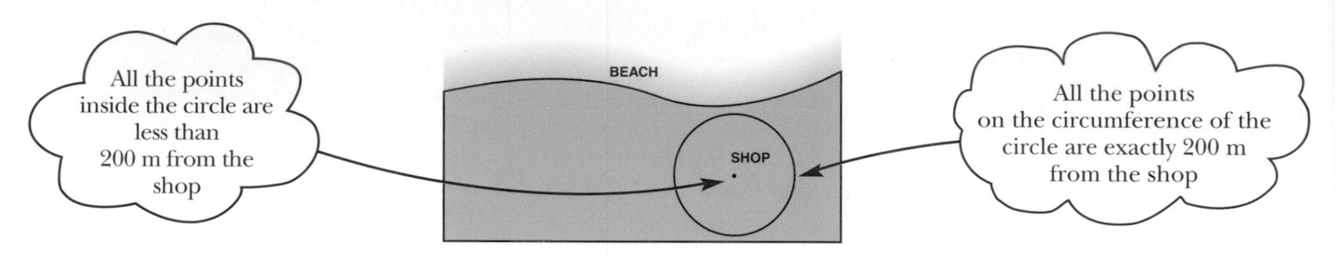

How could you describe the points outside the circle?

The locus of a point a fixed distance, *d*, from a fixed point O forms a circle, centre O and radius *d*

Sarah wants to be less than 100 m from the beach. The map shows the area where she would like to be.

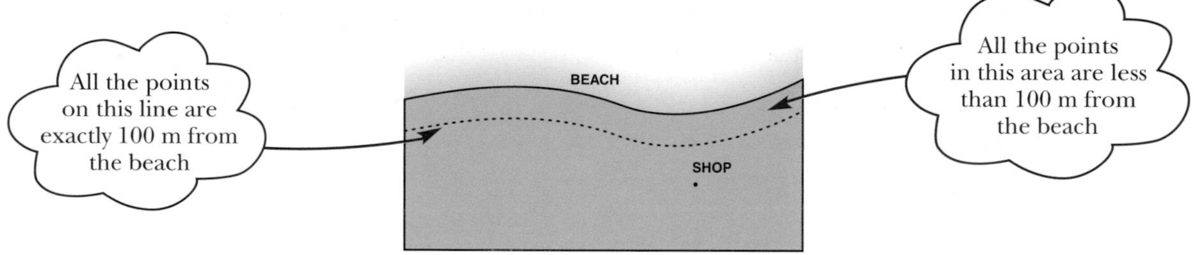

Why is the line in this diagram dotted and not solid?

The locus of a point a fixed distance, *d*, from a line AB forms a line parallel to AB and distance *d* from AB.

This map shows the area satisfying the wishes of both Ainsley and Sarah.

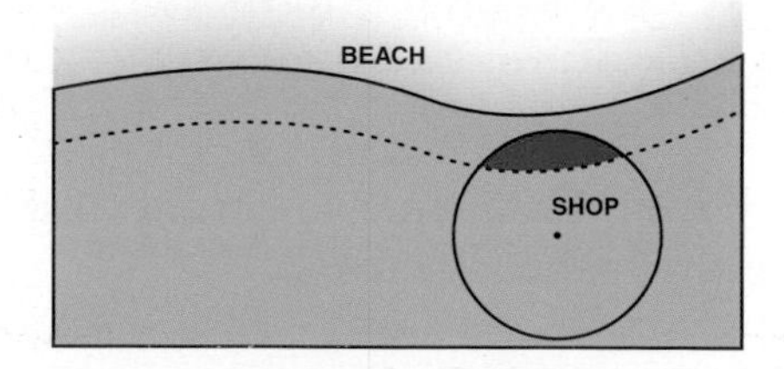

All the points in the shaded area are both no more than 200 m from the shop and less than 100 m from the beach.

Ainsley and Sarah should camp somewhere in this area.

Exercise

1 Mark a point P. Draw the locus of all points 4 cm from P.

2 Draw a line. Assuming the line is of infinite length draw the locus of all points 2 cm from the line.

3 Draw a line AB 5 cm long. Draw the locus of all points 2 cm from AB.

4 This is a diagram of Chloe's garden.

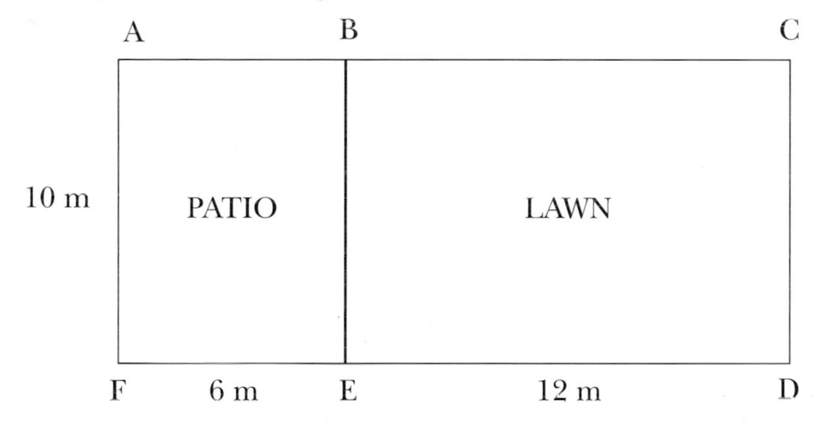

a) Make a scale drawing of Chloe's garden.

b) The house casts a shadow extending 4 m from AF. Draw the locus of the shadow.

c) A tree is to be planted within 2 m of the centre of the lawn. Draw the locus of the points where the tree can be planted.

d) Bulbs are planted in the lawn. They are at least 1 m from any edge of the lawn and no more than 3 m from CD. Draw this locus.

e) Any point on the lawn within 2.5 m of B is turned into a flower bed. Draw this locus.

5 Keith is looking for somewhere to live. He wants to be no more than 3 miles away from the station as he catches a train to work every morning. He is also a keen cinema-goer and would like to be no more than 5 miles away from the local cinema. The station and the cinema are 6.5 miles apart.

Make a scale drawing and show the area where Keith would like to live.

6 Simon and his brother Mark are playing in a rubber dinghy in the sea. The coastguard has told them not to go more than 50 m from the shore. There is a rock 40 m from the beach and he has also told them not to go within 5 m of the rock.

Make a scale drawing and shade the area where they are allowed to go. (Assume that the shoreline is straight.)

Investigate the locus of a point on the circumference of a bicycle wheel as the bicycle moves along.

More loci

A point equidistant from two fixed points

A new road is to pass between two villages A and B. To minimise disturbance to both villages the road is to be equidistant (equal distance) from A and B.

The road lies on the **perpendicular bisector** of AB.

Draw a sketch to show A and B and the route of the road.

The locus of a point equidistant from two points is the perpendicular bisector of the two points.

You can draw the perpendicular bisector of AB as follows:

① Place your compass point on A. Open the compass to a radius more than half the distance from A to B. Draw an arc each side of AB.

② Leave the compass at the same radius. Put the compass point on B and make another arc each side of AB, so that they cross the other arcs.

③ Draw a line joining the two intersections. This is the perpendicular bisector.

A point equidistant from two lines

Two householders each claim ownership of a piece of land between their two plots.

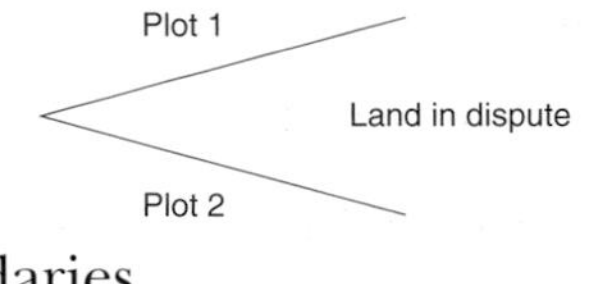

They eventually agree to divide it equally by building a wall equidistant from their plot boundaries.

Draw a sketch to show the position of the wall.

The wall lies on the **angle bisector.**

The locus of a point equidistant from two lines is the angle bisector of the two lines.

You can draw the angle bisector as follows:

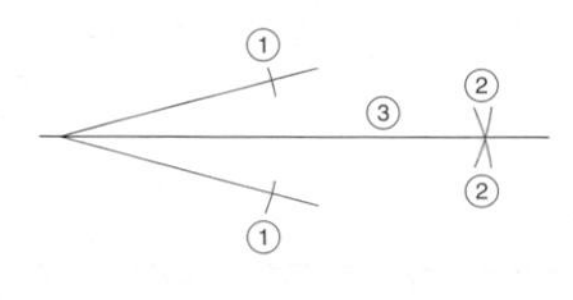

① Put the compass point on the point of the angle and mark off two points as shown.

② Put the compass point on each of the points you have marked off and draw two arcs which meet each other.

③ Draw a line through the point where the arcs intersect to the point of the angle. This line is the angle bisector.

Exercise

1 a) Construct an equilateral triangle ABC of side 6 cm.

b) Construct the perpendicular bisector of BC.

c) Does the perpendicular bisector of BC pass through A?

2 a) Construct an isosceles triangle XYZ with YZ = 5 cm and XY = XZ = 7 cm.

b) Bisect angle X.

The angle bisector meets YZ at W.

c) Is W the midpoint of YZ?

d) Is XW perpendicular to YZ?

3 P is a point on the line AB.

Construct a line through P, perpendicular to AB, as follows.

Step 1. Put the compass point on P and draw arcs on AB as shown.

Step 2. Now bisect CD.

4 Lois is building a fence from a tree, T, to the wall.

This fence must be perpendicular to the wall.

Copy the diagram and construct the line of the fence as follows.

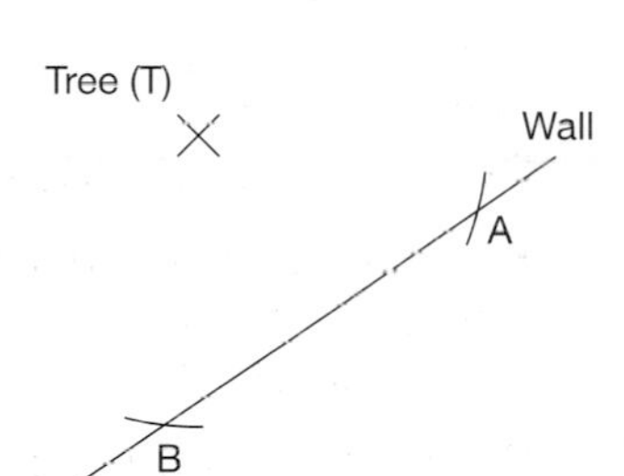

Step 1. Put the compass point on point T and draw arcs on AB as shown in the diagram.

Step 2. Now bisect AB.

5 In each of the following find

(i) the midpoint of PQ

(ii) the equation of the perpendicular bisector of PQ.

a) P(1, 0) Q(3, 0) b) P(0, 2) Q(0, 8) c) P(3, 1) Q(1, 3)

6 This diagram shows a field in which there are some rabbits. The field is surrounded by hedges on sides AB, BC and CD. A fox appears at point F and all the rabbits run to the nearest hedge. Trace the field and divide it up, showing the areas which are nearest to each of the three hedges.

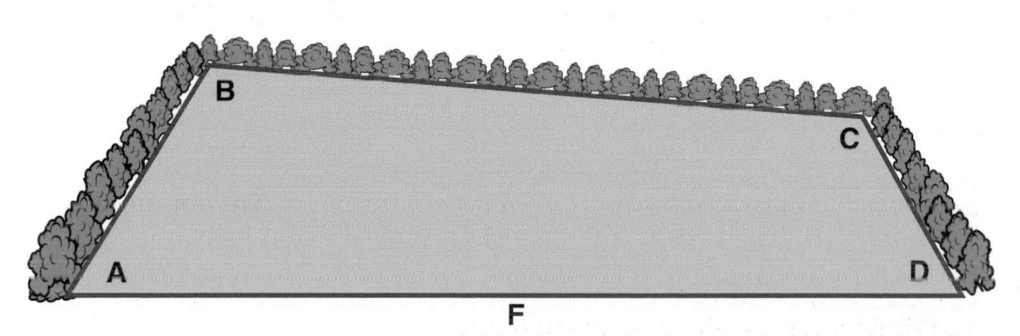

Finishing off

Now that you have finished this chapter you should be able to:

- ★ make and interpret scale drawings
- ★ draw a triangle accurately given
  - – three sides
  - – two sides and an angle
  - – one side and two angles
- ★ use bearings
- ★ solve problems involving loci, including intersecting loci
- ★ construct the perpendicular bisector of a line
- ★ construct the bisector of an angle.

Use the questions in the next exercise to check that you understand everything.

Mixed exercise

1 Karen is writing a book on insects. She wants to include a picture of a flea that is 0.3 cm long in real life. She needs to decide what scale to use for the drawing.

Find the length of the flea in the drawing if Karen uses a scale of

a) 20:1 b) 1:5 c) 500:1

Which scale do you think would be the most suitable?

2 a) Choose a suitable scale and make an accurate scale drawing of each of the triangular sails shown below.

b) Measure on your drawings the angles marked p, q, r and s.

c) Measure on your drawings the sides marked a, b, c and d and use your scale to work out the real lengths

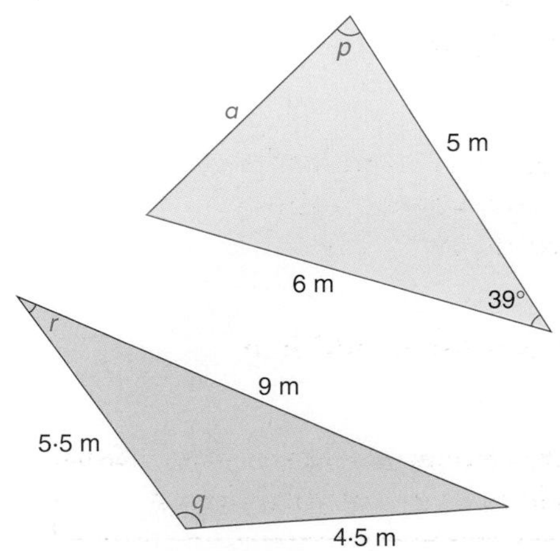

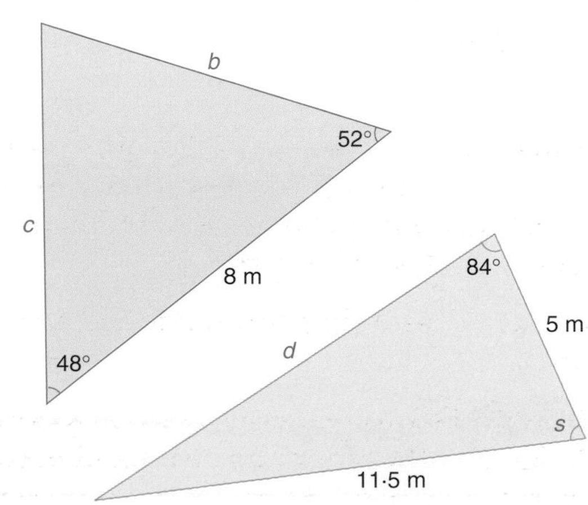

3 The bearing of X from Y is 142°. What is the bearing of Y from X?

4 A ship sails out of port on a bearing of 212°. It covers 10 km before changing course to a bearing of 228°. After a further 12 km it develops engine trouble and drops anchor. A rescue boat is sent from port by the most direct route to help.

a) Make a scale drawing of the ship's journey.

b) What bearing should the rescue boat take from the port?

c) How far will the rescue boat travel before reaching the ship?

5 Draw an angle ABC of size 63°. Draw the locus of all points equidistant from the lines AB and BC.

6 Carla's office is rectangular. It is 9 m long and 7 m wide. She is going to fit a burglar alarm inside. It consists of two motion sensors each with a range of 6 m. She is going to place the motion sensors in opposite corners of her office.

a) Make a scale drawing of Carla's office shading the region not covered by a motion sensor.

b) Show how the sensors can be placed so as to cover every part of the office.

7 Jed is at a rock concert. He wants to be equidistant from the two speakers to get the best stereo effect. He also wants to be less than 10 m from the stage.

Make a drawing and show the possible places where Jed would like to be.

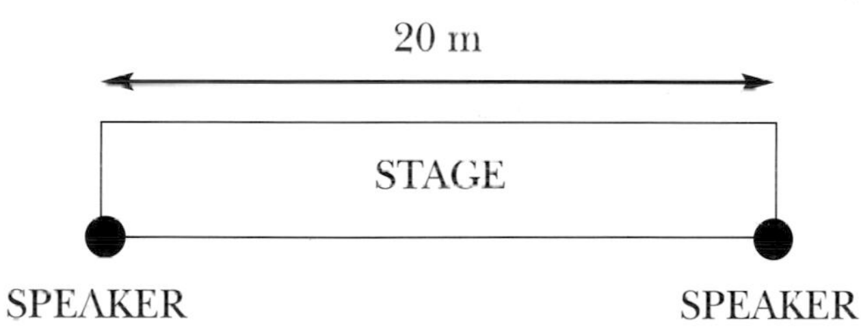

8 Sophie works in town S and her parents live in town T, 50 miles from town S. She is looking for a place to live somewhere between the two towns. She wants to be nearer to S than T but would like to be within 30 miles of her parents.

Make a scale drawing and shade the area where Sophie would like to live.

Design an office for one person and make a scale drawing of your design.

Twenty five

Pythagoras' rule

Finding the hypotenuse

Measure the triangles below in millimetres and find the areas of squares A, B and C in each diagram. What do you notice?

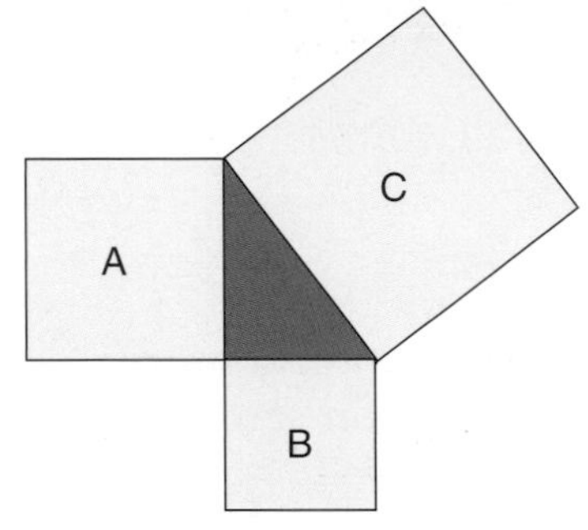

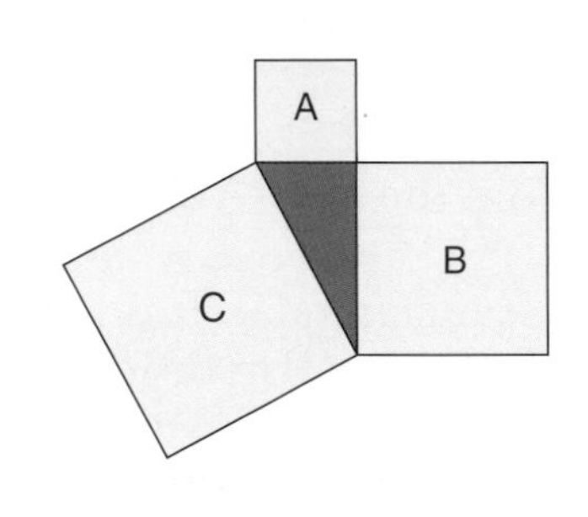

You should have found that the areas of the two smaller squares add up to the area of the largest square in each case.

This rule is called Pythagoras' rule (or theorem). It is true for all right-angled triangles. It is usually written like this:

$$a^2 + b^2 = c^2$$

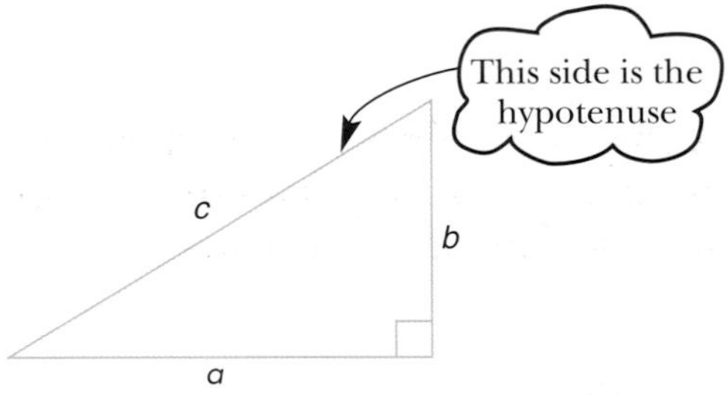

The side labelled *c must* be the hypotenuse (the longest side, always the one opposite the right angle).

Robert is a farmer and is building a gate for one of his fields. He wants to work out how long the diagonal piece of wood needs to be.

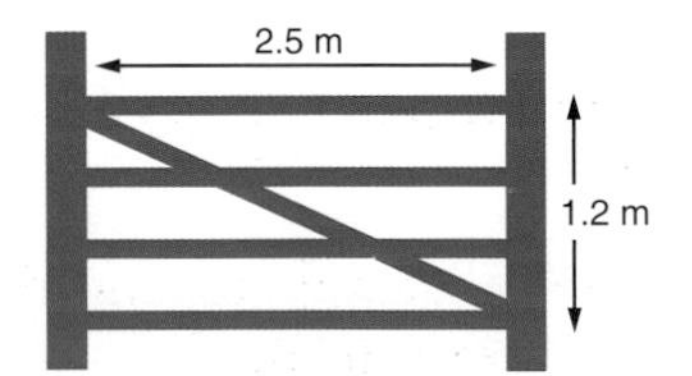

The diagonal bar on Robert's gate can be drawn as the hypotenuse of a right-angled triangle like this:

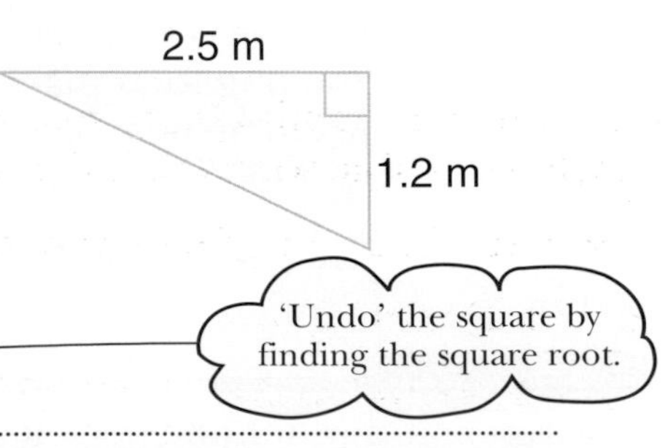

Pythagoras' rule is: $a^2 + b^2 = c^2$

In this case: $2.5^2 + 1.2^2 = c^2$

$6.25 + 1.44 = c^2$

$7.69 = c^2$

$c = 2.77$

So the diagonal piece on Robert's gate needs to be 2.77 m long.

Why must the value of c be greater than 2.5 metres?

Exercise

1 Use Pythagoras' rule to find the length of the hypotenuse in the triangles below.

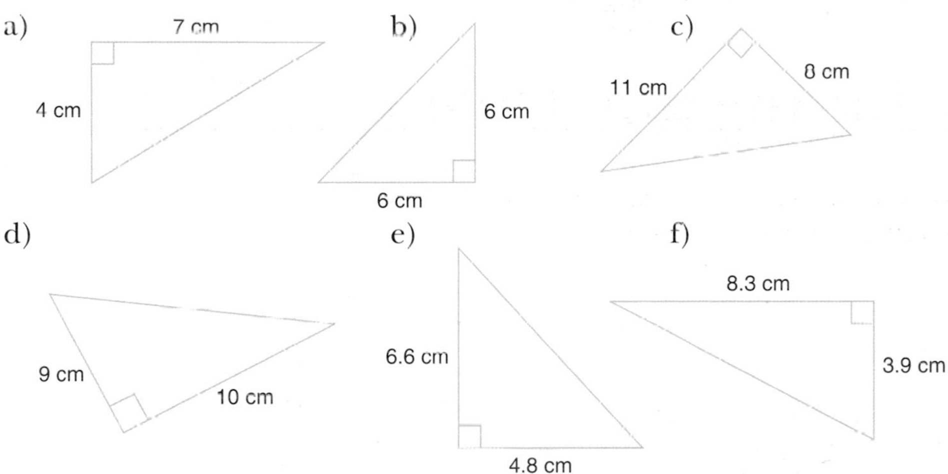

2 A field is 150 metres long and 120 metres wide. A footpath goes diagonally across the field. How long is the footpath?

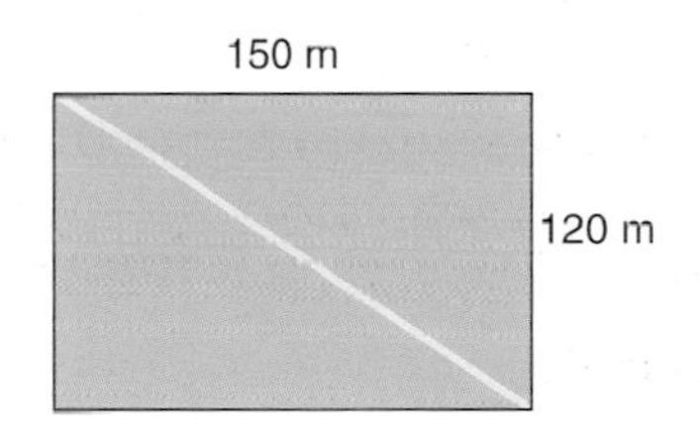

3 A ship sails 23 km due north and then 17 km due east. It then sails back to its starting point in a straight line. How far is the distance back to the starting point?

4 The diagram shows the two points A (1, 2) and B (4, 4).

a) What is the distance from A to the point P?

b) What is the distance from B to the point P?

c) Use Pythagoras' rule to find the distance from A to B.

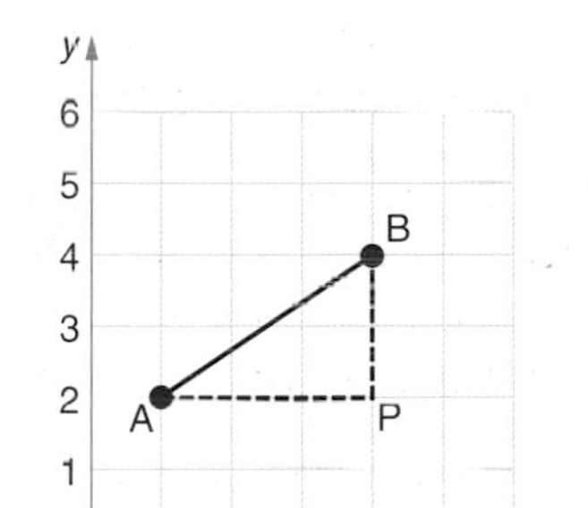

5 Use the same method as in question 4 to find the distances between each of the following pairs of points.

(You will find it helpful to draw diagrams showing the points.)

a) (3, 1) and (5, 6)

b) (4, 2) and (1, 5)

c) (–2, 4) and (3, 5)

d) (0, 3) and (–2, –1)

Finding one of the shorter sides

Terry is a window-cleaner. His ladder is 8 metres long.

For safety reasons he always places the foot of the ladder at least 1.5 metres from the wall. He wants to know how far up the wall he can make his ladder reach.

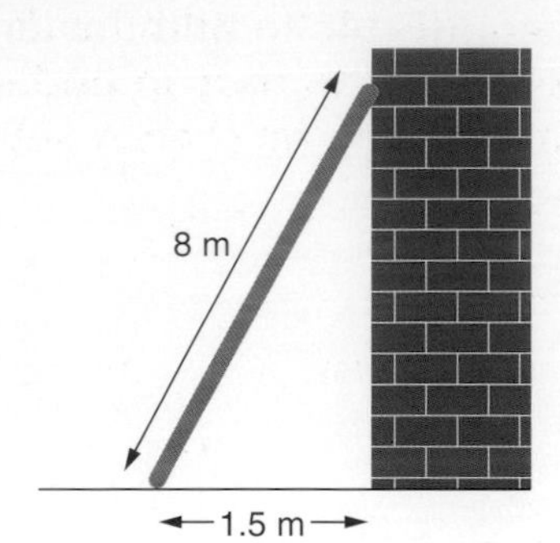

So far you have only been asked to find the length of the hypotenuse in a right-angled triangle. To solve Terry's problem, you need to be able to find one of the two shorter sides. You can use Pythagoras' rule to solve this kind of problem as well.

This is a simplified diagram of Terry's ladder.

y stands for the height up the wall that the ladder reaches.

Pythagoras' rule is:

$$a^2 + b^2 = c^2$$

In this case:

$$1.5^2 + y^2 = 8^2$$

$$2.25 + y^2 = 64$$

To find y^2, you need to subtract 2.25 from both sides of the equation.

$$y^2 = 64 - 2.25$$

$$y^2 = 61.75$$

Now you can find y by taking the square root of 61.75.

$$y = \sqrt{61.75} = 7.86$$

The ladder reaches 7.86 metres up the wall.

Why must the value of y be less than 8 metres?

Remember!

To find the hypotenuse, you have to **add**.

To find one of the shorter sides, you have to **subtract**.

In the calculation above you obtained $y = \sqrt{61.75}$. Your calculator gives the value of y as 7.8581 . . . and you round this to 7.86. Numbers like $\sqrt{61.75}$ which do not work out exactly are called **surds**. Sometimes lengths are given in surd form.

Exercise

1 Find the lengths of the sides marked x in each of these triangles. In some of them you have to find the hypotenuse, in others you have to find one of the shorter sides.

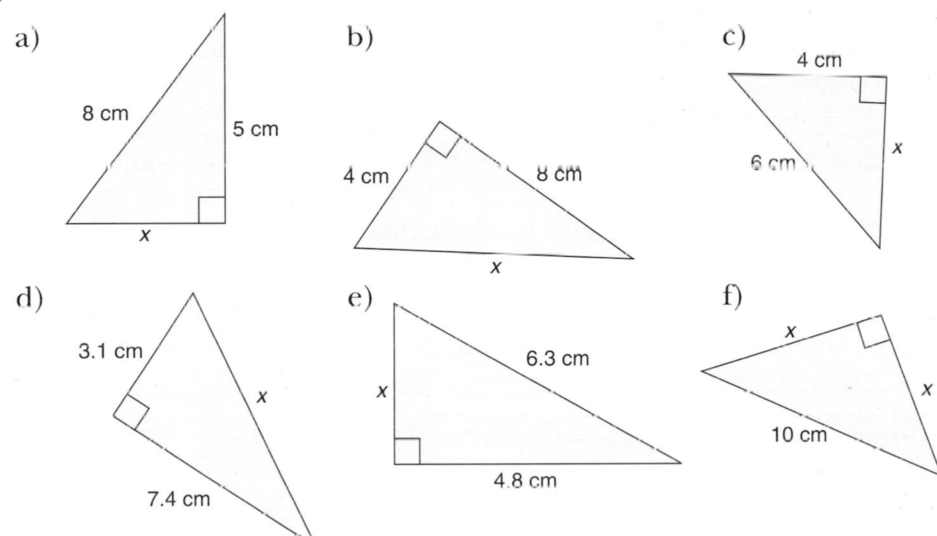

2 A ladder 6.2 metres long is to be placed so that it just reaches a window 5.7 metres from the ground. How far from the wall is the foot of the ladder?

3 The diagram shows an isosceles triangle split into two congruent right-angled triangles.

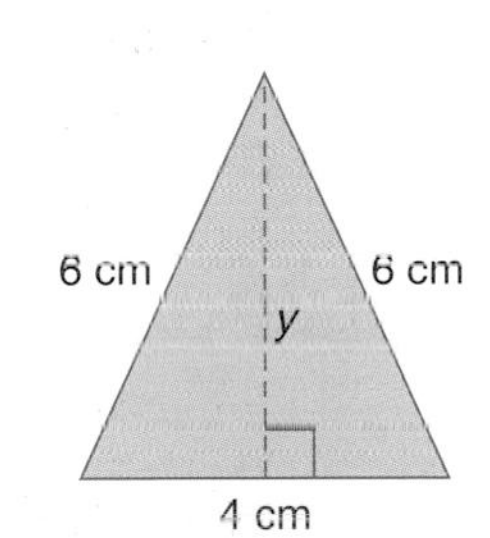

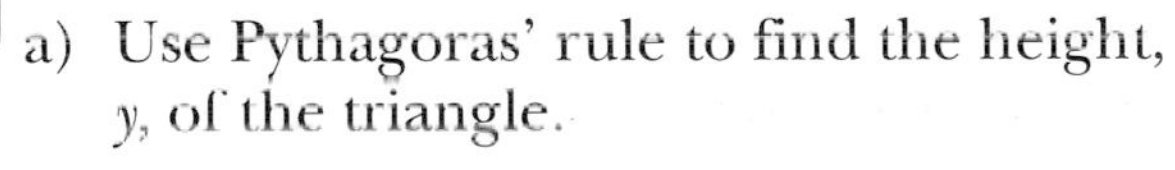

a) Use Pythagoras' rule to find the height, y, of the triangle.

b) Find the area of the triangle.

4 The diagram shows a right-angled triangle ABC. The line BN has been drawn in, splitting the triangle into two smaller right-angled triangles ANB and CNB.

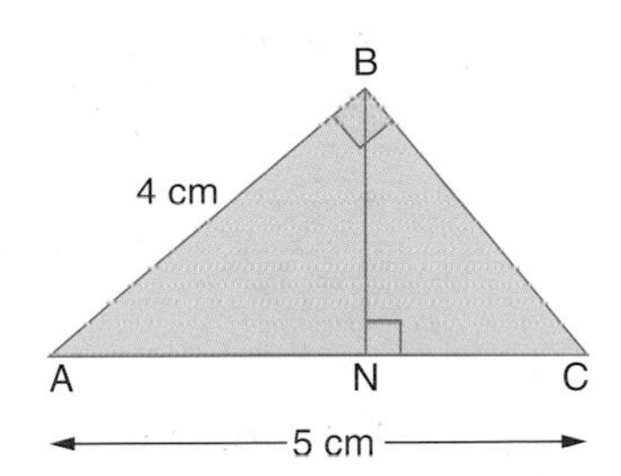

a) Work out the length of the side BC.

b) Using AB as the base of the triangle, work out the area of the triangle.

c) Using AC as the base of the triangle, use your answer to b) to work out the length of BN.

d) Work out the lengths of AN and CN.

5 In triangle ABC, angle B = 90°, AB = 4 cm and AC = 7 cm.

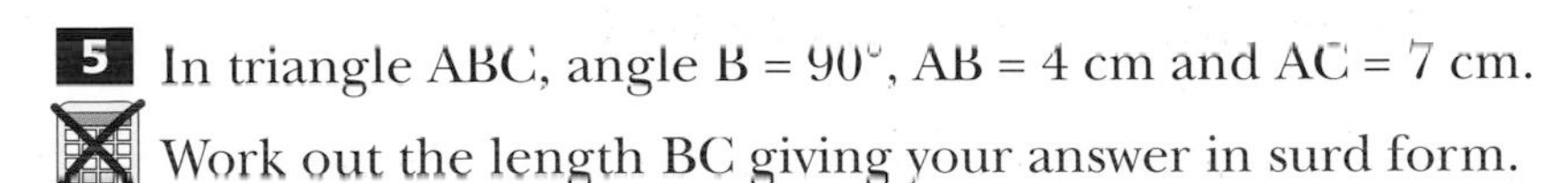

Work out the length BC giving your answer in surd form.

Finishing off

Now that you have finished this chapter you should be able to:

★ use Pythagoras' rule to find the hypotenuse of a right-angled triangle

★ use Pythagoras' rule to find one of the shorter sides of a right-angled triangle.

Use the questions in the next exercise to check that you understand everything.

Mixed exercise

1 Find the lengths of the sides marked with letters in these triangles.

a)

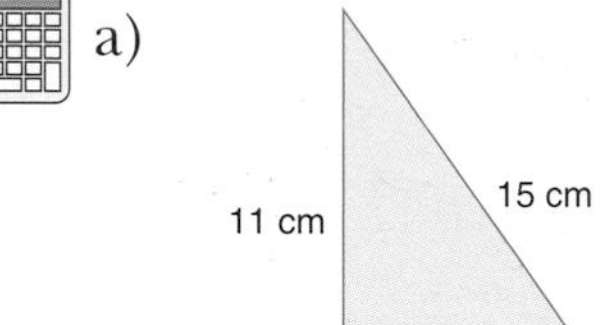

b)

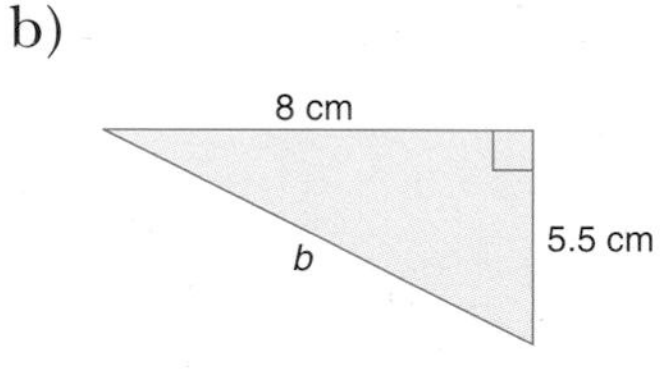

c)

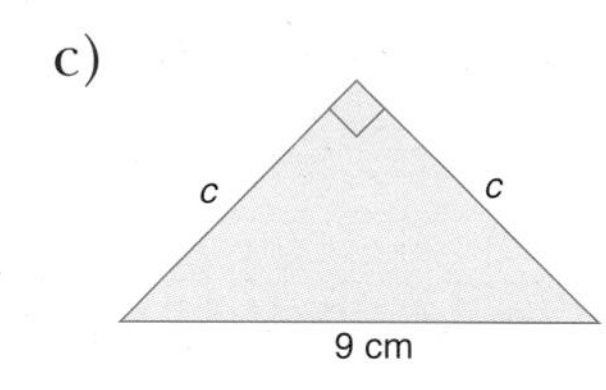

2 Find the distance between each pair of points.
Give your answers in surd form where appropriate.

a) (1, 4) and (4, 0)

b) (–2, 3) and (2, –1)

c) (–3, –4) and (–1, 1)

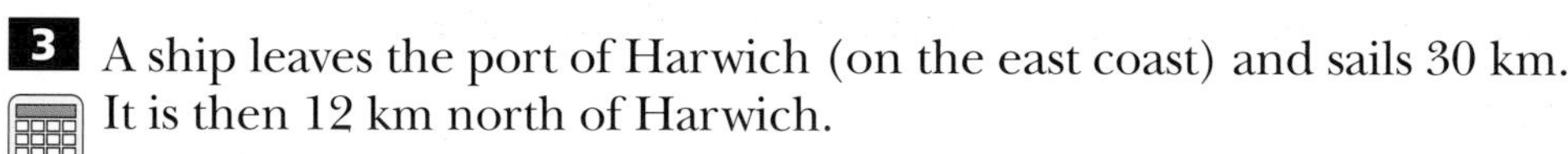

3 A ship leaves the port of Harwich (on the east coast) and sails 30 km. It is then 12 km north of Harwich.

How far east is it from Harwich?

4 Jenny designs this ramp to provide access to a building.

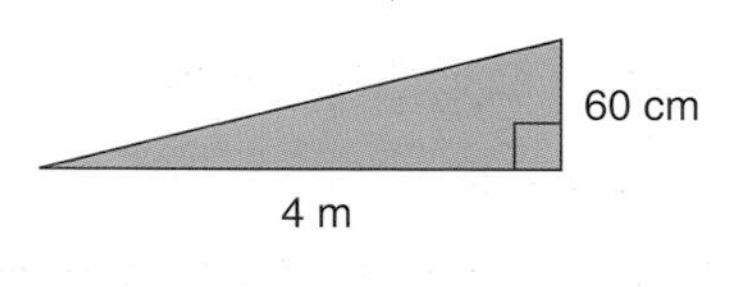

Work out the length of the ramp giving your answer to the nearest centimetre.

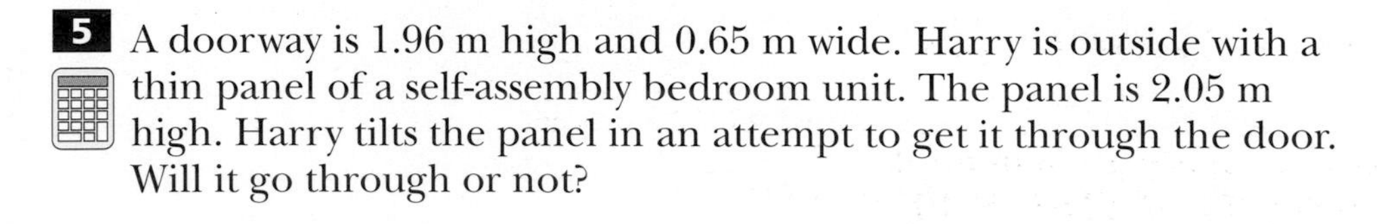

5 A doorway is 1.96 m high and 0.65 m wide. Harry is outside with a thin panel of a self-assembly bedroom unit. The panel is 2.05 m high. Harry tilts the panel in an attempt to get it through the door. Will it go through or not?

6 Mark and Imogen are putting up Christmas decorations (streamers) in their office.

This is a plan of their office.

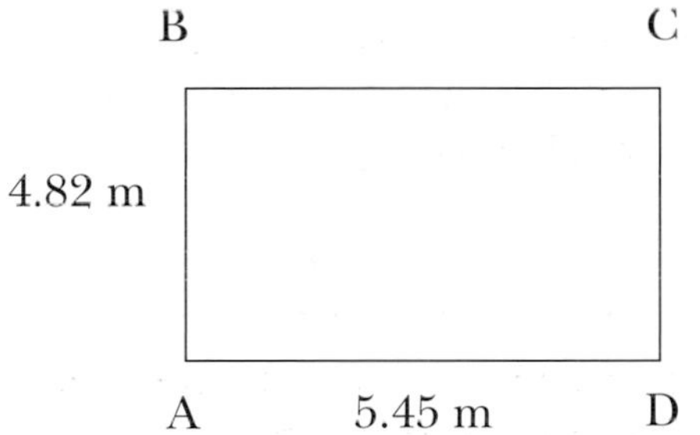

a) How long must a streamer be to go from A to C?

b) Mark pins one end of a 6 m streamer at A. He pins the other end on wall BC at E.

How far is E from C?

c) Imogen pins one end of a 6 m streamer at A. She pins the other end on wall DC at F.

How far is F from C?

d) What length of streamer is needed to go from E to F?

7 This drawing shows Debbie's house.

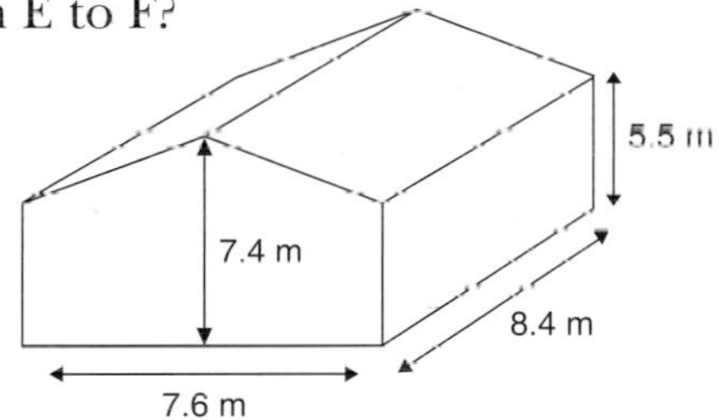

She is building it herself and wants to work out how many roof tiles she will need.

A row of tiles is needed approximately every 25 cm.

a) How many rows of tiles are needed to cover the roof?

b) A tile is 30 cm wide. How many tiles are there in a row?

c) Work out the total number of tiles needed to cover the roof.

8 A thin rod is 50 cm long. It is straight and does not bend.

Will the rod fit into a box 40 cm long, 25 cm wide and 20 cm high?

Explain your answer.

Investigation

This triangle is a right-angled triangle.

$3^2 + 4^2 = 9 + 16 = 25 = 5^2$

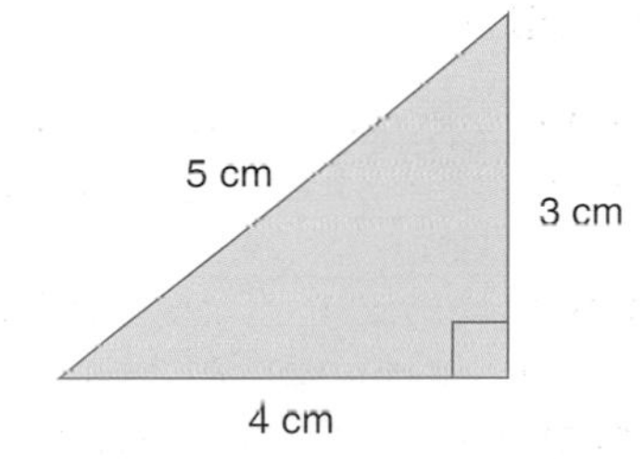

The numbers 3, 4, 5 are called a **Pythagorean triple** because they obey Pythagoras' rule.

Find as many different Pythagorean triples as you can.

(Don't count triples which are just a multiple of one you have already found, like 6, 8, 10 or 9, 12, 15, which are both multiples of 3, 4, 5.)

Twenty six

Trigonometry

Introduction to trigonometry

Trigonometry is the study of triangles. In this chapter you will learn how to calculate sides and angles in right-angled triangles.

You need a scientific calculator for this chapter.

Triangles A, B and C are all right-angled triangles, and they all have an angle of 30°. This means that they are all enlargements of each other, with different scale factors. They are **similar** triangles.

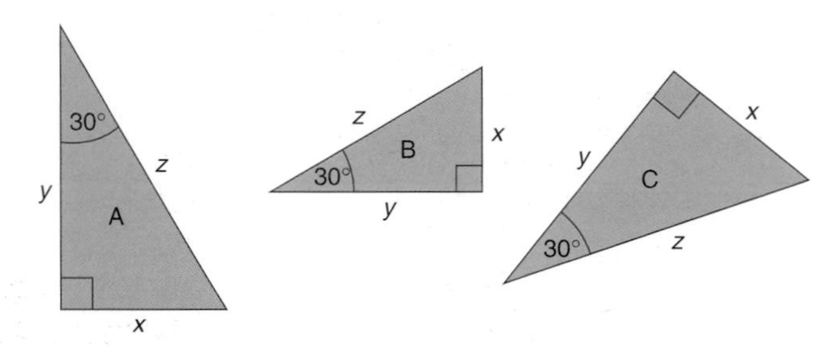

- The sides marked z in the diagram are all opposite to the right-angle. They are the longest side of each triangle. The longest side is called the **hypotenuse**.
- The sides marked x are all **opposite** to the angle of 30°.
- The sides marked y are all **adjacent** to (next to) the angle of 30°.

For any right-angled triangle with one angle marked, you can label the sides hypotenuse, opposite and adjacent. (Some people use the abbreviations **hyp**, **opp** and **adj**.)

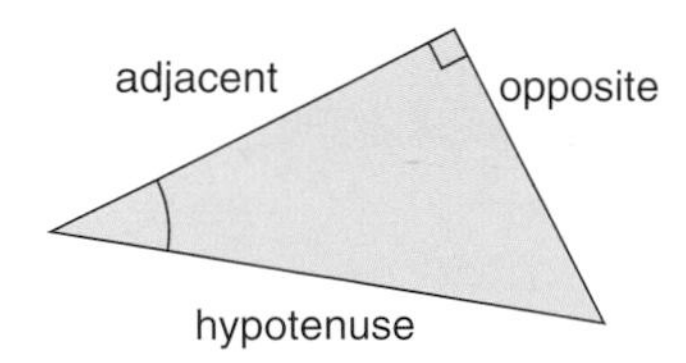

By looking at the diagrams below you can see that the labels 'opposite' and 'adjacent' are not permanently attached to a particular side.

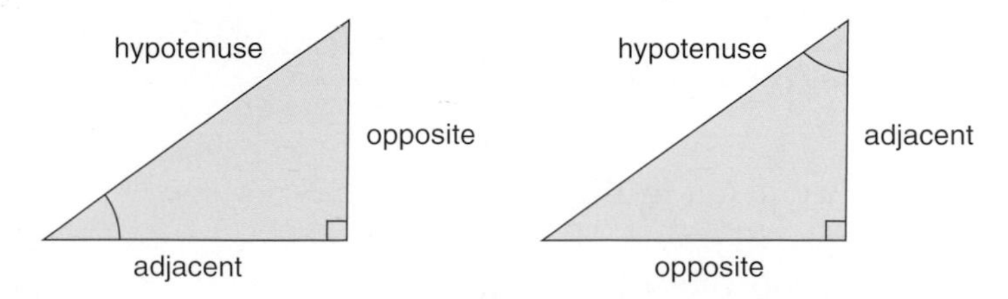

The labelling of opposite and adjacent depends on which angle you are working with.

Exercise

1 Copy each of the triangles below and label the hypotenuse, opposite and adjacent sides.

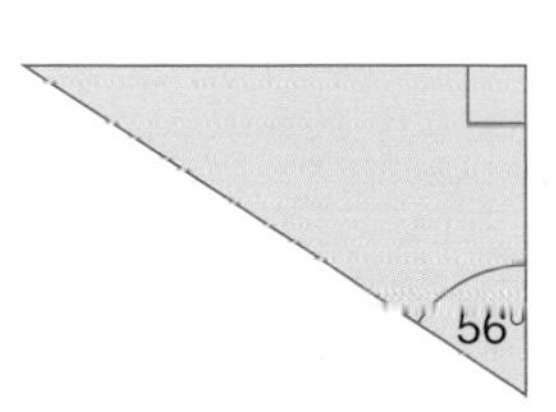

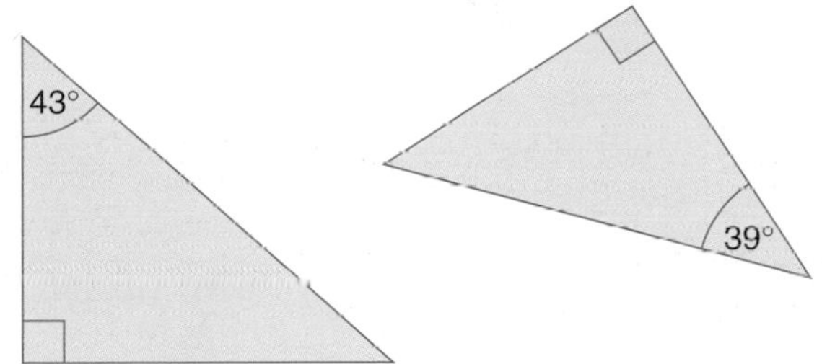

2 Triangles P, Q and R are similar triangles.

They all have an angle of 40°.

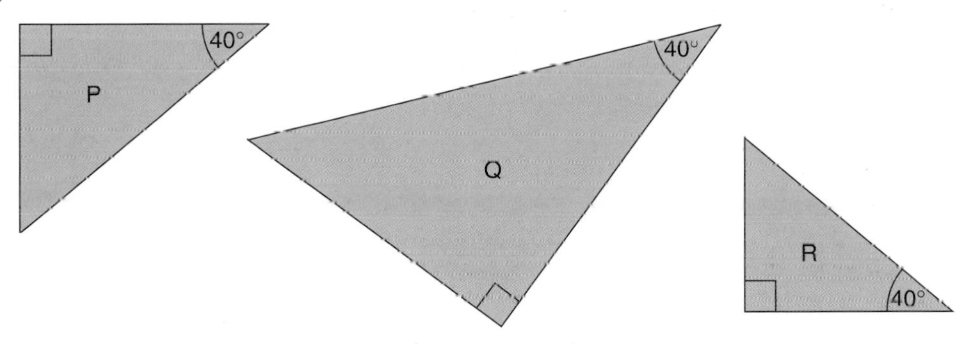

a) Measure the hypotenuse, opposite and adjacent sides of each triangle.

b) Work out the ratio $\frac{\text{opposite}}{\text{adjacent}}$ for each triangle.

c) Work out the ratio $\frac{\text{opposite}}{\text{hypotenuse}}$ for each triangle.

d) Work out the ratio $\frac{\text{adjacent}}{\text{hypotenuse}}$ for each triangle.

e) What do you notice?

3 Repeat question 2 for triangles X, Y and Z.

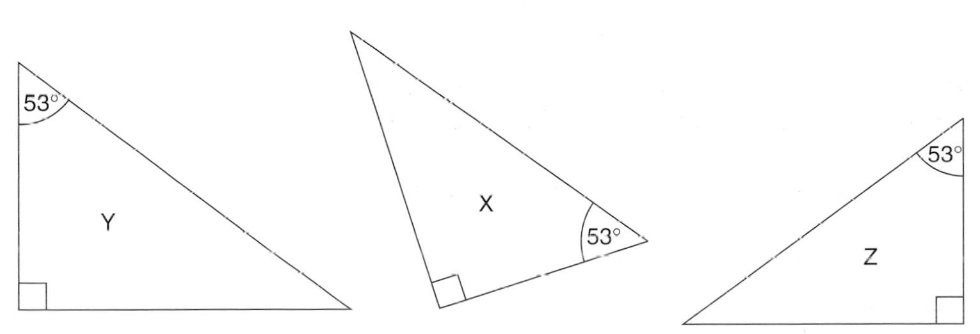

Sine, cosine and tangent

In question 2 on page 245 you found that for each of triangles P, Q and R

$\frac{\text{opp}}{\text{hyp}} = 0.64$ $\quad\quad$ $\frac{\text{adj}}{\text{hyp}} = 0.77$ $\quad\quad$ $\frac{\text{opp}}{\text{adj}} = 0.84$

These ratios are given correct to two decimal places and are true for all right-angled triangles with an angle of 40°.

The ratio $\frac{\text{opp}}{\text{hyp}}$ is called the **sine** (or **sin**) of that angle.

sin 40° = 0.64

The ratio $\frac{\text{adj}}{\text{hyp}}$ is called the **cosine** (or **cos**) of that angle.

cos 40° = 0.77

The ratio $\frac{\text{opp}}{\text{adj}}$ is called the **tangent** (or **tan**) of that angle.

tan 40° = 0.84

For all right-angled triangles with a particular angle these three trigonometrical (trig.) ratios are fixed. They can be written like this.

$$\sin\theta = \frac{\text{opp}}{\text{hyp}}$$

$$\cos\theta = \frac{\text{adj}}{\text{hyp}}$$

$$\tan\theta = \frac{\text{opp}}{\text{adj}}$$

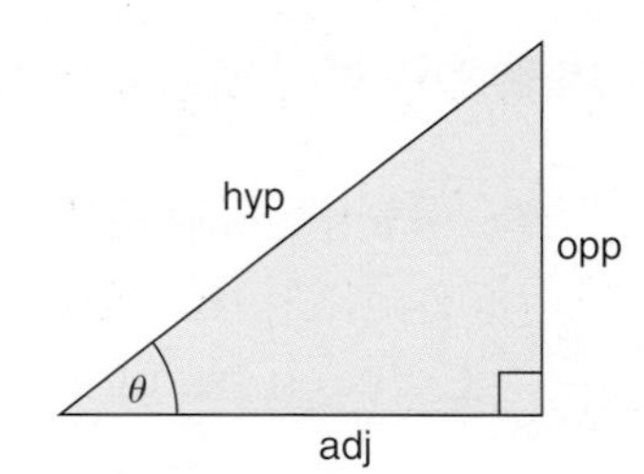

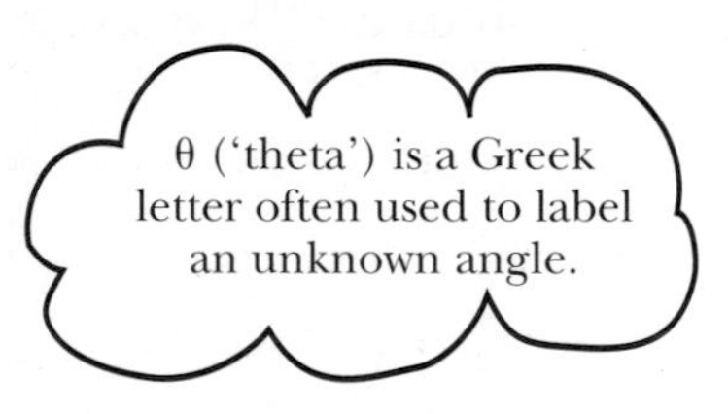

Some people remember these three ratios using the made-up word **SOHCAHTOA**. (**S**in, **O**pp, **H**yp; **C**os, **A**dj, **H**yp; **T**an, **O**pp, **A**dj)

You can find the sin, cos or tan of an angle using a scientific calculator. First make sure that your calculator is in degree (deg) mode. Now check that you know how to use the *sin* button. For some calculators you must enter the angle and then press the *sin* button. For others you must press *sin* first and then key in the angle, followed by ENTER or EXE.

Use you calculator to work out sin 40°, cos 40° *and* tan 40°.
Check that you get the answers shown above (although your calculator will give more decimal places).

In the following pages you will use the trig. ratios to find unknown sides and angles in right-angled triangles.

Exercise

1 Use your calculator to work out each of these trig. ratios. Write down your answer as far as the fourth decimal place. (e.g. sin 32° = 0.5299…)

a) sin 65° b) cos 28° c) tan 83°

d) sin 12° e) tan 45° f) cos 57°

Investigation

Use your calculator to complete this table.

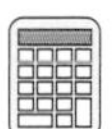

sin 0° = 0	cos 0° = 1	tan 0° = 0
sin 15° = 0.2588…	cos 15° =	tan 15° =
sin 30° =	cos 30° =	tan 30° =
sin 45° =	cos 45° = 0.7071…	tan 45° =
sin 60° =	cos 60° =	tan 60° =
sin 75° =	cos 75° =	tan 75° =
sin 90° =	cos 90° =	tan 90° = error

Use the results in the table to answer these questions.

1 a) What happens to sine as the angle increases?

b) What happens to cosine as the angle increases?

c) Which cosine is the same as (i) sin 15° (ii) sin 30°.

d) Explain how sine and cosine are related.

e) Sketch a triangle with angles 30°, 60° and 90° and explain why your answer to d) is true.

2 a) Work out values for (i) tan 80° (ii) tan 85° (iii) tan 89° (iv) tan 89.5°

b) Use your results from a) to explain why you do not get a value for tan 90°.

3 a) Sketch a triangle with angles 15°, 75° and 90° and use it to explain how tan 15° and tan 75° are related.

b) Find two other angles related in the same way as 15° and 75° are in a).

4 a) Work out $\frac{\sin 15°}{\cos 15°}$ and explain how it is related to tan 15°.

b) Does this work for other angles?

c) Explain why it works.

Finding an unknown side

When you want to find a missing side or angle in a right-angled triangle, the first thing to do is to decide whether to use sin, cos or tan. The example below shows how to do this.

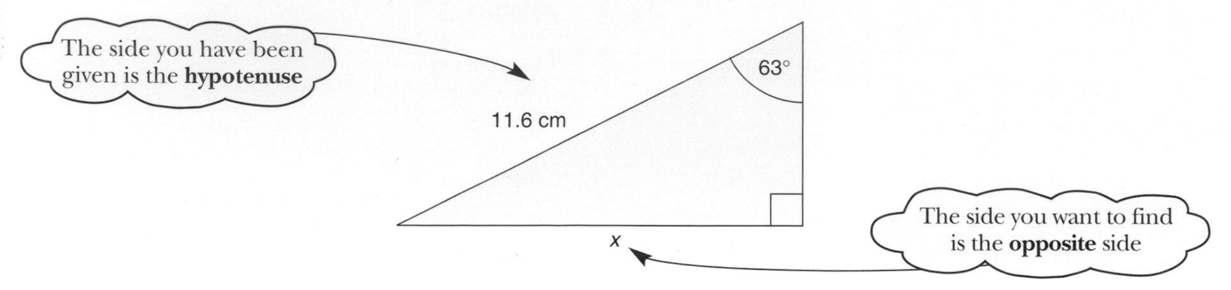

You need the ratio which involves **hypotenuse** and **opposite**. Look at the ratios on page 246 and you will see that the one you need is sin.

$$\sin 63° = \frac{\text{opp}}{\text{hyp}}$$

$$\sin 63° = \frac{x}{11.6}$$

You solve this equation by multiplying by 11.6.

$$x = 11.6 \times \sin 63°$$

$$x = 10.3 \text{ (3 s.f.)}$$

Why must the required side be less than 11.6 cm*?*

The side you want to find is the **hypotenuse**.

x

68°

7.5 cm

The side you have been given is the **adjacent** side.

In the last example, the unknown side was on top in the ratio.

In the next example it is on the bottom

You start off, as before, by deciding which ratio to use.

You need the ratio which involves **hypotenuse** and **adjacent**. It is cos.

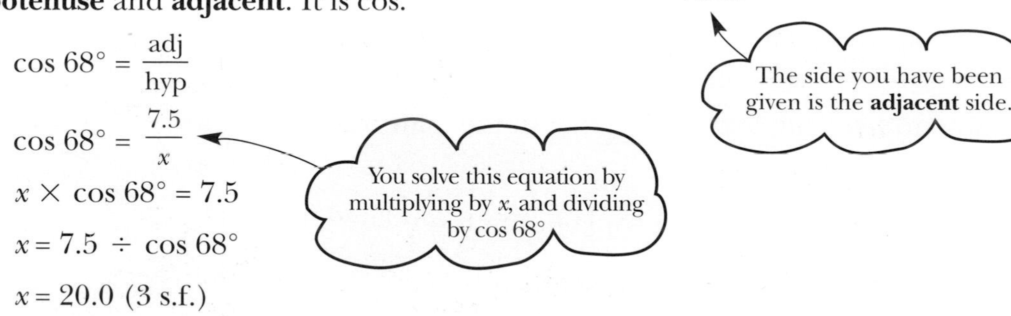

$$\cos 68° = \frac{\text{adj}}{\text{hyp}}$$

$$\cos 68° = \frac{7.5}{x}$$

You solve this equation by multiplying by x, and dividing by cos 68°

$$x \times \cos 68° = 7.5$$

$$x = 7.5 \div \cos 68°$$

$$x = 20.0 \text{ (3 s.f.)}$$

Why must the required side be at least 7.5 cm*?*

Exercise

1 Find the sides marked with letters in each of these triangles.

a)
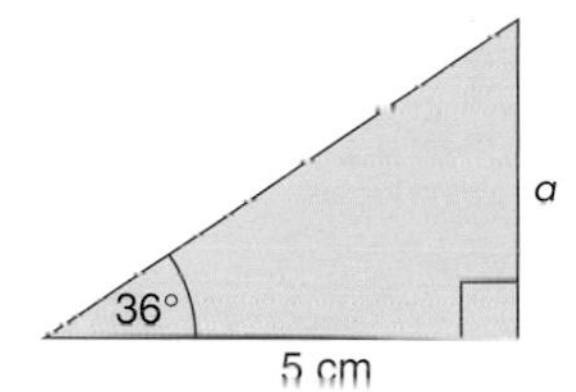

b)
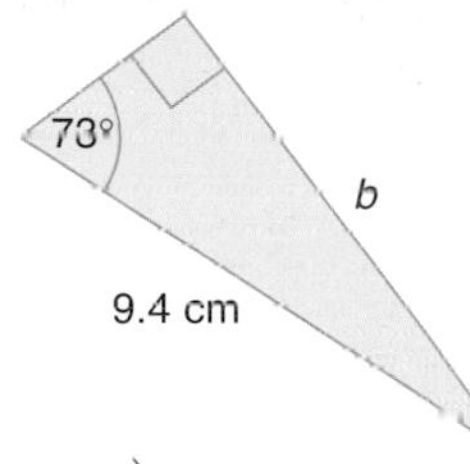

c)
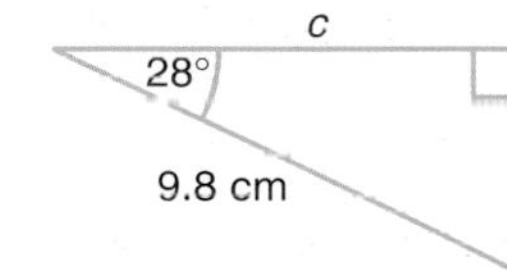

d) 62° 11.4 cm d

e) 37° 8.3 cm e

f) 24° 6.7 cm f

2 Find the sides marked with letters in each of the triangles below.

a) 4.4 cm 51° a

b)
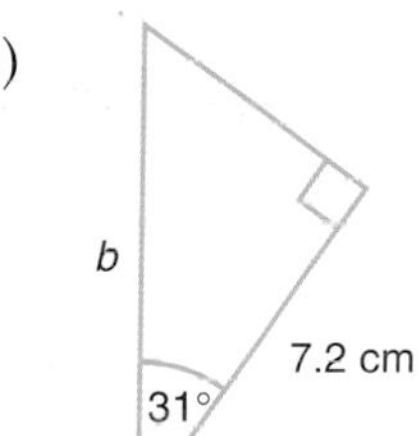

c)
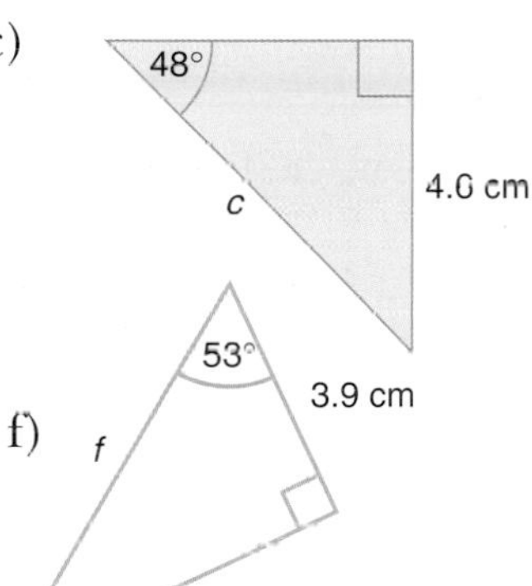

d) d 62° 7.1 cm

e) e 25° 2.3 cm

f)

3 Find the sides marked with letters in these triangles. They are a mixture of the types that you have met so far.

a)
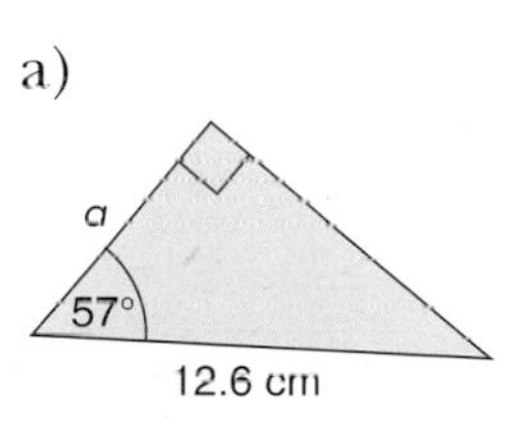

b)
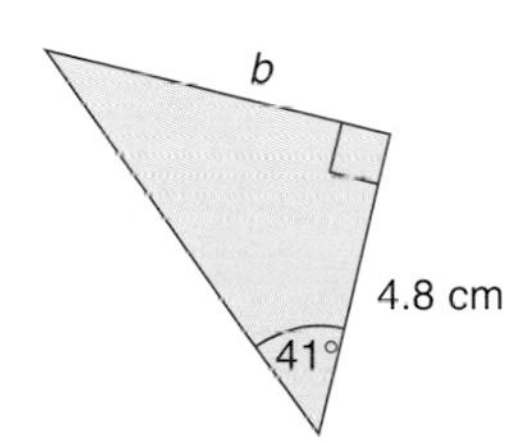

c)
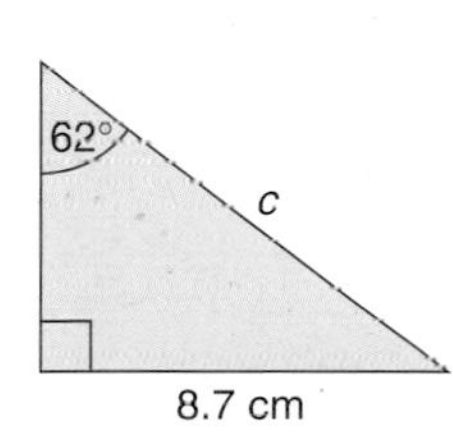

d)
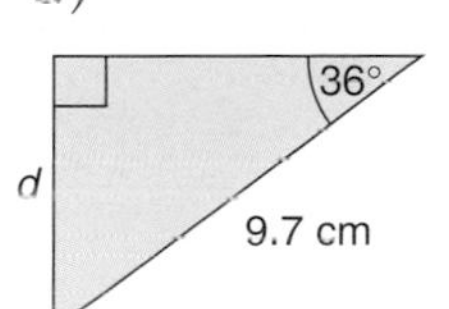

e)
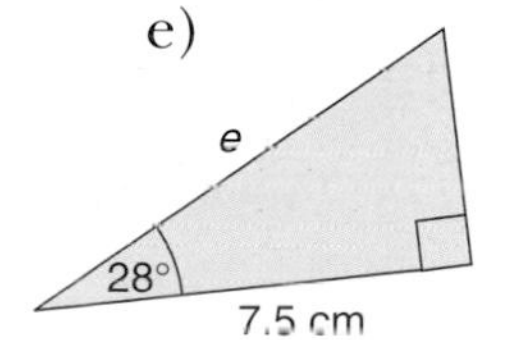

f)
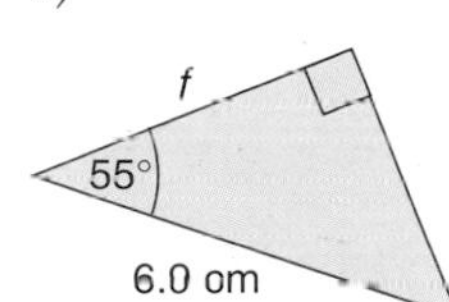

Finding an unknown angle

In question 2 on page 245 you measured sides of right-angled triangles with an angle of 40° and found that

$\sin 40° = 0.64$ $\cos 40° = 0.77$ $\tan 40° = 0.84$

Now suppose that you know two sides and that you want to find the angle.

You can use the two sides to work out a ratio. This leads to

$\sin \theta = 0.64$ (or a similar result involving $\cos \theta$ or $\tan \theta$)

You now need to find out what angle has a sin of 0.64. To do this you need to 'undo' sin. You can do this using a scientific calculator. The key you need may be labelled $\sin^{-1}$ or arcsin or you may need to press INV followed by sin. You may have to press the key before or after entering 0.64.

Use your calculator to check that both $\cos \theta = 0.77$ *and* $\tan \theta = 0.84$ *give the angle* θ *as approximately* 40°.

In the following example you have to find the unknown angle.

You start off, as before, by deciding which ratio to use.

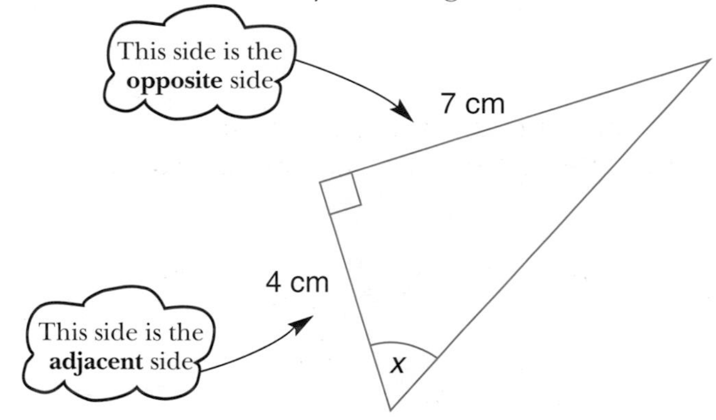

You need the ratio which involves **opposite** and **adjacent**. It is tan.

$$\tan x = \frac{\text{opp}}{\text{adj}}$$

$$\tan x = \frac{7}{4} = 1.75$$

$$x = 60.3° \text{ (to 1 d.p.)}$$

How would you work out the third angle in the triangle?

How would you work out the length of the hypotenuse?

Exercise

1 Use your calculator to work out the angle θ. Give your answer correct to the nearest tenth of a degree.

a) $\sin\theta = 0.72$ b) $\cos\theta = 0.249$ c) $\tan\theta = 0.7625$

d) $\cos\theta = 0.6667$ e) $\sin\theta = 0.8371$ f) $\tan\theta = 1.6$

2 Find the angles marked with letters in these triangles.

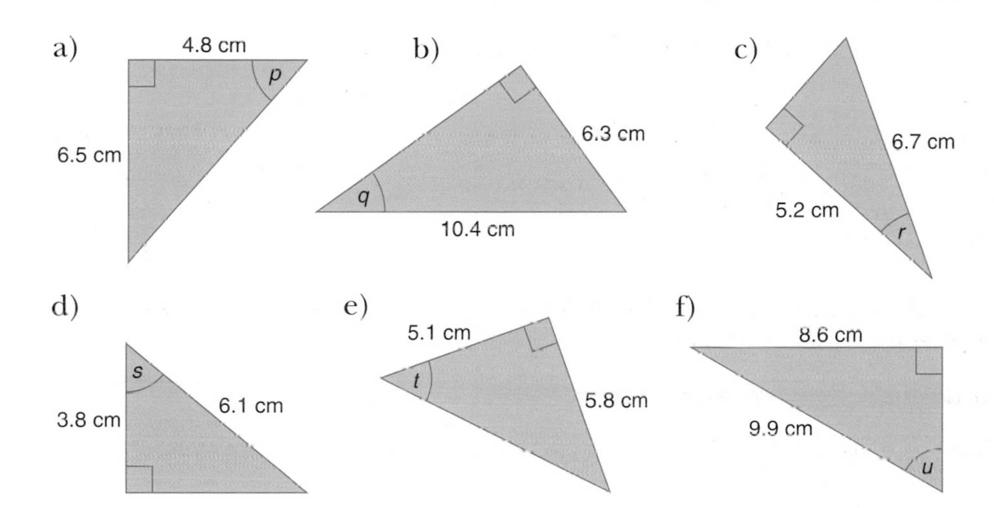

3 Find the sides and angles marked with letters in each of the triangles below. They are a mixture of the types you have met so far

a)

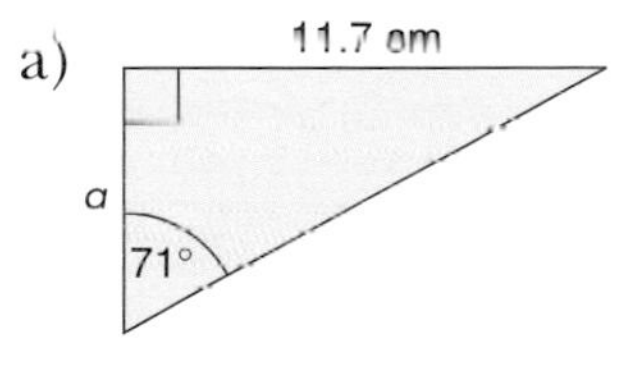

b)

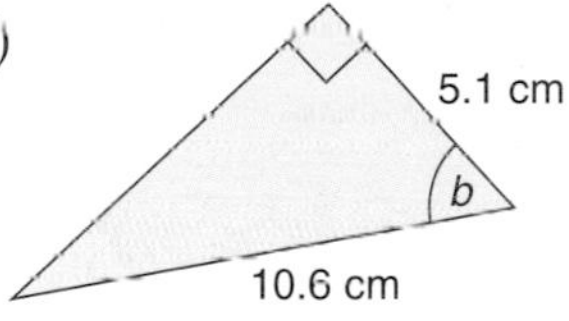

c)

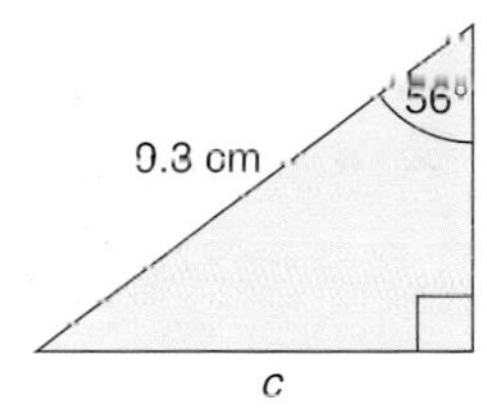

d)

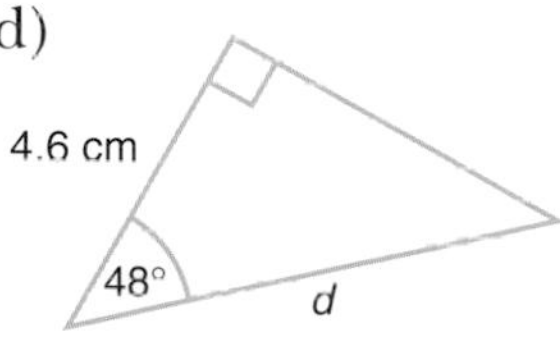

e)

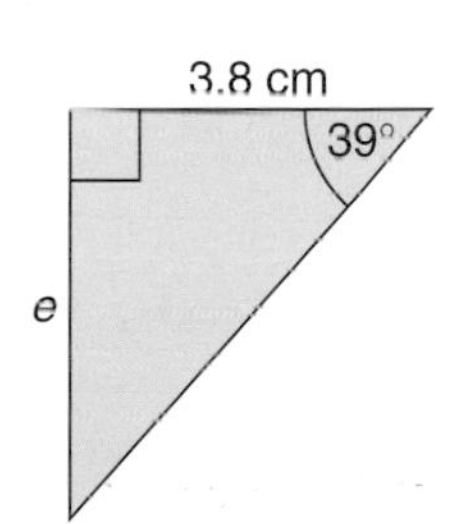

f)

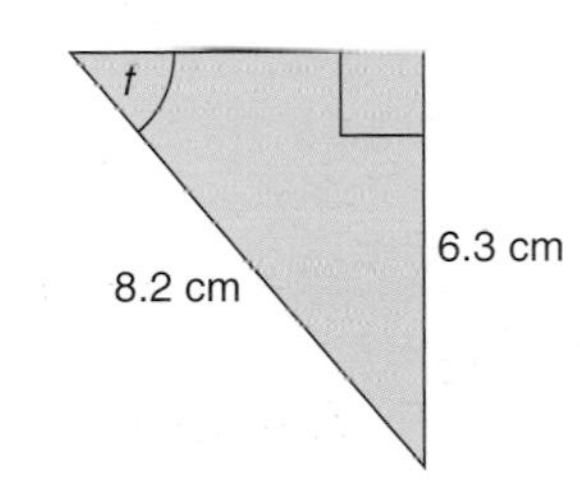

Find out what road gradients such as '1 in 10' mean. Use trigonometry to work out the angle of the slope for different gradients.

Using trigonometry

Trigonometry can be used to solve real life problems.

Example 1

Jenna stands 25 m from the foot of a tall tree. She measures the angle between the ground and a line in the direction of the top of the tree (this is called the **angle of elevation**). She finds that the angle of elevation is 34°. How high is the tree?

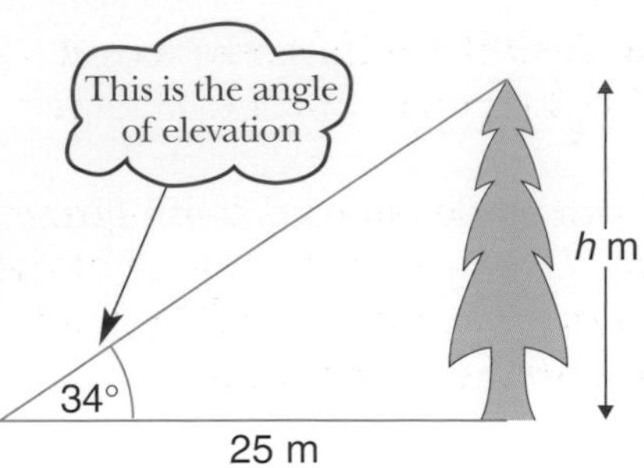

Solution

The two sides involved are the **opposite** and the **adjacent**. So you need to use **tangent** to find the angle.

$\tan 34° = \frac{\text{opp}}{\text{adj}}$ $\tan 34° = \frac{h}{25}$

$h = 25 \times \tan 34° = 16.9$ (to 1 d.p.)

The height of the tree is 16.9 m (1.d.p.)

What difficulties would you expect to meet if you were working out the height of a tree in this way?

Example 2

A ship sails 50 km on a bearing of 140°. How far south and how far east is it from its starting point?

Solution

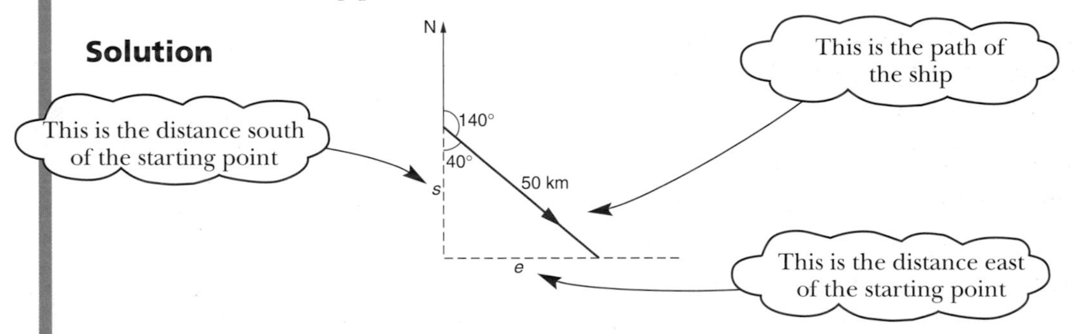

The side marked s is the **adjacent** side, and the side marked 50 km is the **hypotenuse**. So you need to use **cosine** to find s.

$\cos 40° = \frac{\text{adj}}{\text{hyp}}$ $\cos 40° = \frac{s}{50}$

$s = 50 \times \cos 40° = 38.3$

The ship is 38.3 km south of its starting point.

How would you work out e?

Work it out and check that you get 32.1 km.

1 a) An aeroplane flies 300 km on a bearing of 245°.
How far south and how far west is it from its starting point?

b) An aeroplane flies 550 km on a bearing of 318°.
How far north and how far west is it from its starting point?

2 A short flight of steps, 1.2 m high, is to be replaced by a ramp. The slope of the ramp must not be more than 10°. What is the shortest the ramp could be?

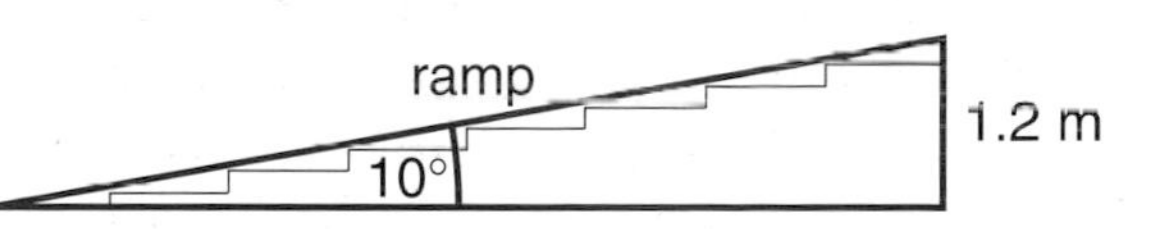

3 a) A path up a hill slopes at 15°. The path is 3.6 km long.
How high is the hill?

b) Another path up the other side of the same hill is 4.3 km long.
What is the angle of the slope of this path?

4 The guy ropes for a tent run from the top of the tent pole, which is 1.5 metres tall, to a point on the ground near the tent. Each rope should ideally be at 45° to the ground.

a) How long should each guy rope be?

b) How far will each rope be from the base of the tent pole?

5 A sailor in a small boat can see a cliff in the distance which he knows is 150 m high. The angle of elevation of the top of the cliff is 8°.
How far is the boat from the foot of the cliff?

6 A ladder is 8 m long.

a) For safety reasons, the angle it makes with the ground should not be more than 75°.

What is the highest the ladder can reach?

b) What angle must it make with the ground to just reach a gutter 6 m above the ground?

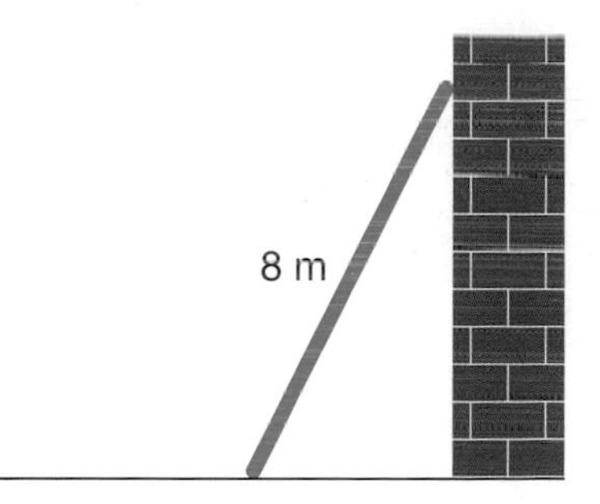

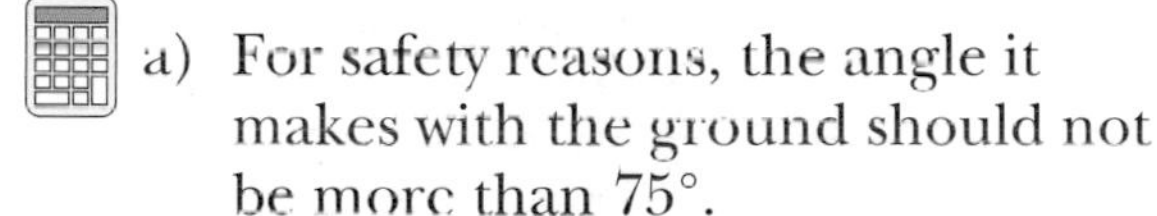

Measure the angle of elevation of a tall tree or building, and use it to work out its height.

Finishing off

Now that you have finished this chapter you should be able to:

★ find sides and angles in right-angled triangles using sine, cosine and tangent

★ use trigonometry to solve problems involving right-angled triangles.

Use the questions in the next exercise to check that you understand everything.

Mixed exercise

1 Find the sides and angles marked with letters in the triangle.

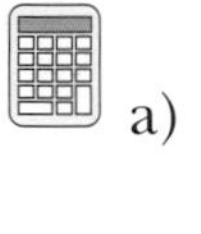

a) 11.7 cm, a, 71°

b) 48°, 6.3 cm, b

c)

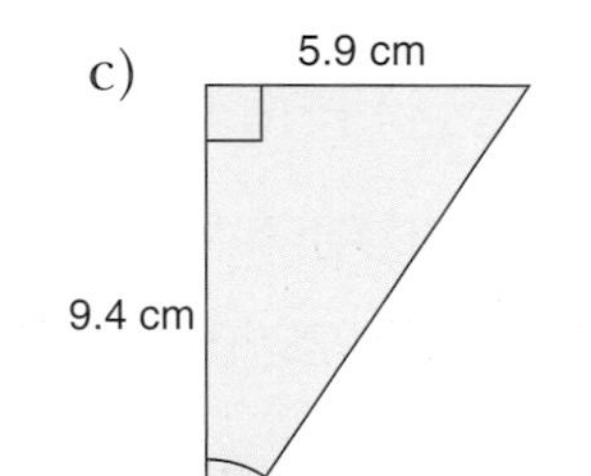

d)

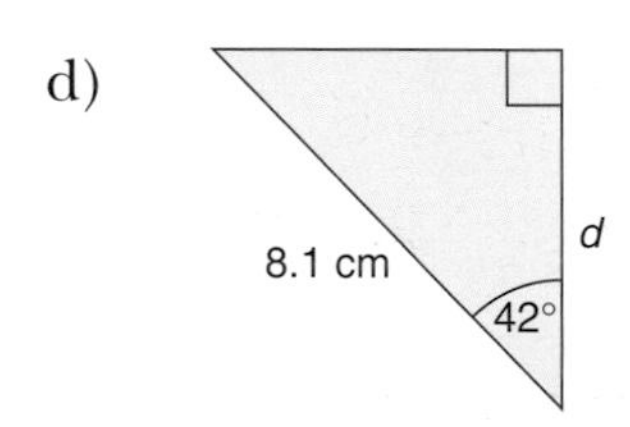

e)

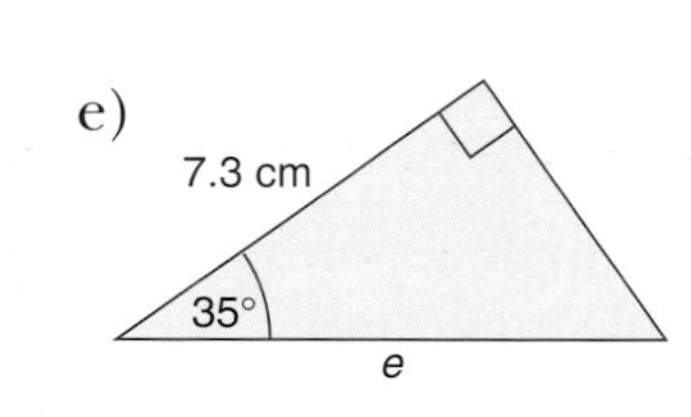

f)

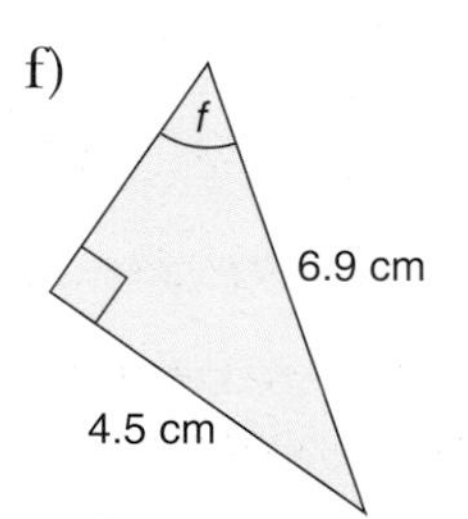

2 Simon and Sue set out from a Youth Hostel for a country walk. They walk 3 km on a bearing of 285°.

a) How far north and how far west are they from the Youth Hostel?

b) They then walk 2 km due north. Find the bearing they need to walk on to get back to the Youth Hostel.

3 A road slopes at 5° to the horizontal for 2 km measured horizontally, as shown in the diagram below. How long is the road?

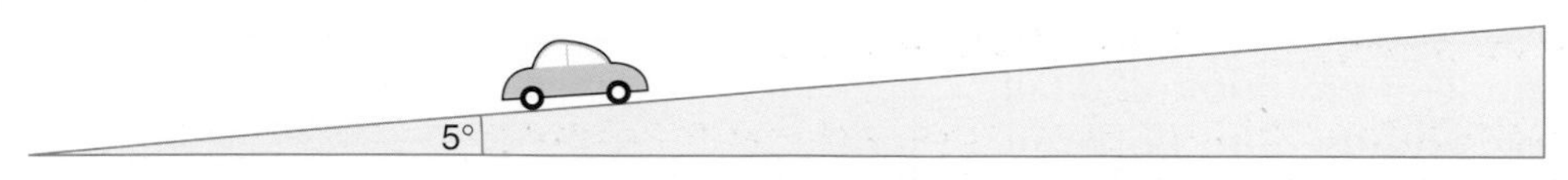

4 Charlotte stands on top of a cliff 35 m high. She looks directly out to sea and sees two small boats. The nearer boat has an angle of depression of 19° while the further boat has an angle of depression of 14°.

Work out the distance between the two boats. (Note: the angle of depression is measured from the horizontal).

5 The diagram shows a beam of light from a spotlight in a concert hall 5.4 m high.

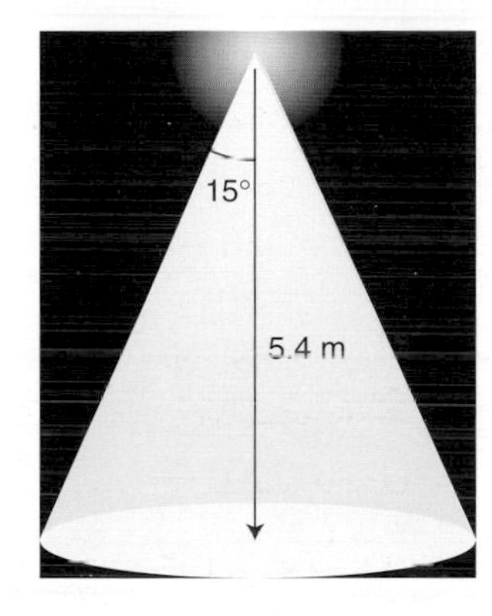

a) Find the radius of the circle of light on the floor.

b) Find the area of the circle of light.

6 a) A ship starts from point A and sails 200 km on a bearing of 115° until it reaches point B. How far south and how far east is it from point A?

b) The ship then sails a further 150 km from point B on a bearing of 230° until it reaches point C. How far south and how far west is it from point B?

c) How far south and how far east is the ship now from point A?

d) What bearing must the ship sail on to get back to point A directly from point C?

7 Daniel is standing at the edge of a river bank. The river is w metres wide. Directly opposite, and also on the edge of the bank, is a tree of height h metres. Daniel measures the angle of elevation of the top of the tree as 35°.

a) Write down a formula connecting h, w and the angle 35°.

b) Write your answer to a) in the form $h = \dots$.

Daniel moves back 10 metres away from the tree. The angle of elevation of the top of the tree is now 23°.

c) Write down a formula connecting h, w and the angle 23°.

d) Write your answer to c) in the form $h = \dots$.

e) Use your results to b) and d) to work out w and h. Give your answers to the nearest whole number.

Twenty seven

Circles and tangents

Shapes in a circle

ABCD is a quadrilateral. It is called a **cyclic quadrilateral** because all 4 points are on the circumference of a circle.

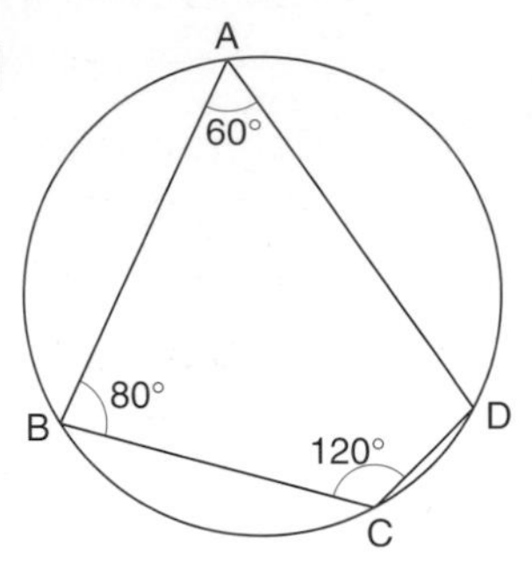

What is the angle sum of a cyclic quadrilateral?

What is the size of the angle at D?

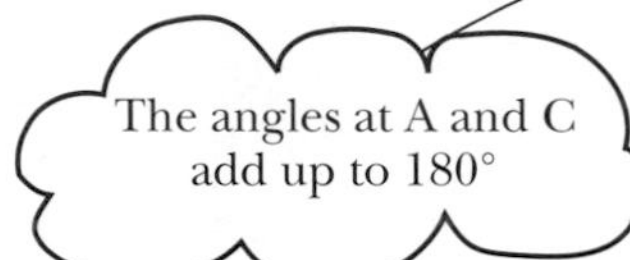

So do the angles at B and D

Draw a circle and a cyclic quadrilateral of your own.
Measure all the angles and add the pairs of opposite angles.

What do you notice?
Does everyone get the same result?

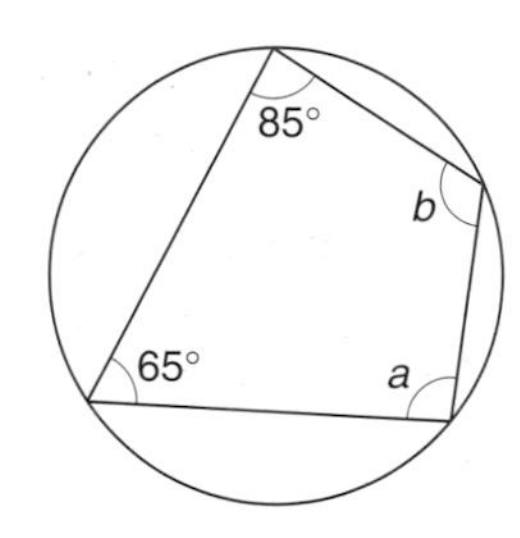

Complete this statement:
'Opposite angles of a cyclic quadrilateral … .'

What is the value of a?
What is the value of b?

Here is a triangle drawn in a circle.

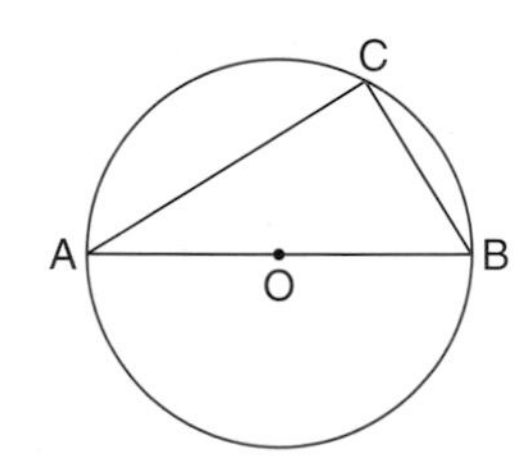

AB is the diameter.

Measure the angle at C.

Draw your own circle and add a diameter.

Complete a triangle by joining the ends of the diameter to any point on the circumference.

Measure the angle at the circumference.

What do you notice?
Does everyone get the same result?

Complete this statement:
'The angle in a semi-circle is … .'

Exercise

1 The diagram shows a circle, with centre O. AB is a diameter of the circle and C is a point on the circumference. OA, OB and OC are all the radii of the circle, so the triangles OAC and OBC are both isosceles.

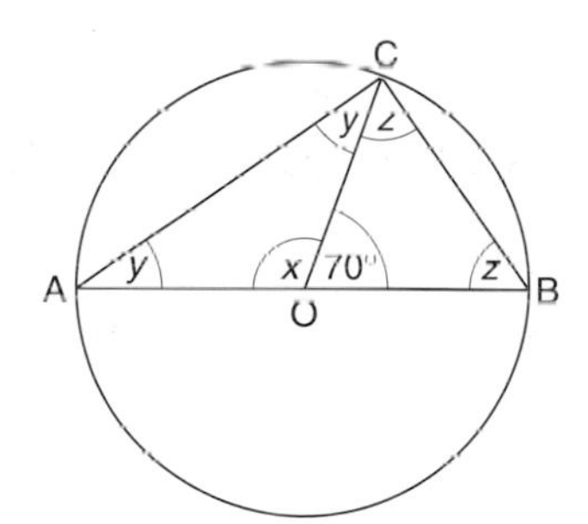

a) (i) Find the angles marked with letters.

(ii) Find the angle ACB by adding together angles y and z.

b) Choose a different angle instead of 70° for angle COB, and repeat part a) of this question. Do this several times.

c) What have you found out about the angle ACB?

2 Check that your answer to question 1 c) is correct before you start this question.

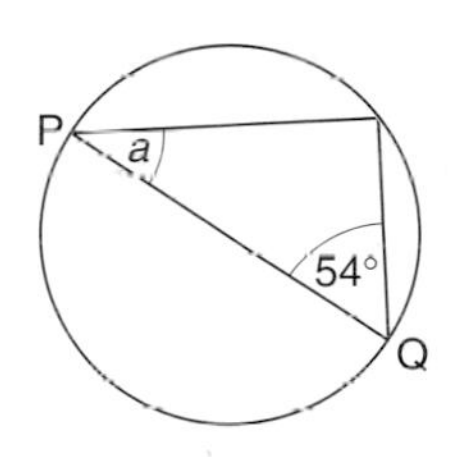

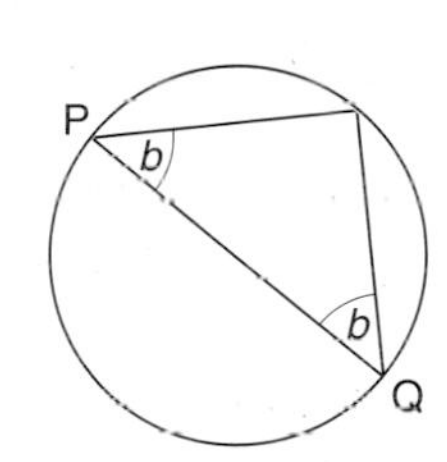

You will need to use the rule that you have found.

In each of these diagrams, PQ is a diameter of the circle.

Find the angles a and b.

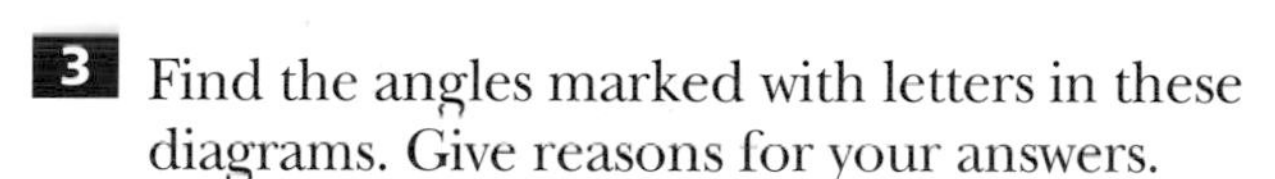

3 Find the angles marked with letters in these diagrams. Give reasons for your answers.

a)

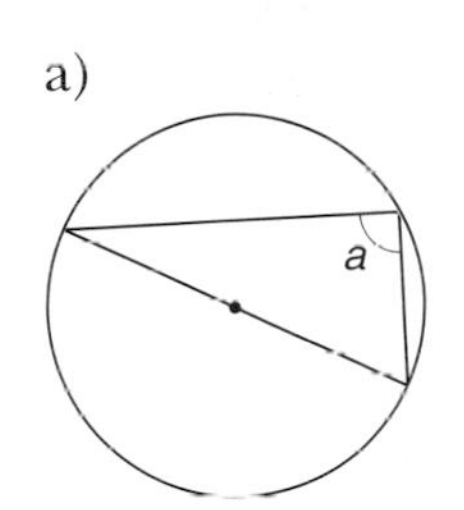

b)

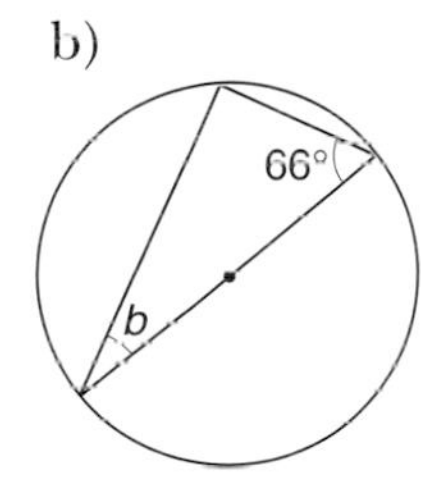

c)

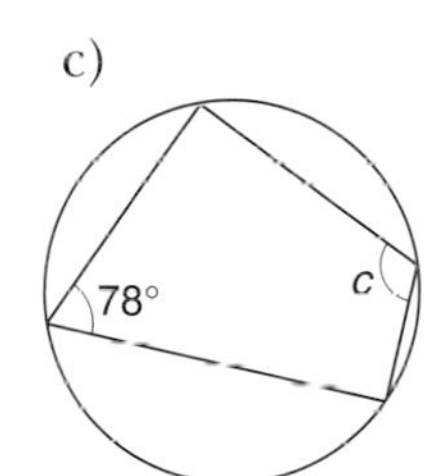

d)

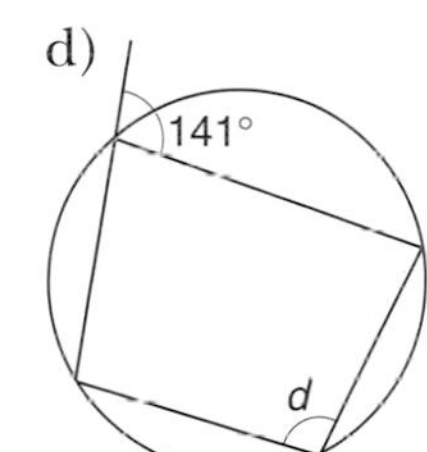

e)

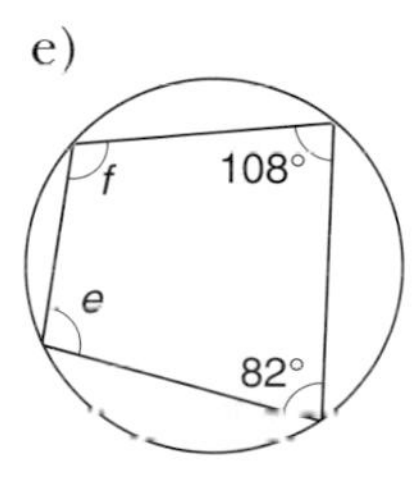

f)

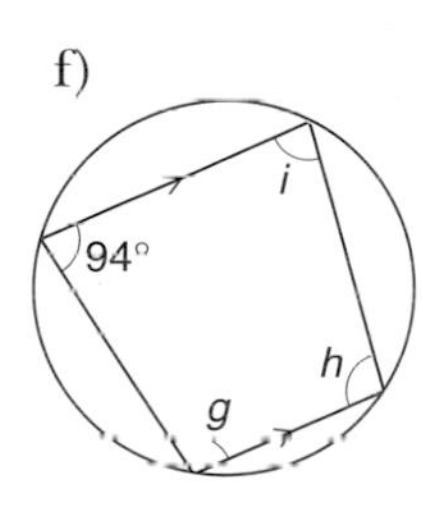

g)

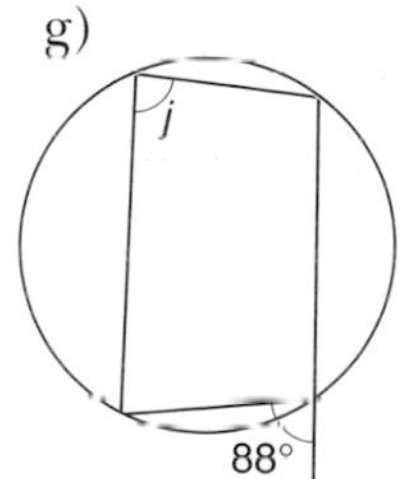

h)

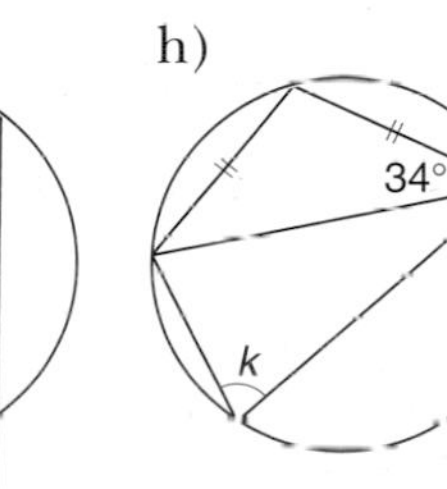

Angles in a circle

These two angles are standing on the same arc AB.

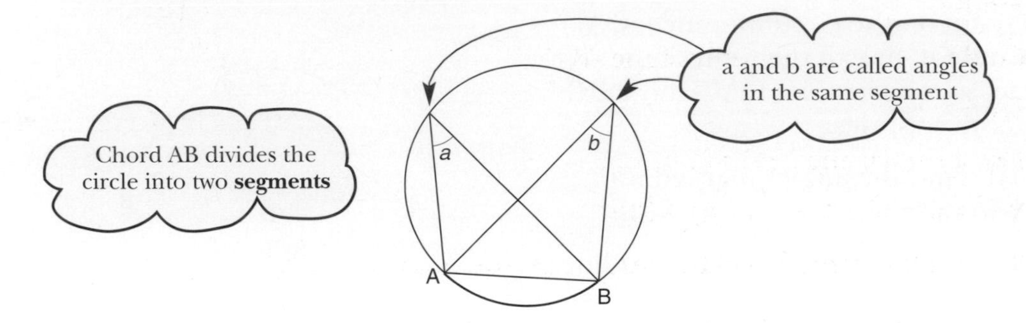

Measure angles a and b.
What do you notice?

Are there any other equal angles in the diagram?

Complete this statement:
'Angles in the same segment … .'

Which angles are equal in this diagram?

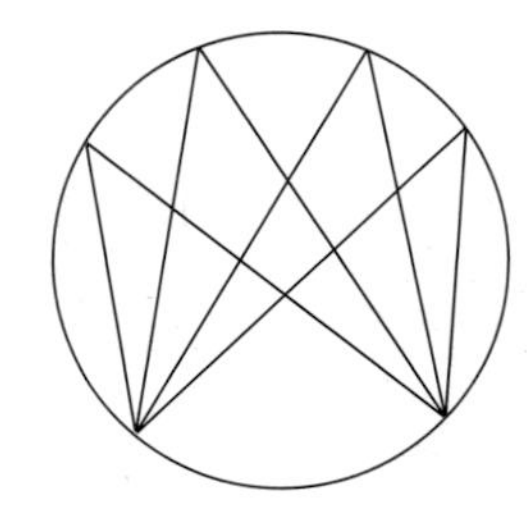

Here are two more angles which are standing on the same arc AB.

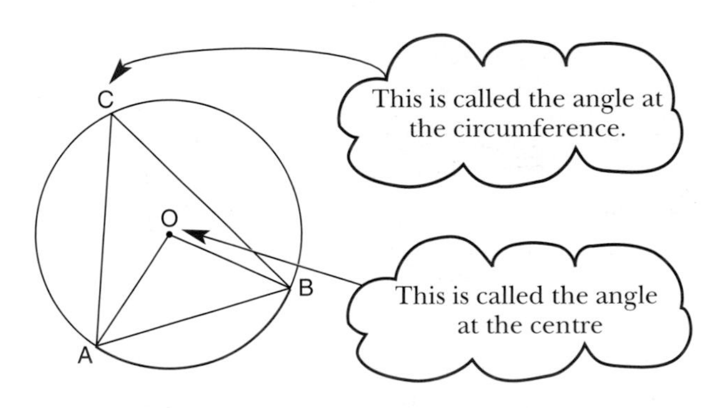

Measure the angle at the centre and the angle at the circumference.

One is twice the other

Draw a similar diagram of your own.
Measure the angle at the centre and the angle at the circumference.

What do you notice?
Does everyone get the same result?

Complete this statement:
'The angle at the centre of a circle is … .'

1 a) Explain why triangle AOC is isosceles.

b) Find a.

c) Find the third angle of triangle AOC.

d) Find x.

e) Use a similar method to find y.

f) What size is (i) AOB (ii) ACB?

g) What is the connection between these two angles?

C 40° 30° O b B x y a A

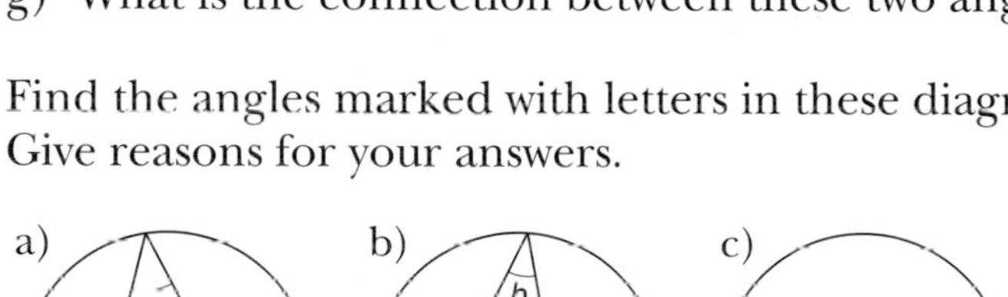

2 Find the angles marked with letters in these diagrams.
Give reasons for your answers.

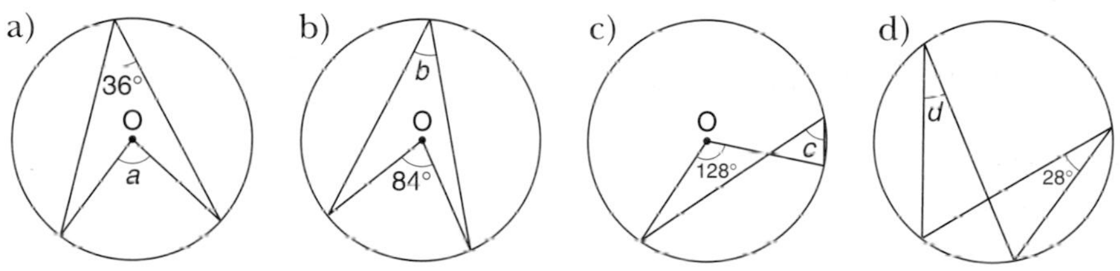

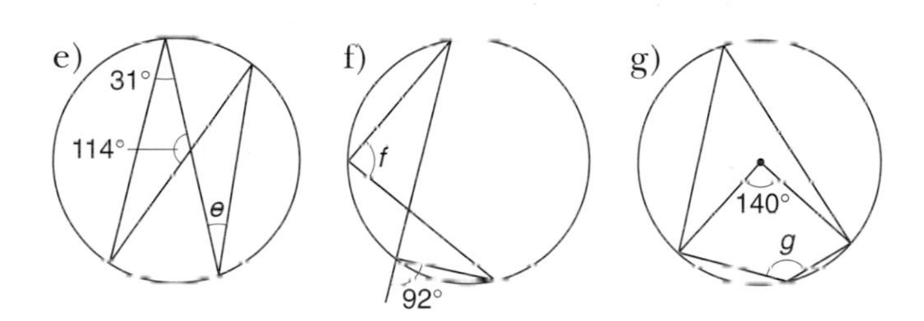

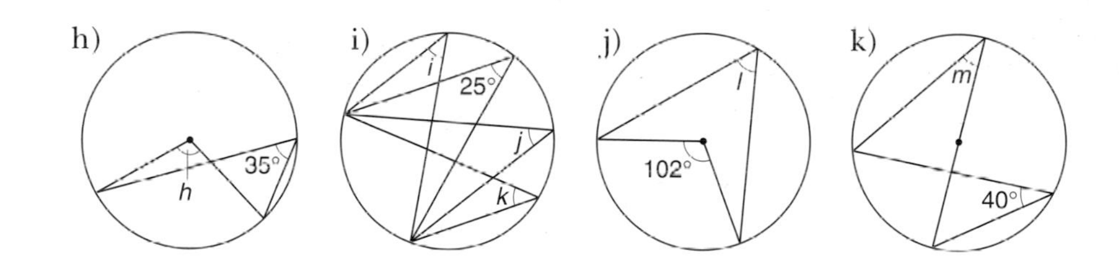

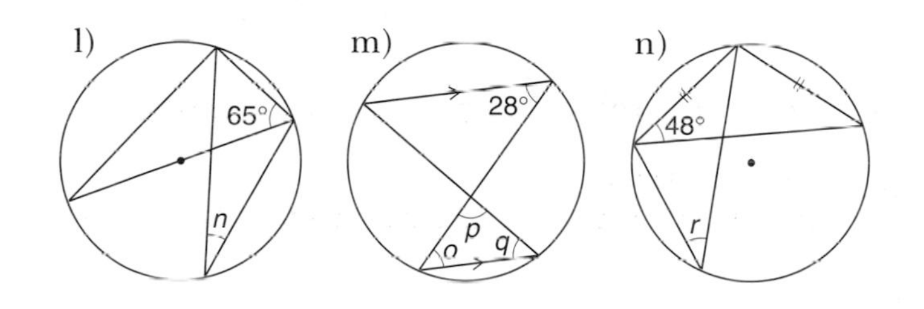

Tangents

A **tangent** is a line which just touches a circle.

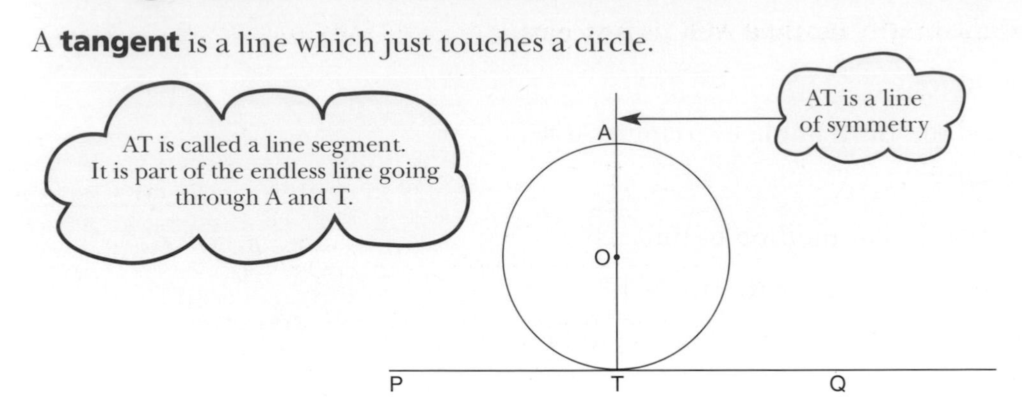

Look below. What size are angles OTP *and* OSP*?*

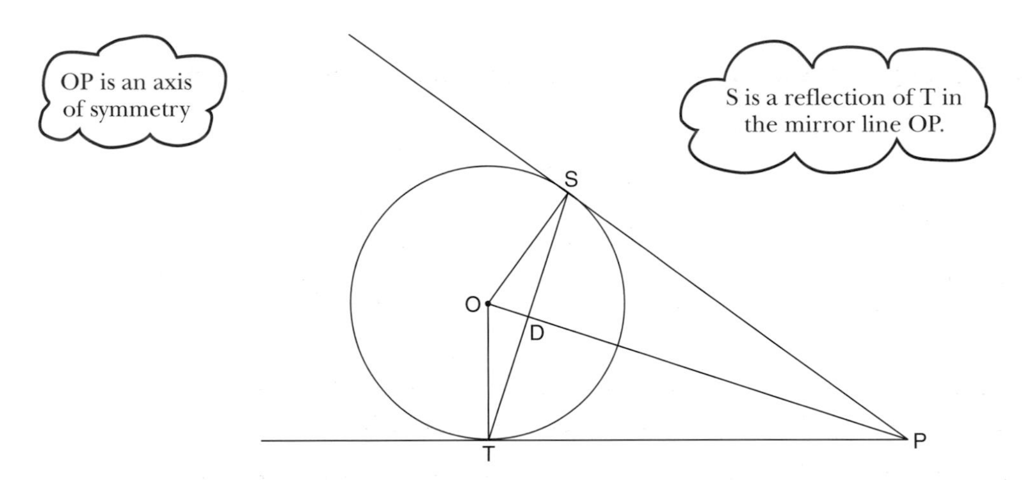

Is SP *equal to* TP*?*

Are tangents drawn from an external point always an equal length?

What can you say about the angles at D*?*

Is a line from the centre of a circle to the mid-point of a chord always perpendicular to the chord?

Does the perpendicular bisector of a chord always pass through the centre of the circle?

1 Find the angles marked with letters, giving reasons for your answers.

a)

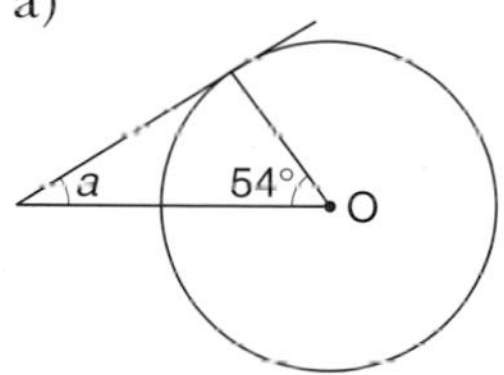

b)

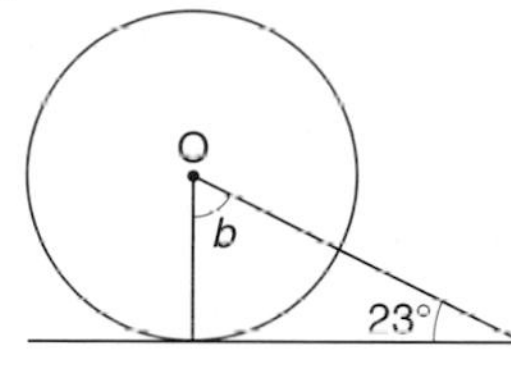

c)

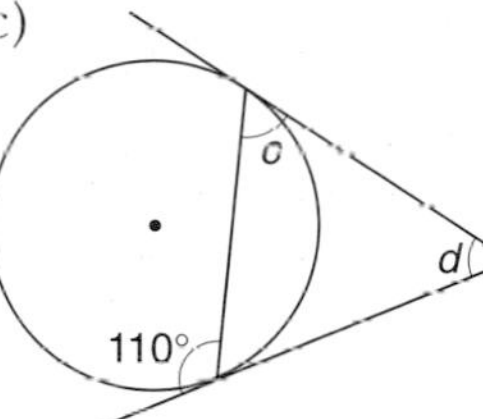

d)

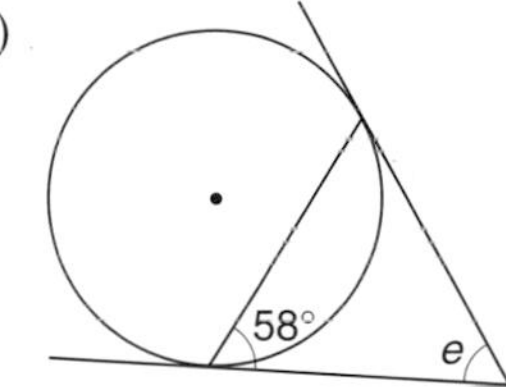

e)

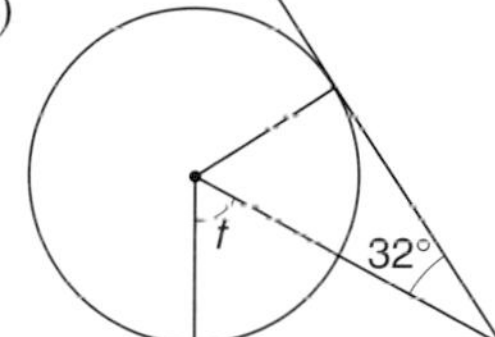

f)

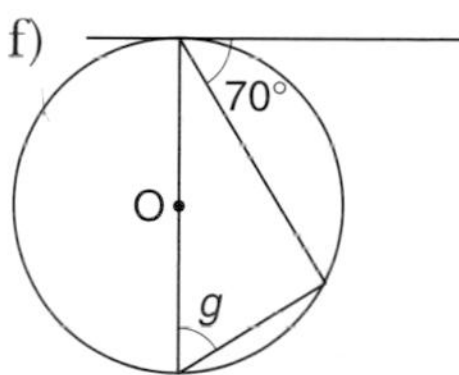

2 Find the angles marked with letters. Give reasons for your answers.

a)

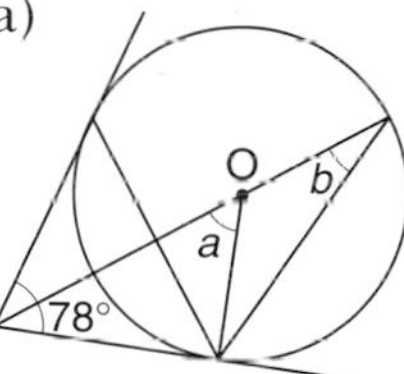

b)

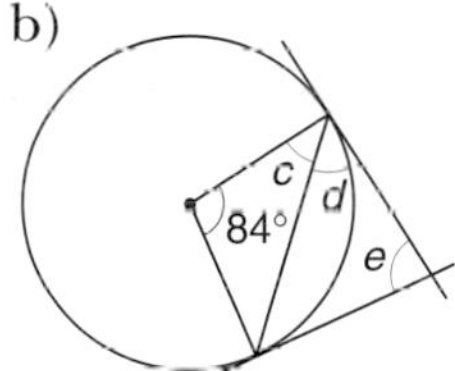

c)

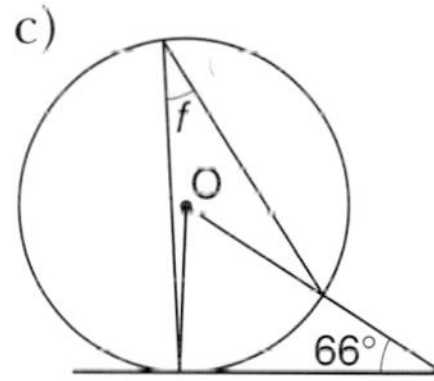

3 The point O is the centre of a circle of radius 8 cm and OP is 17 cm. The tangents from P to the circle touch it at Q and R.

Draw a diagram and use Pythagoras' rule to calculate the length of PQ.

4 The tangents from a point P to a circle are of length 35 cm. P is 37 cm from the centre of the circle.

Draw a diagram and calculate the radius of the circle.

5 A circle, centre O, has a radius of 10 cm. AB is a chord of the circle and AB = 12 cm. Work out the shortest distance from O to the chord AB.

Finishing off

Now that you have finished this chapter you should know that:

- ★ opposite angles of a cyclic quadrilateral add up to 180°
- ★ the angle in a semi-circle is a right angle
- ★ angles in the same segment are equal
- ★ the angle at the centre is twice the angle at the circumference
- ★ the tangent is perpendicular to the diameter at the point of contact
- ★ tangents from a point to a circle are equal in length
- ★ the centre of a circle lies on the perpendicular bisector of any chord.

Use the questions in the next exercise to check that you understand everything.

Mixed exercise

1 Find the angles marked with letters, giving reasons for your answers.

a)

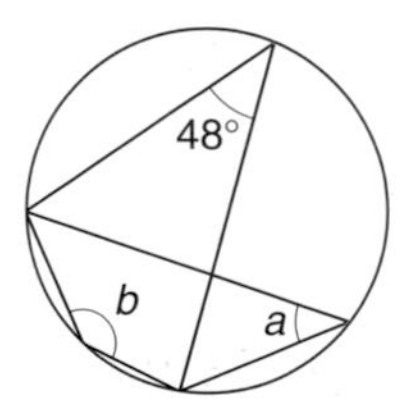

b)

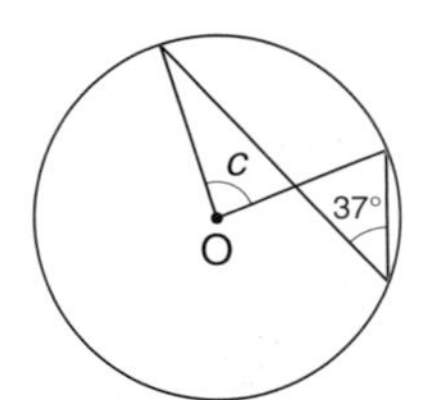

c)

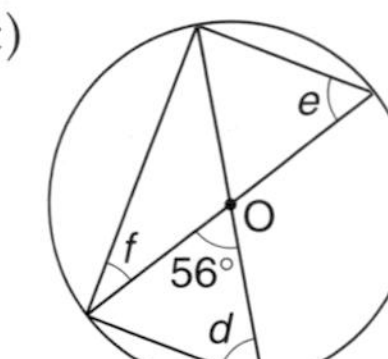

d)

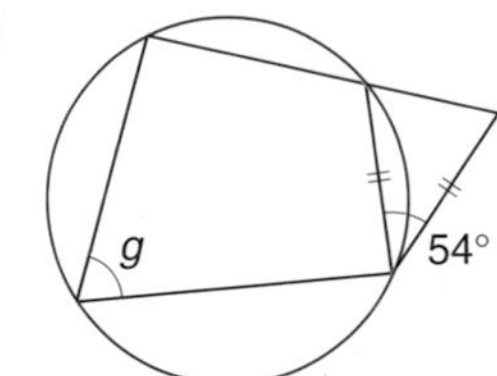

e)

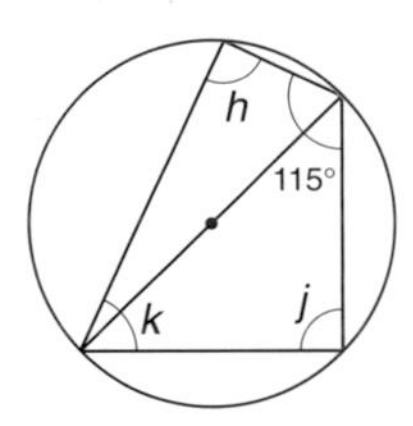

f)

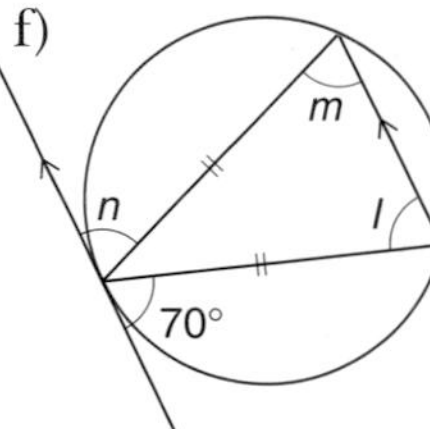

g)

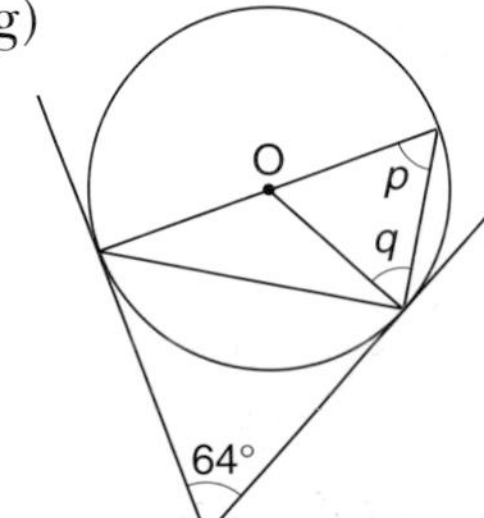

h)

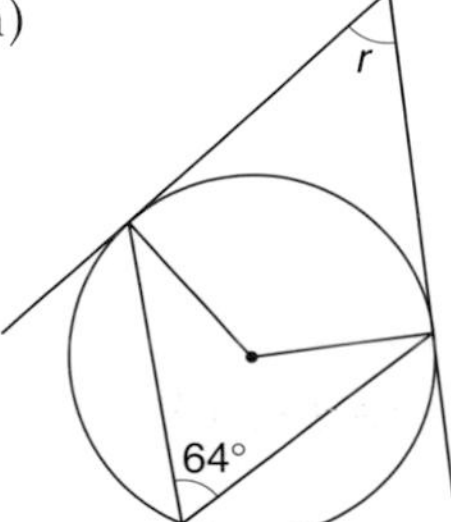

i)

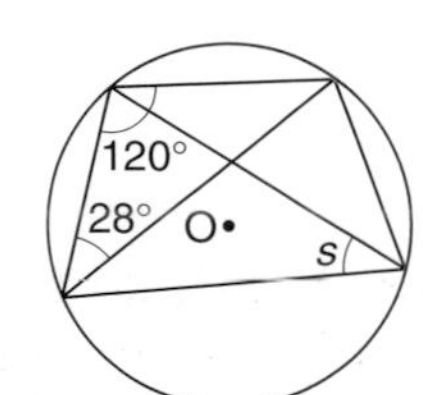

2 In the diagram, O is the centre of the circle and angle ATB is 50°.

TA and TB are tangents.

a) Find angle AOB.

b) Find angle ACB, giving a reason for your answer.

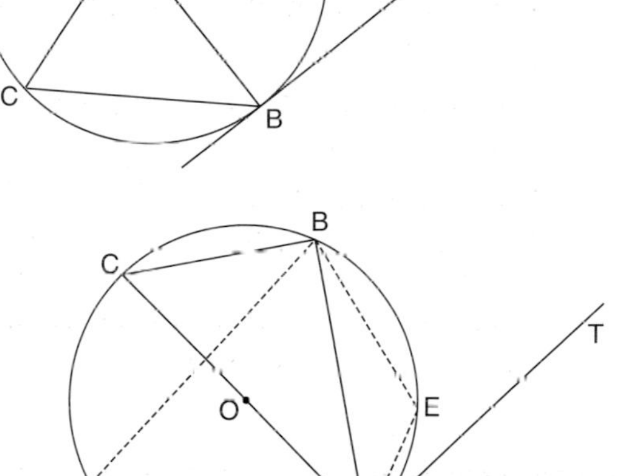

3 In the diagram, the circle through A, B, C, D and E has centre O, which lies on AC.

AT is the tangent to the circle at A.

a) Explain why angles TAC and ABC are right angles.

b) Angle TAB = 58°. Find the following angles, in each case giving a reason for your answer.

(i) BAC (ii) ACB (iii) ADB (iv) AEB

4 O is the centre of a circle. BC is parallel to ED.

Calculate the unknown angles. Give reasons for your answers.

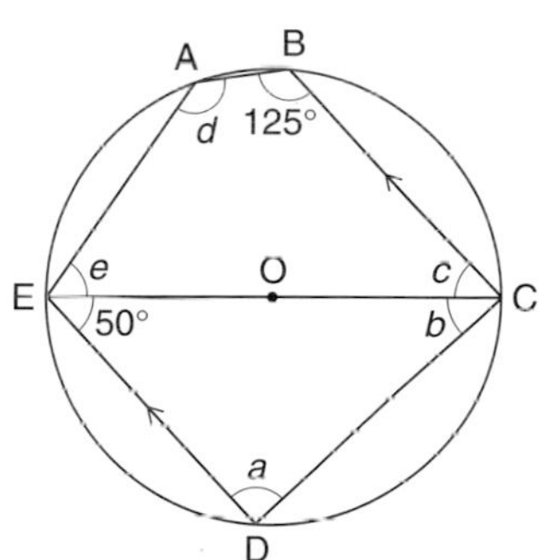

5 PQ is 2 cm and OQ is 3 cm.

Find the lengths of PR and PS, the tangents from P to the circle.

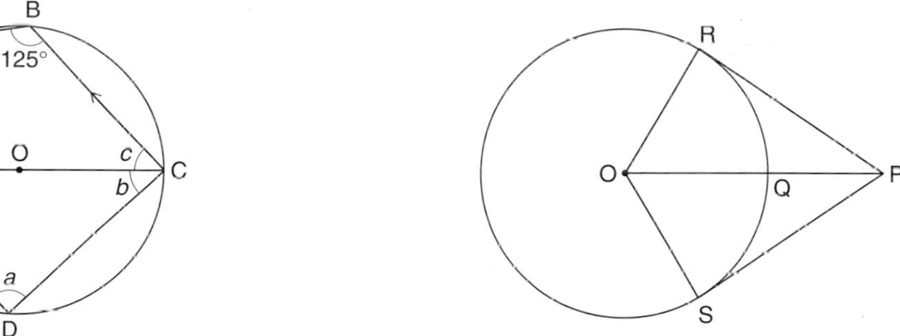

6 A circle centre O and radius 9 cm touches a circle centre C and radius 4 cm as shown.
The line AB is a tangent to both circles.

Find the length of AB.

(Hint: draw a line through C parallel to AB.)

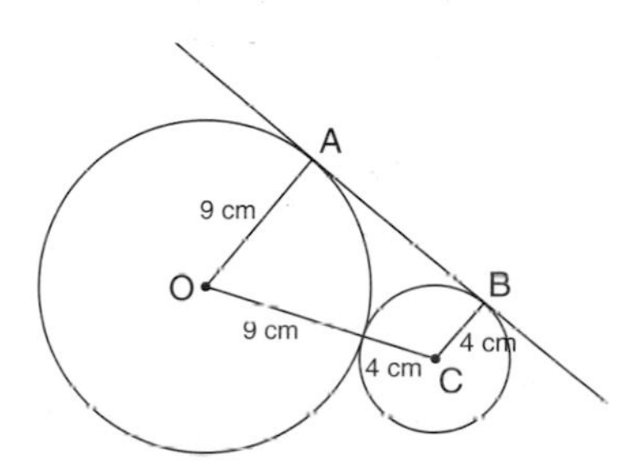

Twenty eight

Collecting data

Conducting a survey

A group of students have been asked by a firm to do some market research on left-handedness.

The firm want to know:

- the proportion of left-handed people in the population
- whether left-handedness is related to scientific, artistic or sporting ability
- whether, because many gadgets are designed for right-handed people, left-handed people find certain everyday tasks more difficult and would like more gadgets designed for their use.

Here are the students' first ideas.

The students seem to be restricting the survey to within their school.

Do you think that this approach is satisfactory or not?

One student suggests that left-handed students are identified and that then only left-handers are asked about which gadgets they would find useful.

Do you think that this approach is satisfactory or not?

Exercise

1 Below is a list of important points to remember when you do a survey. Most of these were brought up by the students.

Who made each point?

Were all the points covered?

Guidelines for surveys

- Always plan before starting to collect lots of data. Identify possible sources of bias and plan to minimise it.
- Don't duplicate effort. Share **primary data** (the data you collect), or use **secondary data** (data that is already available) when you can. Two important sources of secondary data are government publications and the Internet.
- Put questions in a logical order.
- Keep questions simple, with definite answers.

  A **closed question** has only a fixed number of possible answers (such as yes/no/don't know) from which the respondent is asked to select.

  An **open question** has a wide range of possible answers.

  Closed questions produce answers that are easier to process.
- Don't use questions which suggest a particular answer.
- Be sensitive about asking embarrassing questions.
- Check that you have asked everything you want to know.
- Run a **pilot survey** to test your questionnaire (ask just a few people to answer it, and see if any questions cause difficulties).

2 Tim and Jeremy don't like the lunches served in the canteen. They decide to do a survey of customer opinion so that the canteen can provide what people want.

Tim and Jeremy each design a questionnaire for the survey. Tim includes some questions on other matters that interest him.

Compare the two surveys. List your criticisms of each one.

Jeremy's questionnaire

Questionnaire

We are considering altering lunchtime catering arrangements. It would be helpful to know your view.

1. Please tick your preference

   Cooked meal ☐ Light snack ☐
   Salad bar ☐ Sandwiches ☐

2. Please tick if you have special dietary requirements

   Vegetarian ☐ Halal meat ☐
   Vegan ☐ Kosher ☐
   Other ☐

3. Would you prefer

   Coffee ☐ Tea ☐
   Coke ☐ Lemonade ☐

   Other ------------------------------

Thank you for completing my questionnaire.
Please return to reception.

Tim's questionnaire

Questionnaire

1. Do you have a cooked meal or salad or sandwiches for lunch?
2. What do you think of our chances in the World Cup?
3. If you don't have any of them what do you have for lunch?
4. How much telly did you watch last year?
5. Aren't you glad we're going to replace yucky meals with some good nosh?

Recording data

Karen's department at work decide to have an evening out together. They ask Karen to organise it.

Karen goes round asking everyone what they would most like to do.

She records their answers like this:

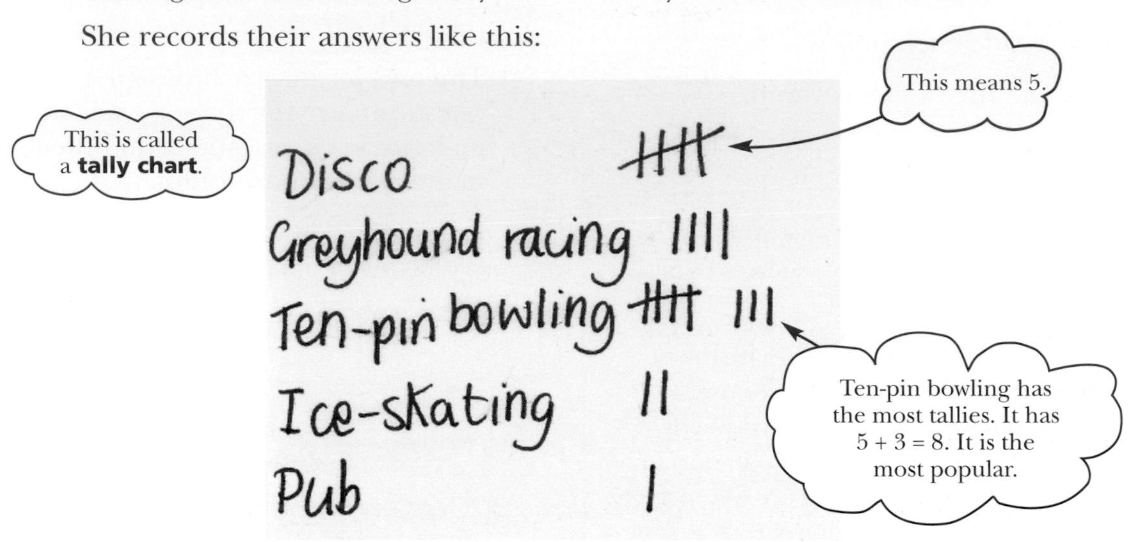

Keeping a tally is an easy way to count things. It is easy to add them up afterwards too, especially if you can count in fives.

When numbers are collected like this they are called **data**.

After she has collected the data, Karen shows the results to everyone as a **frequency table**.

Activity	Disco	Greyhounds	Bowling	Skating	Pub
Frequency	5	4	8	2	1

These are the **frequencies**.

When you add these frequencies up you get 20. This is the number of people who are going out for the evening.

In this case the data are **categorical** (or **qualitative**) because they are grouped by categories such as *ice-skating* and *disco*.

Data which is grouped by numbers, e.g. number of goals scored, heights of people, … is **numerical** (or **quantitative**). Numerical data can be identified as either discrete or continuous data.

The number of goals scored is **discrete**. You can have 0, 1, 2, … goals but you can't have 2.4 goals or $3\frac{1}{2}$ goals. There are gaps.

The heights of people is **continuous**. There are no gaps. (It does not go up in steps.)

Exercise

1 Samantha goes to a theme park with a party of friends. Afterwards she asks them which ride they liked best. Here are their answers.

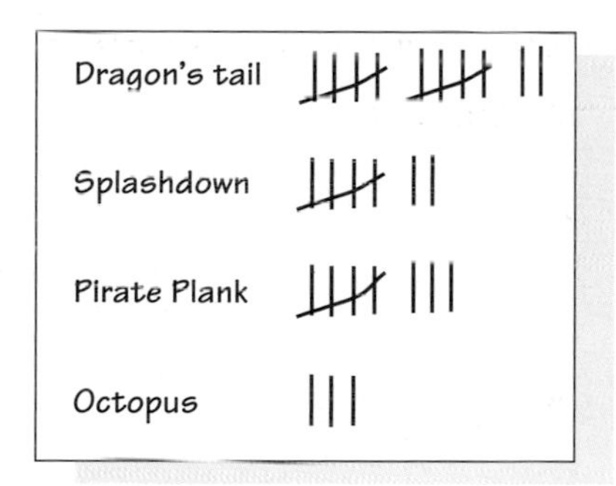

a) Show their answers as a frequency table.

b) How many are in the party?

c) Which is the most popular ride?

2 James manufactures souvenir mugs. He sells them in boxes of 100. Last week James recorded the number of mugs per box that are returned by the customer as unsatisfactory. Here are his results.

0	3	2	1	1	5	2	0	4	2
3	1	0	1	4	7	1	4	2	3

a) Make a frequency table.

b) What percentage of these boxes have 4 or more mugs that are unsatisfactory?

3 For each of these, say whether the data described are discrete or continuous.

a) The number of cars using a car park each day.

b) The weight of food waste from a school kitchen each day.

c) The number of patients admitted to a hospital each week.

d) The length of time that patients wait in an accident and emergency unit.

e) The annual rainfall in Leicester.

Find out the most popular music group among your friends or classmates.

Design a tally chart and use it to record their answers when you question them.

Show the results both as tallies and as a frequency table.

Stem-and-leaf diagrams

Mel and Jane measure the heights of all the girls in their class to the nearest centimetre and record the results.

Mel

Tally chart

140 – 149	II
150 – 159	III
160 – 169	卌 II
170 – 179	III
180 – 189	

Jane

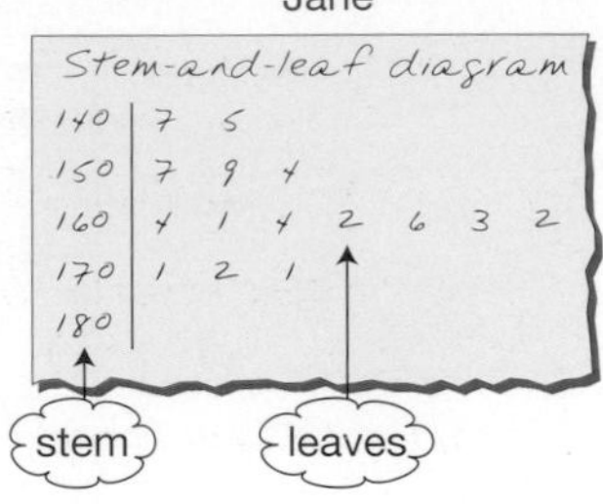

What advantages does Jane's stem-and-leaf diagram have over Mel's tally chart?

Are there any advantages in Mel's tally chart?

Jane orders her stem-and-leaf diagram and adds a key to explain it.

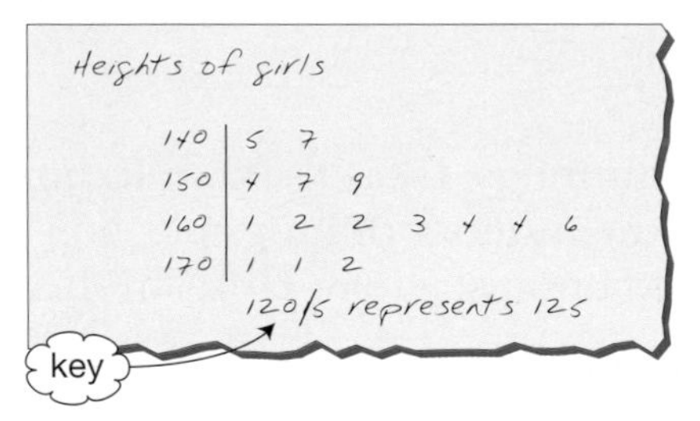

How many girls are 164 cm*?*

Marion's height is 159 cm. *Find the entry corresponding to Marion.*

What is the smallest height?

How tall is the tallest girl?

Mel and Jane also measure the boys and Jane draws a back-to-back diagram to compare the two sets of data. She puts the number of leaves beside the diagram.

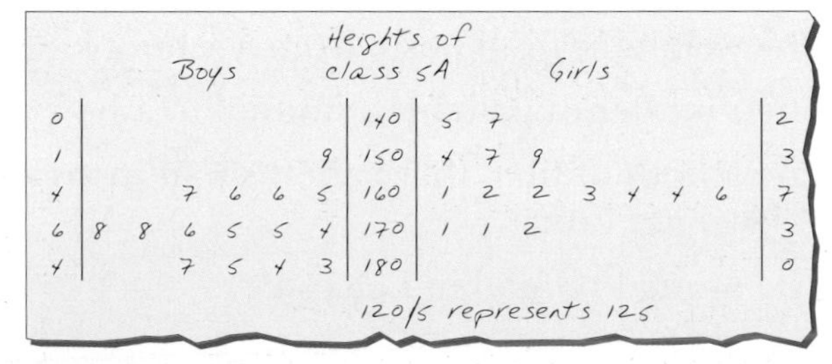

How tall is the tallest student in the class?

Allie is 166 cm. *How many girls are taller than Allie? How many boys are taller than Allie?*

Do you think that, in general, the boys are taller than the girls?

The students are divided into five groups according to their heights.

Which group has the most students? This is called the **modal group**.

Exercise

1 A baker keeps a record of the number of sponge cakes he sells each day.

30 35 61 73 64 62 59 33 42 55
34 36 45 42 39 51 47 38 42 30
43 45 65 53 57 42 45 34 37 65

a) Construct a sorted stem-and-leaf diagram to show the data.
b) What is the smallest number sold?
c) What is the greatest number sold?
d) On how many days are 45 sponge cakes sold?

2 The numbers of miles travelled by students to a conference were as follows.

23 46 63 25 39 47 52 61 69 58
47 53 47 45 63 47 59 47 65 33
29 47 35 47 68 21 25 57 47 47

a) Construct a sorted stem-and-leaf diagram to show the data.
b) What was the smallest number of miles travelled?
c) What was the greatest number of miles travelled?
d) How many students travelled 47 miles?

3 The ages of skaters at a local skating rink are recorded in a survey.

7 24 27 17 9 11 13 25 22 18
37 27 14 17 23 31 16 8 19 6
23 15 14 27 34 26 17 13 14 14

a) Construct a sorted stem-and-leaf diagram to show the data.
b) How old is the youngest skater?
c) How old is the oldest skater?
d) How many 17-year-olds are present?
e) What is the most common age?

4 Jane and Robert live next door to each other. They both planted 20 hollyhocks. Jane used a fertiliser but Robert didn't. The heights, in centimetres, of their hollyhocks were as follows.

Jane: 185 201 256 248 200 254 239 234 196 223
199 243 257 239 246 226 229 254 180 250

Robert: 208 245 150 228 230 215 217 228 215 158
164 179 208 212 226 230 188 196 214 226

a) Construct a sorted back-to-back stem-and-leaf diagram to show the data.
b) State the heights of the tallest and shortest hollyhocks belonging to Jane.
c) State the heights of the tallest and shortest hollyhocks belonging to Robert.
d) Compare the two sets of data and comment on whether the fertiliser is effective.

Construct a stem-and-leaf diagram with the following features.

- The range from lowest to highest value is 50.
- The number of values is 15.
- The highest value is 89.
- The middle value is 53.

Grouping data

Amy manages a supermarket. She does not think that the present checkout system meets customer needs as well as it should so she decides to carry out an investigation. As part of her investigation she records how many items a customer buys. Here are her results.

3 12 6 24 25 7 2 9 11 28 2 5 13 25 27 1 7 23 21 31

16 4 10 26 29 32 1 15 28 3 28 24 1 9 10 4 24 17 5 10

2 22 19 4 10 27 35 26 2 6 26 20 27 5 6 19 4 25 21 9

Amy thinks about making a frequency table with classes 1, 2, 3, 4, … 35.

She realises that this is not very helpful as there are far too many classes.

Instead she groups the data and chooses classes 1–5, 6–10, 11–15, …

Make a tally chart and check that you get this frequency table.

No. of items	1–5	6–10	11–15	16–20	21–25	26–30	31–35
No. of customers	16	12	4	5	10	10	3

You can see that the number of items a customer buys is most commonly between 1 and 10 or between 21 and 30.

Amy also records the times, in minutes and seconds, that these customers wait at the checkout before they are served. She draws up this table.

Waiting time (minutes)	$0 \le t < 2$	$2 \le t < 4$	$4 \le t < 6$	$6 \le t < 8$	$8 \le t < 10$
No. of customers	6	16	22	14	2

Notice that Amy uses different symbols at each end of an interval.

$\le$ means 'is less than or equal to', $<$ means 'is less than'.

Time is continuous so, for example, $0 \le t < 2$ means from 0 up to 2 but not including 2.

Which class does a waiting time of 1 minute 59 seconds go into?

You can see, from the table, that most customers wait between 2 and 8 minutes before getting served. Very few (6 at most) are served immediately.

From these results Amy will consider whether to introduce 'fast' lanes for those with 10 or less items and whether more checkouts are needed.

Note: The notation with $\le$ and $<$ may be a hindrance to a non-mathematician so the groups may be presented more compactly as 0–2, 2–4, 4–6, …

Why not use 0–2, 2–4, 4–6, … *from the outset?*

1 Here are the weights, in kg, of 20 male students all aged 18.

61	98	72	84	63	77	77	81	85	72
90	83	76	82	77	81	80	83	75	68

a) Make a tally chart using groups 60–64, 65–69, 70–74, 75–79, …

b) Make a frequency table.

c) A healthy weight for an 18-year old male is about 70–80 kg.

Comment on the weights of this group of students.

2 Arif wants to know if more babies are born at some times of the year than others. He asks 24 people what month they were born in and gets these responses.

June	Feb	Jan	July	Dec	May	Sept	June	Nov	May
Oct	Jan	Sept	Aug	June	Oct	Aug	Nov	April	Feb
Aug	Sept	June	Sept						

Arif groups the data by season.

a) Copy and complete this frequency table.

Season	Winter Dec, Jan, Feb	Spring March, April, May	Summer June, July, Aug	Autumn Sept, Oct, Nov
Frequency (people)				

b) Do you think Arif has collected enough data to draw any conclusions?

3 Samantha is a receptionist in a doctor's surgery. She monitors the efficiency of the appointments system. At the end of each of 30 surgeries she records by how many minutes the surgery overran. Here are her results.

3	17	12	8	10	24	14	8	9	34
14	9	11	26	3	13	10	38	12	9
22	7	10	14	18	6	15	9	20	5

a) Make a frequency table using groups 1–10, 11–20, 21–30 and 31–40.

b) Comment on this data.

c) Now make a frequency table using groups 1–5, 6–10, 11–15, …

d) How does the second frequency table influence your comments in b)?

Social statistics

This is called the **base year**, shown by 100. All prices are compared with those in 1990.

Avonford Star Financial

What is the £ in your pocket worth?

PRICES KEEP rising so wages and pensions need to increase. But do they increase enough? Our reporter tells you the truth using the Retail Price Index.

Year	1990	1995	2000	2005
RPI	100	127	165	?

The RPI compares the 1995 and 2000 prices with those in 1990.

In 1995 the basket of essential groceries cost 127% of the 1990 price and in 2000 the price is 165% of the 1990 price.

This basket of groceries cost £15 in 1990.
In 1985 the basket cost
In 1995 the basket c

Martha says, 'For every £1 I paid in 1990, I paid £1.27 in 1995 and I have to pay £1.65 in 2000.'

Is Martha right?

What did the basket cost in 1995?

What did the basket cost in 2000?

Martha says 'For every £1 of pension I spent in 1990 I needed to spend £1.27 in 1995 and £1.65 in 2000. What will happen in 2005?'

The table on the right shows how Martha's pension has changed.

Year	Single pension
1990	£46.90
1995	£58.85
2000	£67.50
2005	?

What pension in 1995 would have been equivalent to the 1990 pension?

What pension in 2000 would have been equivalent to the 1990 pension?

Compare Martha's actual pension with those you have just worked out.

Was she better off in 1990, 1995 or 2000?

Exercise

1 A cafe sells an 'All Day Breakfast'. In 1990 the meal cost £6.

Use the RPI values on page 272 to calculate the cost in

a) 1995
b) 2000.

2 A sports shop sells a mountain bike. In 1990 the mountain bike cost £80.

Use the RPI values on page 272 to calculate the cost in

a) 1995
b) 2000.

3 The diagram shows the sales of organic baby food in millions of pounds. These have been calculated at 1993 prices in order to make comparisons.

Year	1993	1994	1995	1996	1997	1998
RPI	100	105	110	113	122	131

Use the RPI values for baby food in the table above to calculate the actual sales in each year.

The table below shows the expected age distribution of the population in England in 2011, given in millions. The total population is expected to be 52 000 000 people.

	0–14	15–29	30–44	45–59	60–74	75 and over	Total
Males	4.7	5.1	5.4	5.4	3.7	1.6	
Females	4.3	4.9	5.2	5.3	4	2.4	
People		10	10.6	10.7	7.7		

Based on figures issued by the Office for National Statistics.

Copy and complete the table.
Are there more males or females in England?
Find the percentage of under-15s who are

a) male
b) female.

Find the percentage difference between the numbers of males and females who are under 15.
Similarly find the percentage difference between the numbers of males and females who are 75 or over.
Comment.

Finishing off

Now that you have finished this chapter you should be able to:

- ★ conduct a survey
- ★ tell the difference between categorical and numerical data
- ★ tell the difference between discrete and continuous data
- ★ make a tally chart
- ★ draw up a frequency table
- ★ draw up a stem-and-leaf diagram
- ★ decide when to group data and how to do it.

Use the questions in the next exercise to check that you understand everything.

Mixed exercise

1 Mohammed is doing a survey about people's use of public transport.

He wants to find out how many people use public transport, what type of people use it, what type of transport they use, for what purpose and how often.

a) Mohammed produces this questionnaire:

1. *Did you travel by train last week?*
2. *On how many days last week did you travel by train?*
3. *Do you use the local bus service?*
4. *Which bus route(s) do you use?*
5. *Do you think that the bus service is more reliable than the train service?*
6. *Do you own a car or a motorcycle?*
7. *Where do you live?*
8. *How old are you?*

Comment on this questionnaire and suggest ways to improve it.

b) Mohammed has to decide where to carry out the survey.
He considered

(i) outside the train station

(ii) on the number 8 bus route

(iii) in the town centre

(iv) outside the cinema

(v) at the superstore on the edge of town

Comment on each of these options.

2 The following data are collected for a group of students. State whether they are categorical or numerical. If they are numerical say also whether they are continuous or discrete.

a) their heights, measured in cm

b) their weights, measured in kg

c) the days of the week on which they were born

d) the number of brothers and sisters they have.

3 The number of hours of sunshine is recorded, to the nearest hour, in a holiday resort one June.

a) Make a tally chart for the number of hours of sunshine per day.

b) Construct a frequency table.

Day	Hours of sun	Day	Hours of sun	Day	Hours of sun
1	10	11	10	21	4
2	9	12	14	22	7
3	12	13	14	23	11
4	11	14	13	24	14
5	3	15	10	25	15
6	8	16	4	26	14
7	5	17	2	27	13
8	0	18	6	28	12
9	1	19	8	29	13
10	6	20	3	30	9

4 Jamie is a production manager. She records the number of components produced in each hour of a day. Here are her results.

39 45 36 45 41 45 38 73 59 70 11 66

26 61 57 59 76 53 41 42 48 19 33 45

a) Decide, for yourself, what groups to form and make a frequency table.

b) Suggest two reasons why the production rate varies throughout the day.

5 Two firms make car batteries and a consumers' association tests 20 batteries from each firm. The battery lives in completed months are:

Longlife Batteries

45 49 47 55 58 40 46 50 51 54
55 61 63 68 51 44 49 58 57 50

Quickstart Batteries

65 45 50 58 60 65 47 68 55 58
62 69 48 62 56 50 58 46 64 66

a) Construct a sorted back-to-back stem-and-leaf diagram to show the data.

b) State the longest and shortest battery life of a 'Longlife' battery.

c) State the longest and shortest battery life of a 'Quickstart' battery.

d) Compare the two sets of data and comment on whether there is a noticeable difference between the two brands.

Twenty nine

Displaying data

Displaying data

People often show their data in charts or diagrams.

Why use charts and diagrams rather than tables of numbers?

Emma has collected this data showing where her friends go for a holiday.

Country	Spain	France	Italy	Greece
No. of friends	8	4	5	6

Pictograms

This pictogram is one way of displaying Emma's data.

Spain

France

Italy

Greece

= 2 people

Bar charts

These bar charts show two more ways of displaying Emma's data.

This is a horizontal bar chart.

(It could have been drawn as a vertical bar chart.)

A bar chart shows clearly the most popular and the least popular categories.

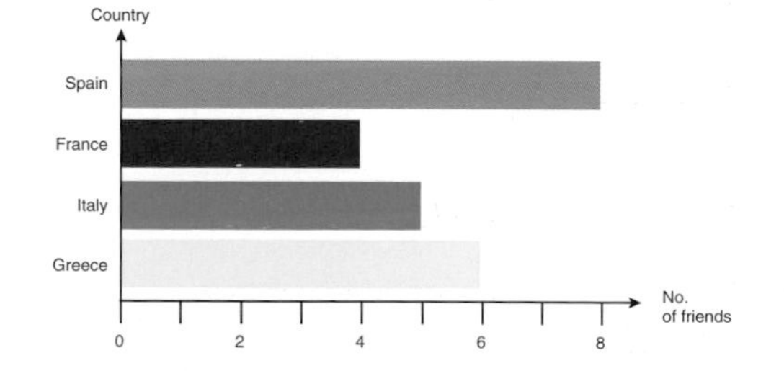

Which is the most popular country?

This is a dual bar chart.

The breakdown between male and female numbers, Spain – 2 male, 6 female, France – 3 male, 1 female, Italy – 4 male, 1 female, and Greece – 2 male, 4 female, is shown on this chart.

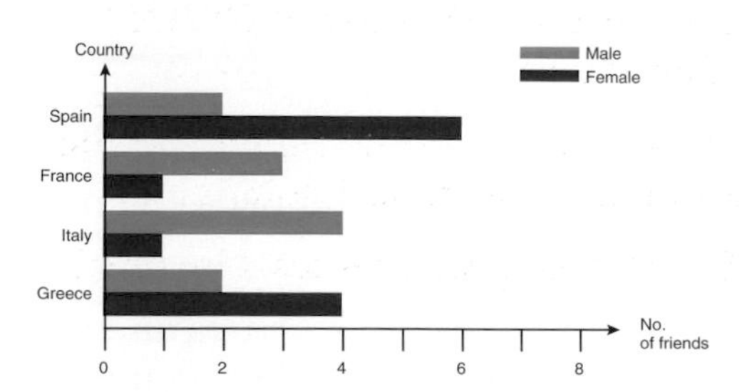

Exercise

1 This pictogram shows the results of a netball team one season.

Win	
Draw	
Lose	
means 4 matches	

a) The symbol means 2 matches.

Draw symbols for 1 match and 3 matches.

b) How many matches did they win, draw and lose?

c) How many matches did they play altogether?

2 This table shows the number of homes sold last year by National Homes plc.

Area	Homes sold
South-east	525
South-west	350
Midlands	275
North	300

a) Choose a suitable symbol and scale and display the data on a pictogram.

b) Draw a bar chart to display the data.

c) Give one reason why your bar chart is preferable to the pictogram.

d) Give one reason why your pictogram is preferable to the bar chart.

e) You have to decide whether to use the pictogram or the bar chart in a brochure to be issued to the public. Which would you use and why?

3 Neil draws this bar chart to show the sales of computers last week.

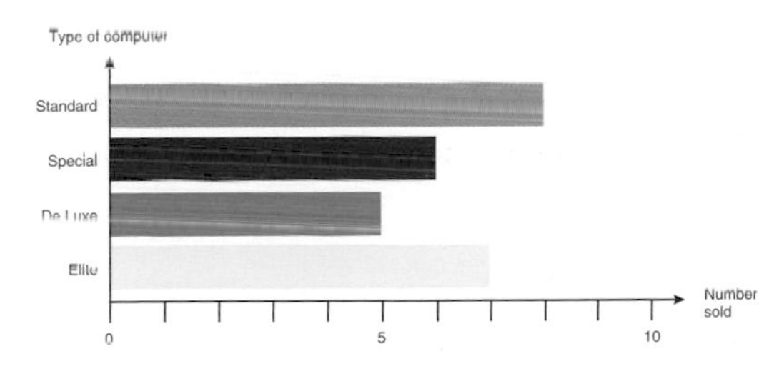

a) How many Standard are sold?

b) How many computers are sold in total?

The prices are Standard £550, Special £700, De Luxe £900 and Elite £1000.

c) Draw a bar chart showing the income from sales for these 4 computers.

d) Which type produces least income?

4 This table shows the numbers of male (M) and female (F) students on particular courses at a college.

Course	M	F
Art and Design	50	40
Business and Finance	45	65
Health and Social Care	10	60
Leisure and Tourism	55	25

a) Draw a vertical bar chart showing the numbers of students on each course.

b) Draw a dual bar chart showing the numbers of males and females on each course.

c) Comment on what your bar charts show.

Pie charts

Each year, many wild animals are killed on the roads. The members of a conservation group keep a record of the bodies they see on the roads in one county.

This table shows the type of road on which these animals died.

Road type	Motorways	A roads	B roads	Other	Total
Deaths	144	1016	560	1160	2880

Mick is designing a pie chart to illustrate these data.

The whole circle is 360°, so Mick works out the angles like this:

Motorways $\frac{144}{2880} \times 360° = 18°$

A roads $\frac{1016}{2880} \times 360° = 127°$

Check that the angle for B roads is 70° *and that the angle for Other is* 145°.

Check that the angles add up to 360°.

When he has worked out all the angles, Mick draws his chart.

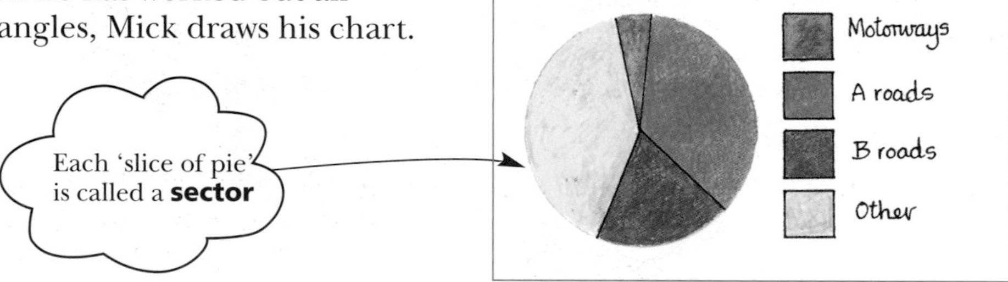

What does Mick's chart tell you?

The chart below illustrates the types of animal involved in the 144 deaths on motorways.

The angle for rabbits is 40°. How many does this represent?

360° is 144

$1°$ is $\frac{144}{360}$

$40°$ is $\frac{144}{360} \times 40 = 16$

The angle of 40° represents 16 rabbits.

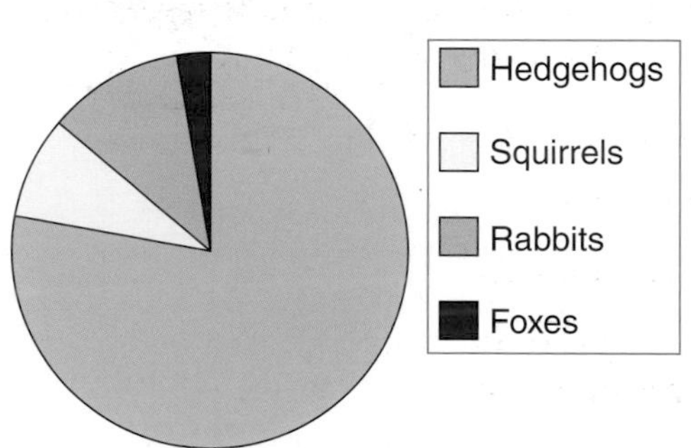

29: Displaying data

Exercise

1 A clothing company makes £360 000 profit one year. The pie chart shows how this was divided between the children's, teenage and adult departments.

a) Measure the angles in the pie chart.

b) How much profit does each department make?

c) What percentage of the profit comes from teenage clothes?

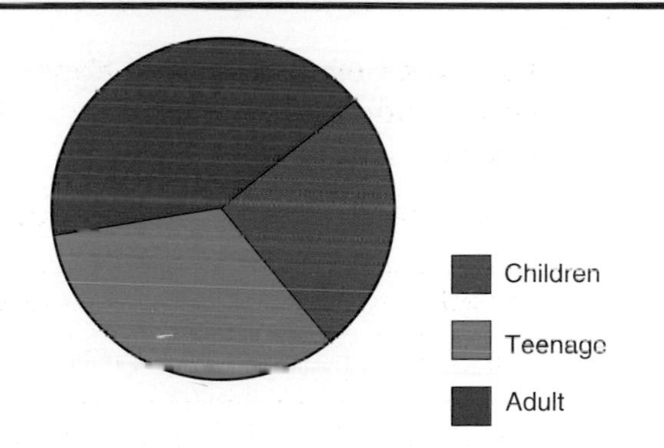

2 Anna says that on a typical day she spends her time like this:

Studying	12 hours
Eating	1 hour
Sleeping	8 hours
Relaxing	2 hours
Housework	1 hour

She wants to show this on a pie chart.

a) How many degrees represent 1 hour?

b) Draw the pie chart.

c) What percentage of her time is spent studying?

3 This pie chart shows the holiday destinations of a number of passengers at Heathrow Airport.

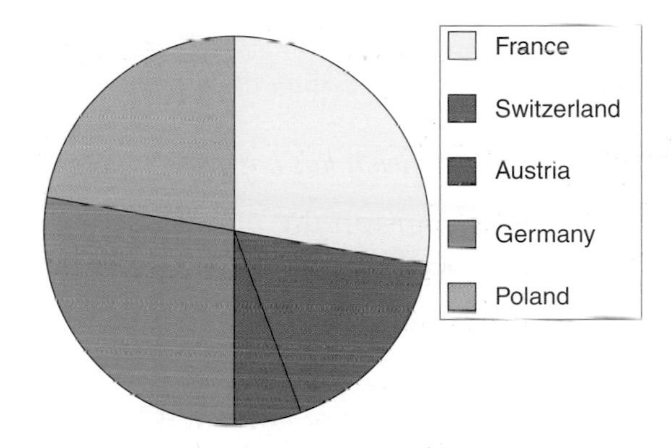

a) The number of people going to Poland is 40. How many passengers are involved altogether?

b) How many people are travelling to each destination?

4 Of 1440 business students, 576 started careers in Finance, 360 in Sales, 288 in Management, 108 in Administration, 72 in Market Research and the rest in Marketing.

Illustrate these data by drawing

a) a pie chart

b) a bar chart.

c) Compare the two displays and identify

(i) one feature that the pie chart shows better

(ii) one feature that the bar chart shows better.

Do the popular colours for cars change from year to year?

Go round a large car park and record the colour of each car and the registration year letter or number.

Draw pie charts to compare the results, year by year.

Line graphs

Nat is in hospital. Every 3 hours his temperature is taken and the points are plotted on a graph.

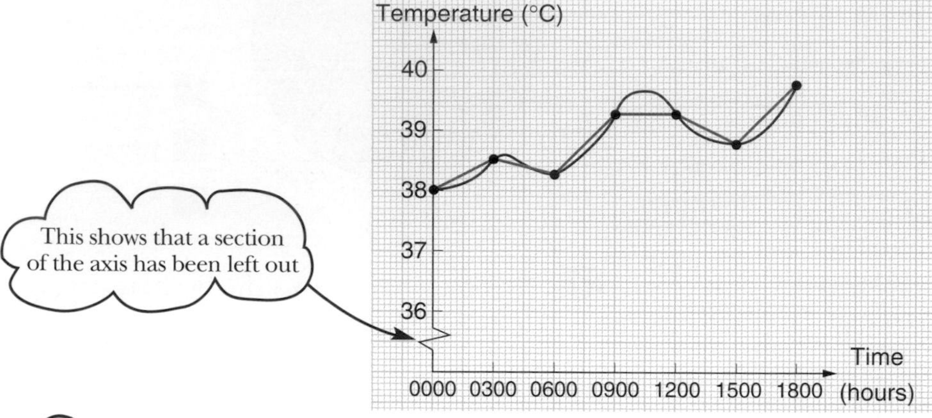

Why has the section of the temperature axis from 0 °C *to* 36 °C *been left out?*

Describe what has happened to Nat's temperature during the day?

In the graph the points have been joined by the blue straight lines. This is usual on a line graph, but be careful: it can be misleading.

You don't know what happened between the points: it could have been the red curve.

When data are collected at time intervals, it is quite usual to show them on a graph like this.

Vertical line charts

A vertical line chart is often used for showing numerical data. This vertical line chart shows the number of people living in the 40 houses on one street.

It is like a bar chart with very thin bars.

In this diagram the zero on the horizontal axis has been offset to make the vertical line there clearer.

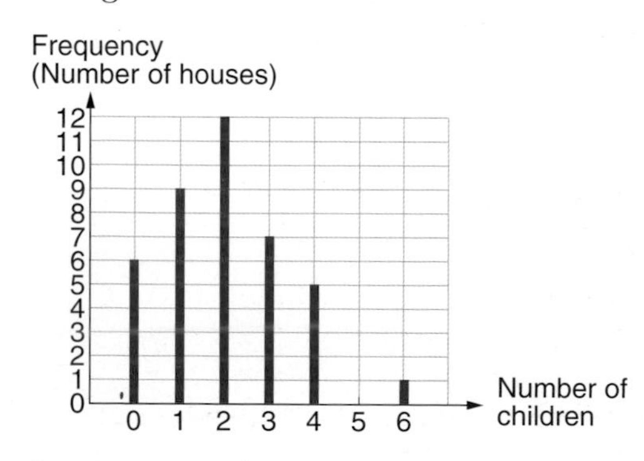

How many houses have no-one living in them?

What percentage of houses have just two people living in them?

1 Jake was born in January weighing 2.1 kg.

This is rather light for a baby so he was weighed every week for the next 10 weeks as a check.

Week	0	1	2	3	4	5	6	7	8	9	10
Weight (kg)	2.1	2.1	2.2	2.3	2.4	2.5	2.3	2.6	2.8	3.0	3.2

a) Plot these figures as a line graph.

b) Jake was ill one week.

Which week do you think it was?

c) Estimate Jake's weight at $9\frac{1}{2}$ weeks.

2 This table shows the population of voles each month on a river bank during 1997.

Month	J	F	M	A	M	J	J	A	S	O	N	D
No. of voles	40	30	28	16	16	20	14	28	56	100	78	60

a) Draw a line graph to illustrate the data.

b) Describe the annual pattern.

c) Over what months did the population increase most rapidly?

3 In one season, Clara's hockey team played 54 matches. She made this frequency table of the number of goals they scored. Show this information on a vertical line chart.

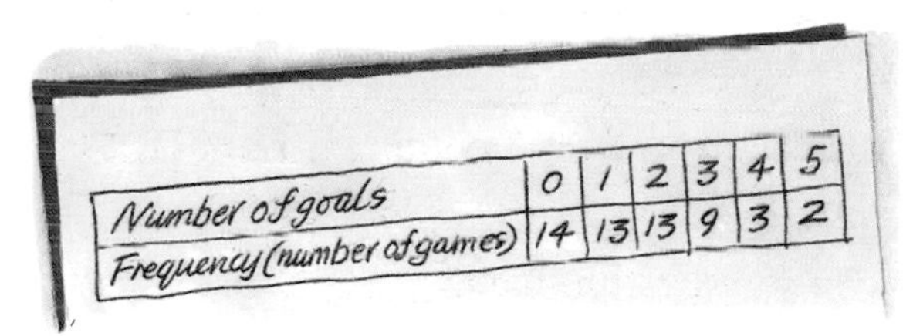

Number of goals	0	1	2	3	4	5
Frequency (number of games)	14	13	13	9	3	2

4 Jenny has kept in touch with some of the girls in her class at school.

This vertical line chart shows the number of children they have had.

a) How many women does the chart represent?

b) What percentage have no children?

c) How many children do they have in total?

Histograms

A bus company plans to start a new long distance service between two cities. They need to know how long the journey will take, so they do 15 trial runs. Each time the driver tells the company how long he took, to the nearest minute.

165	176	176	168	170
166	167	170	179	170
169	174	166	168	186

165 minutes 20 seconds is recorded as 165.
165 minutes 50 seconds is recorded as 166.

Back in the office Tom has to display the data. He starts by making this grouped frequency table.

TIME (MINUTES)	165 – 169	170 – 174	175 – 179	180 – 184	185 – 189
FREQUENCY (NO. OF BUSES)	7	4	3	0	1

In this case the class 165–169 covers $164\frac{1}{2}$ to $169\frac{1}{2}$, the class 170–174 covers $169\frac{1}{2}$ to $174\frac{1}{2}$ and so on. Each class width is five minutes.

Here is Tom's histogram. He has also drawn in a frequency polygon in green, joining the mid-points of the tops of the bars.

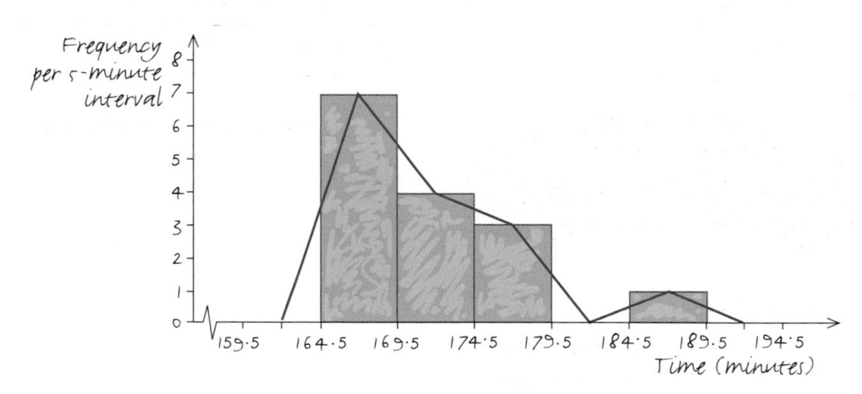

Notice that

- there are no gaps between the bars of a histogram (unless there is an empty class).
- the horizontal scale is continuous
- the vertical scale measures **frequency density**. In this case it is described as 'frequency per 5 minute interval'.

  This means that bars of equal *area* represent equal frequency.

Which display do you find more helpful, the histogram or the frequency polygon?

Exercise

1 The lengths (in minutes, to the nearest minute) of the phone calls made between Jill and Katharine are given in the table.

Time in minutes	1–5	6–10	11–15	16–20
Frequency	15	25	10	5

a) What are the boundary values for the class 6–10?

b) What is the mid-point of the class 6–10?

c) What is the longest possible time for a phone call in this table?

d) What is the shortest possible time for a phone call in this table?

e) Draw a histogram to show these data.

2 The sale price for 30 cars at an auction are as follows.

£1430	£1750	£2430	£4560	£3480
£2520	£4160	£1995	£2460	£2840
£5100	£3275	£2160	£4050	£5120
£3500	£2750	£1850	£3520	£2650
£1200	£4200	£4000	£5800	£4950
£4510	£3840	£2380	£5750	£4800

a) Make out a tally chart to put the data into groups £0 – £999, £1000 – £1999, £2000 – £2999, and so on.

b) Make a frequency table.

c) Draw a histogram and a frequency polygon.

3 The length of time that 60 clients spend in a fashion store is shown in the table below.

Time spent in store	No. of clients
0 m 00 s to 4 m 59 s	9
5 m 00 s to 9 m 59 s	15
10 m 00 s to 14 m 59 s	13
15 m 00 s to 19 m 59 s	10
20 m 00 s to 24 m 59 s	8
25 m 00 s to 29 m 59 s	5

Draw a histogram to illustrate the data.

4 On the opposite page, a frequency polygon has been added to the histogram of times.

a) Find the total area in the rectangles of the histogram.

b) Find the area enclosed by the frequency polygon.

What do you notice?

Finishing off

Now that you have finished this chapter you should be able to:

★ draw and interpret pictograms, bar charts, pie charts, line graphs, vertical line charts, histograms and frequency polygons.

Use the questions in the next exercise to check that you understand everything.

Mixed exercise

1 Fran stands outside her house and counts the number of people in the first 20 cars that pass by.

1, 3, 4, 1, 4   5, 4, 1, 2, 1   2, 3, 1, 1, 2   3, 3, 2, 1, 6

Draw a vertical line chart to illustrate the data.

2 A company is planning to sell red, green, blue and white T-shirts. The marketing manager asks Alison to do a survey to find out the popularity of the colours.

Alison conducts a survey of 270 people and produces this pie chart from her results.

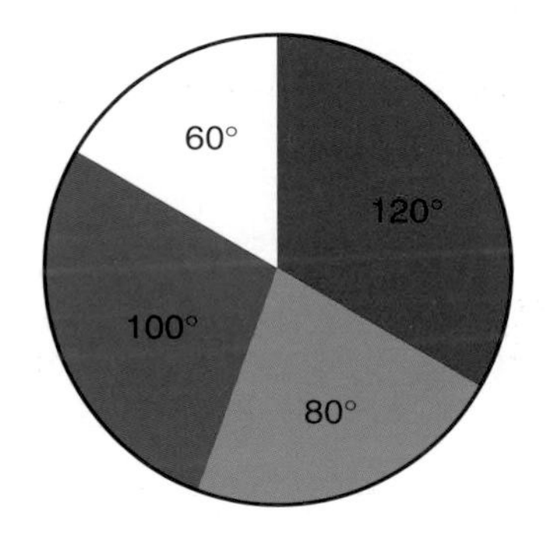

a) How many people in the survey preferred blue?

b) How many preferred green?

c) Draw a bar chart and a pictogram to illustrate Alison's data.

d) Alison has to choose one of these displays for her report. Which would you choose?

3 This table shows the average number of hours of sunshine each month in Malaga.

Draw a line graph to illustrate the data. Then, using your line graph, describe the annual pattern.

Month	**J**	**F**	**M**	**A**	**M**	**J**	**J**	**A**	**S**	**O**	**N**	**D**
Hours of sunshine	6	7	7	9	10	12	12	11	9	7	6	5

4 A manufacturer of children's clothes is changing its sizing policy to be according to height rather than age. Size 1 is to be for children 100 – 108 cm tall, size 2 is to be for children 108 – 116 cm tall, and so on.

The company conducts a survey of the heights of 100 girls and 100 boys to find out the proportion of children of each size. Here are the results for girls between 5 and 10 years of age.

Height (cm)	$100 \le h < 108$	$108 \le h < 116$	$116 \le h < 124$	$124 \le h < 132$	$132 \le h < 140$
Frequency (no. of girls)	12	18	30	25	15

a) Draw a histogram to display the data.

b) Of which size should the company make most girls' clothes?

c) A production run of 5000 garments is planned. How many should be of each size?

5 Scott puts some bread on his bird table and keeps a record of the birds that eat it. His results are shown in the bar chart.

a) Make a frequency table.

b) How many birds does Scott see in total?

c) What percentage of the birds are starlings?

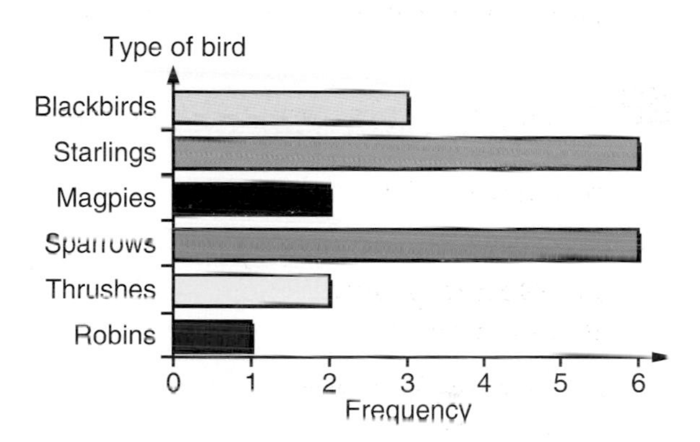

6 Emma has done a survey of the number of items in people's shopping baskets at the checkouts of two different shops. Here are her results.

NUMBER OF ITEMS	1-5	6-10	11-15	16-20	21-25	26-30	31-35
Frequency: shop A	24	49	13	8	4	2	0
Frequency: shop B	5	8	15	33	29	10	0

a) For each shop write down the class with the highest frequency?

b) On the same set of axes, draw the frequency polygon for each shop.

c) Compare the frequency polygons. What do they tell you about the two shops?

d) Suggest what type of shop each might be.

Thirty

Average and spread

Mean, median, mode and range

Sam and Ben are part of a group that go ten-pin bowling. They arrive early to get some practice shots. These are their scores.

Sam: 7, 7, 4, 5, 7   Ben: 10, 1, 10, 3

Who has done better?

Sam's total score of 30 is greater than Ben's 24 but she had more turns.

You need to take an average. Here are three ways of doing it.

The **mean** is the total divided by the number of turns played.

Sam: Mean = 30 ÷ 5 = 6   **Ben: Mean = 24 ÷ 4 = 6**

The **median** is the middle score when the scores are put in order.

Sam: 4, 5, 7, 7, 7   Ben: 1, 3, 10, 10

Ben has no middle score so you add the nearest 2 scores and divide by 2

Sam: Median is 7   **Ben: Median is 6.5**

The **mode** is the most frequent score.

Sam: Mode is 7   **Ben: Mode is 10**

As well as finding an average it is often useful in making comparisons to measure the spread of the scores. This can be done by using the **range**.

range = highest score – lowest score

Sam: Range = 7 – 4 = 3   **Ben: Range = 10 – 1 = 9**

Who do you think is the better player and why?

Sam argues that 'My median (middle) score is better than Ben's and my range is smaller than his so I am a better player and more consistent".

Ben argues that "My modal (most frequent score) is better than Sam's and my greater range of scores make it more likely that I will record the high scores".

Like Sam and Ben you must make a case to support your opinion.

Exercise

1 Find the mean, median, mode and range of

a) 1 1 2 3 8

b) 7 2 4 7

c) 4 6 4 7 4 8

d) 1 1 6 3 0 1 2

2 Mr Doni is planning to start offering boat trips round the bay.

To find out what size of boat will be best he does 10 trial trips for one day. The number of people in them are

3 3 9 2 6 6 4 5 6 5

a) Work out the mean, median and mode of these figures.

b) Advise Mr Doni on how many people his boat should seat.

3 In a survey on TV-watching, a group of boys record how many hours of TV they watch each week.

Boys: 16 23 21 5 12 0 5 13 14 11

a) Find the mean, median, mode and range of these figures.

A group of girls now does the same.

Girls: 9 11 4 16 15

b) From these results do the boys or the girls watch more TV?

4 A hockey coach measures the times it takes members of the team to run 100 m (in seconds).

Hockey team: 15 14 12 11 13 14 11 13 12 11 15 11 13 14 16

a) Calculate the mean, median, mode and range.

The coach then asks the PE staff to run 100 m.

PE staff: 11 18 11 12 13

b) From these results which group is the faster?

5 The pay of employees in a small company is:

£8000 £8000 £8000 £8000 £11 000

£11 000 £11 000 £15 000 £19 000 £35 000

a) Find the mean, median, mode and range.

b) How are the answers in a) affected when the manager's salary of £35 000 is excluded?

6 Greg works out that his mean monthly salary for the past year has been £606. He got £600 for each of the first 11 months.

What was his salary for December?

More discrete data

Rebecca is doing a report on the work undertaken by young adults in Avonford.

Fifty young adults are asked how many days they worked during the last week. The results are summarised in this frequency table.

Number of days	**0**	**1**	**2**	**3**	**4**	**5**	**6**	**7**
Frequency (number of people)	10	0	2	5	9	19	5	0

Rebecca wants a headline

'Young adults work … days per week'

and has to decide what number to put in the space.

Should she use the mode, median or mean?

- The mode is 5 days. This is the number with the greatest frequency.
- The median of 50 numbers is midway between the 25th and 26th numbers.

You can think of them like this:

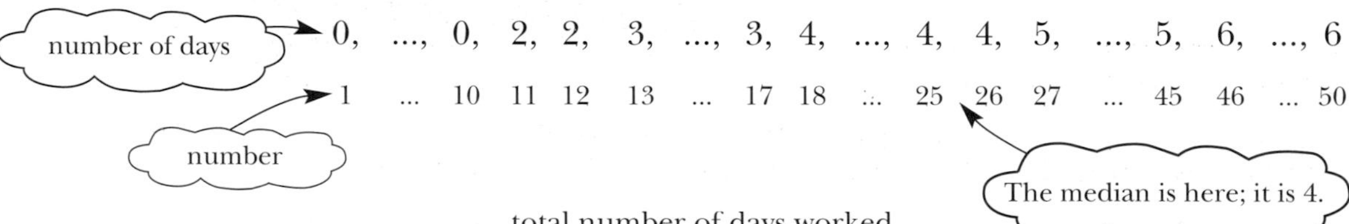

- The mean is $\dfrac{\text{total number of days worked}}{\text{total number of people}}$

$$= \frac{0 \times 10 + 1 \times 0 + 2 \times 2 + 3 \times 5 + 4 \times 9 + 5 \times 19 + 6 \times 5 + 7 \times 0}{10 + 0 + 2 + 5 + 9 + 19 + 5 + 0}$$

$$= \frac{180}{50} = 3.6$$

Would you use mode, median or mean in Rebecca's headline?

You can argue that:

'the mode (5) is an actual result and is given by nearly 40% of the young adults.'

or 'the median (4) is, in this case, an actual result unlike the mean.'

or 'the mean (3.6) uses each of the 50 results in its calculation.'

or '…'

You must make a case to support your opinion.

Exercise

1 a) Work out the mean, median, mode and range of each of these sets of data.

(i)

Value	3	4	5	6
Frequency	1	4	3	2

(ii)

Value	12	13	14	15
Frequency	3	7	7	3

(iii)

Value	1	2	3	4	5
Frequency	8	5	5	4	3

(iv)

Value	1	2	3	4	5
Frequency	2	4	9	3	2

Parts b) to d) must be answered by studying the data and not by calculation.

b) In a) (i) Sean gets a mean of 6.2. Explain why this must be wrong.

c) In a) (ii) Explain how Mark can write down the correct mean immediately.

d) Explain whether the mean in a) (iv) is less than 3, 3 exactly or more than 3.

e) Compare the data in a) (iii) and a) (iv).

2 Tom is a milkman. He delivers to 70 houses in one road. He keeps a record of the numbers of bottles of milk delivered.

a) Write down the mode.

b) Find the median.

c) Calculate the mean number of bottles delivered per house.

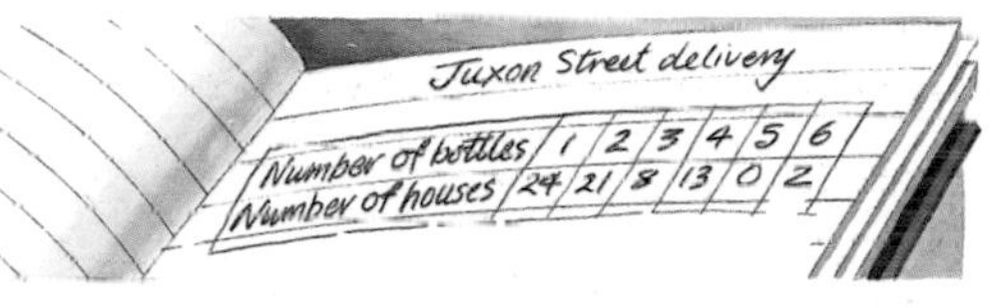
Juxon Street delivery

Number of bottles	1	2	3	4	5	6
Number of houses	24	21	8	13	0	2

d) Explain which of these averages you would use to describe how many bottles of milk Tom delivers to a typical house in Juxon Street.

3 Some households buy a newspaper every evening, others some evenings and others never. These vertical line charts refer to houses in two streets.

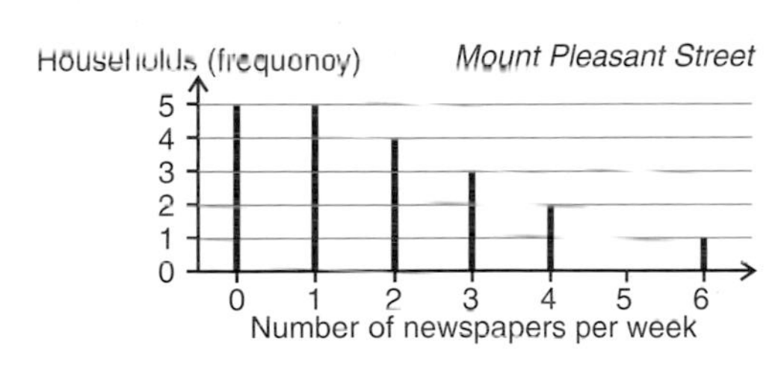

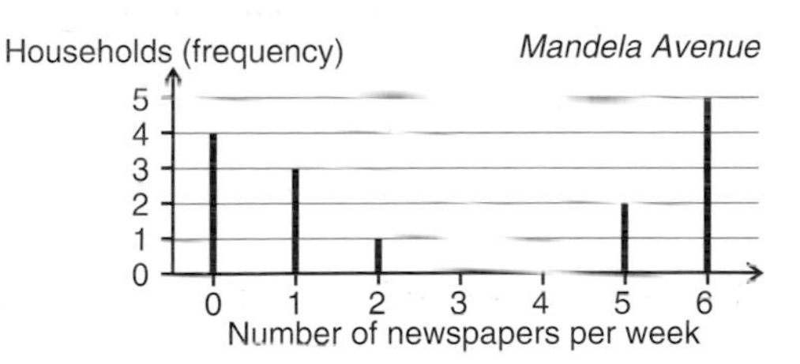

a) Make a frequency table for each street.

b) Find the mean of the number of newspapers per week for each street.

c) Find the range of the number of newspapers per week for each street.

d) Compare the results for the two streets.

Take a tabloid newspaper and count the number of words in each of 20 sentences. Find the mean and range.

Now take a broadsheet newspaper and do the same. Compare the sentences in the two newspapers.

Grouped data

Daniel is investigating the leaf sizes of different plants as part of his science assignment. He measures the width of the leaves of each plant to the nearest millimetre.

This frequency table shows the width of the leaves on a geranium plant.

Width of leaf (mm)	11–20	21–30	31–40	41–50	51–60
No. of leaves	2	6	8	5	4

The **modal class** is the class with the highest frequency.

In this case it is the class 31–40 mm.

How can you work out an estimate of the mean width of the leaves?

In order to carry out a calculation to estimate the mean we need to attach a single value (width) to each of the five classes.

The leaves in the class 11–20 mm could be any width from 10.5 mm to 20.5 mm (as the widths are given to the nearest mm). The fairest single value to use is the mid-point so each of the 6 leaves in the class 11–20 is counted as 15.5 mm.

Similarly the class 21–30 mm is really 20.5 mm to 30.5 mm and the mid-point is 25.5 mm and so on.

An extra line can now be added to the frequency table.

Width of leaf (mm)	11–20	21–30	31–40	41–50	51–60
Mid-point (mm)	15.5	25.5	35.5	45.5	55.5
No. of leaves	2	6	8	5	4

The mid-point is used to calculate an estimate of the mean.

The mean is $\dfrac{\text{total width of leaves}}{\text{number of leaves}}$

$$= \frac{15.5 \times 2 + 25.5 \times 6 + 35.5 \times 8 + 45.5 \times 5 + 55.5 \times 4}{2 + 6 + 8 + 5 + 4}$$

$$= \frac{917.5}{25} = 36.7$$

The estimated mean width of leaves on the geranium plant is 36.7 mm

Why is this only an estimate of the mean?

This is only an estimate of the mean because the actual widths (the raw data) have been replaced by convenient approximations (mid-points). For example, a leaf of width 32 mm is counted in the class 31–40 mm and treated as being 35.5 mm in the calculation.

Exercise

1 Each passenger on a flight to Cairo is allowed one item of hand-luggage. The weights, to the nearest kilogram, of the hand-luggage items are given in the table.

Weight (kg)	1–5	6–10	11–15	16–20
Frequency (no. of items)	12	24	10	4

a) Which is the modal class?
b) What is the mid-point of the 6–10 kg class?
c) Estimate the total weight of hand-luggage on the plane.
d) Estimate the mean weight of the items.

2 Ben works for an estate agency which earns a percentage of the house price for each house it sells. He is working out the mean and mode of the house prices for the last year's sales, in order to prepare a cashflow forecast for the next year. He uses this table.

PRICE (£ THOUSANDS)	31-40	41-50	51-60	61-70	71-140
FREQUENCY (NO. OF HOUSES)	35	42	56	7	3

a) Which is the modal class?
b) What is the mid-point of the class '71–140'?
c) Estimate the mean of these data.

3 In a recent cross-country skiing championship, the times for the first hundred competitors were recorded, to the nearest minute, as follows.

Time (minutes)	80–84	85–89	90–94	95–99	100–104	105–109
Frequency (number of skiers)	8	27	33	20	8	4

a) Which is the modal class?

b) Estimate the mean time taken by these hundred skiers.

The fastest skier in this table was actually disqualified, so another skier whose time was 107 minutes entered the top hundred skiers.

c) Estimate the new mean time for the top hundred.

d) What is the modal class for the top hundred skiers now?

Find the mean height of the people in your class or maths group.

Do it first using the individual heights, then from a grouped frequency table (if possible using a spreadsheet). Compare the answers you get by these two methods.

Moving averages

Dennis has kept his quarterly electricity bills for the last three years.

Year 1				Year 2				Year 3			
W	Sp	Su	A	W	Sp	Su	A	W	Sp	Su	A
76	60	36	64	84	68	44	72	92	76	52	80

He plots the figures as the blue line on the graph shown below.

Look at the figures. You can see that there is **seasonal variation** in them.

In which quarter does Dennis use the most electricity?

In which quarter does he use the least?

Why is this?

Dennis draws another line on the graph in red. This shows the **moving average**. Here is how he worked out the moving average.

76, 60, 36, 64: 4) 236 = 59
60, 36, 64, 84: 4) 244 = 61
36, 64, 84, 68: 4) 252 = 63
. . .
92, 76, 52, 80: 4) 300 = 75

Year 1				Year 2				Year 3			
W	Sp	Su	A	W	Sp	Su	A	W	Sp	Su	A
Moving average	59	61	63	65	67	69	71	73	75		

Notice that the first moving average is placed in the middle of the year, i.e. mid-way between spring and summer.

Where are the moving averages plotted on the graph?

Why is it done like this?

Dennis has averaged his values four at a time so this is called a **4-point** moving average.

Why is the red line much smoother than the blue one?

The red line is called the **trend line**. In this case the trend is increasing. A trend need not be a straight line but it can usually be described as rising, remaining steady, or falling.

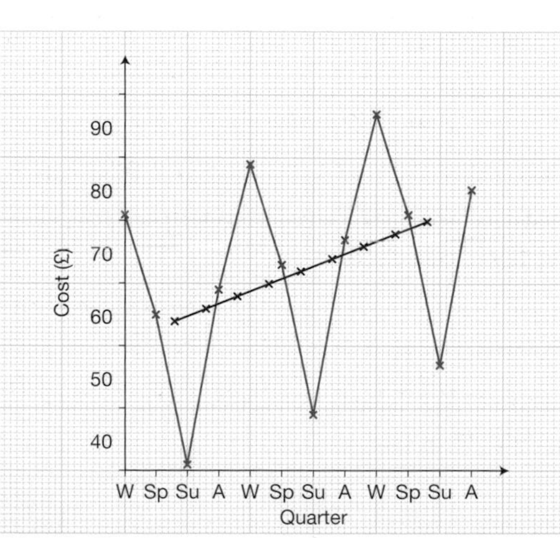

1 a) How many values would you average at a time to find a 3-point moving average?

b) Find the 3-point moving averages for the following figures.

7 5 3 6 4 4 2 7 2 9 6 9 1 5 3 8 2 4

2 How many points would you use in your average

a) for monthly orders in writing an annual report?

b) for daily data when compiling a week-by-week comparison?

c) for a shopkeeper's daily takings if he opens six days each week?

3 Dennis wants to forecast the amounts of his next two bills.

a) State the value expected for the next moving average.

b) Use this value to predict the bill for winter of Year 4.

c) Repeat a) and b) to predict the bill for spring of Year 4.

If you draw a graph you can extend the trend line to find the next moving average.

4 The table shows the sales of ice-creams in thousands.

Year 1				Year 2				Year 3			
W	Sp	Su	A	W	Sp	Su	A	W	Sp	Su	A
1.4	3.2	5.8	2.2	1.8	3.6	6.2	2.6	2.2	4	6.6	3.0

a) Draw a graph to show the data.

b) Calculate the 4-point moving averages.

c) Draw a graph of the moving averages on the same axes.

d) Comment on the general trend.

e) Estimate the expected sales for winter and spring of Year 4.

5 The table shows absences from school for one class over 4 years.

Year 1			Year 2			Year 3			Year 4		
A	Sp	Su	A	Sp	Su	A	Sp	Su	A	Sp	Su
20	50	24	25	60	22	27	90	26	30	50	28

a) Draw a graph to show the data.

b) Calculate the 3-point moving averages.

c) Draw a graph of the moving averages on the same axes.

d) How many moving averages are affected by the '90' in spring of Year 3?

e) Suggest a reason for high absence in the spring term.

A spreadsheet program is ideal for calculating moving averages. Set up a spreadsheet to do the calculations that Dennis did.

Finishing off

Now that you have finished this chapter you should be able to:

- ★ find the mean, median, mode and range of a small data set
- ★ find the mean, median, mode and range of discrete data in a frequency table
- ★ write down the modal class of grouped data
- ★ find an estimate of the mean of grouped data
- ★ calculate moving averages.

Use the questions in the next exercise to check that you understand everything.

Mixed exercise

1 For each of these sets of data find the mean, median, mode and range.

a) 105, 102, 103, 102, 108

b)

Value	0	1	2	3	4
Frequency	7	5	4	3	1

c)

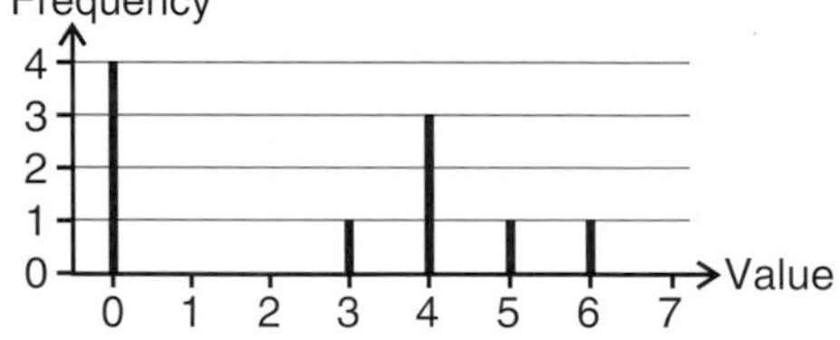

2 Here are the number of goals scored by a football team in their matches this year.

6 0 1 0 4 5 5 1 2 0

a) Find the mean, median, mode and range of these data.

b) What can you say about their performance?

c) How many goals will they have to score in total in their next two matches to increase their mean to 3 goals per match.

3 Kevin does a survey of how many nights his friends went out in the last week.

The data are given below as two frequency tables, one for boys and one for girls.

Boys	Nights out	0	1	2	3	4	5	6	7
	No of boys (frequency)	4	2	1	0	2	2	5	4
Girls	Nights out	0	1	2	3	4	5	6	7
	No of girls (frequency)	0	0	2	4	9	2	3	0

a) For both boys and girls, find the mode and range of the number of nights out.

b) What do the mode and range tell you about the two groups?

c) Draw vertical line charts to illustrate the two groups and comment on what these show.

Mixed exercise

4 Anna is a fitness consultant to the Football Club. She records the resting pulse in beats per minute (b.p.m.) of each footballer and compares them with the pulse rates of members of the Jogging Club. Here is a summary of her results.

Pulse rate (b.p.m.)	50-54	55-59	60-64	65-69	70-74	75-79	80-84
No. of footballers	0	2	7	6	3	2	2
No. of joggers	3	5	8	4	1	1	0

a) For each club work out an estimate of the mean.

b) Compare the fitness of the members of each club.

5 Carl is a keen dog-breeder. He wants to start to breed St Bernards. He visits Ruth, who is already an established breeder of St Bernards, to find out about it.

Carl has written down these questions to ask Ruth.

a) How many litters have you reared?
b) How many pups have you reared?
c) What is the most usual number of pups in a litter?
d) How variable is the size of a litter?
e) What is the average number of pups per litter?

Ruth says, 'All the information you want is on there,' showing Carl this vertical line chart that she keeps on her wall.

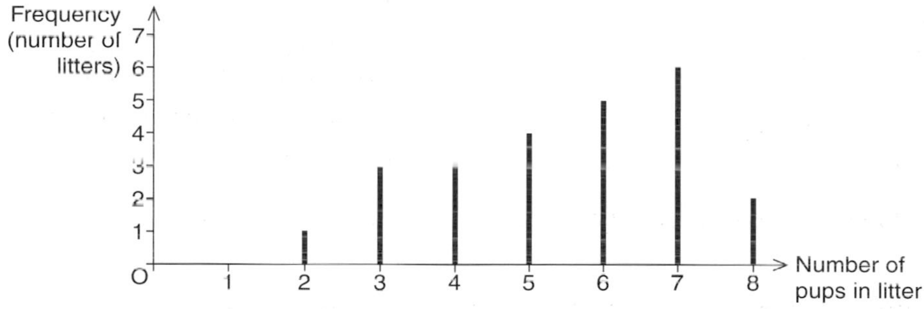

Use the chart to find the answer to each of Carl's questions.

In c), d) and e) write down the mathematical name for the figure he has found.

6 This table shows Lee's sales figures.

	W	Sp	Su	A
Year 1	400	320	340	460
Year 2	410	330	350	470
Year 3	420	340	360	480

a) Draw a graph to illustrate the data.

b) Work out the 4-point moving averages.

c) Draw a graph of the moving averages on the same axes.

d) Comment on the general trend.

e) Estimate the expected sales for winter and spring of the 4th year.

Thirty one

More data handling

Cumulative frequency

This table shows the ages of the members of Avonford Squash Club.

Age	10 – 19	20 – 29	30 – 39	40 – 49	50 – 59
Frequency (No. of members)	15	35	15	8	3

Another way to present these data is to use a **cumulative frequency table**. This gives the total number of members less than a given age.

Age	<10	<20	<30	<40	<50	<60
Cumulative frequency	0	15	50	65	73	76

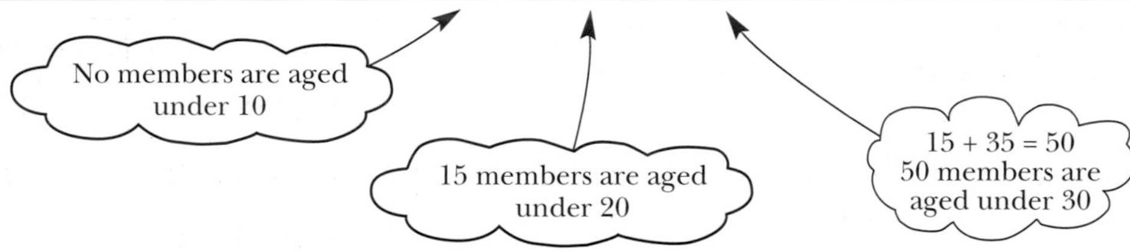

You can plot this data and then join the points by a smooth curve. This graph is called a **cumulative frequency curve**.

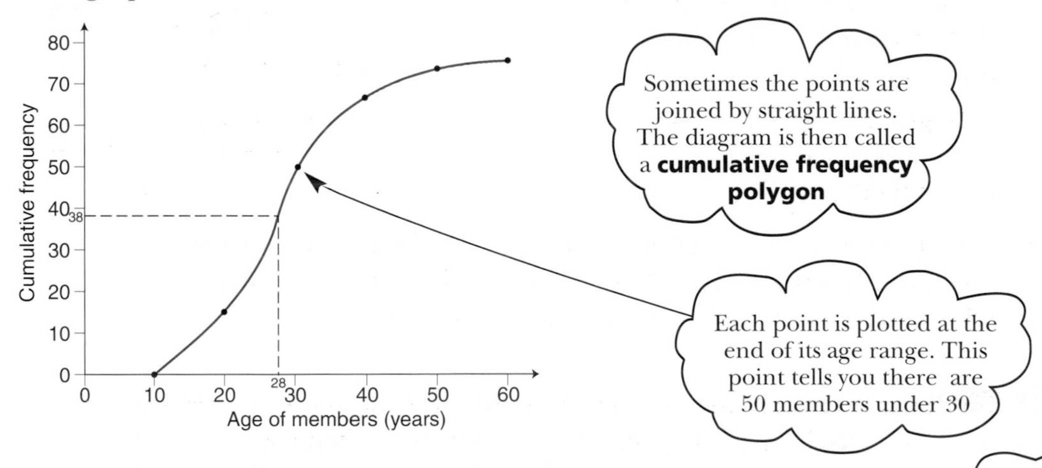

You can estimate the median from this graph.

There are 76 members
76 ÷ 2 – 38

Notice that the median has been taken as the 38th value rather than halfway between the 38th and the 39th. This is often done when the data set is large.

Look at the red line on the graph. It goes along from 38 and down to the median value, which is 28 years.

What percentage of the members are aged under 28 years?

1 Sara is a quality control officer. She is testing the lifetime of torch batteries by switching on 100 new torches and recording, for each torch, the time it takes for the battery to run down.

She produces this table of results.

Lifetime (hours)	<50	<100	<150	<200	<250
Cumulative frequency (no. of torches)	10	35	68	88	100

a) Draw a cumulative frequency diagram for Sara's data.

b) Estimate the median.

c) Estimate the number of torch batteries which exceeded the guaranteed lifetime of 130 hours.

d) The company wants to advertise a battery lifetime that 95% of batteries will reach. What lifetime should they quote?

2 Jordan obtains this table of data showing the annual salaries of the 200 employees at Quickfood plc.

Salary (£)	**No. of workers**
5000–9999	17
10 000–14 999	52
15 000–19 999	59
20 000–24 999	33
25 000–29 999	23
30 000–34 999	12
35 000–39 999	4

a) Draw a cumulative frequency curve illustrating the data.

b) Estimate the median salary.

c) Estimate what percentage of employees have a salary of under £12 000.

d) Estimate what percentage of employees have a salary of over £26 000.

e) Management claim that fewer than 10% of their employees have a salary of less than £12 000. Explain the difference (if any) between management's figure of 10% and your answer to c).

Health visitors, school nurses and other health workers often use growth charts. They check children's heights or weights against the charts to see whether they are within the expected range for their age.

Use a growth chart to find out the median height or weight for boys or girls of a particular age.

If possible, work out from the chart the lower quartile, the upper quartile and the inter-quartile range for that age.

Quartiles

Dino has been keeping a record of the number of people who eat at his diner each evening. He wants to give a clear picture of his business to Mrs Hart his bank manager and has drawn up a frequency table.

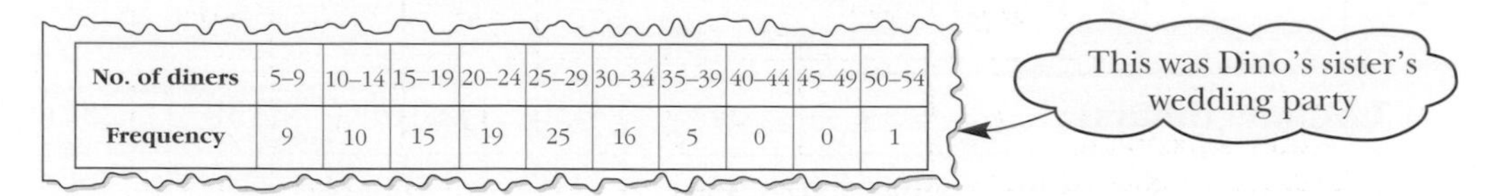

No. of diners	5–9	10–14	15–19	20–24	25–29	30–34	35–39	40–44	45–49	50–54
Frequency	9	10	15	19	25	16	5	0	0	1

The greatest possible range of these data is 54 – 5 = 49.

What is the smallest possible range? Can you tell which of these is right?

Dino decides to make a cumulative frequency table.

Number of diners (less than)	5	10	15	20	25	30	35	40	45	50	55
Cumulative frequency	0	9	19	34	53	78	94	99	99	99	100

Mrs Hart looks at this table and suggests to Dino that he should have used 4.5, 9.5, 14.5, … rather than 5, 10, 15, … .

What choice of boundaries do you think is most appropriate?

He draws the cumulative frequency curve.

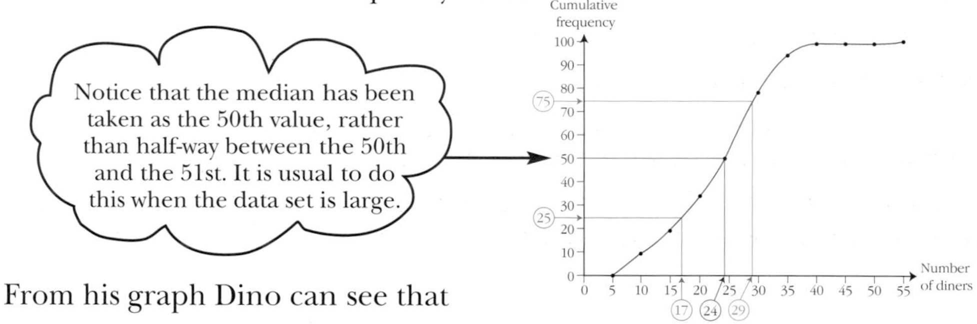

From his graph Dino can see that

- the median number of diners is about 24 (shown by the red line);
- on about 25 evenings (a quarter of the total) there were fewer than 17 diners. 17 is called the **lower quartile** and this is shown in green on the graph;
- on about 75 evenings (three quarters of the total) there were fewer than 29 diners. 29 is called the **upper quartile** and this is shown in blue on the graph;
- the difference between the upper and lower quartiles is 29 – 17 = 12. This is called the **inter-quartile range**.

On what proportion of the evenings did Dino have between 17 *and* 29 *diners?*

Which is the better measure of spread in this case, the inter-quartile range or the range?

Exercise

1 The table shows the weight of breakfast cereal in 50 packets that are labelled as 375 g.

Weight in g	360 to 365	365 to 370	370 to 375	375 to 380	380 to 385	385 to 390	390 to 395
Frequency (no. of packets)	1	3	8	21	12	4	1

a) Plot these data on a cumulative frequency curve.

b) Use your curve to estimate the median and inter-quartile range of the weights. (Give your answers to the nearest whole number.)

c) Use your curve to estimate how many packets contain at least 376 g.

2 The table gives the typing speed of 160 entrants for a secretarial exam.

Speed (words/minute)	47 – 51	52 – 56	57 – 61	62 – 66	67 – 71	72 – 76	77 – 81
Frequency (no. of entrants)	16	50	32	24	22	10	6

a) Draw a cumulative frequency curve.

b) Estimate the median.

c) Estimate the inter-quartile range.

d) Estimate the number of candidates gaining a distinction (with a speed greater than 75 words per minute).

e) A candidate needs to do 55 words per minute to pass. Estimate the number of candidates who failed the exam.

3 The following information on the birth weight of 200 babies born at Avonford General Infirmary was collected as part of a national survey of birth weights.

Weight in kg	1.5–1.9	2.0–2.4	2.5–2.9	3.0–3.4	3.5–3.9	4.0–4.4	4.5–4.9	5.0–5.4
Frequency (no. of babies)	8	18	40	52	38	22	16	6

a) Draw a cumulative frequency curve for the birth weights, and use it to estimate the median and the inter-quartile range.

b) The median weight for babies born in the UK as a whole is 3.25 kg, and the lower and upper quartiles are 2.9 kg and 3.6 kg. How do these Avonford babies compare with the national figures?

Box-and-whisker diagrams (boxplots)

Diana is a biologist. The Forestry Commission have asked her to report on a plantation of pine tree saplings. She measures the heights of the trees and draws a cumulative frequency graph. To present a simple picture in her report, Diana also draws a box-and-whisker diagram.

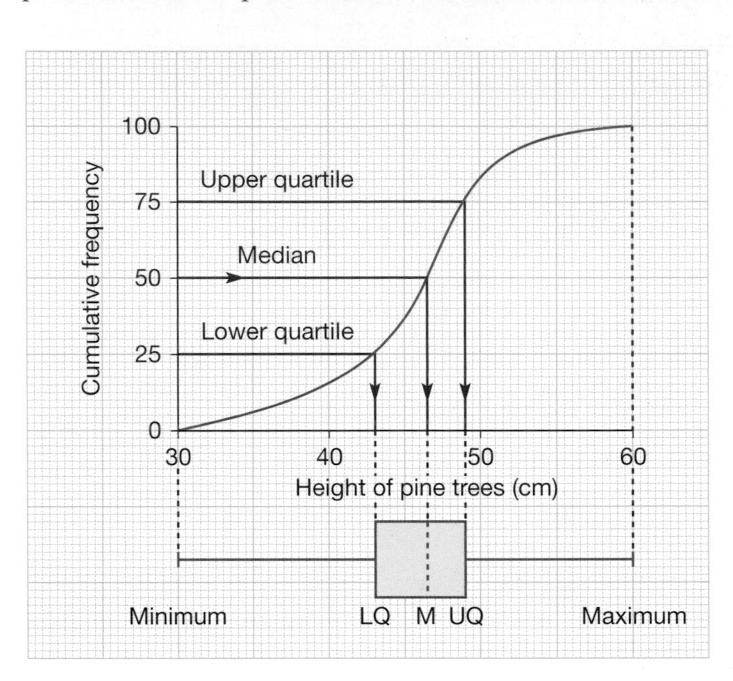

What are the main features of a box-and-whisker diagram?

What does the box show?

What do the whiskers show?

It is very easy to compare two (or more) distributions using boxplots. Look at these two. They show the marks in English and Maths of one year group in a school.

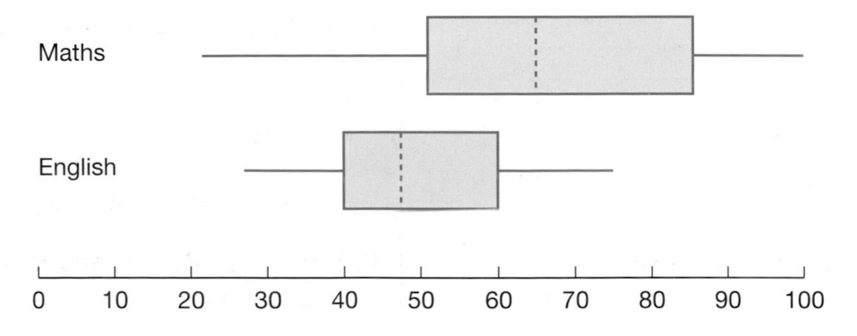

In which subject do most pupils get a higher mark?

In which subject do the marks cover a wider range?

Exercise

1 Find the median, mode and quartiles of the following values.

3 9 12 5 17 21 4 9 15 9 18

Draw a boxplot for these data.

2 A farm records the daily milk yield, in litres, of each cow in its herd. For one particular day the yields were as given below.

41 35 24 17 12 36 28 29 31 43 37
26 40 34 39 11 26 34 25 39 23 29
24 36 25 39 33 19 42 35 32

a) Draw an ordered stem-and-leaf diagram to show the data.

b) State the lower quartile, the median and the upper quartile.

c) Draw a boxplot for the data.

3 Jason carried out a survey of the ages of cars in a supermarket car park. Here is a frequency table of his results.

Age x (years)	$1 \leqslant x < 2$	$2 \leqslant x < 3$	$3 \leqslant x < 4$	$4 \leqslant x < 5$	$5 \leqslant x < 6$	$6 \leqslant x < 7$
Frequency	34	30	20	10	5	1

a) Compile a cumulative frequency table for these data.

b) Draw a cumulative frequency diagram.

c) Find the median and quartiles from your graph.

d) Draw the associated boxplot.

4 A company wanted to evaluate the training programme in its factory. They gave the same task to trained and untrained employees and timed each one in seconds.

Trained: 121 137 131 135 130 128 130 126 132
127 129 120 118 125 134

Untrained: 135 142 126 147 145 156 152 153 149
145 144 134 139 140 142

a) Draw a back-to-back stem-and-leaf diagram to show the two sets of data.

b) Find the medians and quartiles for both sets of data.

c) On the same scale draw the two boxplots.

d) Comment on the results.

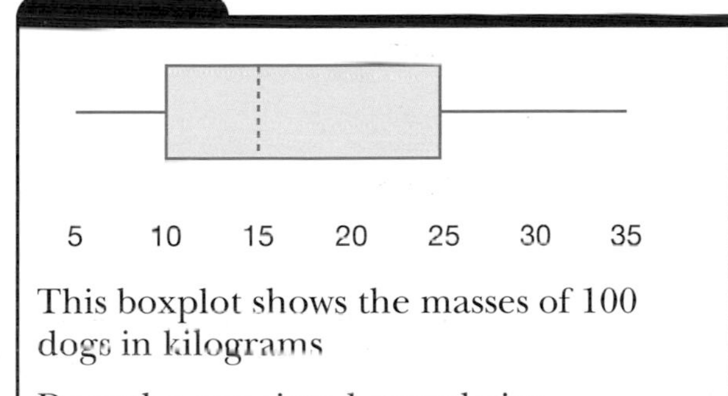

This boxplot shows the masses of 100 dogs in kilograms

Draw the associated cumulative frequency graph.

Bivariate data

Fatima has weighed and measured the people in her maths group, and drawn up this table.

	Angela	Brian	Colin	David	Elsa	Fatima	Gary	Helen
Weight (in stones)	7	10	13	14	10	11	15	9
Height (in inches)	65	69	72	71	64	66	76	60

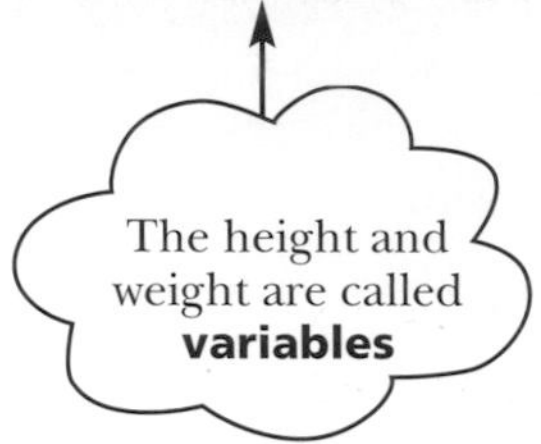

Data like these are called **bivariate** ('bi' for 'two')

Fatima suspects that height and weight are related, so she draws this **scatter diagram**.

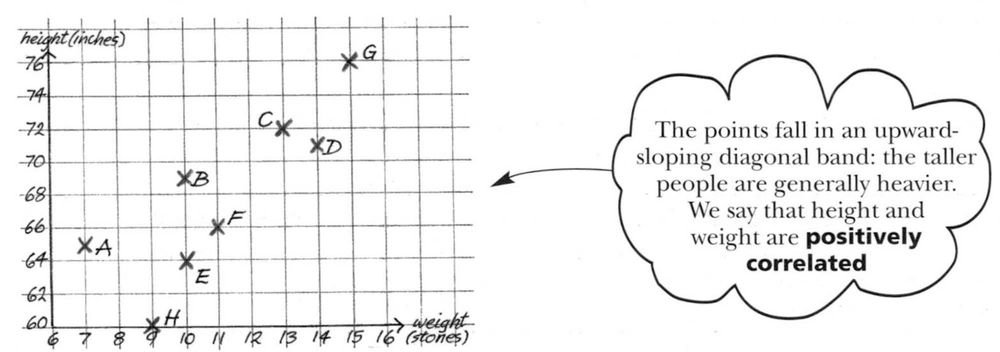

- This scatter diagram shows a strong correlation because the points are close to a line. If the points are not clustered in a band there is little or no correlation.
- The correlation is positive in this case because the slope is upwards.

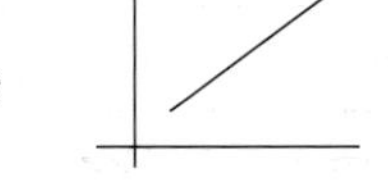

- A downward slope shows negative correlation.

Here are some more scatter diagrams.

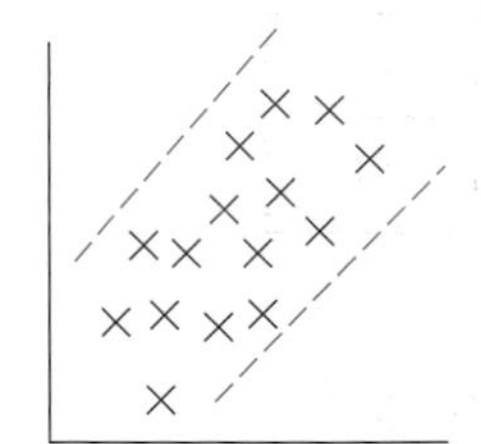

moderate positive correlation

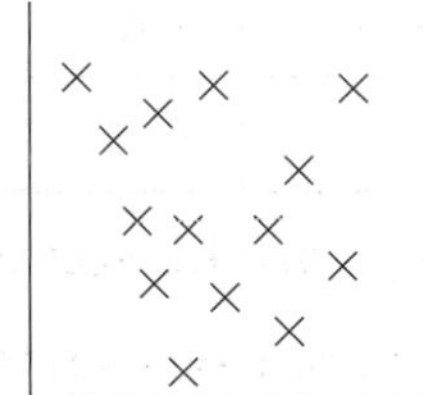

little or no correlation

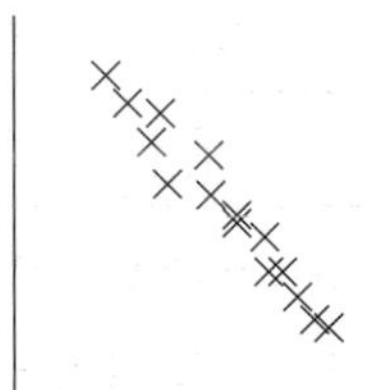

strong negative correlation

No correlation implies no linear relationship – there may be some other relationship.

What data would you expect to show each of these types of correlation?

1 The table shows the number of hours of sunshine at a seaside resort one week in June, together with the number of people visiting the aquarium there.

Day	Mon	Tues	Wed	Thur	Fri	Sat	Sun
Hours of sunshine	2.5	5.5	16	13	10	0	7.5
No. of visitors	90	70	14	28	44	96	64

a) Draw a scatter diagram.

b) Describe the correlation and explain what it means.

2 The table shows the amount of vitamin B present in 100 g samples of spinach from crops treated with different amounts of fertiliser.

Fertiliser applied (kg per ha)	0	30	60	90	120	150	180	210
Vitamin B content (mg)	0.6	0.98	1.49	1.95	1.9	1.83	1.6	1.51

Draw a scatter diagram and comment on the correlation.

3 A hospital keeps a record of the birth-weight and length of each baby born there. Here are the records for one day.

Name	**Sex**	**Weight in kg**	**Length in cm**	**Name**	**Sex**	**Weight in kg**	**Length in cm**
Faulkner	Boy	3.64	55	Wolf	Boy	4.18	63
Jones	Girl	3.49	53	Holt	Boy	2.7	59
Patel	Girl	4.46	55	Deane	Girl	3.47	56
Berry	Boy	3.84	53	Worth	Boy	2.98	53
Haynes	Boy	3.38	51	Murphy	Girl	3.81	61
Adams	Boy	3.69	58	McPhee	Boy	1.96	46
Gardner	Boy	2.61	47	Hughes	Girl	2.89	52
Roper	Girl	3.13	56	la Rue	Girl	3.98	57
Holmes	Boy	3.55	51	Marsh	Girl	2.55	49
McGrace	Boy	4.33	58	Mehta	Girl	2.6	50

a) Draw a scatter diagram and comment on the correlation.

b) Mark girls and boys in different colours. Does a pattern emerge?

Line of best fit

If the points on a scatter diagram lie in a narrow band there is a strong correlation. The stronger the correlation, the narrower the band. Provided there is some correlation, you can add a line of best fit to your diagram as in the example below.

Adding a line of best fit to your scatter diagram helps you to estimate the value of one variable, given the value of the other.

This table shows the mean height of parents and the mean height of their adult offspring.

Surname	Mean height of parents (cm)	Mean height of adult offspring (cm)
Wilson	161	163
Allan	164	167
Gupta	166	168
Smith	169	170
Lipton	174	173
Morris	179	177
Spicer	183	179
Gibson	183	183

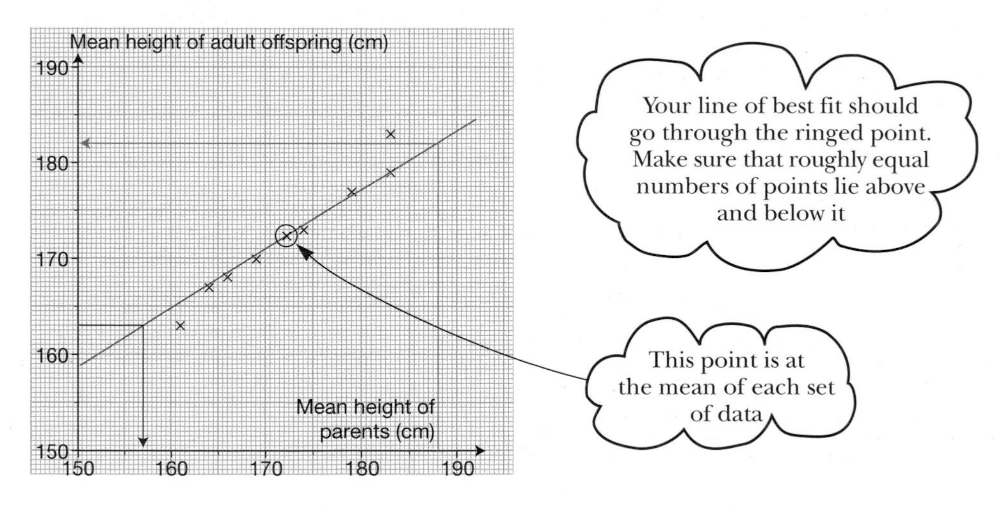

What does the diagram tell you?

A group of brothers and sisters found that their mean height was 163 cm. *Use the scatter diagram to estimate the mean height of their parents. (The red line will help.)*

A couple whose mean height is 188 cm *are worried their children will be giants. They draw the green line on the diagram.*

Do you think their children will all be 182 cm *tall?*

Exercise

1 This table records the force P newtons required to move an object of mass M kg on a rough surface.

M	1	2	3	4	5	6	7	8
P	7.5	14.4	22.5	30.5	37.4	45.4	53.5	61.4

a) Draw a scatter diagram and add the line of best fit.

b) Estimate the force required to move a mass of 2.6 kg.

c) Estimate the mass you could move with a force of 40 newtons.

2 This table shows the height in metres above sea level and the temperature in Celsius, on one day at nine different places in Europe.

Height (m)	1400	400	280	790	370	590	540	1250	680
Temperature (°C)	6	15	16	10	14	14	13	7	13

a) What is the mean height?

b) What is the mean temperature?

c) Plot a scatter diagram and describe the correlation.

d) Add a line of best fit.

e) Use your diagram to estimate the temperature at a height of 500 m.

f) Use your diagram to estimate the height of an area with a temperature of 8 °C.

3 The people in Gil's chemistry class have done an experiment in which an acid reacts with a solid. They recorded the volume of gas given off during the reaction, and the loss in mass of the reactants. Here are their results.

Name	Alan	Bridie	Callie	Donna	Ed	Flick	Gil	Helen
Loss in mass (g)	0.060	0.032	0.120	0.083	0.090	0.160	0.140	0.107
Volume of gas (cm³)	45	24	90	62	35	120	105	80

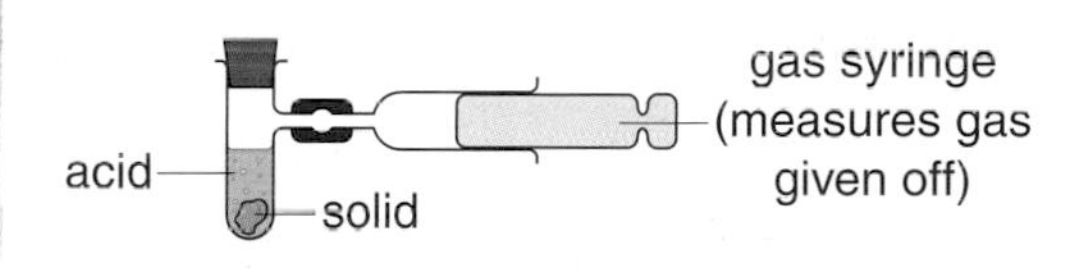

a) Draw a scatter diagram of their results.

b) Comment on what your diagram shows.

c) One student made a mistake during the measuring. Who do you think it was?

d) Ignoring the data from the person who made a mistake, draw a line of best fit (by eye) for the data.

e) Use your line to estimate the mass of 100 cm^3 of the gas.

Finishing off

Now that you have finished this chapter you should be able to:

★ draw and interpret cumulative frequency curves

★ estimate the median and the inter-quartile range of a set of data from its cumulative frequency curve

★ draw and interpret box-and-whisker diagrams

★ interpret bivariate data by drawing scatter diagrams

★ recognise positive, negative and little or no correlation

★ draw (by eye) and use lines of best fit.

Use the questions in the next exercise to check that you understand everything.

Mixed exercise

1 This cumulative frequency curve shows the profits recorded by 100 companies.

a) From the graph, estimate the median.

b) From the graph, estimate the profit exceeded by 80% of the companies.

c) Estimate the number of companies recording profits between £12 million and £22 million.

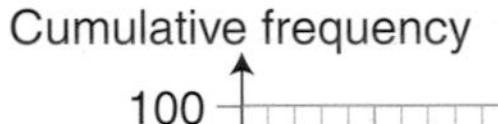

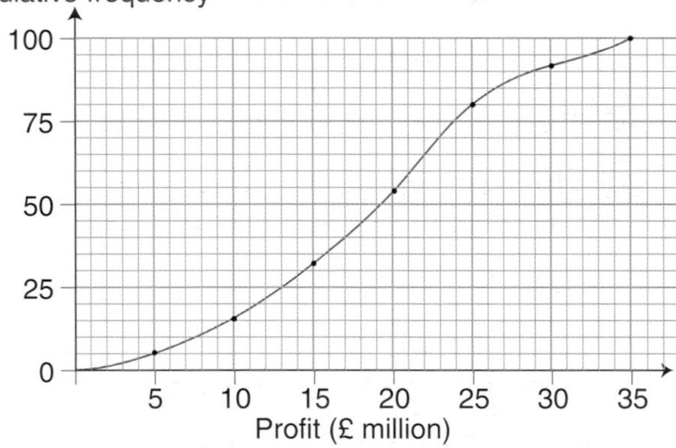

2 A new fertiliser is being tried out on some apple trees. The data give the amount of fertiliser used and the fruit yield of each tree.

Fertiliser (g)	Yield (kg)	Fertiliser (g)	Yield (kg)
5	52	12	32
8	40	8	44
6	48	6	46
10	38	5	50
4	50	10	40

a) Plot these data on a scatter diagram. Use the horizontal axis for the amount of fertiliser. Use the vertical axis for the yield.

b) What sort of correlation is there?

c) Do you think it is a good thing to use a lot of fertiliser?

3 This table shows the results of a survey that was done to check the phasing of the traffic lights at the site of some major roadworks. The time spent waiting by 200 cars travelling in each direction was recorded one morning.

Delay, *t* (minutes)	$0 \le t < 2$	$2 \le t < 4$	$4 \le t < 6$	$6 < t < 8$	$8 \le t < 10$	$10 \le t < 12$	$12 \le t < 14$	$14 \le t < 16$
North-bound	50	80	60	10	0	0	0	0
South-bound	10	25	35	55	45	20	10	0

a) On the same set of axes, draw a cumulative frequency curve for the delay times in each direction.

b) Estimate the median and the inter-quartile range for each direction.

c) Draw a boxplot for the data for each direction.

d) Do you think the traffic lights are sensibly phased?

4 A company uses two types of car in its fleet (A and B).

The fleet manager keeps these records of the distances travelled and the running costs for 20 of the cars.

Type A

Distance (thousands of miles)	5	5.6	6.9	7.7	9	9.9	9.9	11.1	11.8	13.1
Costs (hundreds of pounds)	6.3	7.7	8.5	9.8	9	10.1	11.5	11	12.7	13.4

Type B

Distance (thousands of miles)	4.5	5.6	6.3	7	8.2	9.4	11.1	12.1	12.5	13.3
Costs (hundreds of pounds)	7.9	8.6	8.9	9.3	9.8	10.2	10.6	11.3	11.1	12.3

a) For each type of car calculate the mean distance and the mean cost.

b) Plot these figures on a scatter diagram using one colour for Type A and a different colour for Type B.

c) This year the company expects to expand and distances will be 50% higher. From the graph, do you think the company should buy new cars of Type A or Type B?

Thirty two

Surveys

Planning a survey

Yvette is a keen swimmer.

One morning she gets this letter.

It is from her swimming club.

AVONFORD SWIMMING CLUB

Dear Yvette

The club committee met last night to discuss how to attract new members.

As a first step we would like a survey done in your school. Could you carry it out for us?

Step 1

Choose a topic that you find interesting.

Yvette agrees to do the survey.

She makes some notes.

My swimming club wants me to find out whether people like swimming and if they are good at it.
THINGS TO DO
Write some questions to ask people
Try them out on friends
Get 60 copies made
Get to school early + give to first 30 girls + 30 boys
Make data collection sheet
Write a report

Step 2

Write down why you are doing the survey and what you want to find out, and plan how to do the survey.

Yvette writes these questions.

SWIMMING QUESTIONS
1 Can you swim?
2 Why do you like swimming?
3 How long do you take to do 1 length?
4 What stroke do you like best?

Step 3

Write the questions for your survey.

Yvette tries the questions out on some of her friends. Here are some of their answers.

I don't know the names of the strokes.

I can swim 1 length of Uncle Pete's pool in 5 seconds.

I don't like it. I hate it.

I like every stroke except butterfly.

Anyone can swim - even my baby brother can.

Step 4

Try out your questions on a few people.

Exercise

Discussion

1 Look at the answers that Yvette's friends have given.

Which question is each person answering?

Do you think Yvette is getting the information she wants?

2 Yvette decides to write some better questions so that people will give her useful answers.

Is this new set of questions better?

Step 5

Write your new questions.

SWIMMING QUESTIONS

1 Boy ☐ Girl ☐ (tick one box)

2 Is it important for everyone to be able to swim?
Yes ☐ No ☐ Don't know ☐

3 Can you swim at least 1 width in a swimming pool?
Yes ☐ No ☐

4 Which swimming stroke do you prefer?
Breast stroke ☐ Front crawl ☐ Back stroke ☐ Butterfly ☐

5 How many seconds do you take to swim 25 metres? ________ seconds

Please return to Yvette Lee.

3 Why do you think Yvette asks question 1?

4 Why does Yvette need a data collection sheet?

Yvette has started to design the data collection sheet for her survey.

It looks like this:

Data Collection Sheet

1 Boys ________ Girls ________ Totals ☐ ☐

2 Is it important for everyone to be able to swim?

Yes ________ Totals ☐ ☐

No ________ Totals ☐ ☐

Don't know ________ Totals ☐ ☐

Yvette has decided to record the boys' answers in blue and the girls' in red.

Design the rest of the data collection sheet for Yvette's survey.

The survey report

Yvette collects the answers from 20 boys and 20 girls as they arrive at school.

Do you think that this method of collecting data gives Yvette a set of answers which are representative of the whole school?

Here is her data collection sheet.

Step 6

Collect your data.

Data Collection Sheet

	Boys	Girls	Totals (boys)	Totals (girls)
1 Boys / Girls	𝍸 𝍸 𝍸 𝍸	𝍸 𝍸 𝍸 𝍸	20	20
2 Is it important for everyone to be able to swim?				
Yes	𝍸 𝍸 𝍸 𝍸	𝍸 𝍸 𝍸 𝍸	20	20
No			0	0
Don't know			0	0
3 Can you swim at least 1 width of a swimming pool?				
Yes	𝍸 𝍸 |||	𝍸 𝍸 𝍸	13	15
No	𝍸 ||	𝍸	7	5
4 Which stroke do you like best?				
Breast stroke	|||	𝍸 ||||	3	9
Front crawl	𝍸 |||	||||	8	4
Back stroke	||			2	1
Butterfly				0	1

5 How many seconds do you take to swim 25 metres

Boys: 22 33 35 34 35 61 51 55 43 25 47 32 46

Girls: 31 24 48 43 32 66 52 38 38 24 41 35 36 24 27

Yvette talks to the swimming pool manager about her survey.

He gives her a copy of a recent magazine article.

90% of young people can swim

Step 7

Find and use information from other sources.

Do you think Yvette can use this in her report?

Now Yvette must write her report. This is her plan:

Step 8

Write your report.

- start with an *executive summary* explaining in a few lines the main points to emerge from my survey.
- write an *introduction* explaining what I wanted to find out and why.
- explain my *approach* detailing how I collected my data, how many people I asked, how I selected them and what I asked them (see questionnaire).
- present my *findings* using tables, charts and diagrams, and work out measures of average and spread.
- finish up with my *conclusions and recommendations.*

Exercise

1 Look at the answers from question 2 of the survey.

Everyone said they thought it was important to be able to swim.

Do you think Yvette should draw a chart to illustrate this?

2 Look at the answers from question 3 of the survey.

a) Show the girls' answers in a bar chart and a pie chart.

b) Show the boys' answers in a bar chart and a pie chart.

c) Which kind of chart do you think Yvette should use, and why?

3 Look at the answers from question 4 of the survey.

Why do you think the numbers do not add up to 20 for boys and 20 for girls?

4 Look at the times taken to swim 25 metres.

a) What is the mean time for the boys and for the girls?

b) What is the range of times for the boys and for the girls?

c) What conclusions can you draw from your answers to a) and b)?

5 a) Copy and complete this frequency table of the time taken by the girls to swim 25 metres.

b) Make a similar table for the boys' times.

c) Compare the two charts.

Girls	
Seconds	Frequency
20-29	4
30-39	6
40-49	
50-59	
60-69	

Discussion

What should Yvette write in her conclusions?

Now you are ready to do a survey of your own.

Follow the 8 steps given in this chapter.

Don't be afraid to ask other people what they think about the way you are doing it.

Here are some possible topics. Use one of these, or (even better) think up a topic of special interest to you.

- *Favourite TV programmes.*
- *Do people believe in horoscopes?*
- *Attitudes to smoking.*
- *Health and fitness.*

Thirty three

Probability

Calculating probabilities

Su throws the die in a game of ludo. (Note: die is the singular of dice.)

How likely is she to throw a 4?

When she throws the die there are six possible outcomes: 1, 2, 3, 4, 5 or 6. They are all equally likely.

In everyday English you say

'There is 1 in 6 chance of getting a 4.'

In mathematics you say

'The probability of getting a 4 is $\frac{1}{6}$.'

You can write this as $P(4) = \frac{1}{6}$.

short for probability

Probability is defined as $\dfrac{\textbf{number of favourable outcomes}}{\textbf{number of possible outcomes}}$

(This is only true when each outcome is equally likely.)

Here are other examples of calculating probabilities.

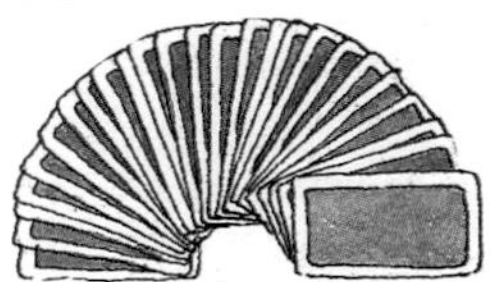

Tossing a fair coin:

$P(\text{head}) = \frac{1}{2}$

Choosing a card at random from a pack:

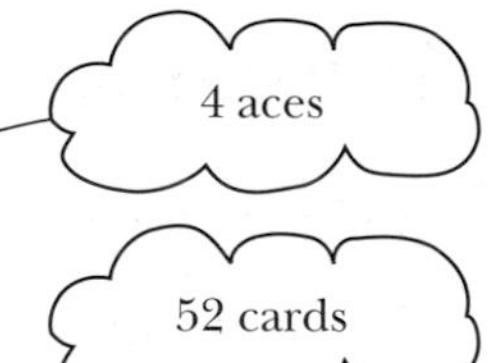

$P(\text{ace}) = \frac{4}{52} = \frac{1}{13}$

What do the words 'fair' and 'at random' mean in the above contexts?

A fair coin means an unbiased one. It is equally likely to come down heads as tails. At random means that each card has an equal chance of being chosen.

Probability is a number on a scale between 0 (impossible) and 1 (certain). It can be written as a fraction, a decimal or percentage.

Impossible	Unlikely	Evens or '50–50'	Likely	Certain
0		$\frac{1}{2}$		1

What is the probability that a card chosen at random from a pack is not an ace?

Exercise

1 You throw a die. What is the probability that you get

a) a 6? b) an even number? c) an odd number?

d) a 7? e) less than 3? f) a 3 or higher?

2 You select a card at random from a standard pack of 52 cards.

Find the probability that you have selected

a) a heart

b) a red card

c) a court card (king, queen or jack)

d) an ace

e) the queen of hearts.

3 There are 28 dominoes in a standard set. This one is 2–4.

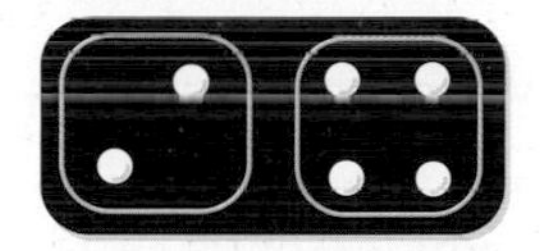

List the full set of dominoes.

You choose a domino at random from the complete set. What is the probability that

a) it has a 6 on it?

b) it is a double?

c) the two numbers on it are different?

4 In a raffle 400 tickets have been sold. Alex has bought 5 of them.

The tickets are put in a hat and one is selected for first prize.

What is the probability that

a) Alex wins first prize?

b) Alex does not win first prize?

In fact Ron wins first prize. His ticket is not put back.

Another ticket is now drawn for second prize.

c) What is the probability that Alex wins second prize?

5 Bella is playing Scrabble.

These 20 tiles are in the bag and she chooses one without looking.

What is the probability that she chooses:

a) the letter A?

b) a tile which scores 1 point?

c) a tile which scores more than 1 point?

d) a tile which scores 5 points?

e) a tile which is not an N?

Estimating probability

In some situations, like throwing dice, tossing coins and selecting cards, you can calculate probability exactly. This is what you did on the last page.

There are other situations where you want to know a probability but cannot work it out from theory. In that case you have to estimate it from data.

Shona spent all of August last year in a small village in Cornwall. On her calendar she put S for sunny or R for rainy against every day.

She says

There were 8 rainy days in August

'If I choose a day at random then P(rainy) = $\frac{8}{31}$.'

There are 31 days in August

Shona has estimated the probability from the **relative frequency**: the number of times the outcome happened divided by the number of times it could have happened.

If Shona collects August rainfall data this year and uses it to do the same calculation, will she get the same result?

There are not very many situations where you can work out probability from theory. It is more common to estimate it, but you need more data than Shona collected to have any confidence in your answers.

How many years of August weather data would you need to feel confident?

Do you think you would have got the same results 100 *years ago?*

People's estimates of probability are not always based on data. Sometimes they are just guesswork, like predicting the result of a sporting fixture

During Festival Week the village football team are, for the first time, playing a showbiz team. What is the probability that the showbiz team will win?

Exercise

1 Diana works in a travel agency. She notes that of the last 20 holiday bookings 13 were for Europe and the rest for America.

Estimate the probability that the next holiday booking will be for

a) Europe b) America c) Australia

2 Last year there was no rain in Avonford on 30 out of 90 days in January, February and March.

a) Estimate the probability that it will rain on 6 March next year.
b) Estimate the probability that it will not rain on 6 March next year.
c) Do you trust your answers?

3 Angus is a bird watcher; he keeps a record of all the birds he sees. Over a long period he sees 161 cuckoos, 23 of which are female and the rest are male. Use these figures to estimate:

a) the probability that a cuckoo chosen at random is female
b) the probability that a cuckoo chosen at random is male
c) how many females Angus expects to find among 1000 cuckoos.

4 Ela keeps a record of how many letters she receives each day (except Sunday) for 10 weeks and then draws this line chart.

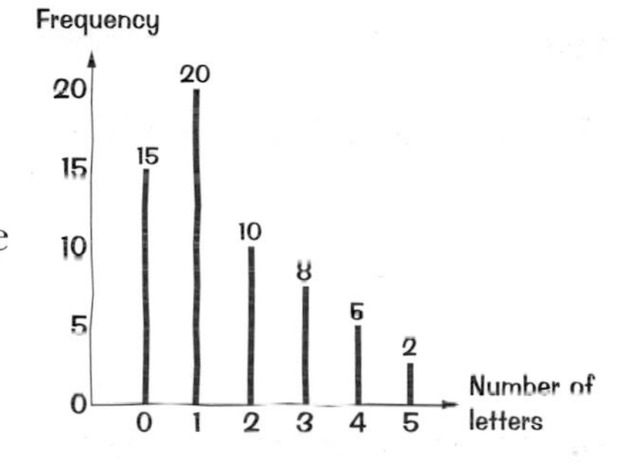

Use Ela's data to estimate the probability that on Thursday next week she will receive

a) 1 letter b) 3 letters
c) 3, 4 or 5 letters d) 12 letters.

Do you believe your answer to part d)?

5 Roy plants a packet of 40 seeds. They all grow into plants.

10 of these plants have pink flowers, the rest white.

Next year Roy plants 5 similar packets.

a) Estimate the probability that a seed chosen at random from these will grow into a pink-flowering plant.
b) About how many of the seeds would Roy expect to grow into plants with pink flowers?
c) Estimate the probability that a seed chosen at random will grow into a white-flowering plant.
d) About how many of the seeds would Roy expect to grow into plants with white flowers?

Solitaire is a well known card game that you can play on many computers. Play enough games for it to come out completely several times and so estimate the probability that this happens.

Two outcomes: 'either, or'

Avonford Car Auctions holds a monthly Company Draw. One employee is picked (at random) by the company computer to win £200.

This week several people are off work because of a flu epidemic, and some people are away on courses.

Absences 15th June

flu	7
courses	12
working	31
TOTAL	50

What is the probability that the employee who wins the draw is absent, either with flu or on a course?

You can see that there are 50 employees in total, so there are 50 possible winners. Of those 50:

7 have flu: $P(\text{flu}) = \frac{7}{50}$

12 are away on a course: $P(\text{course}) = \frac{12}{50}$

Nobody with flu is on the course. They are all in bed.

Of the 50 employees, a total of 19 are away either with flu or on a course, so you can write

$$P(\text{flu or course}) = \frac{19}{50}$$

Notice that in this case,

$$P(\text{flu or course}) = P(\text{flu}) + P(\text{course})$$

To find the probability that the winner either has flu or is on a course you can add the probability that the winner has flu and the probability that the winner is on a course

Of the 50 employees, 31 are in the sports club and 10 are in the fell-walking club. What is the probability that the winner is in one of these clubs?

You may have realised that to answer the question you need more information. Some of the fell-walkers may also be in the sports club: the outcomes are not **mutually exclusive**.

In fact, Ted, Liz and Nina are in both clubs. You want to add the probabilities without counting these people twice, so you can write

$$P(\text{sports or fell}) = P(\text{sports}) + P(\text{fell but not sports}) = \frac{31}{50} + \frac{7}{50} = \frac{38}{50}$$

There are 10 people in the fell-walking club but 3 have already been counted in the sports club. $10 - 3 = 7$

You can add the probabilities in 'either, or' situations, provided you make sure that the outcomes do not overlap.

How many people, altogether, are involved in the sports club or the fell-walking club?

Exercise

1 When you select a card at random from an ordinary pack, what is the probability that it is either

a) a king or a queen?

b) a king or a heart?

2 Dick plants mixed crocuses: 5 purple, 7 yellow and 8 white.

Work out the probability that the first one to flower in the Spring will be

a) purple

b) yellow or white

c) purple or white

d) neither purple nor white.

3 Paul is a keen bird-watcher. One June he keeps a close watch on 65 nests of house-martins. He records the number of hatchlings that survive and fly from each nest. This is his table of the results.

Number of hatchlings flying the nest	0	1	2	3	4	5	6
Frequency (number of nests)	1	5	12	18	24	3	2

Using Paul's data, estimate the probability that a nest of house-martins will produce

a) 5 hatchlings that fly the nest

b) at least 4 hatchlings that fly the nest

c) fewer than 3 hatchlings that fly the nest.

4 Make a table to show all the possible scores when two fair dice (one red, one blue) are thrown.

Find the probability that

a) the scores are the same (it is a double)

b) the score on the red die is greater than the score on the blue one

c) both dice show odd numbers

d) the total score is more than 7.

You are going to check whether real dice behave as your answers to question 4b) suggest. Find two dice of different colours. Draw a table like the one you drew for question 4 to record your results.

The relative frequency each time is the number of times red has scored more than blue, divided by the total number of throws so far.

You would expect the relative frequency to get closer and closer to your calculated probability. Is this noticeable after 10 throws? Extend the table, and see if it is closer after 20 throws, or 50 throws.

Two outcomes: 'first, then'

When you select a playing card at random from an ordinary pack, then replace it, then select another, the probability of an ace is $\frac{4}{52}\left(=\frac{1}{13}\right)$ each time.

(i) What is the probability that both cards are aces?

(ii) What is the probability of getting just one ace?

It is helpful to work out a table showing the probabilities of all possible outcomes. When you work out the probability of first one outcome, then another happening, you multiply the probabilities.

First card	Second card	Probability
ace	ace	$\frac{1}{13}\times\frac{1}{13}=\frac{1}{169}$
ace	not ace	$\frac{1}{13}\times\frac{12}{13}=\frac{12}{169}$
not ace	ace	$\frac{12}{13}\times\frac{1}{13}=\frac{12}{169}$
not ace	not ace	$\frac{12}{13}\times\frac{12}{13}=\frac{144}{169}$

Since the first card is replaced the outcome of the second draw is **independent** of the first card drawn

What do these four probabilities add up to?

(i) You can see in the table that the probability that both cards are aces is $\frac{1}{169}$.

(ii) You can see that there are two outcomes in the table with just one ace. These are 'either, or' events so you add the probabilities.

$$P(\text{just one ace}) = P(\text{ace, not ace}) + P(\text{not ace, ace}) = \frac{12}{169}+\frac{12}{169}=\frac{24}{169}$$

Now suppose that the original card selected is not replaced. The probability of getting an ace with the first selection is $\frac{4}{52}$ and assuming an ace is selected then there are only 3 aces left among the remaining 51 cards so

$$P(\text{ace, ace}) = \frac{4}{52}\times\frac{3}{51}=\frac{1}{13}\times\frac{1}{17}=\frac{1}{221}$$

In 'first, then' situations, you multiply the probabilities, but you need to think carefully about those probabilities when the events are not independent.

Exercise

1 Kim selects two cards from an ordinary pack. The first is replaced before the second is taken. Find the probability that

a) both cards are spades

b) just one card is a spade.

2 Repeat question 1 with the first card not being replaced before the second is taken.

3 China mugs are packed in individual boxes at the pottery. 5% of the mugs are damaged in transit to gift shops.

Rachel buys two boxes.

Find the probability that

a) both mugs are damaged

b) neither mug is damaged

c) just one mug is damaged.

4 Harry and Jessica plan to meet at a restaurant at midday. The probability that Harry is late is 0.1. The probability that Jessica is late is 0.2.

Find the probability that

a) Harry and Jessica are both late

b) Harry is late but Jessica is not

c) neither is late.

5 Rowena does a survey of pedestrians in Avonford town centre. She asks them how they have travelled there, and the main purpose of their visit. She presents the results in these pie charts.

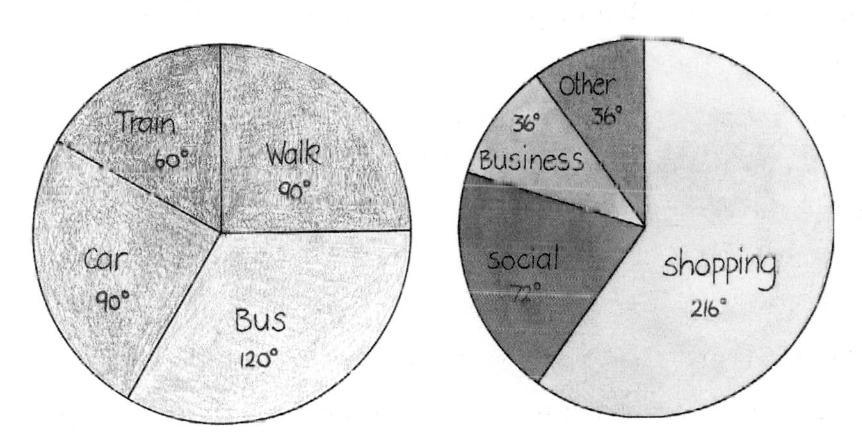

Using Rowena's pie charts, work out the probability that a pedestrian chosen at random

a) has travelled into Avonford by car

b) is there mainly for a social event.

Assuming that the events are independent, what is the probability that a pedestrian chosen at random

c) arrived by car for a social event?

d) arrived by train for business?

e) arrived by bus or train for a shopping trip?

Probability trees

The probability that a person (chosen at random) is left-handed is $\frac{1}{10}$. The probability that a person wears glasses is $\frac{1}{4}$.

What is the probability that a person chosen at random is left-handed and wears glasses?

What is the probability that the person is right-handed and wears glasses?

You can work out these probabilities using the ideas from the previous pages. Alternatively you can represent the situation in a **tree diagram**, which shows the probabilities of all possible outcomes.

Any person is taken to be either left-handed (LH) or right-handed (RH): two alternatives. You show this as two branches of a 'tree', and write the probability on each branch.

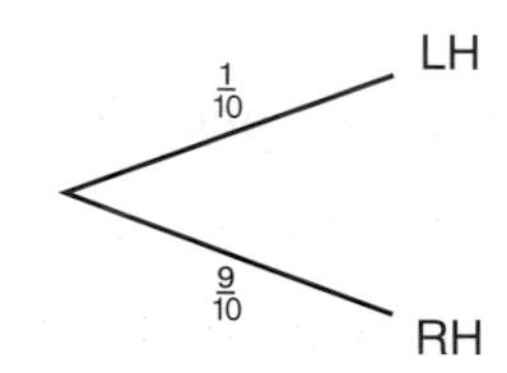

Any person either wears glasses (G) or doesn't (not G). Again there are two alternatives. You extend the diagram like this.

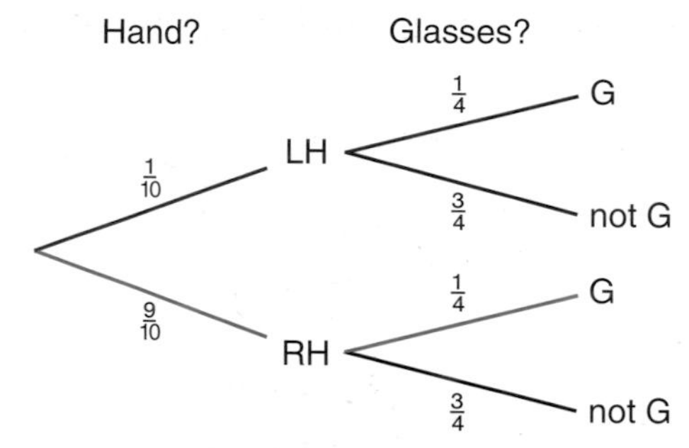

The probability that a person is left-handed and wears glasses is given by multiplying the probabilities along the red branches:

$$P(\text{LH and G}) = \frac{1}{10} \times \frac{1}{4} = \frac{1}{40}$$

The probability that a person is right-handed and wears glasses is given by

$$P(\text{RH and G}) = \frac{9}{10} \times \frac{1}{4} = \frac{9}{40} \text{ (the green branches)}$$

There are four possible outcomes. The probability of two of them have been found above. Find the probability of the other two outcomes, and add all four probabilities together.

What does this tell you?

Probability trees are particularly helpful when there are more than two outcomes from each event, and when the events are not independent.

Exercise

1 A student survey finds that 60% of students own a bicycle and 70% own a CD player. Assuming these to be independent, draw a probability tree and find the probability that a randomly chosen student owns

a) a bicycle but no CD player

b) a CD player but no bicycle

c) neither a CD player nor a bicycle

d) both a CD player and a bicycle.

2 Fred claims to be able to tell the colour of a Smartie by taste alone. A test is organised with Fred blindfolded. He has to pick 30 Smarties out of a big bowl containing equal numbers of green (G), red (R) and brown (B) Smarties. He eats each one and says what colour he thinks it is.

a) How many out of the 30 would you expect Fred to get right if

(i) his claim is false and he is just guessing?

(ii) his claim is true and he really can tell them apart?

b) In fact Fred can tell the green ones but guesses between the red and brown.

Copy and complete this tree diagram for the situation.

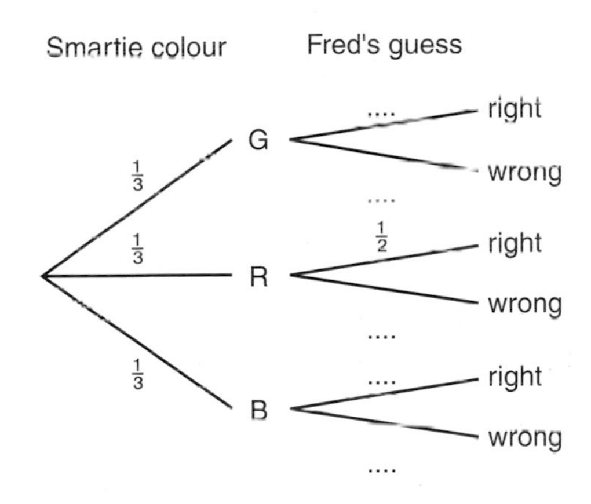

c) What is the probability that Fred gets a Smartie right?

d) In fact Fred gets 21 right. Is that about what you would expect?

Set up and carry out an experiment like the one in question 2. Write a short report on your findings, including tree diagrams to show what the probabilities appear to be.

Finishing off

Now that you have finished this chapter you should be able to:

- ★ work out simple probability
- ★ work out the probability of a particular outcome in an 'either, or' situation
- ★ work out the probability of a particular outcome in a 'first, then' situation
- ★ use a tree diagram to work out the probabilities of different outcomes for two or more events.

Use the questions in the next exercise to check that you understand everything.

Mixed exercise

1 a) Write down the probability that:

(i) when a coin is tossed it comes up heads

(ii) when a die is thrown the number on it is more than 2

(iii) someone chosen at random was born on a Thursday

(iv) a card chosen at random from an ordinary pack is a heart

(v) when a die is thrown the number on it is 7.

b) Draw a probability scale and mark your answers (i) to (v) on it.

2 In a survey of 500 homes it was found that 253 had a microwave, 398 had a washing machine and 147 had a dishwasher.

By rounding these numbers in a suitable way, write down an estimate of the probability that a home chosen at random has

a) a microwave   b) a washing machine   c) a dishwasher.

3 a) Copy and complete this table to show the larger score when 2 dice, one blue and one red are thrown.

b) Work out the probability of each of the possible outcomes, 1, 2, 3, 4, 5 and 6.

c) Show that the probabilities add up to 1.

d) The dice are thrown 360 times. How many times would you expect the outcome to be 1?

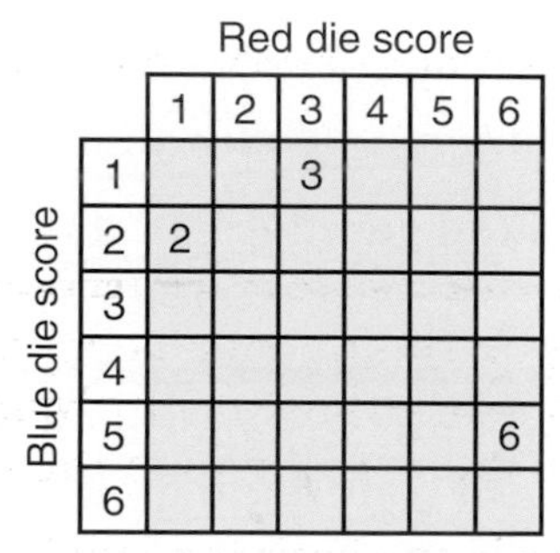

Red die score (columns), Blue die score (rows)

	1	2	3	4	5	6
1			3			
2	2					
3						
4						
5						6
6						

4 The carnation plants at a garden centre have lost their labels and become mixed up. The manager knows that 40% are red, 20% are white, 30% are pink and 10% are yellow.

Amy buys two of the carnation plants. What is the probability that

a) they are both red?

b) they are both white?

c) they are both the same colour?

d) they are different colours?

e) one is red and one is white?

f) one is red and one is not?

5 The probability that a new baby will be a boy is about 0.51.

a) Draw a tree diagram to show the probabilities for a woman who has two children.

b) What is the probability that she has two boys?

c) What is the probability that she has two girls?

d) What is the most likely combination?

6 On Marie's route to work she drives through a junction with traffic lights. The whole traffic light sequence at this junction takes 2 minutes, and it runs as shown in the diagram.

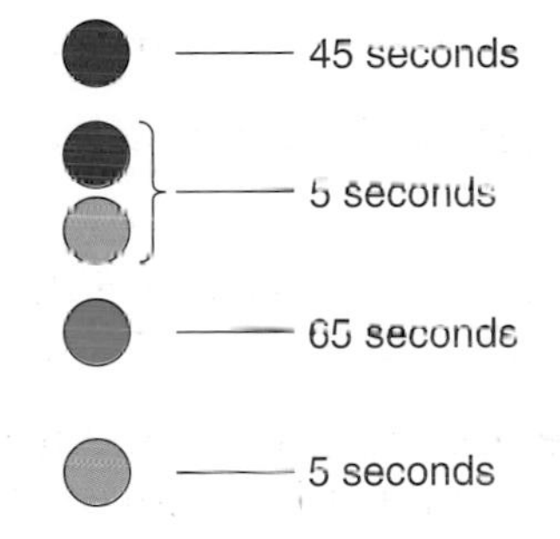

What is the probability that on any journey Marie arrives at the junction when the lights are

a) green? b) red? c) red and amber?

One day, Marie drives through the traffic lights 3 times. What is the probability that she arrives

d) at a green light every time? e) at a red light every time?

Look up Gregor Mendel in an encyclopedia or on the Internet. Read about his experiments involving sweet-peas, and write a short description of them, explaining the importance of his findings.

Mendel's work is the basis for much of modern genetics.

Answers

Chapter 1: Types of number

Page 3: Squares and square roots

1 a) 49 b) 100 c) 225 d) 40 000
2 a) 6 b) 9 c) 12 d) 20
3 a) 64 b) 125 c) 1000 d) 343
4 a) 4 b) 6 c) 5 d) 10
5 a) 36 b) 100
6 a) 10 b) 14×14 c) 4
7 a) 64 b) 512

Page 5: Multiples

1 a) 12 b) 14
2 a) 10 b) 15 c) 20
3 a) 6, 12, 18, 24, 30, 36, 42, 48, 54, 60
b) Multiples of 6
4 a) Grids of 9×2, 6×3, 3×6, 1×18 and 18×1
b) 1, 2, 3, 6, 9, 18
5 a) 1, 3, 5, 15 b) 1, 2, 4, 8
c) 1, 2, 4, 5, 10, 20 d) 1, 7
e) 1, 2, 4, 8, 16 f) 1, 5, 7, 35
g) 1, 3, 5, 9, 15, 45 h) 1, 2, 3, 4, 6, 9, 12, 18, 36
i) 1, 2, 3, 4, 5, 6, 10, 12, 15, 20, 30, 60
j) 1, 2, 3, 4, 6, 8, 9, 12, 18, 24, 36, 72
k) 1, 2, 4, 5, 10, 20, 25, 50, 100
l) 1, 2, 3, 4, 6, 8, 9, 12, 16, 18, 24, 36, 48, 72, 144
6 23, 29, 31, 37

Page 7: Prime factorisation

1 a) 2×7 b) 3×5 c) $2 \times 2 \times 7$
d) $2 \times 2 \times 3 \times 3$ e) $2 \times 3 \times 5$
f) $3 \times 3 \times 3$ g) $2 \times 3 \times 3 \times 5$
h) $2 \times 3 \times 3 \times 7$ i) $2 \times 3 \times 5 \times 5$
j) $2 \times 3 \times 5 \times 7$ k) $7 \times 7 \times 11$
l) $2 \times 2 \times 5 \times 7 \times 11$
2 a) 2 b) 3 c) 6 d) 4
e) 5 f) 1 g) 9 h) 2
i) 7 j) 11 k) 9 l) 14
m) 6 n) 5 o) 8 p) 20
3 a) 20 b) 30 c) 8 d) 36
e) 30 f) 21 g) 54 h) 16
i) 70 j) 40 k) 60 l) 90
m) 12 n) 30 o) 72 p) 30
4 a) 8 b) 5 c) 7
5 a) every 6 weeks b) every 4 weeks
c) every 12 weeks

Pages 8–9: Finishing off

1 a) 81 b) 10 c) 216 d) 3
e) 7 f) 1728 g) 625 h) 100
2 8
3 a) 27 b) 64 c) 125
4 a) 32 b) 60 c) No
5 a) 5, 10, 15, 20, 25, 30, 35, 40, 45, 50
b) Yes c) No
d) a multiple of 5 has a final digit of 0 or 5.
6 a) Grids of 8×3, 6×4, 4×6, 2×12, 12×2, 1×24 and 24×1
b) 1, 2, 3, 4, 6, 8, 12, 24
7 a), c) and e)
8 a) 3, 9, 27 b) 36 c) 3, 9, 27, 29
d) 9, 36 e) 27 f) 3, 29
9 a) 12 b) 24 and 25 May c) 2
10 a) $2 \times 3 \times 3$ b) $2 \times 2 \times 2 \times 2 \times 3$
c) $2 \times 2 \times 5 \times 5$ d) $2 \times 2 \times 2 \times 3 \times 5$
11 a) 5 b) 4 c) 10 d) 12
12 a) 12 b) 24 c) 100 d) 90

Chapter 2: Using numbers

Page 11: Brackets

1 a) 19 b) 27 c) 36
d) 3 e) 8 f) 11
2 a) 17 b) 23 c) 10 d) 10
3 a) 15 b) 10 c) 22 d) 3
4 a) $4 + 5 - 3 = 6$ b) $(7 + 5) \div 3 = 4$
c) $(2 \times 6) + 3 = 15$ d) $(5 + 2) \times 2 = 14$
e) $3 \times (5 - 1) = 12$ f) $(3 \times 2) + 2 = 8$
g) $(1 \times 6) - 2 = 4$ h) $(6 \times 6) \div 4 = 9$
i) $8 + 2 - 3 = 7$ j) $(10 + 8) \div 2 = 9$

Page 13: Using the number line

1 a) 4 b) –4 c) –6 d) 1
e) 7 f) –7 g) 3 h) –3
i) 3 j) –3 k) 7 l) –7
2 a) –1 b) 100 c) –20 d) –1000
3 London 4 °C; Manchester –1 °C; Leeds –3 °C; Inverness –6 °C; Accra 30 °C.
4 –£8 (or £8 DR)
5 a) £13 000 profit b) £5000 loss
c) £11 000 loss
6 a) –2 °C b) –1 °C c) –4 °C

Page 15: Using negative numbers

1 a) 7 b) 5 c) 4 d) 8
e) 4 f) −3 g) −5 h) −4
i) −8 j) 2 k) 3 l) 8

2 a) 20 b) −20 c) −20 d) 20
e) −21 f) −32 g) 8 h) 56
i) 60 j) −30 j) −24 l) −125

3 a) 6 b) −6 c) −6
d) 6 e) −3 f) −5
g) 4 h) 7 i) −4
j) −8 k) 6 l) 11

4 a) 3 b) 3 c) −2
d) −4 e) 4 f) 6
g) −9 h) −3 i) −16

Pages 16–17: Finishing off

1 a) 10 b) 39 c) 32 d) 1
e) 20 f) 4 g) 16 h) 10

2 a) −5 b) 4 c) −22
d) −3 e) −11 f) 0

3 a) 1 °C b) −4 °C

4 a) up 1 floor b) down 1 floor
c) down 3 floors d) down 2 floors
e) up 4 floors f) down 3 floors

5 £60; £93; £21; −£107 (or £107 DR); −£92 (or £92 DR); £458

6 a) 1215 b) 1705

7 a) 2 b) −5 c) −1
d) −15 e) 8 f) 9
g) −4 h) −9 i) 15
j) −4 k) 4 l) 1

Chapter 3: Fractions

Page 19: Equivalent fractions

1 a) $\frac{13}{16}$ b) $\frac{12}{16} = \frac{3}{4}$ c) $\frac{25}{32}$

2 a) 2 b) 2 c) 4 d) 3
e) 3 f) 3 g) 12 h) 3
i) 4 j) 14 k) 5 l) 30

3 $\frac{2}{4}$, $\frac{3}{6}$, $\frac{4}{8}$ (and many other possibilities)

4 a) $\frac{1}{2}$ b) $\frac{3}{4}$ c) $\frac{3}{8}$ d) $\frac{1}{3}$
e) $\frac{2}{3}$ f) $\frac{1}{5}$ g) $\frac{3}{4}$ h) $\frac{1}{5}$
i) $\frac{1}{4}$ j) $\frac{5}{7}$ k) $\frac{2}{3}$ l) $\frac{2}{5}$

5 a) $\frac{7}{8}$ b) $\frac{1}{2}$ c) $\frac{5}{8}$ d) $\frac{13}{16}$
e) $\frac{1}{2}$ f) $\frac{7}{10}$ g) $\frac{3}{2}\left(=1\frac{1}{2}\right)$ h) $\frac{7}{8}$

6 a) $\frac{3}{8}$" b) $\frac{5}{8}$" c) $\frac{11}{16}$" d) $\frac{7}{16}$"

Page 21: Improper fractions and mixed numbers

1 a) $4\frac{1}{2}$ b) $1\frac{5}{8}$ c) $2\frac{2}{5}$
d) $3\frac{2}{3}$ e) $3\frac{3}{4}$ f) $2\frac{1}{6}$

2 $4\frac{1}{2}$

3 $3\frac{1}{4}$

4 a) $\frac{7}{2}$ b) $\frac{35}{8}$ c) $\frac{23}{16}$
d) $\frac{11}{4}$ e) $\frac{16}{3}$ f) $\frac{59}{16}$

5 5

6 11

7 a) $1\frac{1}{4}$ miles b) $2\frac{3}{4}$ miles c) 2 miles d) $5\frac{3}{4}$ miles

8 a) $3\frac{1}{2}$ b) $5\frac{1}{8}$ c) $1\frac{5}{16}$ d) $1\frac{9}{16}$
e) $1\frac{3}{16}$ f) $6\frac{1}{8}$ g) $2\frac{7}{8}$ h) $4\frac{1}{6}$
i) $2\frac{1}{3}$ j) $7\frac{1}{2}$ k) $1\frac{1}{2}$ l) $7\frac{17}{20}$

Page 23: Multiplying fractions

1 a) $\frac{1}{6}$ b) $\frac{3}{16}$ c) $\frac{3}{20}$ d) $\frac{5}{8}$
e) $\frac{1}{4}$ f) $\frac{3}{5}$ g) $\frac{15}{64}$ h) $7\frac{1}{2}$

2 a) $3\frac{1}{2}$ b) $4\frac{1}{2}$ c) $2\frac{2}{3}$ d) $1\frac{3}{5}$
e) $12\frac{1}{2}$ f) $9\frac{1}{3}$ g) $3\frac{3}{4}$ h) $7\frac{1}{2}$

3 a) $3\frac{1}{4}$ b) $3\frac{3}{8}$ c) $\frac{7}{8}$ d) $1\frac{3}{4}$
e) $1\frac{3}{4}$ f) 4 g) $3\frac{3}{4}$ h) $7\frac{7}{8}$
i) $5\frac{1}{2}$ j) 20 k) $4\frac{13}{16}$ l) 12

4 $27\frac{1}{2}$

5 £15 6 £1.44

7 a) 54 b) 20 8 $36\frac{1}{4}$ lbs

9 a)(i) $\frac{1}{5}$ (ii) 5 1 b) 1 c) No

Page 25: Dividing fractions

1 a) $1\frac{1}{4}$ b) $\frac{1}{10}$ c) $\frac{2}{5}$ d) $\frac{1}{9}$
e) $\frac{1}{2}$ f) $\frac{3}{4}$ g) $\frac{5}{16}$ h) $\frac{1}{4}$

2 a) 12 b) 6 c) 16 d) 30
e) $\frac{7}{8}$ f) $1\frac{4}{5}$ g) 10 h) $1\frac{3}{4}$
i) $\frac{3}{4}$ j) 7 k) $2\frac{1}{2}$ l) $3\frac{3}{5}$

3 a) 10 b) 20 4 40

5 28 6 a) 39 b) 26 c) 18

Pages 26–27: Finishing off

1 a) $\frac{5}{8}$ b) $\frac{3}{4}$ c) $\frac{7}{8}$ d) $\frac{1}{2}$
2 $\frac{1}{2}$
3 18
4 21
5 a) 2 miles b) $3\frac{1}{8}$ miles c) $5\frac{1}{8}$ miles
6 a) $\frac{11}{3}$, 4, $4\frac{3}{16}$, $\frac{17}{4}$ b) $\frac{11}{4}$, $2\frac{13}{16}$, $\frac{23}{8}$, 3
7 a) (i) $\frac{1}{8}$ (ii) 4 (iii) $\frac{2}{3}$ b) 1
8 a) 16 b) 12
9 a) £23.33 b) No (dearer by £2.09)
10 a) $4\frac{5}{8}$ b) $\frac{13}{16}$ c) $4\frac{1}{2}$ d) $9\frac{1}{2}$
e) $\frac{2}{3}$ f) $10\frac{1}{8}$ g) $5\frac{25}{32}$ h) 5
i) $\frac{3}{4}$ j) 11 k) $\frac{1}{4}$ l) $3\frac{3}{4}$
11 a) $\frac{5}{8}$ b) $\frac{1}{10}$ c) 8
d) 5 e) $10\frac{5}{12}$ f) $4\frac{4}{9}$

Chapter 4: Decimals

Page 29: Tenths and hundredths

1 a) $\frac{3}{10}$ b) $\frac{43}{100}$ c) $9\frac{2}{10}$ ($=9\frac{1}{5}$)
d) $7\frac{9}{100}$
2 a) 0.7 b) 2.3 c) 0.47 d) 3.07
3 a) 36.4 °C b) 36.7 °C c) 35.9 °C
4 a) 1.66 m b) 1.73 m
5 a) 13.9 b) 23.56 c) 5.908
d) 5.5 e) 4.1 f) 17.04
6 £2.39
7 a) 1.08 m b) 1.78 m
8 0.009, 0.08, 0.1, 0.8, 0.81

Page 31: Multiplying and dividing

1 a) 84 b) 390 c) 5.6 d) 0.073
e) 720 f) 1750 g) 24 h) 30
i) 600 j) 0.05 k) 437.5 l) 0.082
2 a) 4.35 b) 4.8 c) 3 d) 3.408
e) 4.2 f) 12.48 g) 1.04 h) 6.25
3 a) 20 b) 14 c) 44 d) 6.25
e) 8 f) 3.6 g) $4.\dot{6}$ h) 0.3
4 Malta 81 °F Cyprus 84 °F Tunisia 97 °F
5 a) 16 litres b) 41 litres
c) in a) £11.40; in b) £29.30
6 a) £8000 b) £416 000
7 £1 600 000
8 Total is £528.00

Page 33: Fractions to decimals

1 a) 0.6 b) 0.375 c) 0.45 d) 1.25
e) 0.32 f) 0.0625 g) 0.175 h) 1.22
2 a) $1.\dot{3}$ b) $0.\dot{1}$ c) $0.8\dot{3}$ d) $0.41\dot{6}$
3 0.07, 0.1, $\frac{3}{5}$, 0.66, $\frac{2}{3}$

Pages 34–35: Finishing off

1 a) 1.65 m b) 1.78 m
2 a) $\frac{7}{10}$ b) $5\frac{4}{10}$ ($= 5\frac{2}{5}$) c) $1\frac{83}{100}$ d) $6\frac{371}{1000}$
3 a) 2.3 b) 3.17 c) 6.07 d) 0.141
4 2, 2.04, $2\frac{3}{10}$, $\frac{7}{3}$, 2.35 **5** 12.7 cm
6 a) 0.6 °C fall b) 0.4 °C fall c) 0.1 °C rise
7 a) £28 (Pat's Printery) b) £66.25 (Fast Print)
c) over 300 copies
8 a) 10 b) 15
9 a) 76.8 cm b) 22.6 cm
10 a) £1 500 000 b) £13
11 a) 28.09 b) 2.9 c) 46.24 d) 3.9
12 a) 0.8 b) 0.625 c) 0.35
d) $1.1\dot{6}$ e) 0.3125 f) $0.\dot{2}8571\dot{4}$

Chapter 5: Percentages

Page 37: 25%, 50% and 75%

1 a) (i) B (ii) A (iii) C
b) 10% approximately
2 C
3 57%
4 a) Knight 20%; Diamond 44%; Pathways 36%
b) Knight $\frac{1}{5}$; Diamond $\frac{11}{25}$; Pathways $\frac{9}{25}$
c) Knight 0.2; Diamond 0.44; Pathways 0.36

Page 39: Percentage calculations

1 a) 240 b) 150 c) 315.5
d) 87 e) 25 f) 27.9
2 75
3 a) £1500 b) £187.50
4 a) £8240 b) £8569.60
5 6089
6 £396
7 a) £1575 b) £1638 c) unchanged
8 a) 105 000 b) 110 250
c) Population after 10 years d) will exceed

Page 41: Further percentage problems

1 a) £160 b) £127.66 c) £59.57
2 a) £12 500 b) £13 520
3 a) West
b) NW 14 182, NE 14 947, SW 14 554, SE 14 785
c) + 2.3%

4 a) £280 000 b) £189 000
c) £378 000, £408 240

Page 43: Fractions to percentages

1 a) 80% b) 36% c) 12.5%
d) 42.5% e) 30% f) 66.6%
2 a) 47.5% b) 31.25% c) 21.25%
3 a) 125 b) 56% c) 52%
4 a) 12.5% b) 7.5%
5 a) 20% b) 52.5% c) 25%
6 a) May + 15%; June – 4%; July + 5%; August + 3.5%
b) 4.3%

Page 45: Making comparisons

1 A
2 a) A 4.5%, B 4.9%, C 4.6%
b) A
c) Breakdown or older machine
3 a) Oakington b) Waterbeach
c) Waterbeach
4 a) Northhill 79.5%; Heartland 54.9%; Southdown 78.8%
b) Heartland's success rate is much lower.

Pages 46–47: Finishing off

1 a) 40% b) 30% c) 30%
2 a) 16% b) 80% c) 4%
3 300 ml
4 Sporting Life £49.50
5 a) £7100 b) £8547.50 c) 20.4%
6 £74 million
7 House £66 000; car £10 300; computer £1710; calculator £10
8 a) £95 400 b) £107 200 c) £181 100
9 a) 80% b) Parkside (71.7%)
c) Eastwood (62.9%)
10 64

Chapter 6: Units

Page 49: Length, mass, capacity and time

1 a) 200 cm b) 3000 m c) 4 cm d) 72 inches
e) 1300 mm f) 54 feet
2 a) 1200 g b) 140 pounds
c) 0.25 kg d) 8 ounces
3 a) 1500 ml b) $1\frac{1}{4}$ gallons
c) 0.7 litres d) 4 pints
4 a) 2 h b) 180 s
c) $1\frac{1}{2}$ minutes d) 3600 s
5 a) 30 minutes b) 10 minutes
c) Tennis highlights d) 35 minutes
6 a) Emily by 10 ml
b) Ask your teacher to check your answer

7 Two 500 g or four 250 g or one 500 g and two 250 g
8 Yes (8 cm spare)

Page 51: Imperial and metric units

1 a) 270 l b) 30 or 30.5 cm
c) 50 miles d) 20 kg
e) 6 or 7 gallons f) 6 feet 6 inches
g) 5 ounces h) $\frac{1}{4}$ inch
2 a) 208 km b) 232 km
3 a) 61.8 kg b) 72.7 kg
4 a) 22 lbs b) 8 or 9 pints
c) 117 or 118 inches d) 7 ounces
5 a) Yes b) 5 feet 7 inches
c) 11 gallons d) 22.5 mpg
6 a) 340 g, 11 or 12 ounces
b) 450 g, 1 lb or 16 oz

Pages 52–53: Finishing off

1 a) 24 mm b) 0.75 kg
c) 880 yards d) 13 600 m
e) 9 stone 9 lb f) 150 cl
2 a) 10.7 cm b) 107 mm
3 a) 6 m b) £27
c) 59 inches
4 a) 2000 kg b) 2 tonnes
c) 5.5 tonnes d) £30 609
5 a) 5 b) 25
6 1830
7 a) 15 cm b) 8 gallons
c) 96 km d) 121 pounds
e) 120 g f) 1.14 litres
8 a) 3.18 kg b) 20 cm
c) 2.28 litres
9 675
10 a) 64 km per hour b) 68.8 mph
11 a) 90 000 m b) 25 m
c) 25 m/s

Chapter 7: Ratio and proportion

Page 55: Ratio

1 a) 2:1 b) 3:4 c) 3:2 d) 1:3
e) 2:5 f) 2:3 g) 5:2 h) 8:3
i) 3:4:5 j) 2:3:4 k) 2:4:5 l) 4:5:9
2 a) 1:8 b) 3:2 c) 3:8
d) 4:1 e) 1:5 f) 3:20
3 a) 100 units b) 6 units
4 a) 60 ml b) 1:50
5 a) £360, £240 b) £120, £420
c) £500, £750, £1000 d) £1080, £1600, £320
6 a) Rosie £44, Natasha £66
b) Rosie £39, Natasha £52
7 a) 400 g b) Copper 350 g, zinc 150 g

8 a) Charlotte £150, Thomas £250
b) £640
c) Charlotte £234, Thomas £390

Page 57: Unitary method

1 a) 6 pack b) 6 pack c) 500 g pack
2 a) 96.5 g b) 579 g
3 a) £4.80 b) £182.40 c) 42
4 a) 255 miles b) 14 c) No
5 a) 0.69 milligrams b) 1.725 milligrams
6 a) 100 g flour, 40 g sugar, 40 g butter
b) 500 g c) 10
7 54 words per minute

Page 59: Changing money

1 a) 6800 b) 16 150 c) £109 d) £47.06
e) £16.50
2 a) 795 b) £306 c) £69.81 £245.28 d) £49
3 a)

Rand	£		Rand	£
10	0.74		150	11.03
20	1.47		200	14.71
50	3.68		500	36.76
100	7.35		1000	73.53

b) (i) 70 = 50 + 20
(ii) 450 = 500 – 50
c) (i) 70 rand = £5.15
(ii) 450 rand = £33.08

Page 61: Distance, speed and time

1 a) 100 km b) 105 km c) 1980 km d) 20 km
2 a) 56 km/h b) 64.8 km/h
c) 84 km/h
3 a) $2\frac{1}{2}$ h b) $3\frac{3}{4}$ h c) 1 h 40 min
d) 2 h 6 m 36 s
4 a) 1240 b) 80 km/h
c) 64 km/h d) 35 minutes

Pages 62–63: Finishing off

1 a) 5:8 b) 8:5 c) 2:4:1
d) 9:40 e) 1:20 000 f) 10:1
2 a) Andrew £5000; Harriet £3000;
Justin £7200; Mel £4800
b) 5:3 c) 36% d) $\frac{6}{25}$
3 350 g jar
4 a) sand 3 m^3; gravel 6 m^3; cement 1.5 m^3
b) 8.75 m^3
5 a) 174 b) £18.99 c) £129.11
6 a) 6 g b) 30.8 g
c) 35 g iron and 15 g oxygen

7 a) 22.5 km b) 720 km/h c) 45 minutes
8 a) 75 km/h b) 15 minutes
c) 80 km/h d) 5 hours
e) 63 km/h

Chapter 8: Money

Page 65: Simple and compound interest

1 a) £30 b) £330 c) £937.50 d) £17
2 a) £40.80 b) £382.03 c) £238.82 d) £2991.94
3 a) Rosie, Southern Building Society, £72
Scott, Western Building Society, £125
b) £12.50 c) £8671.77
4 a) £168.75 b) £84.38 c) 4.50%
d) Mohan £33.75, Alka £16.87

Page 67: Income and tax

1 a) £185 b) £160 c) £243.60 d) £286
2 a) July £1400, August £1436, September £1382
b) £2360
3 a) £8600 b) £2150 c) £10 350
4 a) £13 292.50 b) £367.88
5 £14 760
6 a) £21 262.40 b) Worse off by £62.40

Page 69: Bills

1 a) £79.86 b) £94.66
2 a) £50.46 b) £62.96 c) 500
3 a) £105.96 b) Yes
4 a) £26 b) £34 c) £32 d) £32

Page 71: Hidden extras

1 a) £927, £178 b) £1426.64, £231.64
2 a) £950.40 b) £1440
c) £330 d) £546.75
3 a) £152.75 b) £434.75
4 a) A by £9.17 b) B by £22.33
5 £6.60

Page 73: Profit and loss

1 a) Profit 52% b) Profit 37.5%
c) Loss 30% d) Loss 14.4%
2 a) £72 b) £86.80 c) £68 d) £14.25
3 a) £180 b) 30%
4 a) Jackets £6, Trousers £4
b) Jackets 12%, Trousers 20% c) £67.50
5 a) £330 b) 49.3%
6 a) £115 b) £25 c) 27.8%

Pages 74–75: Finishing off

1 a) £440 b) £20
2 £1071.91
3 a) £207.20 b) £240.80 c) 44
4 a) £21 600 b) £150
5 a) £12 080 b) £180
6 a) £210 b) £437.50
7 £91.78
8 a) £1294.70 b) £299.70
9 a) £46.90 c) £308.20
10 a) 35% b) £15 c) £360
11 a) 40% b) £18 c) £880
d) £1240.70 e) 41.0%

Chapter 9: Indices and standard form

Page 77: Powers

1 a) 36 b) 125 c) 81 d) 64
e) 256 f) 243 g) 343 h) 10 000
i) 121 j) 5 k) 512 l) 3.375
2 a) $\frac{1}{16}$ b) $\frac{1}{1000}$ c) $\frac{1}{25}$ d) $\frac{1}{8}$
e) $\frac{1}{27}$ f) $\frac{1}{36}$ g) $\frac{1}{100}$ h) $\frac{1}{81}$
i) $\frac{1}{4}$ j) $\frac{1}{16}$ k) $\frac{1}{216}$ l) $\frac{1}{10\,000}$
3 a) 49 b) $\frac{1}{9}$ c) $\frac{1}{10}$ d) 64
e) $\frac{1}{81}$ f) 1 g) 32 h) $\frac{1}{125}$
i) 1 000 000 j) 9 k) $\frac{1}{5}$
l) 1 m) 216 n) $\frac{1}{64}$ o) 4
p) 625
4 a) 10^6 b) 10^{-2} c) 10^2
d) 10^{-3} e) 10^{-1} f) 10^3

Page 79: Using index form

1 a) 5^6 b) 2^{11} c) 6^4 d) 10^3
e) 2^7 f) 10^6 g) 3^8 h) 4^2
i) 10^4 j) 5^{-2} k) 3^3 l) 2
m) 4^2 n) 3^{-2} o) 2^0 p) 10^3
2 a) 520 b) 8.3 c) 0.69 d) 2300
e) 0.47 f) 6400 g) 7 h) 5.28
i) 9345 j) 8 k) 0.0912 l) 573 000
3 a) 280 000 000 000 b) 57 200 000 000 000
4 Yes. Their answers have been rounded because calculators have limited capacity and display space. Tom gets 800 000 100 000 003

Page 81: Standard form

1 a) 600 b) 30 000
c) 0.007 d) 0.000 04
e) 4 500 000 f) 0.0054
g) 9400 h) 0.000 875
i) 0.016 j) 2 750 000
k) 0.000 083 l) 10 500
m) 7300 n) 0.000 000 008
o) 0.4 p) 82 500 000 000
2 a) 4×10^3 b) 8×10^5
c) 3×10^{-3} d) 9×10^{-4}
e) 2.6×10^4 f) 2.5×10^{-2}
g) 7.5×10^6 h) 3.7×10^{-5}
i) 8.1×10^2 j) 5.43×10^{-3}
k) 9.3×10^{-1} l) 6.4×10^4
m) 1.6×10^{-2} n) 1.47×10^8
o) 5.07×10^{-1} p) 9.04×10^3
3 a) 0.005 b) 4 600 000 000 000
c) 942 000 000 d) 0.000 0075
4 Pluto (smallest), Mercury, Mars, Venus, Earth, Neptune, Uranus, Saturn, Jupiter (largest)

Page 83: Calculations using standard form

1 a) 8.4×10^4 b) 6.8×10^6
c) 1.5×10^{12} d) 1.6×10^7
e) 1.5×10^9 f) 7.02×10^{-2}
g) 1.44×10^{-17} h) 1.6×10^{15}
2 Mercury 7.86×10^{22} kg Venus 4.88×10^{24} kg
Mars 2.05×10^{23} kg Jupiter 2.04×10^{27} kg
3 1.8×10^{-16} J
4 a) Europe b) 9 million
c) 30 million km^2
5 a) 2.06×10^{10} m^3 b) 3596 km/m^3

Pages 84–85: Finishing off

1 a) 64 b) 1000 c) 1 d) 128
e) $\frac{1}{16}$ f) 144 g) $\frac{1}{125}$ h) 10
i) $\frac{1}{7}$ j) 1 k) $\frac{1}{32}$ l) 1024
2 a) 2^8 b) 7^3 c) 5^6 d) 4^8
e) 6^3 f) 4^4 g) 3^4 h) 3
3 a) 6 400 000 b) 0.0002
c) 0.000 007 5 d) 13 600
4 a) 2×10^5 b) 1.5×10^{-3}
c) 8.3×10^4 d) 2.6×10^{-5}
5 a) 8.5×10^5 b) 2.4×10^{-2}
c) 3.88×10^{22} d) 4×10^7
e) 2.4×10^6 f) 1.75×10^{16}
g) 2.025×10^{15} h) 4×10^6
6 4.02×10^7 (or 40 200 000)
7 a) 31 536 000 b) 9.46×10^{15}
c) 9.46×10^{12} d) 3.97×10^{13}

Chapter 10: Approximations

Page 87: Decimal places

1 a) 3.1 b) 3.14 c) 3.142
2 a) 4.5 b) 4.47 c) 4.472
3 a) 1 b) 2
4 a) Approximately 5.38 to the nearest hundredth; 5.4 to the nearest tenth
b) Approximately 8.72 to the nearest hundredth; 8.7 to the nearest tenth
5 a) 0.167 b) 0.083 c) 0.667 d) 0.429
6 5.52
7 a) 5.8 cm, 2.3 cm b) 13.3 cm^2
8 7.86
9 a) 45.8 m b) 167.3 m^2

Page 89: Significant figures

1 a) 77 000 b) 42.2 c) 5400 d) 758 000
e) 4000 f) 62.0 g) 6.7 h) 50 000
i) 33.8 j) 0.005
2 a) Approximately 37.73 to 4 significant figures; 37.7 to 3 significant figures
b) Approximately 0.2156 to 4 significant figures; 0.216 to 3 significant figures
c) Approximately 1.444 to 4 significant figures; 1.44 to 3 significant figures
3 a) 38 b) 0.22 c) 1.4
4 26 000
5 a) 370 m b) 8300 m^2
6 a) £90 000 b) £360 000
7 a)

Day	Thursday	Friday	Saturday
Tickets sold	257	319	348
Income	£1090	£1360	£1480

b) £3900

Page 91: Rough checks

1 a) 280 cm^2 b) 400 cm^2 c) 48 cm^2
2 a) 200 b) £2000
3 a) 200 b) 25 c) 180 d) 200
e) 250 f) 600 g) 300 h) 360
4 a) £6000 b) £3000 c) £2000 d) £1000
5 a) 300 b) 25 c) 200 d) 40
e) 180 f) 250 g) 150 h) 60
6 a) 24 m; 48 m^2 b) 12 m; 12 m^2
c) 25 m; 38 m^2
7 a) £12 b) £26 c) £45

Page 93: Errors

1 a) 2 weeks, 20% b) £200, 50%
c) £3200, 80%
2 a) 18.5 cm, 17.5 cm b) 23.5 cm, 22.5 cm
c) 606.5 cm^2, 577.5 cm^2
3 a) 247.5 miles b) 157.5 miles
4 Smallest 344.375 m^3, largest 569.625 m^3

Pages 94–95: Finishing off

1 a) 5.3 b) 5.29 c) 5.292
2 a) 3.33 b) 3.88 c) 4.64
3 a) 15.9 m^2 b) 32.6 m^2 c) 102.1 m^2
4 a) 8000 b) 765 000 c) 14.8 d) 0.065
5 a) Rutherford £2800, Churchill £1700 (or £1800) Hillary £1200, Sargent £2400
b) £8100 or £8200
6 a) £8 b) £16 c) £15
7 £60
8 a) 300 b) 12 c) 100
d) 400 e) 8 f) 18 000
g) 210 h) 600 i) 740
9 £400
10 a) Smallest 26 m, Largest 30 m
b) Smallest 41.25 m^2, largest 55.25 m^2
11 a) £840 approx. b) £175 approx.
c) £560 approx.

Chapter 11: Algebra 1

Page 97: Sequences

1 a) 9, 11, 13; add 2
b) 13, 10, 7; subtract 3
c) 10 000, 100 000, 1 000 000; multiply by 10
d) 23, 27, 31; add 4
e) 30, 25, 20; subtract 5
f) 4, 2, 1; divide by 2
g) 4.2, 5, 5.8; add 0.8
h) 3, 0, –3; subtract 3
i) 81, 243, 729; multiply by 3
j) 5, –6, 7; count and alternate signs
2 a) EVENS 26-32, EVENS 34-40, EVENS 42-48
b) ODDS 49-63, ODDS 65-79, ODDS 81-95
c) ODDS 65-79
3 2000, 2004, 2008
4 14 Sept, 5 Oct, 26 Oct, 16 Nov, 7 Dec
5

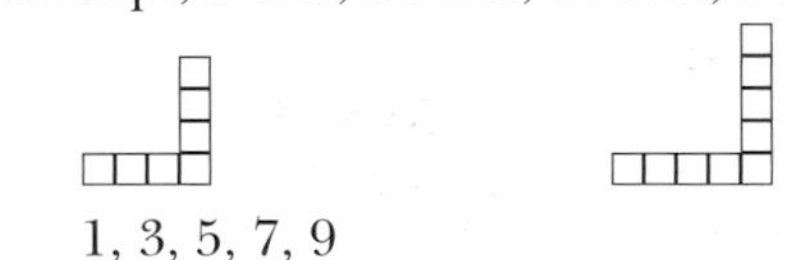

1, 3, 5, 7, 9
6
(i) 1, 2, 3, 4, 5 (ii) 4, 6, 8, 10, 12
7

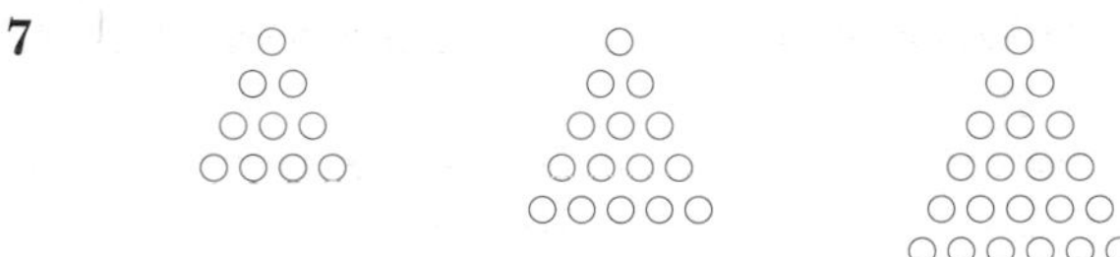

1, 3, 6, 10, 15, 21
8 a) 0945, 1000 b) 1230, 1400 c) 0915, 1000

Page 99: Finding the *n*th term

1 a) $4n + 1$ b) $3n - 1$ c) $2n + 5$ d) $6n$
e) $10 - 2n$ f) $23 - 3n$ g) $10 + n$ h) $1 + 0.5n$

i) $101 - n$ j) $9 + 7n$ k) $-8n$ l) $7 - 3n$
2 a) 128 b) 124 c) 52
3 a) £$(120 + 30n)$
4 a) $(10 + 20n)$ pence b) £4.10
5 a) $2n$ b) $2n - 1$

Page 101: Simplifying expressions

1 $2n + 1$
2 a) $2n + 1$ b) $4n + 1$ c) $5n + 2$ d) $3n + 1$
3 a) $6n$ b) $4n$ c) $4n$ d) $2n$
e) $6x$ f) $-5x$ g) $-y$ h) 0
4 a) $2n + 7$ b) $8n - 1$ c) $7n + 3$ d) $3n + 2$
e) $11 + a$ f) $3b - 3$ g) $7c + d$ h) $5f - e$
i) $11a + 5b$ j) $8p - 5q$ k) $7x + y$ l) $4r - 3s$
m) $2k + 1$ n) $4e - 3f + 7$

Page 103: Using brackets

1 $4n + 2$
2 a) $3n + 7$ b) $3n + 5$
c) $5n + 2$ d) $3n + 4$
3 a) $3n + 12$ b) $5n - 10$
c) $12 + 4n$ d) $6 - 6n$
e) $-2x - 8$ f) $8x + 4$
g) $15 - 10y$ h) $-21p + 3q$
i) $45x + 36y$ j) $-a + 5$
k) $-32h - 20$ l) $5x + 10y - 15z$
4 a) $3n + 2$ b) $3n + 7$
c) $17n - 5$ d) $15 - 12n$
e) $7x - 20$ f) $5p + 2q$
g) $13a + 14b$ h) $15c - 2d$
i) $19x - 25y$ j) $17r - 19s$
k) $a + 32b$ l) $3e - 14f + 13$
m) $7k - 19m$ n) $-35x$

Page 104–105: Finishing off

1 a) 15, 18; add 3 b) 22, 27; add 5
c) 26, 20; subtract 6 d) 3.9, 4.3; add 0.4
2 a) 2002, 2006, 2010 b) No
3 a) $4n - 1$ b) $3n + 11$
c) $8 - n$ d) $0.75n + 1$
4 a) (i) 4 (ii) 7 (iii) 10
b) $3n + 1$
5 a) (i) 94 cm (ii) 88 cm (iii) 82 cm
b) $(100 - 6n)$ cm c) -2, only 4 cm of bar left
6 $4n + 4$
7 a) $2n + 3$ b) $3n + 6$
8 a) $9n$ b) $3n - 2$
c) $5n$ d) $3n + 2$
e) $10a + 7b$ f) $5c + 6d$
g) $8r - 3s + 2$ h) $8 - 4t - u$
i) $5x - 8y + 2z$ j) $6a - b - 5c$
9 a) $6n + 18$ b) $15n - 6$
c) $6x + 8y$ d) $5x + 20y + 10z$
e) $7a + 8$ f) $3 - 2b$

g) $8c - 3d$ h) $7p + 5q$
i) $22e + 14f$ j) $16y - 3x$
k) $6k - 11m$ l) $34a - b$
m) $26x + 4$ n) $32t - 4s - 23u$

Chapter 12: Algebra 2

Page 107: Using powers

1 a) a^2 b) a^3 c) $5a^2$ d) $3p^2$
e) $4q^2$ f) $6r^2$ g) $2t^2$ h) $30c^2$
i) $16d^2$ j) e^3 k) $12f^3$ l) x^5
m) $36h^2$ n) $-15p^2$ o) $14q^3$ p) $4t^2$
q) $-10a^2$ r) $10d^4$ s) $4e^4$ t) $9f^3$
u) $8a^3$ v) a^6 w) 1 x) t^{13}
2 a) a^3 b) a c) $3a$ d) 2
e) $\frac{x}{4}$ f) $\frac{3}{y}$ g) $2p$ h) $3q$
i) $\frac{1}{2}$ j) $\frac{c}{4}$ k) $\frac{2}{d}$ l) $\frac{3e^2}{4}$
m) $\frac{3}{2}$ n) $-\frac{3}{5k}$ o) $\frac{2}{5m^2}$ p) -1
q) $\frac{3}{2b}$ r) $\frac{5}{a}$ s) $\frac{p}{2}$ t) y^2
u) $-\frac{3}{4e}$ v) r^5 w) x^4 x) $\frac{1}{a}$
3 a) 45 b) 22 c) 50 d) 100
4 a) -12 b) 80 c) -64 d) -1728
5 a) 2 b) -3 c) 2 d) 3
6 a) $2n$ b) $4n^2$ c) $8n^3$

Page 109: Multiplying out brackets

1 a) $n^2 + 6n$ b) $n^2 - 2n$
c) $n^2 + n$ d) $n^2 - n$
e) $2n^2 + 10n$ f) $-8n - 4$
g) $2x^2 - 3x$ h) $-7p^2 + p$
i) $15a^2 + 6a$ j) $-2c^2 - 5c$
k) $12k + 20k^2$ l) $25x^3 - 5x$
m) $-a^3 + 4a^2$ n) $-4b^4 - 28b$
o) $3x^4 - 18x^2$
2 a) $n^2 + 6n + 8$ b) $n^2 + 7n + 6$
c) $n^2 + 2n - 15$ d) $n^2 + 3n - 28$
e) $n^2 - 10n + 16$ f) $2n^2 + 11n + 12$
g) $3x^2 + 19x + 6$ h) $4x^2 + 3x - 10$
i) $3x^2 - 5x - 28$ j) $5a^2 + 14a - 3$
k) $4b^2 - 11b + 6$ l) $6c^2 + 29c + 28$
m) $16d^2 - 2d - 3$ n) $6t^2 - 35t + 36$
o) $20y^2 - 9y - 77$ p) $1 + 5y + 6y^2$
q) $12 + 5s - 2s^2$ r) $14 - 19u - 3u^2$
s) $4b^2 - 5b - 6$ t) $-12d^2 + 31d - 20$
u) $25 - 9f^2$
3 $x^2 + 8x + 16$
4 a) $x^2 + 10x + 25$ b) $x^2 - 12x + 36$
c) $9x^2 + 6x + 1$ d) $16x^2 - 24x + 9$

5 a) $n^2 + 6n$ b) $13n^2 - 3n$
c) $10x^2 + 30x + 25$ d) $6p^2 + 5p$
e) $2x^2 + 11x - 6$ f) -9
g) $3c^2 - 2$ h) $-n + 10$
i) $49x^2 - 34x + 22$ j) $24x$

6 a) $n(n + 1)$ b) $(n + 3)n$
c) $(n + 1)(n + 2)$ d) $(n + 1)(n - 1)$
e) $n(n + 1)(n + 2)$ f) $n(n + 2) + (n + 1)$

Page 111: Factorising

1 a) $2(3n + 4)$ b) $5(n - 3)$
c) $4(5 + 2n)$ d) $3(2 - n)$
e) $2(x + 1)$ f) $6(2x + 5y)$
g) $4(3 - 8y)$ h) $9(4 - 3q)$
i) $10(2s + 5t)$ j) $16(3a - 5b)$
k) $15(3a + 5b)$ l) $6(16x + 9)$
m) $6(7m - 11n)$ n) $13(4p - 3q)$
o) $28(3c + 1)$ p) $11(2 + 7t)$
q) $12(12x - 5y)$ r) $17(2a - 3b)$

2 a) $4(x + 2y + 3)$ b) $5(m - 2n + 4)$
c) $3(4t - 2u + 1)$ d) $2(1 + 2a + 5b)$
e) $6(4p - 5q - 7)$ f) $12(3x + 6y - 11z)$

3 a) $n(n + 5)$ b) $n(8n - 3)$
c) $2n(n + 3)$ d) $2a(3a - 2)$
e) $y(5 - 2y)$ f) $3x(5 + x)$
g) $4x(5x + 2)$ h) $p(3p - 1)$
i) $x(25 + 4x)$ j) $4a(7a - 3)$
k) $4(q^2 + 2)$ l) $3y(4y - 7)$
m) $t(2 + 5t^2)$ n) $2n(3 - 5n^2)$
o) $2x^2(1 - 2x)$ p) $16u^2(6u^2 - 5)$
q) $3d^2(9d + 8)$ r) $25t^3(3 + 5t)$

4 a) $x(x^2 + 5x + 7)$ b) $2x(2 - 3x - 4x^2)$
c) $3(a^2 + 5a + 3)$ d) $8p^2(4p^2 + 3p + 2)$

5 a) $3(x + 5)$ b) $5(n - 1)$
c) $5(x + 4)$ d) $2(x + 7)$
e) $2(3x + 1)$ f) $x(x + 15)$
g) $5(x - 4)$ h) $4(x + 8)$
i) $x(x + 8)$ j) $5(2x - 5)$

Page 113: Two unknowns

1 a) 150 pence b) $6x$ pence
c) nx pence

2 a) 120 b) $40m$ c) mp

3 a) $10mn$ b) $12ab$ c) $5xy$ d) $-6rs$
e) $35tu$ f) $18g^2h$ g) $32a^3b$ h) $3x^2y^2$
i) $14p^3q^2$ j) $-30r^4s^2$ k) $24c^2d^3$ l) $25a^2b^2$

4 a) £50 b) £$(\frac{200}{n})$ c) £$(\frac{x}{n})$

5 a) £2 b) £$(\frac{10}{p})$ c) £$(\frac{c}{p})$

6 a) $\frac{3a}{b}$ b) $\frac{2x}{5y}$ c) $\frac{4x}{3}$ d) $-\frac{p}{4}$
e) $\frac{r}{s}$ f) $\frac{2m}{3n}$ g) $4b$ h) $\frac{1}{5tu}$
i) $-\frac{2x^2}{5}$ j) $\frac{5}{2a^2b}$ k) $\frac{e^2}{c^2}$ l) $5t$

7 a) $2x + xy$ b) $p^2 - 4pq$
c) $6ab - 3a^2$ d) $4r^2 - 12rs$
e) $6m^2 + 2mn$ f) $35ef - 15e^2$
g) $x^2 + 5xy + 6y^2$ h) $x^2 + 2xy - 35y^2$
i) $10p^2 - 7pq + q^2$ j) $14c^2 + 13cd + 3d^2$
k) $20t^2 + 23tu - 21u^2$ l) $12x^2 - 47xy + 45y^2$

8 a) $x(4 + 3y)$ b) $b(5a - 7)$
c) $2c(3 + 4d)$ d) $5q(p - 2)$
e) $x(6x - y)$ f) $2g(1 + 5h)$
g) $4r(2r - s)$ h) $ab(b + a)$
i) $pq(1 - 9p)$

Pages 114–115: Finishing off

1 a) $15ax$ b) $36b^2$ c) $10cd^2$
d) $21e^2f^4$ e) $-36g^3$ f) $18h^7$

2 a) $\frac{p}{2}$ b) $\frac{4}{q}$ c) $\frac{2r^2}{7}$
d) $\frac{3s}{2t}$ e) $\frac{5}{3uv}$ f) $\frac{3x}{w^2}$

3 a) 16 b) 48 c) 576 d) 128

4 a) 17 b) 9 c) –54 d) –729

5 a) $2n - 1$ b) $4n^2 - 4n + 1$

6 a) 38 b) $3w + d$

7 a) $15 + 5x$ b) $6p^2 - 24p$
c) $3a - 7a^3$ d) $-3m - n$
e) $-8t^2 + 12tu$ f) $21c^2d - 15cd^2$

8 a) $x^2 + 10x + 16$ b) $a^2 - 2a - 15$
c) $6p^2 - 13p + 5$ d) $12f^2 + 13fg + 3g^2$
e) $28 + 17y - 3y^2$ f) $16x^2 - 25y^2$
g) $a^2 - 6a + 9$ h) $9n^2 + 30n + 25$
i) $64p^2 - 48pq + 9q^2$

9 a) $n^2 + 9n$ b) $9a^2 + 8a$
c) $3x^2 - x$ d) $4y^2 + 7y$
e) $7ab + 14a$ f) $-6a - 8$
g) $d^2 + 6d$ h) $k^2 + 39n^2$

10 a) $n(n + 2)$ b) $(n + 2)(n + 1) - (n - 1)$
c) $n(n + 1) + (n + 2)$ d) $n(n + 2)(n + 4)$

11 a) $4(n + 3)$ b) $4(2t - 5)$
c) $3(5x + 9y)$ d) $2(3a - 1)$
e) $3(q^2 - 5)$ f) $5(5x + 7y + 8)$

12 a) $p(4 - p)$ b) $3x(2 + x)$
c) $8n(3n + 1)$ d) $3r(2r^2 + 7s)$
e) $14f(5f - 2g)$ f) $16yz(7z - 3y)$
g) $3(8a + 4b - 5)$ h) $x(x^2 + 3x + 10)$
i) $7r(5p - 3q + 8r)$

13 a) 108 pence b) $27n$ pence
c) np pence

14 a) £6 b) £$(\frac{30}{x})$
c) £$(\frac{c}{x})$

Chapter 13: Graphs

Page 117: Co-ordinates

1 A (2, 4), B (4, 3), C (3, –1), D (–2, 1) E (1, 0), F(–3, –2), G (2, –3), H (–4, 3), I (–1, –4), J (0, 2)

2 Ask your teacher to check your diagram
a) Square b) Rectangle c) Triangle

3 a) Ask your teacher to check your diagram
b) y increases by 2 each time
c) $y = 2x + 1$

4 Ask your teacher to check your diagrams
A: b) y is one more than x
c) $y = x + 1$
d) One answer is (4, 5), (5, 6)
B: b) y is five subtract x
c) $y = 5 - x$
d) One answer is (4, 1), (5, 0)
C: b) y is three times x
c) $y = 3x$
d) One answer is (3, 9), (4, 12)

Page 119: Straight line graphs

1 Ask your teacher to check your diagram
g) (4, 8) h) (3, 1) i) (0, 4)

2 a) + 1, 5, $y = x + 5$ b) 2, – 2, $y = -2x - 2$
c) $-\frac{2}{3}$, 2, $y = -\frac{2}{3}x + 2$ d) $-\frac{1}{2}$, –1, $y = -\frac{1}{2}x - 1$

3 a) b) Ask your teacher to check your graphs.
c) (i) – 1 (ii) 2 (iii) 2 (iv) – 1
d) (i) and (iv), (ii) and (iii)
e) by comparing the coefficients of x (i.e. the values of m)

Page 121: Curved graphs

1 a)

x	–3	–2	–1	0	1	2	3
x^2	+9	+4	+1	0	+1	+4	+9
–5	–5	–5	–5	–5	–5	–5	–5
y	+4	–1	–4	–5	–4	–1	+4

b) Ask your teacher to check your graph.
c) –3.56
d) +2.65, –2.65

2 a)

x	0	1	2	3	4	5
–4	–4	–4	–4	–4	–4	–4
+5x	0	5	10	15	20	25
$-x^2$	0	–1	–4	–9	–16	–25
y	–4	0	2	2	0	–4

b)

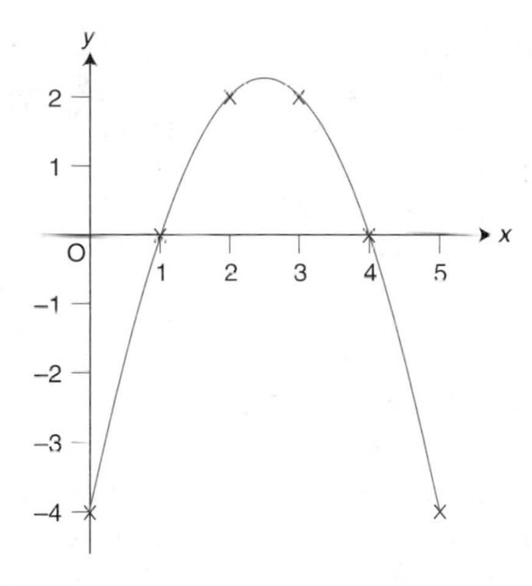

c) 2.25
d) 1.4, 3.6

3 a)

x	–2	–1	0	1	2
x^3	–8	–1	0	1	8
2x	+4	2	0	–2	–4
y	–4	1	0	–1	4

b)

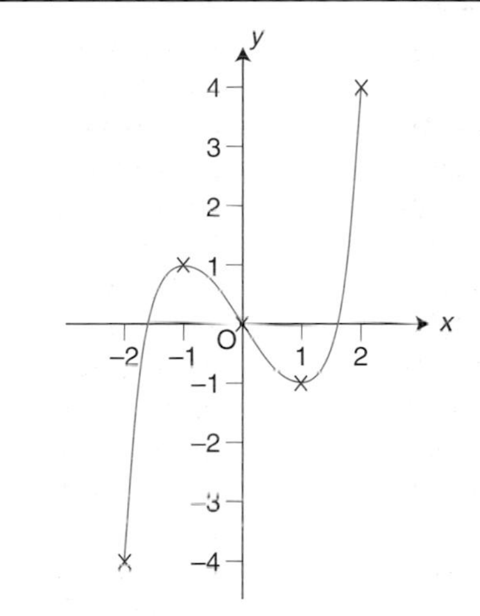

c) 1.4, 0, –1.4

4 72, 32, 18, 8, 4.5, 2.88, 2
Ask your teacher to check your answer

Page 123: Obtaining information

1 a) 12.30 p.m., $1\frac{1}{2}$ hours b) 15 minutes
c) 2 hours, she stopped for 1 hour
d) the same speed

2 a) 25p b) £2.50 c) £1.25

3 a) Ask your teacher to check your graph
b) 50 miles per hour, 48 miles per hour, 70 miles per hour
c) In Reading at 1242

4 a) £750
b) licence, insurance, MOT service, depreciation
c) 20p per mile
d) petrol, parking, wear and tear

Pages 124–125: Finishing off

1 (i) Ask your teacher to check your graphs
(ii) (3.5, 5.5)
2 Ask your teacher to check your graphs
(iii) a) $\frac{1}{2}$ b) –3 (iv) a) 4 b) 7
3 a)

t	0	1	2	3	4
$20t$	0	20	40	60	80
$-5t^2$	0	–5	–20	–45	–80
h	0	15	20	15	0

b) Ask your teacher to check your graphs
c) 0.6 seconds and 3.4 seconds
d) The ball meets the ground
4 a) and c)
5 a) £25 b) £350
c) A (1, 500) B (6, 1250); AC = 5 thousand, BC = £750
d) £150 e) £1850
6 a) £100 b) £35 per day
c) $C = 100 + 35x$ d) £800
7 In these sketches the vertical axis measures the distance from home.

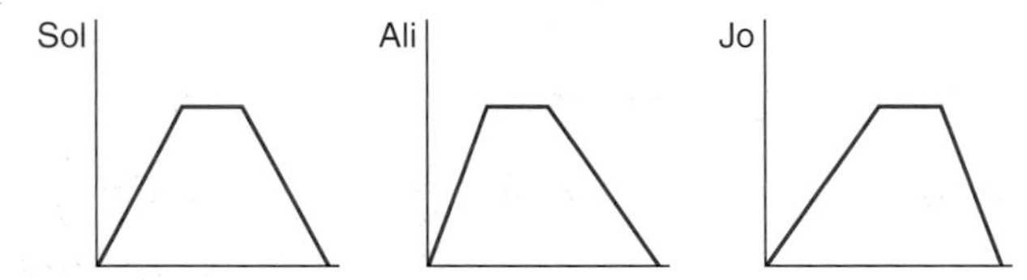

Chapter 14: Equations

Page 127: Finding unknowns

1 $3x = 21$; 7
2 a) $x = 4$ b) $x = 6$ c) $x = 9$ d) $x = 8$
3 $x - 5 = 9$; 14
4 a) $x = 19$ b) $x = 3$ c) $x = 20$ d) $x = 12$
5 $3x + 2 = 26$; 8
6 a) $x = 6$ b) $x = 7$ c) $x = 4$ d) $x = 2$
e) $x = 6$ f) $x = 9$ g) $x = 4$ h) $x = 1.5$
i) $x = 11$ j) $x = -1$ k) $x = 2$ l) $x = 4$
m) $x = 5.6$ n) $x = 4$ o) $x = -0.6$
7 $2x + 100 = 250$, $x = 75$
8 $6x = 84$, $x = 14$

Page 129: More equations

1 a) $n = 3$ b) $n = 13$ c) $x = 2$
d) $x = -2$ e) $y = 2.5$ f) $y = 6$
g) $a = 4$ h) $x = -3$ i) $n = -2.5$
2 a) $x = 4$ b) $x = 6$ c) $n = 13$
d) $n = 9$ e) $a = -10$ f) $y = -9$
g) $t = 2$ h) $x = 1$ i) $n = 4$
j) $d = 2.4$ k) $u = -4$ l) $v = 4.5$
3 a) $x = 6$ b) $x = 4$ c) $y = 10$ d) $m = -3$
e) $a = 2$ f) $n = 5$ g) $p = 4$ h) $q = 1\frac{1}{3}$
4 a) $c = 3\frac{3}{4}$ b) $y = 1\frac{2}{7}$ c) $x = \frac{4}{15}$
d) $t = \frac{3}{7}$ e) $n = 1\frac{6}{7}$ f) $k = 1\frac{3}{22}$
5 a) 77 b) $11.\dot{6}$ c) –40

Page 131: Equations with fractions

1 a) $x = 10$ b) $x = 24$ c) $x = 10\,000$
d) $x = 4$ e) $x = -4$ f) $x = -50$
g) $x = 8$ h) $x = 20$
2 b) 36
3 a) $x = 24$ b) $x = 3$ c) $x = 12$ d) $x = 11$
4 a) $x = 11$ b) $x = -2$ c) $x = 1$ d) $x = -9$
e) $x = -2$ f) $x = 7$ g) $x = 13$ h) $x = 10$
5 a) $x = 15$ b) $x = 5$ c) $x = 0.5$ d) $x = -9$
6 a) 40 b) 200

Page 133: Using equations to solve problems

1 a) (i) $80x$ (ii) $780 + 65x$ b) 52
2 a) (i) $12x + 16 = 100$ (ii) $x = 7$
b) (i) $3w + 6 = 72$ (ii) $w = 22$
c) (i) $2(3w + w) = 600$ (ii) $w = 75$
d) (i) $4A + 60 + A = 180$ (ii) $A = 24$
e) (i) $3y = y + 24$ (ii) $h = 12$
3 a) A-serve: £15
Beeline: £$(5 + 2x)$
Comic: £$(9 + 1.5x)$
When $x = 1$, Beeline is cheapest
When $x = 10$, A – serve is cheapest
b) $x = 5$ c) $x = 8$ d) $x = 4$
e) A-serve is cheapest for a total of more than 10 hours otherwise use Beeline.

Page 135: Using graphs to solve equations

1 a) Ask your teacher to check your graph
b) (i) $x = -2, -0.4, 2.4$ approximately
(ii) $x = 2.7$ approximately
2 a) Ask your teacher to check your graph
b) (i) $x = -0.7, 0.8, 3.9$ approximately
(ii) $x = -0.9, 1.2, 3.7$ approximately
3 3.141
4 1.325
5 a) 5.6 b) 1.66 c) 5.62
6 a)

x	1	2	3	4
y	5	–2	$-\frac{1}{3}$	5

b) Ask your teacher to check your graph
c) 1.4, 3.1
d) 1.43, 3.08

Pages 136–137: Finishing off

1 a) $a = 3$ b) $b = 1.4$ c) $c = 0.4$ d) $d = 2.4$
e) $e = 5$ f) $f = 5$ g) $g = 2.5$ h) $h = 3.1$

2 a) $z = 3$ b) $y = 4$ c) $x = 0.5$ d) $w = 7$
e) $v = 4.5$ f) $u = 1.5$ g) $t = 6$ h) $s = 1$

3 a) $x = 7$ b) $x = 3.75$ c) $x = 1.4$
d) $x = 11$ e) $x = -4$ f) $x = 4$

4 a) $4(n - 1) - 10$ b) 7

5 a) $x = 18$ b) $x = 10$ c) $x = 50$ d) $x = 5$
e) $x = 3$ f) $x = 1$ g) $x = 7$ h) $x = 8$

6 a) Ask your teacher to check your graph
b) $x = -0.9, 1.3, 2.5$ approx
c) $x = -0.5, 0.7, 2.9$ approx

7 a) Ask your teacher to check your graph
b) $x = 2.3$ c) 2.289 d) $\frac{12}{x} = x^2$ when $x^3 = 12$

8 a) 2.2056 b) 2.2056

9 a) £$(12 + 6x)$ b) £$(36 + 4.5x)$
c) $x = 16$ d) Silver price is cheaper

Chapter 15: Formulae

Page 139: Substituting into a formula

1 a) (i) 88 (ii) 100 (iii) 106
b) (i) 48 (ii) 54

2 a) 25 b) 8 c) $\frac{1}{4}$ d) 0.04

3 a) 5 b) 20 c) 1.25 d) 11.25

4 a) (i) £30 (ii) £75 (iii) £525
b) (i) 30p (ii) 7.5p (iii) 5.25p
c) $p = \frac{2500 + 5n}{n}$ or equivalent

5 a) 5, moving upwards b) 3, moving upwards
c) 0, at the top d) –5, moving downwards

Page 141: Making up formulae

1 a) $3n$ b) $\frac{1}{2}bh$ c) $6a^2$
e) $\frac{20}{r^2}$ e) $3h + 6d$

2 a) $m = 12y$ b) $m = 12y = 120$
c) $y = \frac{m}{12}$

3 a) $C = 100\ M$ b) $M = \frac{c}{100}$

4 a) 10 miles b) $m = 10l$
c) $l = \frac{m}{10}$

5 a) (i) $P = 30A$ (ii) $\frac{p}{30}$
b) (i) $C = 2.54I$ (ii) $I = \frac{C}{2.54}$

6 a) $M = D + 12$ b) $D = M - 12$

Page 143: Changing the subject of a formula

1 a) $x = y - 4$ b) $x = y - a$
c) $x = y - 3$ d) $x = y - c$
e) $x = y + 5$ f) $x = y + b$
g) $x = 6 - y$ h) $x = d - y$

2 a) $x = \frac{1}{2}y$ b) $x = \frac{y}{a}$
c) $x = 4y$ d) $x = by$
e) $x = \frac{4}{3}y$ f) $x = \frac{b}{a}y$
g) $x = \frac{5}{4}y$ h) $x = \frac{b}{a}y$

3 a) $t = \frac{x+3}{2}$ b) $t = \frac{y-4}{3}$ c) $t = \frac{p-6}{2}$
d) $t = 4 - c$ e) $t = \frac{6-z}{2}$ f) $t = \frac{s-a}{2}$
g) $t = \frac{x+c}{5}$ h) $t = \frac{n+3x}{7}$ i) $t = \frac{v-u}{a}$

4 a) $u = v - at$ b) $l = \frac{p-2b}{2}$ c) $x = \frac{V+9y}{4}$

5 a) $x = \frac{p-2y}{2}$ or $\frac{p}{2} - y$
b) $x = \frac{V-12r}{12}$ or $\frac{V}{12} - r$
c) $x = \frac{8-s}{4}$ or $2 - \frac{s}{4}$
d) $x = \frac{4a-y}{4}$ or $a - \frac{y}{4}$

6 a) $l = \frac{A}{w}$ b) $h = \frac{V}{lw}$ c) $R = \frac{V}{I}$
d) $d = \frac{C}{\pi}$ e) $r = \frac{C}{2\pi}$ f) $P = \frac{100I}{r}$
g) $T = \frac{100I}{PR}$ h) $R = \frac{100I}{PT}$

7 a) $x = \sqrt{\frac{y}{6}}$ b) $x = \sqrt{3y}$ c) $x = 5y^2$
d) $x = \frac{1}{2}y^2$

8 a) $c = \sqrt{\frac{E}{m}}$ b) $V = \sqrt{PR}$ c) $u = \sqrt{v^2 + 2gs}$
d) $h = \frac{v^2}{2g}$ e) $a = c^2 - 2$ f) $A = \frac{\pi d^2}{4}$

9 a) $x = \frac{5k}{6}$ b) $x = \frac{t}{4}$ c) $x = \frac{b}{c-a}$
d) $x = 2y + 7z$ e) $x = \frac{nd}{n-1}$ f) $x = \frac{u-sy}{r-t}$

Pages 144–145: Finishing off

1 Athens 59 °F, Buenos Aires 84 °F, Moscow –2 °F, Prague 32 °F

2 a) 10 b) 50 c) 64

3 a) $3\frac{3}{4}$ hours b) 3 hours 43 mins; yes

4 a) $A = 283$, $V = 314$ b) $A = 628$, $V = 1005$

5 a) (i) $A = 0.25\ g$ (ii) $g = \frac{A}{0.25}$ or $g = 4A$
b) $d = \frac{k}{5}$ c) $C = \frac{150}{N}$
d) $T = 1.8 + 0.5x$

6 a) $x = y - 7$ b) $x = 5 - y$ c) $P = \frac{k}{V}$
d) $F = PA$ e) $h = \frac{A}{b}$ f) $V = \frac{m}{d}$
g) $t = \frac{v - u}{10}$ h) $W = \frac{1}{2}P - L$ or $W = \frac{P - 2L}{2}$
i) $P = \frac{100I}{RT}$

7 a) $d = 60$ b) £425 c) $p = \frac{10d}{n}$

8 a) $x = \sqrt{2y}$ b) $t = \sqrt{\frac{2s}{a}}$ c) $r = \sqrt[3]{\frac{3V}{4\pi}}$
d) $c = \frac{b}{a^2}$ e) $x = \frac{y^2}{9}$ f) $l = \frac{gT^2}{4\pi^2}$
g) $x = \frac{d - b}{a - c}$ h) $t = \frac{n}{a - 2}$ i) $h = \frac{6n}{2n - 1}$

Chapter 16: Inequalities

Page 147: Using inequalities

1 a) $x > 3$ b) $x \geq 0$ c) $x < 5$ d) $x \leq -2$
2 a) $s \leq 30$ b) $f \geq 79$ c) $g \geq 20$
d) $p \leq 5$ e) $c < 250$ f) $n \geq 50$
3

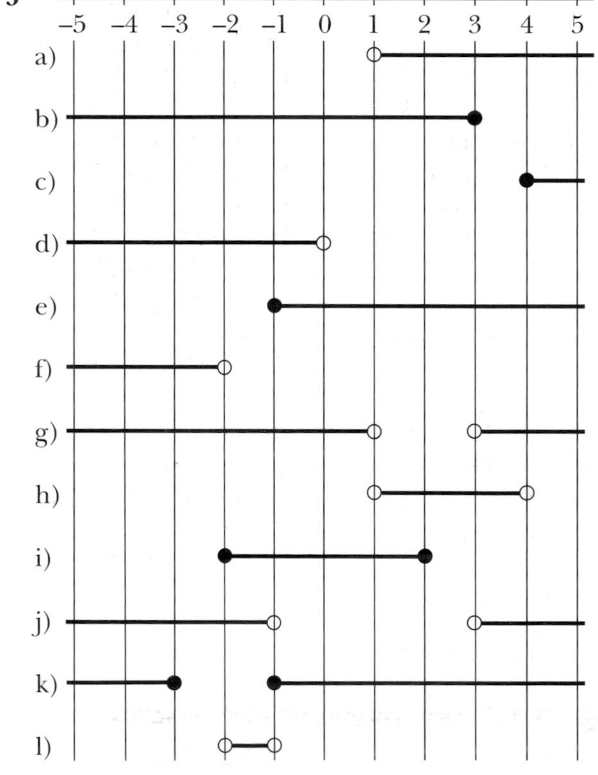

4 a) $x \leq 2$ b) $x > 0$
c) $-2 < x \leq 3$ d) $x < -1$ or $x > 4$

Page 149: Solving inequalities

1 a) $x < 6$ b) $x \leq 5$ c) $x \geq 8$ d) $x \geq 4$
e) $x < 3$ f) $x > 6$ g) $x \leq 3.6$ h) $x \leq -3$
i) $x > 2$ j) $x \leq 2$ k) $x \geq 3$ l) $x \leq -3$

2 a) 2, 3, 5, 7, 11, 13, 17, 19
b) 25, 36 c) 4, 6 d) 6, 9
3 a) $5x - 200$ b) $5x - 200 > x$; $x > 50$
4 $w + 52 > 14w$, $w < 4$
5 a) $x \geq 11$ b) $x \geq 2$
c) $x > 7$ d) $x < 6$
e) $1 \leq x \leq 8$ f) $10 \geq x > 3$
g) $6\frac{1}{2} < x < 9\frac{1}{2}$ h) $2 \leq x \leq 20$
i) $2 < x \leq 11$ j) $x > 11$
k) $x \geq 16$ l) $x > 1$
6 $5\,000 \leq m \leq 9999$

Page 151: Inequalities and graphs

1 a) $-1 \leq x \leq 2$, $0 \leq y \leq 2$ b) $x > 2$, $y > 1$
c) $0 \leq x \leq 3$, $0 \leq y \leq 1$ d) $3 < x < 10$, $5 \leq y \leq 10$
e) $x \geq 4$, $y \geq 3$ f) $1 \leq x \leq 4$, $y > -1$
2 Ask your teacher to check your graphs
3 a) $20 \leq t \leq 120$, t is time in minutes
$5 \leq d \leq 100$, d is distance in miles
b) Ask your teacher to check your graph
4 Ask your teacher to check your graphs

Page 153: Solution sets

1 Ask your teacher to check your graphs
2 a) $4s + 2w \geq 10$ or $2s + w \geq 5$
b) $w \leq 5$
c) $s \leq 2$
d) Ask your teacher to check your graph
3 a) $8x + 11y \leq 100$ b) $x > y$ c) $y \geq 2$
d) Ask your teacher to check your graph
e) Yes $x = 7$, $y = 4$ lies on $8x + 11y = 100$
4 a) (i) $10 \leq d \leq 15$ (ii) $13 \leq s \leq 28$.
Ask your teacher to check your graph
b) There are two people in a double room and one in a single room.
c), d) Ask your teacher to check your graph

Pages 154–155: Finishing off

1 Ask your teacher to check your answers
2 M $20 \leq a < 36$
15 20 25 30 35 40
F $30 \leq a < 40$
25 30 35 40

3 a) 16 b) One of 11, 13, 17
c) 14 d) 12.7 for example
4 a) $x < -3$ or $x > 4$
b) $x \geq -1$ and $x \leq 3$ (or $-1 \leq x \leq 3$)
5 a) $x \leq 5$ b) $x \geq 13$ c) $x > 8$
d) $x < 5$ e) $x \leq 8$ f) $x > -4$
g) $x > 7$

6 a) $1 < x < 4, 2 < y < 6$ b) $-3 \le x \le 0$
c) $x < 2$ d) $2 \le x \le 5, y \ge 1$
7 Ask your teacher to check your graphs
8 a) 395, b) $10x + 5y$ c) $10x + 5y \le 430$
d) The number of starter and bonus questions cannot be negative. In order to get a bonus question, the starter question must be answered correctly and so there can never be more correct bonus questions than starter questions.
e) Ask your teacher to check your graph

Chapter 17: Simultaneous equations

Page 157: Using simultaneous equations

1 a) $x = 7$ b) $14 + y = 15$ so $y = 1$ c) $7 + 1 = 8$
2 a) 4, 9 b) 10, 1 c) 6, –1
d) 2, 2 e) 5, 3 f) 1, 6
g) 5, 1 h) 10, 20 i) –2, 3
3 a) 35p b) 28p
4 White costs £2, blue £2.50
5 a) 7, 3 b) 3, $\frac{1}{3}$ c) 9, 11
d) 1, 0 e) 2, –1 f) 5, 2
g) 6, 2 h) 4, 3 i) 17, 3
6 a) 3 b) 1

Page 159: More simultaneous equations

1 a) 10, 2 b) 18, 3 c) 8, 5
d) 9, 5 e) 7, –1 f) $11\frac{1}{2}, 1\frac{1}{2}$
2 11, 2
3 a) 4, 1 b) $\frac{1}{2}$, 1 c) 10, 5
d) 5, 2 e) 5, 1 f) $1\frac{1}{2}$, 4
g) 8, $3\frac{1}{2}$ h) 10, 2 i) 50, 13
4 Jam doughnuts cost 18p, ring doughnuts 12p
5 a) £1.10 b) £0.75
6 a) $6x + 10y = 58$ and $25x - 10y = 35$
b) $31x = 93$ c) 3, 4
7 a) 9, 2 b) 3, 1
c) 10, 2 d) 5, –1 e) 0, 5
f) $\frac{1}{2}$, 1 g) 9, 7 h) 10, 20

Page 161: Other methods of solution

1 a) 4, 6 b) 6, 10 c) 4, 3
d) 5, 1 e) 6, –5 f) 1, 2
2 a) $x + 2y = 12$ b) $x + y = 8$
c) 4 200 g jars and 4 400 g jars
3 5 in a packet, 20 in a box
4 a) 3, 7 b) 2, 5 c) $2\frac{1}{2}, \frac{1}{2}$
d) –1, 5 e) 1, –1 f) 6, 1

Pages 162–163: Finishing off

1 a) 8, 5 b) 8, –1 c) 1, 3 d) 2, $1\frac{1}{2}$
e) 120, 180 f) 5, 2 g) 10, –2 h) 6, 1
i) $9\frac{1}{2}$, 3 j) 5, $\frac{1}{9}$ k) 20, 10 l) 11, 10
m) –1, –12 n) $\frac{1}{2}, \frac{1}{2}$ o) 3, 3 p) 10, 4
q) 4, 2 r) –1, 2 s) 5, 2 t) 1, 3
u) 2, –3
2 $x + y = 21$; $x - y = 5$; 13, 8
3 Oranges cost 40p, apples cost 30p
4 Adult £9, child £6
5 Basic rate is £3.50, rate after 11 p.m. is £6 per hour
6 a) 90 pence b) 50 pence
7 70 of A, 30 of B
8 a) 6, 12 b) 10, 4 c) 7, 3
9 a) 1, 5 b) $1\frac{1}{2}, 8\frac{1}{2}$ c) 0, –3

Chapter 18: Quadratics

Page 165: Factorising quadratic expressions

1 a) 1, 5 b) $(x + 1)(x + 5)$
2 a) 3, 5 b) $(y + 3)(y + 5)$
3 a) $(x + 2)(x + 7)$ b) $(x + 1)(x + 14)$
c) $(a + 2)(a + 5)$ d) $(a + 10)(a + 1)$
e) $(n + 4)(n + 1)$ f) $(p + 2)(p + 4)$
g) $(x + 2)(x + 2)$ h) $(x + 3)(x + 6)$
i) $(y + 4)(y + 6)$ j) $(x + 5)(x + 7)$
k) $(x + 5)(x + 12)$ l) $(x + 14)(x + 2)$
4 a) –1, –5 b) $(x - 1)(x - 5)$
5 a) –3, –5 b) $(y - 3)(y - 5)$
6 a) $(x - 1)(x - 2)$ b) $(x - 3)(x - 1)$
c) $(x - 5)(x - 1)$ d) $(x - 11)(x - 1)$
e) $(x - 7)(x - 2)$ f) $(x - 2)(x - 4)$
g) $(p - 3)(p - 6)$ h) $(a - 4)(a - 11)$
i) $(x - 44)(x - 1)$ j) $(t - 10)(t - 2)$
k) $(x - 5)(x - 5)$ l) $(y - 4)(y - 9)$
7 a) $(x - 3)(x - 5)$ b) $(a + 2)(a + 11)$
c) $(x - 3)(x - 7)$ d) $(p - 2)(p - 8)$
e) $(t - 2)(t - 6)$ f) $(y + 3)(y + 4)$
g) $(a + 1)^2$ h) $(x - 3)^2$
i) $(x - 20)(x - 5)$

Page 167: More quadratic factorisation

1 a) 6, 1 b) $(x + 6)(x - 1)$
c) $(x - 6)(x + 1)$
2 a) 3, 5 b) $(y + 5)(y - 3)$
c) $(y - 5)(y + 3)$
3 a) $(x + 11)(x - 1)$ b) $(x - 11)(x + 1)$
c) $(x + 7)(x - 1)$ d) $(x - 7)(x + 1)$
e) $(x + 5)(x - 1)$ f) $(x - 5)(x + 1)$
g) $(x + 7)(x - 2)$ h) $(x - 7)(x + 2)$
i) $(x + 11)(x - 8)$

4 a) $(a+9)(a-2)$ b) $(a-9)(a+2)$
c) $(y+10)(y-1)$ d) $(y+2)(y-5)$
e) $(p-6)(p+3)$ f) $(x+3)(x-4)$
g) $(x+5)(x-4)$ h) $(a-2)(a+10)$
i) $(t+6)(t-2)$

5 a) $(x+4)(x-4)$ b) $x(x-8)$
c) $(y+7)(y-7)$ d) $t(t-4)$
e) $(x+1)(x-1)$ f) $p(p+1)$

6 a) $(x+1)(x+6)$ b) $(x+2)(x+4)$
c) $(r-4)(r-1)$ d) $(x-9)(x+1)$
e) $(y+4)(y-1)$ f) $(x+4)(x-3)$
g) $(t-4)(t+3)$ h) $(x-2)(x-9)$
i) $(p+6)(p-2)$ j) $(y-9)(y+9)$
k) $(b-5)(b+4)$ l) $a(a-11)$

7 a) $4(x+4)(x-3)$ b) $3(a-1)(a-2)$
c) $3(x-2)(x+2)$ d) $3(x+1)^2$
e) $10(x-10)(x+10)$ f) $5(x-8)(x+10)$

Page 169: Quadratic equations

1 a) $x = 2$ or 6 b) $x = 4$ or -3
c) $x = -7$ or 2 d) $y = -8$ or -5
e) $y = 0$ or 3 f) $t = 1$ or -6

2 a) 2, 7 b) 2, 5
c) 2, –7 d) –1, –4
e) 2, –3 f) 6, –2
g) 5, –10 h) 9, –7
i) 11, 1 j) –2, –12,
k) 6, –5 l) 3
m) 4, –4 n) 0, 5
o) 9, –2

3 a) –5, 4 b) 2, –10
c) 1, 4 d) 4, –1
e) –2, –9 f) 4, –3

4 a)

x	0	1	2	3	4	5
x^2	0	1	4	9	16	25
$-5x$	0	–5	–10	–15	–20	–25
$+5$	+5	+5	+5	+5	+5	+5
y	5	1	–1	–1	1	5

b) Ask your teacher to check your graph
c) $x = 1.38$ or $x = 3.62$

5 a) Ask your teacher to check your graph
b) $x = -0.14$ or $x = 3.64$

Page 171: Rational functions

1 a) x b) x c) x^2 d) x^2 e) x f) $\frac{1}{x}$

2 a) (i) $x(3x+1)$ (ii) $x^2(4x+1)$ (iii) $x^2(5+x)$
b) (i) $3x+1$ (ii) $4x+1$ (iii) $5+x$

3 a) $x-5$ b) $x-2$ c) $x+3$ d) $x+1$
e) $x-2$ f) $x-2$

4 a) $\frac{x-5}{x+5}$ b) $\frac{x-2}{x+2}$ c) $\frac{x+3}{x-3}$

5 a) (i) $x(x+3)$ (ii) $(x-1)(x+1)$ (iii) $x(x-9)$
(iv) $x(3x+2)$ (v) $x^2(x+1)$
b) (i) x (ii) $x-1$ (iii) x (iv) $\frac{x}{3x+2}$ (v) $\frac{x}{x+1}$

6 a) (i) $(x+2)(x+5)$ (ii) $(x-2)(x-3)$
(iii) $(x+3)(x-1)$
b) (i) $x+5$ (ii) $x-2$ (iii) $x+3$

Pages 172–173: Finishing off

1 a) $(x+12)(x+1)$ b) $(a-1)(a-2)$
c) $(z-2)(z-4)$ d) $(n+13)(n-2)$
e) $(t+2)(t-6)$ f) $(x+3)(x-10)$

2 a) $(x+1)^2$ b) $(n-5)^2$
c) $(r+4)^2$ d) $(y-6)^2$
e) $(x-4)(x+4)$ f) $(p-7)(p+7)$

3 a) 0 b) –2 c) 0, 2
d) 0, –15 e) 5, –6 f) –7, –11
g) 3, –3 h) $\frac{3}{2}$, 3 i) 5, $\frac{2}{3}$

4 a) 2, 3 b) –3, –4 c) –9
d) 3, –6 e) –3, 7 f) –2, 12
g) 0, –2 h) –2, 2 i) –2, 2

5 a) 8, –7 b) 9, –10 c) –3, 4
d) 2, 3 e) 2, –7 f) 3, –6
g) –4, 5 h) 0, 5 i) 3, 7

6 a) x b) $x-7$ c) 5
d) $3x-4$ e) $x+4$ f) $\frac{3}{x+2}$

7 Ask your teacher to check your answer

8 a)

x	–4	–3	–2	–1	0	1	2	3	4
x^2	20	12	6	2	0	0	2	6	12

b) –2, 3 c) –2, 3
d) $x^2 - x = 6$ when $y = 6$ on the graph
e) 3.4, –2.4
It will not factorise

9 a)

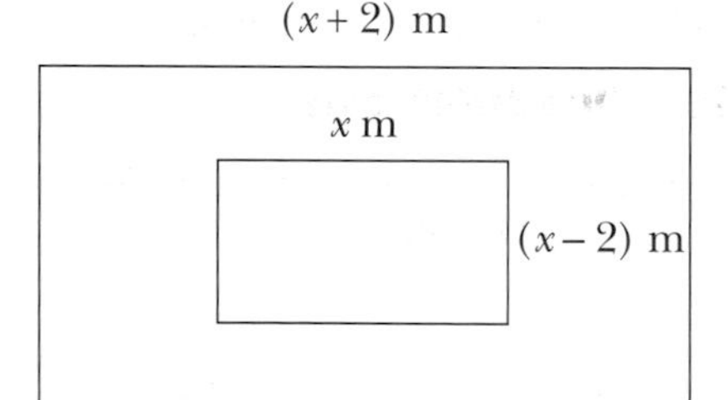

b) Area = $x(x-2)$ m^2
c) $x(x-2) = 120$
d) $x = 12$
12 m by 14 m

10 a) (i) $w^2 + 3w - 54 = 0$ (ii) $w = 6$
b) (i) $8l^2 - 80l - 3000 = 0$ (ii) $l = 25$
c) (i) $2x^2 + 8x - 90 = 0$ (ii) $x = 5$

Chapter 19: Angles and shapes

Page 175: Angles

1 a) 15° b) 150°
2 a) 720° b) 540° c) 432°
3 a) Half b) One quarter
c) Three quarters
4 a) Obtuse b) Reflex c) Acute d) Reflex
5 Ask your teacher to check your reasons
$a = 127^\circ$ $b = 125^\circ$ $c = 76^\circ$ $d = 123^\circ$
$e = 88^\circ$ $f = 31^\circ$ $g = 71^\circ$ $h = 109^\circ$
$i = 109^\circ$ $j = 60^\circ$ $k = 72^\circ$ $l = 45^\circ$
$m = 68^\circ$ $n = 68^\circ$ $p = 62^\circ$

Page 177: Parallel lines

1 $d = 108^\circ$, $e = 72^\circ$, $f = 108^\circ$, $g = 108^\circ$,
$h = 72^\circ$, $i = 92^\circ$, $j = 92^\circ$, $k = 88^\circ$,
$l = 88^\circ$, $m = 92^\circ$, $p = 127^\circ$, $q = 53^\circ$,
$r = 127^\circ$, $s = 53^\circ$, $t = 127^\circ$, $u = 53^\circ$
2 a) m and p, n and o, q and s, r and t
b) n and r, p and s, o and t, m and q
c) p and q, n and t
3 a) $a = 69^\circ$ $b = 62^\circ$ $c = 49^\circ$ b) 180°
4 Ask your teacher to check your reasons
$a = 54^\circ$ $b = 54^\circ$ $c = 42^\circ$ $d = 42^\circ$
$e = 131^\circ$ $f = 66^\circ$ $g = 114^\circ$ $h = 50^\circ$
$i = 130^\circ$ $j = 50^\circ$ $k = 115^\circ$

Page 179: Triangles

1 a) F b) E c) B d) A
e) G f) D g) C
2 $p = 30^\circ$ $q = 45^\circ$ $r = 56^\circ$ $s = 71^\circ$
$t = 60^\circ$ $u = 35^\circ$ $v = 38^\circ$ $w = 36^\circ$
3 Ask your teacher to check your reasons
$a = 105^\circ$ $b = 42^\circ$ $c = 33^\circ$ $d = 42^\circ$
$e = 43^\circ$ $f = 69^\circ$ $g = 27^\circ$ $h = 66^\circ$
$i = 66^\circ$ $j = 59^\circ$ $k = 59^\circ$ $l = 55^\circ$

Page 181: Quadrilaterals

1 a) $x = 115^\circ$ b) $y = 74^\circ$ c) $z = 254^\circ$
2 $\angle B = 104^\circ$, $\angle D = 53^\circ$
3 34°
4 60°
5 a) Trapezium b) 140° c) 20°
6 a) 1 is a square, 2 is a parallelogram, 3 is a trapezium, 4 is a rhombus, 5 is a kite, 6 is a rectangle
b) Square: A, B, F; Parallogram: B, C, E;
Trapezium: D; Rhombus: A, B, C, F;
Kite: E, F; Rectangle: B, E

Pages 182–183: Finishing off

1 Ask your teacher to check your reasons
$a = 82^\circ$ $b = 96^\circ$ $c = 62^\circ$ $d = 30^\circ$
$e = 103^\circ$ $f = 77^\circ$ $g = 103^\circ$ $h = 111^\circ$
$i = 69^\circ$ $j = 111^\circ$ $k = 111^\circ$ $l = 69^\circ$
$m = 58^\circ$ $n = 62^\circ$ $p = 67^\circ$ $q = 113^\circ$
$r = 74^\circ$ $s = 106^\circ$ $t = 74^\circ$ $u = 89^\circ$
$v = 89^\circ$ $w = 47^\circ$ $x = 48^\circ$
$y = 60^\circ$ $z = 57^\circ$
2 60° 3 30° 4 141°
5 a) 45° b) 80° c) 125°
6 a) (i) ABF or CDE (ii) BEC or FBE
(iii) FBCD, ABEF, BCDE or ACEF
(iv) FBCE
b) $x = 55^\circ$, $y = 110^\circ$, $z = 55^\circ$
7 $a = 55^\circ$, $b = 35^\circ$, $c = 35^\circ$, $d = 70^\circ$, $e = 20^\circ$

Chapter 20: Polygons

Page 185: Polygons

1 A – regular pentagon
B – irregular octagon
C – irregular hexagon
D – irregular quadrilateral
E – irregular pentagon
F – regular octagon
2 a) 120° b) 90° c) 72°
d) 60° e) 45° f) 36°
3 a) A: (i) 91° (ii) 89° (iii) 91°
B: (i) 61° (ii) 119° (iii) 61°
C: (i) 118° (ii) 62° (iii) 118°
b) They are always equal
4 a) 24
b) Ext $\angle = \frac{360}{n}$, $\frac{360}{n} = 25$ gives $n = 14.4$ which is not a whole number

Page 187: Angle sum of a polygon

1 a) 720° b) 1080° c) 1260° d) 1440°
2 a) 120° b) 135° c) 140° d) 144°
3 a) 60° b) 45° c) 40° d) 36°
4 a) Hexagon (120°) b) Octagon (135°)
c) Nonagon (140°) d) Decagon (144°)
6 a) 18
b) Ext $\angle = 50^\circ$ and $\frac{360}{n} = 50$ gives $n = 7.2$ which is not a whole number.

Page 189: Congruent shapes

1 D
2 Ask your teacher to check your answer
3 Triangle CDA
4 When AD = BC
5 Ask your teacher to check your answer

Pages 190–191: Finishing off

1 a) 72° b) 54° c) 108°
2 a) 900° b) 1260°

3 a) ext ∠ = 36°, int ∠ = 144°
b) ext ∠ = 30°, int ∠ = 150°
4 a) 6 b) 15 c) 8
5 a) (i) square (ii) rhombus
b) Shape A
6 a) There are many possibilities.
One answer is (i) OCDE (ii) EDCF (iii) EDA
b) (i) 60° (ii) 60° c) equilateral
7 a) 60° b) 3
c) Ask your teacher to check your answer
8 a) 140° b) No
9 a) $n \times$ int ∠ (135°) = 360° gives $n = 2.\dot{6}$ which is not a whole number
b) 90° c) Square

Chapter 21: Perimeter and area

Page 193: Rectangles

1 a) Area : 63 cm^2; Perimeter: 32 cm
b) Area : 60 cm^2; Perimeter 34 cm
c) Area : 176 m^2; Perimeter 54 m
d) Area : 30 m^2; Perimeter 23 m
2 a) 14 cm^2 b) 27 cm^2 c) 20 cm^2
d) 33 cm^2 e) 14.5 cm^2 f) 38 cm^2

Page 195: Other areas

1 a) 15 cm^2 b) 14 cm^2 c) 6 cm^2 d) 92 cm^2
e) 60 cm^2 f) 120 cm^2
2 19.2 m
3 10000 cm^2 or 1m^2
4 a) 2.5 km^2 b) 3 km^2 c) 3 km^2

Page 197: Using similarity

1 a) 4/3 or 1.33 b) 2.25 c) 0.375
2 a) B, D, E b) (i) 36 cm (ii) 1:3
3 a) x = 4.5 cm, y = 6 cm
b) a = 2 cm, b = 7.5 cm
c) p = 3.2 cm, q = 2 cm
d) d = 4.8 cm, e = 3.33 cm, f = 3.75 cm
e) x = 2.5 cm, y = 10.5 cm
f) p = 5.33 cm, q = 6.67 cm

Page 199: Circumference and area of a circle

1 a) 15.7 cm b) 25.1 cm c) 81.7 cm
2 a) 12.6 cm^2 b) 153.9 cm^2 c) 95.0 cm^2
3 a) 7.96 cm b) 3.82 cm c) 10.82 cm
4 a) 3.09 cm b) 4.22 cm c) 5.97 cm
5 25.1 cm
6 a) 3.82 cm b) 45.84 cm^2
7 a) 18.3 cm^2 b) 38.6 cm^2 c) 21.5 cm^2
8 a) 41.9 m b) 40 m c) 1.9 m
9 a) 18π m b) 81π m^2

Pages 200–201: Finishing off

1 a) 22 m b) £371.25
2 a) 20 cm^2 b) 112.5 cm^2 c) 126 cm^2
d) 50 cm^2 e) 54 cm^2
3 a = 5, b = 4.5, c = 30, d = 22.5.
4 a) 1 : 500 b) 5 cm
c) (i) 10 cm^2 (ii) 250 m^2
d) 250 m^2 is 2500 000 cm^2 so ratio of answers in c) is 1 : 250 000 or 1: 500^2

5

A0	1188	840
A1	840	594
A2	594	420
A3	420	297
A4	297	210
A5	210	148
A6	148	105

6 450 7 6400 km 8 20 cm
9 1178 mm^2 10 8 inch

Chapter 22: Three dimensions

Page 203: Cubes and cuboids

1 a) 270 cm^3 b) 180 cm^3 c) 300 cm^3
2 a) 2 340 000 cm^3 b) 2.34 m^3
c) Divide by 1 000 000 (=100^3)
3 72
4 a) 40 cm b) 19 cm
5 162 cm^3
6 a) 480 cm^3 b) 336 g
7 75 (= 5 × 5 × 3)
8 50 m

Page 205: Volume of a prism

1 a) 45 cm^3 b) 100 cm^3 c) 62.8 cm^3
d) 30 cm^3 e) 35.3 cm^3
2 870 m^3
3 795
4 90π cm^3

Page 207: More about prisms

1 a = 1.99 cm b = 4 cm c = 2 cm
d = 2.44 cm e = 15.92 cm f = 6 cm
2 a) 150.5 cm^2 c) 108 cm^2 d) 98.9 cm^2 e) 81.8 cm^2
3 a) 8.84 cm b) 3.26 cm
4 2.78 m

Page 209: Drawing solid objects

1 A, D
2 a) Ask your teacher to check your answer
b) 52 cm^2

3 House a) 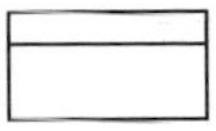 b)  c) 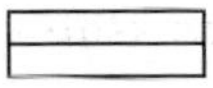

Tent a) 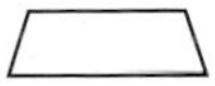 b) 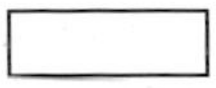 c) 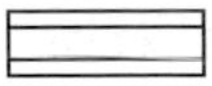

4 a) (i) 4 (ii) 6 (iii) 4
b) Ask your teacher to check your drawing
5 Ask your teacher to check your answers

Page 211: Using dimensions

1 (i) a) 3 dimensions; x, x, y
b) volume
c) C
(ii) a) 1 dimension; $x + y$
b) length
c) A
(iii) a) 2 dimensions; y, y
b) area
c) E
(iv) a) 2 dimensions; x, y
b) area
c) A
(v) a) 2 dimensions; x, $x + y$
b) area
c) D
(vi) a) 3 dimensions; x, x, y
b) volume
c) F
(vii) a) 2 dimensions; x, y
b) area
c) B
(viii) a) 1 dimension; y
b) length
c) E
2 a) volume
b) area
c) length
d) area
e) not a real formula
f) length
g) not a real formula
h) volume

Pages 212–213: Finishing off

1 a) 96 cm^3 b) 260 cm^3 c) 150.8 cm^3
2 a) 254 cm^3 b) 5.3 cm
3 2.52 cm
4 $7\frac{1}{2}$ feet
5 a) Ask your teacher to check your drawing
b) 32 cm^2
6 a) 16 cm^2 b) (i) 5 (ii) 8 (iii) 5
7 a) Ask your teacher to check your drawing
b) 84 cm^2 c) 36 cm^3
8 a) volume b) area
c) none of these d) length
e) none of these f) volume
g) area
9 22.1 g

Chapter 23: Transformations

Page 215: Reflection symmetry

1 Equilateral triangle – 3 lines of reflection symmetry
rotational symmetry order 3
Rectangle – 2 lines of reflection symmetry
rotational symmetry order 2
Regular octagon – 8 lines of reflection symmetry
rotational symmetry order 8
Parallelogram – 0 lines of reflection symmetry
rotational symmetry order 2
Regular pentagon – 5 lines of reflection symmetry
rotational symmetry order 5
Kite – 1 line of reflection symmetry
no rotational symmetry
Isosceles triangle – 1 line of reflection symmetry
no rotational symmetry
Square – 4 lines of reflection symmetry
rotational symmetry order 4
2 Ask your teacher to check your patterns
3 Ask your teacher to check your pattern

Page 217: Translations

1 a) $\begin{pmatrix} 5 \\ 1 \end{pmatrix}$ b) $\begin{pmatrix} -5 \\ 1 \end{pmatrix}$ c) $\begin{pmatrix} 2 \\ -3 \end{pmatrix}$
d) $\begin{pmatrix} 7 \\ 0 \end{pmatrix}$ e) $\begin{pmatrix} -7 \\ 4 \end{pmatrix}$ f) $\begin{pmatrix} -5 \\ -3 \end{pmatrix}$

2 Discuss your answers with your teacher
3 a) reflection in y-axis b) reflection in $y = 1$
c) reflection in $x = 1$ d) (3, 1), (5, 1), (5, 2)
e) translation $\begin{pmatrix} 2 \\ -2 \end{pmatrix}$
4 a) (1, 3), (–1, 4), (–2, 2)
c) (3, 1), (4, –1), (2, –2)
e) (–3, –1), (–4, 1), (–2, 2)

Page 219: Rotations

1 Ask your teacher to check your drawings
2 a) Rotation through 90°, clockwise, about (0, 0)
b) Rotation through 180°, about (–2, 4)
c) Rotation through 90°, anticlockwise about (2, –4)
d) Rotation through 180°, about (–3, 1)
e) Rotation through 90°, clockwise about (3, 2)
f) Rotation through 180°, about (2, –4)
3 Ask your teacher to check your answer.
4 a) Reflection in $y = x$
b) Rotation through 90° clockwise about (2, –1)
c) Reflection in $x = \frac{1}{2}$

d) Translation $\begin{pmatrix} -1 \\ 7 \end{pmatrix}$
e) Rotation through 180° about (–5, 0)
f) Reflection in $y = -x$
g) Translation $\begin{pmatrix} -5 \\ -6 \end{pmatrix}$
h) Reflection in $y = 2$
i) Rotation through 90° clockwise about (0, 3)

Page 221: Enlargements

1 a) 54 cm b) 4.5 cm c) 27 cm d) 12 cm
2 Ask your teacher to check your answers
3 Ask your teacher to check your answers
4 a) Scale factor 2, centre (1, 8)
b) Scale factor 4, centre (4, 8)

Page 223: Combining transformations

1 a) (i) Reflection in y axis
(ii) Reflection in x axis
(iii) Rotation, centre (0, 0), 180°
b) (i) Reflection in y axis
(ii) Reflection in $x = 1.5$
(iii) Translation $\begin{pmatrix} 3 \\ 0 \end{pmatrix}$
c) (i) Yes (ii) Yes (iii) No
2 a) (i) Translation $\begin{pmatrix} 3 \\ 1 \end{pmatrix}$
(ii) Translation $\begin{pmatrix} 1 \\ 3 \end{pmatrix}$
(iii) Translation $\begin{pmatrix} 4 \\ 4 \end{pmatrix}$
b) (i) Yes (ii) No (iii) No
3 a) (i) Enlargement, centre (3, –2), scale factor 2
(ii) Rotation, centre (–1, 4), 90° anticlockwise
b) (i) Rotation, centre (0, 0), 90° anticlockwise
(ii) Enlargement, centre (–1, –2), scale factor 2
c) No
4 a) A → C → E b) A → H → A

Pages 224–225: Finishing off

1 a) (i) 0 (ii) 4 b) (i) 1 (ii) no
c) (i) 2 (ii) 2 d) (i) 0 (ii) 3
2 a) – c) Ask your teacher to check your answer
d) (i) Reflection in y axis
(ii) Reflection in $y = x$
(iii) Reflection in $y = -x$
3 a) True
b) False. Length is not preserved.
4 a) Reflection in y axis
b) Reflection in $y = 3$
c) Rotation through 90°, anticlockwise about (0, 0)
d) Translation $\begin{pmatrix} 0 \\ -6 \end{pmatrix}$
e) Rotation through 180°, about (4, 1)
f) Reflection in $y = x$
g) Translation $\begin{pmatrix} 8 \\ -4 \end{pmatrix}$
h) Rotation through 90°, clockwise, about (2, –6)
i) One possibility is a translation $\begin{pmatrix} 0 \\ -6 \end{pmatrix}$ followed by a reflection in $y = x$
j) One possibility is a translation $\begin{pmatrix} -8 \\ 4 \end{pmatrix}$ followed by a reflection in $y = 3$
5 Ask your teacher to check your answer
6 a) Centre (8, 0), scale factor 2
b) L → Q (4, 0), L → R (0, 0)
c) No

Chapter 24: Drawings and loci

Page 227: Scale drawings

1 a) 20 cm by 8 cm
b) Length 4 cm, height 2.5 cm
2 b) 19.6 cm, 3.92 m in reality
c) (i) 10.8 m (ii) 54 cm
d) About 4.2 cm, 84 cm in reality
3 a) 0.0075 mm
b) 40 cm

Page 229: Drawing triangles

1 a) AC = 5.3 cm, BC = 6.5 cm
b) PR = 9.2 cm, QR = 6.4 cm
2 a) 94°, 56°, 30° b) 87°, 30°, 63°
c) 94°, 36°, 50° d) 102°, 56°, 22°
3 b) $2 + 4 < 7$
4 a) 45° b) 81° c) BC = 5.7 cm
5 BC = 4.5 cm
6 a) GH = 7.4 cm, ∠ GFH = 84°
b) YZ = 8.9 cm

Page 231: Using bearings

1 a) A 030°, B 135°, C 315°, D 260°
b) A 15 km, B 20 km, C 30 km, D 25 km
2 a) 225° b) 45° c) 280°
d) 100° e) 20° f) 200°
g) There is a difference of 180° between each pair of bearings
3 1.8 or 1.9 km and 5.0 or 5.1 km

Page 233: Simple loci

1 Locus is a circle, radius 4 cm, centre P
2 Locus is a pair of parallel lines each 2 cm from the original line
3 Locus is a pair of parallel lines each 5 cm long and 2 cm from the original line joined by semicircles of radius 2 cm
4 Ask your teacher to check your drawing
5 Ask your teacher to check your drawing
6 Ask your teacher to check your drawing

Page 235: More loci

1 Ask your teacher to check your drawing
2 Ask your teacher to check your drawing
3 Ask your teacher to check your drawing
4 Ask your teacher to check your drawing
5 a) (i) (2, 0) (ii) $x = 2$
b) (i) (0, 5) (ii) $y = 5$
c) (i) (2, 2) (ii) $y = x$
6 Ask your teacher to check your diagram

Pages 236–237: Finishing off

1 a) 6 cm b) 0.06 cm c) 150 cm
20:1 would be the most suitable
2 b) $p = 85°$ $q = 128°$ $r = 23°$ $s = 70°$
c) $a = 3.8$ m $b = 6.0$ m
$c = 6.4$ m $d = 10.9$ m
3 322°
4 a) Ask your teacher to check your drawing
b) 221° c) 21.8 km
5 Ask your teacher to check your diagram
6 a), b) Ask your teacher to check your diagram
7 Ask your teacher to check your diagram
8 Ask your teacher to check your diagram

Chapter 25: Pythagoras' rule

Page 239: Finding the hypotenuse

1 a) 8.06 cm b) 8.49 cm c) 13.60 cm
d) 13.45 cm e) 8.16 cm f) 9.17 cm
2 192 m **3** 28.6 km
4 a) 3 units b) 2 units c) 3.61 units
5 a) 5.39 units b) 4.24 units
c) 5.10 units d) 4.47 units

Page 241: Finding one of the shorter sides

1 a) 6.24 cm b) 8.94 cm c) 4.47 cm
d) 8.02 cm e) 4.08 cm f) 7.07 cm
2 2.44 m
3 a) 5.66 cm b) 11.3 cm^2
4 a) 3 cm b) 6 cm^2 c) 2.4 cm
d) AN = 3.2 cm, CN = 1.8 cm
5 $\sqrt{33}$ cm

Pages 242–243: Finishing off

1 a) $a = 10.2$ cm b) $b = 9.71$ cm
c) $c = 6.36$ cm
2 a) $\sqrt{25} = 5$ units b) $\sqrt{32}$ (or $4\sqrt{2}$) units
c) $\sqrt{29}$ units
3 27.5 km
4 4.04 m or 404 cm
5 Yes
6 a) 7.28 m b) 1.88 m c) 2.31 m d) 2.98 m
7 a) 34 (17 each side) b) 28 c) 952
8 Yes. Diagonal of box is 51.2 cm

Chapter 26: Trigonometry

Page 245: Introduction to trigonometry

1

Your answers to questions 2 and 3 may be slightly different (up to about 0.05 either way) from those given. This is because it is not possible to measure accurately.
2 b) 0.84 c) 0.64 d) 0.77
3 b) 1.33 c) 0.80 d) 0.60

Page 247: Sine, cosine and tangent

1 a) 0.9063 . . . b) 0.8829 . . .
c) 8.1443 . . . d) 0.2079 . . .
e) 1.0000 f) 0.5446 . . .

Page 249: Finding an unknown side

1 a) $a = 3.63$ cm b) $b = 8.99$ cm
c) $c = 8.65$ cm d) $d = 5.35$ cm
e) $e = 5.00$ cm f) $f = 2.98$ cm
2 a) $a = 3.56$ cm b) $b = 8.40$ cm
c) $c = 6.19$ cm d) $d = 8.04$ cm
e) $e = 4.93$ cm f) $f = 6.48$ cm
3 a) $a = 6.86$ cm b) $b = 4.17$ cm
c) $c = 9.85$ cm d) $d = 5.70$ cm
e) $e = 8.49$ cm f) $f = 3.96$ cm

Page 251: Finding an unknown angle

1 a) 46.1° b) 75.6° c) 37.3°
d) 48.2° e) 56.8° f) 58.0°
2 a) $p = 53.6°$ b) $q = 37.3°$ c) $r = 39.1°$
d) $s = 51.5°$ e) $t = 48.7°$ f) $u = 60.3°$
3 a) $a = 4.03$ cm b) $b = 61.2°$
c) $c = 7.71$ cm d) $d = 6.87$ cm
e) $e = 3.08$ cm f) $f = 50.2°$

Page 253: Using trigonometry

1 a) 127 km south, 272 km west
b) 409 km north, 368 km west
2 6.9 m
3 a) 932 m b) 12.5°
4 a) 2.12 m b) 1.5 m
5 1067 m
6 a) 7.73 m b) 48.6°

Pages 254–255: Finishing off

1 a) $a = 8.66$ cm b) $b = 8.48$ cm
c) $c = 32.1°$ d) $d = 6.02$ cm
e) $e = 8.91$ cm f) $f = 40.7°$

2 a) 0.78 km north, 2.90 km west b) 134°
3 2.008 km 4 38.7 m
5 a) 1.45 m b) 6.6 m^2
6 a) 84.5 km south, 181.3 km east
b) 96.4 km south, 114.9 km west
c) 180.9 km south, 66.4 km east
d) 340°
7 a) $\tan 35° = \frac{h}{w}$ b) $h = w \tan 35°$
c) $\tan 23° = \frac{h}{w+10}$ d) $h = (w + 10) \tan 23°$
e) $w = 15$, $h = 11$

Chapter 27 Circles and tangents

Page 257: Shapes in a circle

1 a) (i) $x = 110°$, $y = 35°$, $z = 55°$ (ii) ACB = 90°
c) The angle ACB is always 90°
2 $a = 36°$, $b = 99°$
3 Ask your teacher to check your reasons
a) $a = 90°$ b) $b = 24°$ c) $c = 102°$ d) $d = 141°$
e) $e = 72°$, $f = 98°$ f) $g = 86°$, $h = 86°$, $i = 94°$
g) $j = 88°$ h) $k = 68°$

Page 259: Angles in a circle

1 a) OA = OC b) 30° c) 120° d) 60°
e) 80° f) (i) 140° (ii) 70° g) $\angle ABC = \frac{1}{2} \angle AOB$
2 Ask your teacher to check your reasons
a) $a = 72°$ b) $b = 42°$ c) $c = 64°$ d) $d = 28°$
e) $e = 35°$ f) $f = 88°$ g) $g = 110°$ h) $h = 70°$
i) $i = j = k = 25°$ j) $l = 51°$ k) $m = 40°$ l) $n = 25°$
m) $o = 28°$, $p = 124°$, $q = 28°$ n) $r = 48°$

Page 261: Tangents

1 Ask your teacher to check your reasons
a) $a = 36°$ b) $b = 67°$ c) $c = 70°$, $d = 40°$
d) $e = 64°$ e) $f = 58°$ f) $g = 70°$
2 Ask your teacher to check your reasons
a) $a = 51°$, $b = 25.5°$ b) $c = 48°$, $d = 42°$, $e = 96°$
c) $f = 12°$
3 15 cm 4 12 cm 5 8 cm

Pages 262–263: Finishing off

1 Ask your teacher to check your reasons
a) $a = 48°$, $b = 132°$ b) $c = 74°$ c) $d = e = 62°$, $f = 28°$
d) $g = 63°$ e) $h = 90°$, $j = 90°$, $k = 65°$
f) $l = 70°$, $m = 70°$, $n = 70°$ g) $p = q = 58°$
h) $r = 52°$ i) $s = 32°$
2 a) 130° b) 65°
3 a) Angle TAC: tangent is perpendicular to the diameter at the point of contact
Angle ABC: angle in a semi-circle is 90°
b) (i) 32° (ii) 58° (iii) 58° (iv) 122°
4 Ask your teacher to check your reasons
$a = 90°$, $b = 40°$, $c = 50°$, $d = 130°$, $e = 55°$
5 PR = PS = 4 cm 6 12 cm

Chapter 28: Collecting data

Page 265: Conducting a survey

1 Ask your teacher to check your answer
2 Ask your teacher to check your answer

Page 267: Recording data

1 a)

Ride	Dragon's Tail	Splashdown	Pirate Plank	Octopus
Frequency	12	7	8	3

b) 30 c) Dragon's Tail
2 a)

No. of mugs per box returned	0	1	2	3	4	5	6	7
No. of boxes	3	5	4	3	3	1	0	1

b) 25%
3 a) discrete b) continuous
c) discrete d) continuous e) continuous

Page 269: Stem-and-leaf diagrams

1 b) 30 c) 73 d) 3
2 b) 21 c) 69 d) 9
3 b) 6 c) 37 d) 3 e) 14
4 b) 257 cm, 180 cm c) 245 cm, 150 cm
d) Jane's flowers are taller: fertilizer is effective.

Page 271: Grouping data

1 b)

Weight (kg)	No. of male students
60–64	2
65–69	1
70–74	2
75–79	5
80–84	7
85–89	1
90–94	1
95–99	1

c) More information is needed e.g. heights. Half the students are 80 kg or more suggesting that the group may be overweight.
2 a)

Season	Frequency (people)
Winter Dec, Jan, Feb	5
Spring March, April, May	3
Summer June, July, August	8
Autumn Sept, Oct, Nov	8

b) No

3 a)

Time (minutes)	1–10	11–20	21–30	31–40
No. of surgeries	14	11	3	2

b) One answer is that the majority of surgeries overrun by between 1 and 20 minutes

c)

Time (minutes)	No. of surgeries
1–5	3
6–10	11
11–15	8
16–20	3
21–25	2
26–30	1
31–35	1
36–40	1

d) The majority of surgeries overrun by between 6 and 15 minutes (rather than between 1 and 20 minutes)

Page 273: Social statistics

1 a) £7.62 b) £9.90
2 a) £101.60 b) £132.00
3 1993 £1.7m, 1994 £3.045m, 1995 £5.28m, 1996 £7.458m, 1997 £9.882m, 1998 £12.707m

Pages 274–275: Finishing off

1 Ask your teacher to check your answers
2 a) numerical, continuous
b) numerical, continuous
c) categorical
d) numerical, discrete
3 b) Ask your teacher to check your table
4 a) Ask your teacher to check your answer
b) Number of workers varies, machine breakdown, . . .
5 b) 68 months, 40 months
c) 69 months, 45 months
d) These results suggest that Quickstart last longer.

Chapter 29: Displaying data

Page 277: Displaying data

1 a) ; b) 20; 6; 10 c) 36
2 Ask your teacher to check your answers
3 a) 8 b) 26
c) Ask your teacher to check your bar chart
d) Special
4 Ask your teacher to check your bar charts

Page 279: Pie charts

1 a) 90°, 120°, 150°
b) Children £90 000, Adult £150 000, Teenage £120 000
c) $33\frac{1}{3}\%$
2 a) 15°
b) Ask your teacher to check your pie chart
c) 50
3 a) 180
b) France 50, Switzerland 30, Austria 10, Germany 50 and Poland 40
4 Ask your teacher to check your charts

Page 281: Line graphs

1 a) Ask your teacher to check your line graph
b) Week 6
c) 3.1 kg
2 a) Ask your teacher to check your line graph
b) It increases between July and October and then gradually decreases
c) July to October
3 Ask your teacher to check your vertical line chart
4 a) 20 b) 15 c) 40

Page 283: Histograms

1 a) 5.5 and 10.5 b) 8
c) Just under 20.5 minutes
d) 0.5 mins
e) Ask your teacher to check your histogram
2 a) Ask your teacher to check your tally chart
b)

Price (£)	No. of cars
0–999	0
1000–1999	5
2000–2999	8
3000–3999	5
4000–4999	8
5000–5999	4

c) Ask your teacher to check your histogram
3 Ask your teacher to check your histogram
4 a) 15 sq units (each sq unit represents 1 bus)
b) 15 sq units.
The area of the histogram is the same as the area under the frequency polygon.

Pages 284–285: Finishing off

1 Ask your teacher to check your chart
2 a) 75 b) 60
c) Ask your teacher to check your diagrams
d) Ask your teacher to check your answer
3 Ask your teacher to check your line diagram. There is the least sunshine in December. It increases gradually through the months until June. There is the most sunshine in June and July. It decreases gradually through the months until December.
4 a) Ask your teacher to check your histogram
b) Size 3
c) 600 size 1, 900 size 2, 1500 size 3, 1250 size 4, 750 size 5
5 a)

Type of bird	Frequency
Blackbird	3
Starling	6
Magpie	2
Sparrow	6
Thrush	2
Robin	1

b) 20 c) 30
6 a) Shop A, 6–10. Shop B, 16–20
b) Ask your teacher to check your frequency polygons.
c) Fewer items are bought in shop A than in shop B (lower modal class) but the range is the same.
d) A is probably a smaller shop because most people just buy a few items. B is probably a supermarket.

Chapter 30: Average and spread

Page 287: Mean, median, mode and range

1 a) Mean 3, median 2, mode 1, range 7
b) Mean 5, median 5.5, mode 7, range 5
c) Mean 5.5, median 5, mode 4, range 4
d) Mean 2, median 1, mode 1, range 6
2 a) Mean 4.9, median 5, mode 6
b) Ask your teacher to check your answer
3 a) Mean 12 hours; median 12.5 hours; mode 5 hours; range 23 hours.
b) Mean 11 hours; median 11 hours; no mode. The mean and the median are both higher for the boys than for the girls, so the boys watch more TV than the girls.
4 a) Mean 13 s; median 13 s; mode 11 s; range 5 s.
b) Mean 13 s; median 12 s; mode 11 s; range 7 s. The PE staff are perhaps slightly faster: their mean time is the same as the students but their median time is 1 s less. However they are less consistent, with one very slow member.
5 a) Mean £13 400, median £11 000, mode £8000, range £27 000
b) Mean reduced to £11 000, median and mode unchanged, range reduced to £11 000
6 £672

Page 289: More discrete data

1 a) (i) Mean 4.6, median 4.5, mode 4, range 3
(ii) Mean 13.5, median 13.5, no single mode, range 3
(iii) Mean 2.56, median 2, mode 1, range 4
(iv) Mean 2.95, median 3, mode 3, range 4
b) It is above the highest value
c) Use the symmetry of the frequencies
d) Less than 3 because there are four 2s and only three 4s
e) (iii) has a mode of 1 and the frequencies decrease as the value increases
(iv) has a mode of 3 and the frequencies are nearly symmetrical about this value.
2 a) 1 b) 2 c) 2.2
d) Ask your teacher to check your answer
3 a) Ask your teacher to check your answer
b) Mount Pleasant Street 1.8, Mandela Avenue 3
c) Mount Pleasant Street 6, Mandela Avenue 6
d) Ask your teacher to check your answer

Page 291: Grouped data

1 a) 6 – 10 kg b) 8 kg c) 430 kg d) 8.6 kg
2 a) £51 000 – £60 000 b) £105 500
c) $\frac{£7036500}{143} = £49\,206$
3 a) 90 – 94 mins b) 92.25 mins
c) 92.5 mins (approx 25 mins added to total)
d) modal class still 90 – 94 mins

Page 293: Moving averages

1 a) 3
b) 5, 4.67, 4.33, 4.67, 3.33, 4.33, 3.67, 6, 5.67, 8, 5.33, 5, 3, 5.33, 4.33, 4.67
2 a) 12 b) 7 c) 6
3 a) 77 b) £100 c) 79, £84
4 b) 3.15, 3.25, 3.35, 3.45, 3.55, 3.65, 3.75, 3.85, 3.95
d) Rising e) 2600, 4400
5 b) 31.33, 33, 36.33, 35.67, 36.33, 46.33, 47.67, 48.67, 35.33, 36 d) 3 e) Illness

Pages 294–295: Finishing off

1 a) Mean 104, median 103, mode 102, range 6
b) Mean 1.3, median 1, mode 0, range 4
c) Mean 2.6, median 3.5, mode 0, range 6
2 a) Mean 2.4, median 1.5, mode 0, range 6
b) Varied performance, some high scores and some low scores
c) 12
3 a) Boys : mode 6 nights, range 7 nights
Girls : mode 4 nights, range 4 nights

b) Boys higher average nights out, but more varied
c) Ask your teacher to check your answer

4 a) Footballers 67.5 b.p.m., joggers 61.5 b.p.m.

5 a) 24 b) 131 c) 7, mode
d) 6, range e) 5.46, mean

6 b) 380, 382.5, 385, 387.5, 390, 392.5, 395, 397.5, 400
d) Rising
e) 430, 350

Chapter 31: More data handling

Page 297: Cumulative frequency

1 a) Ask your teacher to check your answer
b) 123 hours approximately
c) 100 – 55 = 45
d) 30 hours approximately

2 a) Ask your teacher to check your answer
b) £17 500 approximately
c) 17% approximately
d) 17% approximately
e) The cumulative frequency curve assumes a spread of values across each interval.
It may be that very few of the 52 employees in the £10 000 – £14 999 interval are earning less than £12 000 and that management claims are correct.

Page 299: Quartiles

1 a) Ask your teacher to check your cumulative frequency curve
b) Median 378 g or 379 g, IQR 6 g
c) 34

2 a) Ask your teacher to check your cumulative frequency curve
b) Median 59 words per minute
c) IQR 12 words per minute (66–54)
d) 8
e) 52

3 a) Median 3.3 kg, IQR 1.1 kg (3.9 – 2.8)
b) Avonford babies are slightly heavier but their weights are more variable

Page 301: Box-and-whisker diagrams

1 Median 9, mode 9, quartiles 5 and 17

2 b) LQ 25, median 32, UQ 37

3 c) LQ 1.75, median 2.55, UQ 3.575

4 b) Trained: LQ 125, median 129, UQ 132
Untrained: LQ 139, median 144, UQ 149
d) Trained employees can do the job in less time but their range of times is wider.

Page 303: Bivariate data

1 a) Ask your teacher to check your scatter diagram
b) Strong negative correlation. As the number of hours of sunshine increases the number of visitors decreases.

2 Ask your teacher to check your scatter diagram. There is a strong positive correlation up to 100 kg per ha, then a strong negative correlation if fertilizer is increased.

3 a) Ask your teacher to check your scatter diagram
There is a strong correlation between weight and length.
There is no obvious difference between girls and boys.

Page 305: Line of best fit

1 a) Ask your teacher to check your scatter diagram
b) 19.5 N c) 5.3 kg

2 a) 700 m b) 12 °C
c), d) Ask your teacher to check your scatter diagram
e) 14 °C f) 1150 m approximately

3 a) Ask your teacher to check your scatter diagram
b) Strong positive correlation
c) Ed is wrong. Should be 67.5
e) 0.133 g

Pages 306–307: Finishing off

1 a) £19 million
b) £11.5 million
c) 43 (65 – 22)

2 a) Ask your teacher to check your scatter diagram
b) Moderate negative correlation
c) No

3 a) Ask your teacher to check your cumulative frequency curve
b) Northbound: median 3.25 mins, IQ range
4 mins 40 secs – 2 mins = 2 mins 40 secs
Southbound: median 7 mins, IQ range
9 mins – 5 mins = 4 mins
c) Ask your teacher to check your answers
d) Southbound traffic has longer delays which could be helped by setting traffic lights with longer at green for Southbound traffic.

4 a) Type A : Mean distance 9000 miles, mean cost £1000
Type B : Mean distance 9000 miles, mean cost £1000
b) Ask your teacher to check your scatter diagram
c) Type B

Chapter 33: Probability

Page 313: Calculating probabilities

1 a) $\frac{1}{6}$ b) $\frac{3}{6}=\frac{1}{2}$ c) $\frac{3}{6}=\frac{1}{2}$

d) 0 e) $\frac{2}{6}=\frac{1}{3}$ f) $\frac{4}{6}=\frac{2}{3}$

2 a) $\frac{13}{52}=\frac{1}{4}$ b) $\frac{26}{52}=\frac{1}{2}$ c) $\frac{12}{52}=\frac{3}{13}$

d) $\frac{4}{52}=\frac{1}{13}$ e) $\frac{1}{52}$

3 a) $\frac{7}{28}=\frac{1}{4}$ b) $\frac{7}{28}=\frac{1}{4}$ c) $\frac{21}{28}=\frac{3}{4}$

4 a) $\frac{5}{400}=\frac{1}{80}$ b) $\frac{395}{400}=\frac{79}{80}$ c) $\frac{5}{399}$

5 a) $\frac{2}{20}=\frac{1}{10}$ b) $\frac{10}{20}=\frac{1}{2}$ c) $\frac{10}{20}=\frac{1}{2}$

d) $\frac{2}{20}=\frac{1}{10}$ e) $\frac{18}{20}=\frac{9}{10}$

Page 315: Estimating probability

1 a) $\frac{13}{20}$ b) $\frac{7}{20}$ c) 0

2 a) $\frac{60}{90}=\frac{2}{3}$ b) $\frac{30}{90}=\frac{1}{3}$

c) Not really, January probably had most of the rainy days

3 a) $\frac{23}{161}=\frac{1}{7}$ b) $\frac{138}{161}=\frac{6}{7}$ c) 143

4 a) $\frac{20}{60}=\frac{1}{3}$ b) $\frac{8}{60}=\frac{2}{15}$ c) $\frac{15}{60}=\frac{1}{4}$ d) 0

Zero is probably a good approximation but it is not impossible that she'll get 12; it could be her birthday.

5 a) $\frac{10}{40}=\frac{1}{4}$ b) 50 c) $\frac{30}{40}=\frac{3}{4}$ d) 150

Page 317: Two outcomes: 'either, or'

1 a) $\frac{8}{52}=\frac{2}{13}$ b) $\frac{16}{52}=\frac{4}{13}$

2 a) $\frac{5}{20}=\frac{1}{4}$ b) $\frac{15}{20}=\frac{3}{4}$ c) $\frac{13}{20}$ d) $\frac{7}{20}$

3 a) $\frac{3}{65}$ b) $\frac{29}{65}$ c) $\frac{18}{65}$

4 a) $\frac{6}{36}=\frac{1}{6}$ b) $\frac{15}{36}=\frac{5}{12}$ c) $\frac{9}{36}=\frac{1}{4}$ d) $\frac{15}{36}=\frac{5}{12}$

Page 319: Two outcomes: 'first, then'

1 a) $\frac{1}{16}$ b) $\frac{3}{8}$

2 a) $\frac{1}{17}$ b) $\frac{13}{34}$

3 a) $\frac{1}{400}$ b) $\frac{361}{400}$ c) $\frac{38}{400}=\frac{19}{200}$

4 a) 0.02 b) 0.08 c) 0.72

5 a) $\frac{90}{360}=\frac{1}{4}$ b) $\frac{72}{360}=\frac{1}{5}$ c) $\frac{1}{20}$

d) $\frac{1}{60}$ e) $\frac{3}{10}$

Page 321 Probability trees

1 a) 0.18 b) 0.28 c) 0.12 d) 0.42

2 a) (i) 10 (ii) 30

b) Smartie colour Fred's guess

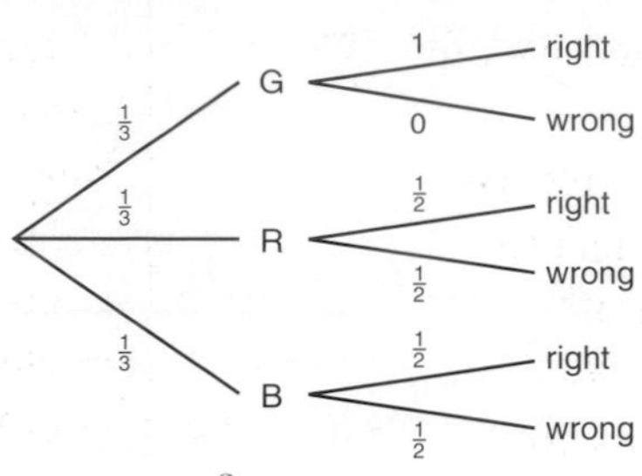

c) $P(\text{right})=\frac{2}{3}$ d) Yes

Pages 322-323: Finishing off

1 a) (i) $\frac{1}{2}$ (ii) $\frac{4}{6}=\frac{2}{3}$ (iii) $\frac{1}{7}$ (iv) $\frac{1}{4}$ (v) 0

b) Ask your teacher to check your drawing

2 a) $\frac{1}{2}$ b) $\frac{4}{5}$ c) $\frac{3}{10}$

3 a)

Red die score (columns); Blue die score (rows)

	1	2	3	4	5	6
1	1	2	3	4	5	6
2	2	2	3	4	5	6
3	3	3	3	4	5	6
4	4	4	4	4	5	6
5	5	5	5	5	5	6
6	6	6	6	6	6	6

b) $P(1)=\frac{1}{36}$, $P(2)=\frac{3}{36}$, $P(3)=\frac{5}{36}$,

$P(4)=\frac{7}{36}$, $P(5)=\frac{9}{36}$, $P(6)=\frac{11}{36}$

c) $\frac{1+3+5+7+9+11}{36}=\frac{36}{36}=1$

d) 10

4 a) 0.16 b) 0.04 c) 0.3

d) 0.7 e) 0.16 f) 0.48

5 b) 0.2601 c) 0.2401 d) 1 boy and 1 girl

6 a) $\frac{65}{120}=\frac{13}{24}$ b) $\frac{45}{120}=\frac{3}{8}$

c) $\frac{5}{120}=\frac{1}{24}$ d) $\left(\frac{13}{24}\right)^3=0.159$

e) $\left(\frac{3}{8}\right)^3=0.0527$

Index